Human Anatomy Laboratory Manual

WITH CAT DISSECTIONS

Ninth Edition

Elaine N. Marieb, R.N., Ph.D.
Holyoke Community College

Lori A. Smith, Ph.D.
American River College

Pearson

Editor-in-Chief: Serina Beauparlant
Senior Courseware Portfolio Manager: Lauren Harp
Managing Producer: Nancy Tabor
Content & Design Manager: Michele Mangelli, Mangelli Productions, LLC
Courseware Editorial Assistant: Lidia Bayne
Production Supervisor: David Novak
Proofreader: Betsy Dietrich
Compositor: SPi Global
Interior & Cover Designer: Hespenheide Design
Illustrators: Imagineering STA Media Services, Inc.
Rights & Permissions Management: Ben Ferrini
Senior Anatomy & Physiology Specialist: Derek Perrigo
Manufacturing Buyer: Stacey Weinberger
Product Marketing Manager: Wendy Mears
Director of Product Marketing, Science: Allison Rona

Cover Photo Credit: K H FUNG/Science Source

www.pearson.com

10 2024

ISBN 10: 0-13-516803-1
ISBN 13: 978-0-13-516803-5

Contents

iv   Contents

Preface for the Instructor

The philosophy behind the revision of this manual mirrors that of all earlier editions. It reflects a still-developing sensibility for the way teachers teach and students learn, engendered by years of teaching the subject and by listening to the suggestions of other instructors as well as those of students enrolled in multifaceted healthcare programs. *Human Anatomy Laboratory Manual with Cat Dissections* was originally developed to facilitate and enrich the laboratory experience for both teachers and students. This edition retains those same goals.

This manual, intended for students in one-term introductory human anatomy courses, presents a wide range of anatomical laboratory experiences for students concentrating in nursing, physical therapy, dental hygiene, pharmacology, respiratory therapy, and health and physical education, as well as biology and premedical programs. This manual studies anatomy of the human specimen in particular, but the cat and isolated animal organs are also used in the dissection experiments.

Basic Approach and Features

Although the main textbook, *Human Anatomy,* Ninth Edition (Elaine N. Marieb, Patricia Brady, and Jon Mallatt, Pearson Education ©2020), provided the impetus for this revision, the laboratory manual, as in previous editions, is based largely on exercises developed for use independent of any textbook. It contains all the background discussion and terminology necessary to perform all manipulations effectively and eliminates the need for students to bring a textbook into the laboratory. The laboratory manual is comprehensive and balanced enough to be flexible and is carefully written so that students can successfully complete each of its 30 exercises with little supervision.

Features

- Each exercise begins with learning outcomes, followed by a Pre-lab Quiz that poses questions on basic information students should know before doing the lab.

- Laboratory Review Sheets at the end of each exercise require students to label diagrams and answer multiple-choice, short-answer, and essay questions. We have strived to achieve an acceptable balance between questions that require students to recognize structures and those that ask students to explain important concepts.

- Illustrations are large and of exceptional quality. Full-color photographs and drawings highlight and differentiate important structures and focus student attention on them.

- All laboratory instructions and procedures incorporate the latest precautions as recommended by the Centers for Disease Control and Prevention (CDC); these are reinforced by the laboratory safety procedures described inside the front cover and in the front section of the *Instructor's Guide*. These procedures can be easily photocopied and posted in the lab.

- Three icons alert students to special features or instructions:

The **dissection scissors icon** appears at the beginning of activities that entail the dissection of the cat as well as isolated animal organs.

The **homeostatic imbalance icon** appears where a clinical disorder is described to indicate what happens when there is a structural abnormality or physiological malfunction, that is, a loss of homeostasis.

A **safety icon** notifies students that they are to observe specific safety precautions when they are using certain equipment or conducting particular lab procedures. For example, when handling body fluids such as blood, urine, and saliva, they are to wear gloves.

New to the Ninth Edition

- **Dozens of new, full-color illustrations and photos** replace many black and white line drawings to help students differentiate among structures and more easily interpret diagrams.

- **New Clinical Application Questions** have been added to the Exercise Review Sheets that challenge students to apply lab concepts and critical thinking skills to real-world clinical scenarios.

- **Improved interior design** incorporates more saturated colors in headings and exercise tabs to improve readability.

- **Content and illustration updates** have been made throughout the Ninth Edition. Please contact your Pearson representative for more details.

Customization Options

For information on creating a custom version of this manual, visit pearsoncollections.com or contact your Pearson sales representative for details.

Instructor Supplements

Instructor's Guide

The *Instructor's Guide* that accompanies the *Human Anatomy Laboratory Manual* contains a wealth of information, including answers to the updated review sheets. The guide includes help for anticipating pitfalls and problem areas, directions for lab setup, a complete materials list for each lab, and answers to all questions in the manual. (ISBN 0-13-520215-9)

Student Supplements

A Photographic Atlas for Anatomy & Physiology

by Nora Hebert, Ruth Heisler, Karen Krabbenhoft, Olga Malakhova, and Jett Chinn
(ISBN 0-321-86925-7)

The *Photographic Atlas for Anatomy & Physiology* helps students learn and identify key anatomical structures. Featuring photos from *Practice Anatomy Lab*™ and other sources, the *Atlas* includes over 250 cadaver dissection photos, histology photomicrographs, and cat dissection photos plus over 50 photos of anatomical models from leading manufacturers such as 3B Scientific®, SOMSO®, and Denoyer-Geppert Science Company. The *Atlas* is composed of 13 chapters, organized by body system, and includes a final chapter with cat dissection photos.

Practice Anatomy Lab™ 3.1
(ISBN 0-321-68211-4)

Practice Anatomy Lab™ (PAL) 3.1 is now accessible on all mobile devices to give students 24/7 access to the most widely used laboratory specimens including human cadaver, cat, and fetal pig as well as anatomical models and histology images that are used in the lab.

NEW! Customizable PAL 3.1 flashcards allow students to create a personalized, mobile-friendly deck of flashcards and quizzes using images from PAL 3.1. Using the checklist, students can generate flashcards using only the structures that their instructor has emphasized in lecture or lab. Visit practiceanatomylab.com for more information.

Practice Anatomy Lab 3.1 Lab Guide

by Ruth Heisler, Nora Hebert, Karen Krabbenhoft, Olga Malakhova, and Jett Chinn
without PAL DVD (ISBN 0-321-84025-9)
with PAL DVD (ISBN 0-321-85767-4)
Written to accompany PAL™, the *Practice Anatomy Lab Lab Guide* contains lab exercises that direct students to select images and features in PAL 3.1, and then assess their understanding by completing labeling, matching, short-answer, and fill-in-the-blank questions. Exercises cover human cadaver, anatomical models, and histology.

The Anatomy Coloring Book, Fourth Edition

by Wynn Kapit and Lawrence M. Elson
(ISBN 0-321-83201-9)
For more than 35 years, *The Anatomy Coloring Book* has been the #1 best-selling human anatomy coloring book! A useful tool for anyone with an interest in learning anatomical structures, this concisely written text features precise, extraordinary hand-drawn figures that were crafted especially for easy coloring and interactive study.

Acknowledgments

Many thanks to the Pearson Education team: Serina Beauparlant, Editor-in-Chief; and Lauren Harp, Senior Acquisitions Editor. Thanks also to Wendy Mears, product marketing manager. Kudos as usual to Michele Mangelli and her team: David Novak for his production work and Gary Hespenheide for an updated interior design and a beautiful new cover. And last, but not least, a special thanks to Susan Mitchell for her authorial contribution to this lab manual over the years.

As always, we invite users of this edition to send us their comments and suggestions for subsequent editions.

Elaine N. Marieb
and Lori A. Smith
Pearson Education
50 California Street,
San Francisco, CA 94111

A Word to the Student

We hope you will enjoy your laboratory experiences. As with any unfamiliar experience, it really helps to know in advance what you can expect and what will be expected of you.

Laboratory Activities

The laboratory exercises in this manual are designed to help you gain a broad understanding of anatomy. You can anticipate examining models, dissecting a specimen and isolated animal organs, and using a microscope to look at tissue slides (anatomical approaches). You will also investigate a limited selection of physiological phenomena (conduct visual tests, plot the distribution of sweat glands, and so forth) to make your anatomy studies more meaningful.

Icons/Visual Mnemonics

We have tried to make this manual easy for you to use, and to this end two colored section heads and three different icons (visual mnemonics) are used throughout:

The **Dissection** head is purple and is accompanied by the **dissection scissors icon** at the beginning of activities that require you to dissect the cat as well as isolated animal organs.

The **Activity** head is blue. Because most exercises provide some explanatory background before the experiment(s), this visual cue alerts you that your lab involvement is imminent.

The **homeostatic imbalance icon** appears where a clinical disorder is described to indicate what happens when there is a structural abnormality or physiological malfunction, that is, a loss of homeostasis.

The **safety icon** notifies you that you are to observe specific safety precautions when using certain equipment or conducting particular lab procedures (for example, using a hood when you are working with ether, or wearing gloves when you are handling body fluids, such as blood, urine, or saliva).

Hints for Success in the Laboratory

Most students can use helpful hints and guidelines to ensure that they have successful lab experiences.

1. Perhaps the best bit of advice is to attend all your scheduled labs and to participate in all the assigned exercises. Learning is an *active* process.

2. *Before* going to lab, complete the pre-lab quiz, and scan the scheduled lab exercise and the questions in the Review Sheet at the end of the exercise.

3. Be on time. Most instructors explain what the lab is about, pitfalls to avoid, and the sequence or format to be followed at the beginning of the lab session. If you are late, you will miss this information and also risk annoying the instructor.

4. Review your lab notes after completing the lab session to help you focus on and remember the important concepts.

5. Keep your work area clean and neat. This reduces confusion and accidents.

6. Assume that all lab chemicals and equipment are sources of potential danger to you. Follow directions for equipment use, and observe the laboratory safety guidelines provided inside the front cover of this manual.

7. Keep in mind the real value of the laboratory experience—a place for you to observe, manipulate, and experience hands-on activities that will dramatically enhance your understanding of the lecture presentations.

We really hope that you enjoy your anatomy laboratories and that this lab manual makes learning about intricate structures and functions of the human body a fun and rewarding process. We're always open to constructive criticism and suggestions for improvement in future editions. If you have any, please write to us in care of Pearson Education.

Elaine N. Marieb

Lori A. Smith

Elaine N. Marieb
and Lori A. Smith
Pearson Education
50 California Street,
San Francisco, CA 94111

The Language of Anatomy

MATERIALS

- Human torso model (dissectible)
- Human skeleton
- Demonstration: sectioned and labeled kidneys [three separate kidneys uncut or cut so that (a) entire, (b) transverse sectional, and (c) longitudinal sectional views are visible]
- Gelatin-spaghetti molds
- Scalpel

LEARNING OUTCOMES

☐ Describe the anatomical position, and explain its importance.

☐ Use proper anatomical terminology to describe body regions, orientation and direction, and body planes.

☐ Name the body cavities, and indicate the important organs in each cavity.

☐ Name and describe the serous membranes of the ventral body cavities.

☐ Identify the abdominopelvic quadrants and regions on a torso model or image.

PRE-LAB QUIZ

1. Circle True or False. In the anatomical position, the body is lying down.
2. Circle the correct underlined term. With regard to surface anatomy, <u>abdominal</u> / <u>axial</u> refers to the structures along the center line of the body.
3. The term *superficial* refers to a structure that is:
   a. attached near the trunk of the body   c. toward the head
   b. toward or at the body surface   d. toward the midline
4. The _____ plane runs longitudinally and divides the body into right and left parts.
   a. frontal   c. transverse
   b. sagittal   d. ventral
5. Circle the correct underlined terms. The dorsal body cavity can be divided into the <u>cranial</u> / <u>thoracic</u> cavity, which contains the brain, and the <u>sural</u> / <u>vertebral</u> cavity, which contains the spinal cord.

A student new to any science is often overwhelmed at first by the terminology used in that subject. The study of anatomy is no exception. But without this specialized terminology, confusion is inevitable. For example, what do *over, on top of,* and *behind* mean in reference to the human body? Anatomists have an accepted set of reference terms that are universally understood. These allow body structures to be located and identified with a minimum of words and a high degree of clarity.

This exercise presents some of the most important anatomical terminology used to describe the body and introduces you to basic concepts of **gross anatomy,** the study of body structures visible to the naked eye.

Anatomical Position

When anatomists or doctors refer to specific areas of the human body, the picture they keep in mind is a universally accepted standard position called the **anatomical position.** In the anatomical position, the human body is erect, with the feet only slightly apart, head and toes pointed forward, and arms hanging at the sides with palms facing forward (Figure 1.1). It is also important to remember that "left" and "right" refer to the sides of the individual, not the observer.

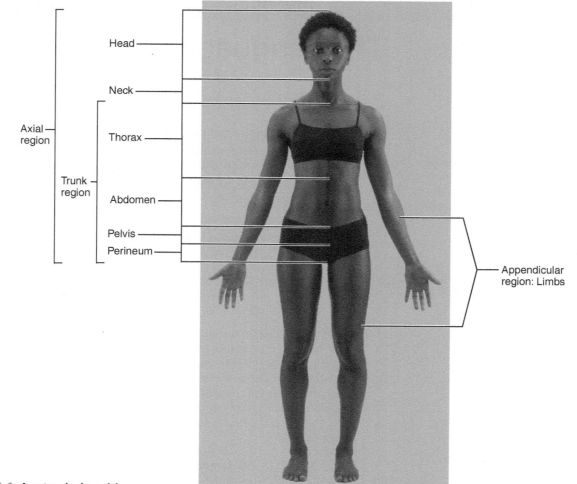

Figure 1.1 Anatomical position.

☐ Assume the anatomical position, and notice that it is not particularly comfortable. The hands are held unnaturally forward rather than with the palms toward the thighs.

Check the box when you have completed this task.

Regional Anatomy

The body is divided into two main regions, the axial and appendicular regions (Figure 1.1). The **axial region** includes the head, neck, and trunk; it runs along the vertical axis of the body. The **appendicular region** includes the limbs, which are also called the appendages or extremities. The body is also divided up into smaller regions within those two main divisions. These anterior and posterior body regions are summarized below and they are illustrated in **Figure 1.2**.

Anterior Body Landmarks

Note the following regions in Figure 1.2a:

Abdominal: Located below the ribs and above the hips

Acromial: Point of the shoulder

Antebrachial: Forearm

Antecubital: Anterior surface of the elbow

Axillary: Armpit

Brachial: Arm (upper portion of the upper limb)

Buccal: Cheek

Carpal: Wrist

Cephalic: Head

Cervical: Neck

Coxal: Hip

Crural: Leg (lower portion of the lower limb)

Digital: Fingers or toes

Femoral: Thigh

Fibular (peroneal): Side of the leg

Frontal: Forehead

Hallux: Great toe

Inguinal: Groin

Mammary: Breast

Manus: Hand

Mental: Chin

Nasal: Nose

Oral: Mouth

Orbital: Bony eye socket (orbit)

Palmar: Palm of the hand

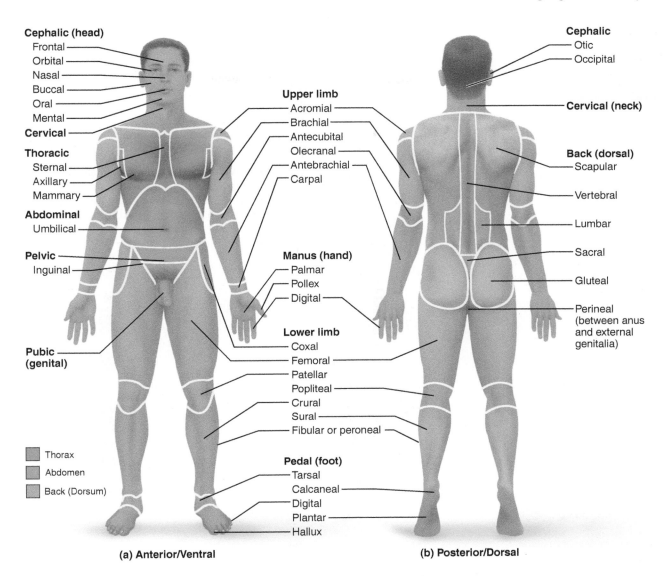

Figure 1.2 Anatomical position and regional terms. Heels are raised to illustrate the plantar surface of the foot, which is actually on the inferior surface of the body.

Patellar: Kneecap

Pedal: Foot

Pelvic: Pelvis

Pollex: Thumb

Pubic: Genital

Sternal: Breastbone

Tarsal: Ankle

Thoracic: Chest

Umbilical: Navel

Posterior Body Landmarks

Note the following body surface regions in Figure 1.2b:

Acromial: Point of the shoulder

Brachial: Arm

Calcaneal: Heel of the foot

Cephalic: Head

Dorsal: Back

Femoral: Thigh

Gluteal: Buttocks

Lumbar: Lower back

Manus: Hand

Occipital: Back of the head

Olecranal: Posterior aspect of the elbow

Otic: Ear

Pedal: Foot

Perineal: Between the anus and external genitalia

Plantar: Sole of the foot

Popliteal: Back of the knee

Sacral: Posterior region between the hip bones

Scapular: Shoulder blade

Sural: Calf

Vertebral: Spine

Locating Body Regions

Locate the anterior and posterior body regions on yourself, your lab partner, and a human torso model. ▬

Body Planes and Sections

The body is three-dimensional, and to observe its internal structures, it is often helpful and necessary to make use of a **section,** or cut. When the section is made through the body wall or through an organ, it is made along an imaginary surface or line called a **plane.** A section is named for the plane along which it is cut. Anatomists commonly refer to three planes **(Figure 1.3)**, or sections, that lie at right angles to one another.

Sagittal plane: A plane that runs longitudinally and divides the body into right and left parts is referred to as a sagittal plane. If it divides the body into equal parts, right down the midline of the body, it is called a **median,** or **midsagittal, plane.**

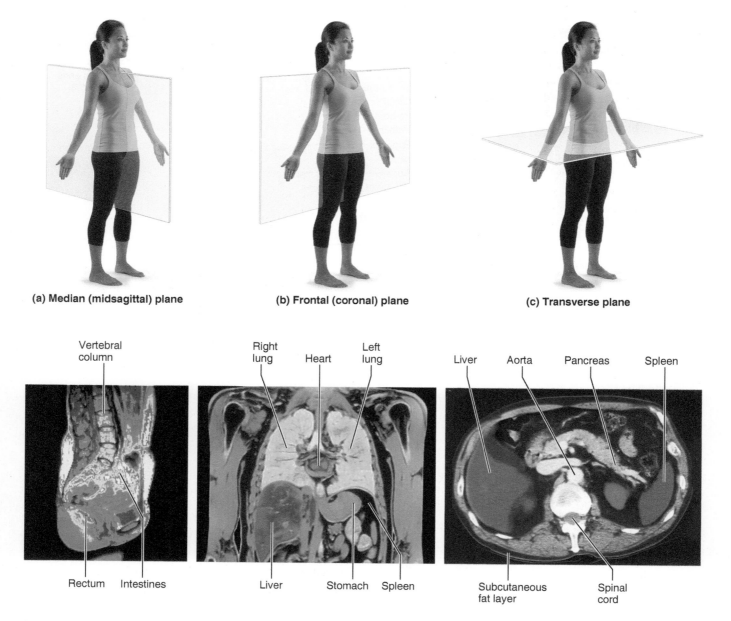

(a) Median (midsagittal) plane   (b) Frontal (coronal) plane   (c) Transverse plane

Vertebral column

Right lung   Heart   Left lung

Liver   Aorta   Pancreas   Spleen

Rectum   Intestines

Liver   Stomach   Spleen

Subcutaneous fat layer   Spinal cord

Figure 1.3 Planes of the body, with corresponding magnetic resonance imaging (MRI) scans. Note that the transverse section is an inferior view.

Frontal plane: Sometimes called a **coronal plane,** the frontal plane is a longitudinal plane that divides the body (or an organ) into anterior and posterior parts.

Transverse plane: A transverse plane runs horizontally, dividing the body into superior and inferior parts. When organs are sectioned along the transverse plane, the sections are commonly called **cross sections.**

On microscope slides, the abbreviation for a longitudinal section (sagittal or frontal) is l.s. Cross sections are abbreviated x.s. or c.s.

A median or frontal plane section of any nonspherical object, be it a banana or a body organ, provides quite a different view from a cross section (Figure 1.4).

ACTIVITY 2

Observing Sectioned Specimens

1. Go to the demonstration area and observe the transversely and longitudinally cut organ specimens (kidneys).

2. After completing instruction 1, obtain a gelatin-spaghetti mold and a scalpel, and take them to your laboratory bench. (Essentially, this is just cooked spaghetti added to warm gelatin, which is then allowed to gel.)

3. Cut through the gelatin-spaghetti mold along any plane, and examine the cut surfaces. You should see spaghetti strands that have been cut transversely (x.s.) and some cut longitudinally (a median section).

4. Draw the appearance of each of these spaghetti sections below, and verify the accuracy of your section identifications with your instructor.

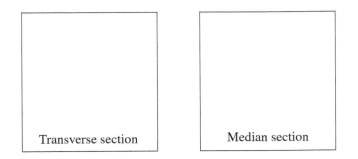

| Transverse section | Median section |

Directional Terms

Study the terms that follow, referring to Figure 1.5. Notice that certain terms have different meanings depending on whether they refer to a four-legged animal (quadruped) or to a human (biped).

Superior/inferior *(above/below):* These terms refer to placement of a structure along the long axis of the body. For example, the nose is superior to the mouth, and the abdomen is inferior to the chest.

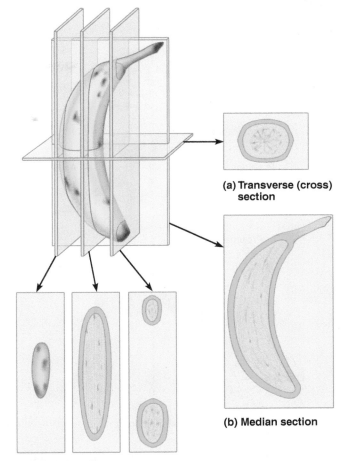

(a) Transverse (cross) section

(b) Median section

(c) Frontal sections

Figure 1.4 Objects can look odd when viewed in section. This banana has been sectioned in three different planes **(a–c)**, and only in one of these planes **(b)** is it easily recognized as a banana. If one cannot recognize a sectioned organ, it is possible to reconstruct its shape from a series of successive cuts, as from the three serial sections in **(c)**.

Anterior/posterior *(front/back):* In humans, the most anterior structures are those that are most forward—the face, chest, and abdomen. Posterior structures are those toward the backside of the body. For instance, the spine is posterior to the heart.

Medial/lateral *(toward the midline/away from the midline or median plane):* The sternum (breastbone) is medial to the ribs; the ear is lateral to the nose.

These terms of position assume the person is in the anatomical position. The next four term pairs are more absolute. They apply in any body position, and they consistently have the same meaning in all vertebrate animals.

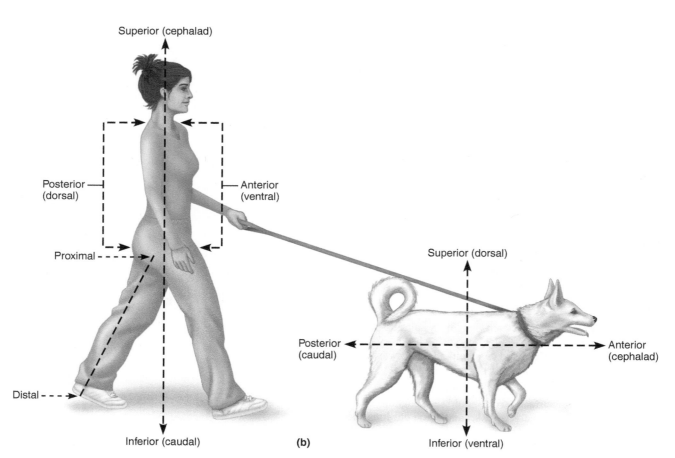

Figure 1.5 Directional terms. (a) With reference to a human. **(b)** With reference to a four-legged animal.

Cephalad (cranial)/caudal *(toward the head/toward the tail):* In humans, these terms are used interchangeably with *superior* and *inferior,* but in four-legged animals they are the same as *anterior* and *posterior,* respectively.

Ventral/dorsal *(belly side/backside):* These terms are used chiefly in discussing the comparative anatomy of animals, assuming the animal is standing. In humans, the terms *ventral* and *dorsal* are used interchangeably with the terms *anterior* and *posterior,* but in four-legged animals, *ventral* and *dorsal* are the same as *inferior* and *superior,* respectively.

Proximal/distal *(nearer the trunk or attached end/farther from the trunk or point of attachment):* These terms are used primarily to locate various areas of the body limbs. For example, the fingers are distal to the elbow; the knee is proximal to the toes. However, these terms may also be used to indicate regions (closer to or farther from the head) of internal tubular organs.

Superficial (external)/deep (internal) *(toward or at the body surface/away from the body surface):* For example, the skin is superficial to the skeletal muscles, and the lungs are deep to the rib cage.

ACTIVITY 3

Practicing Using Correct Anatomical Terminology

Use a human torso model, a human skeleton, or your own body to practice using the regional and directional terminology.

1. The wrist is _____ to the hand.
2. The trachea (windpipe) is _____ to the spine.
3. The brain is _____ to the spinal cord.
4. The kidneys are _____ to the liver.
5. The nose is _____ to the cheekbones.
6. The thumb is _____ to the ring finger.
7. The thorax is _____ to the abdomen.
8. The skin is _____ to the skeleton. ▬

Body Cavities

The axial region of the body has two large cavities that provide different degrees of protection to the organs within them (Figure 1.6).

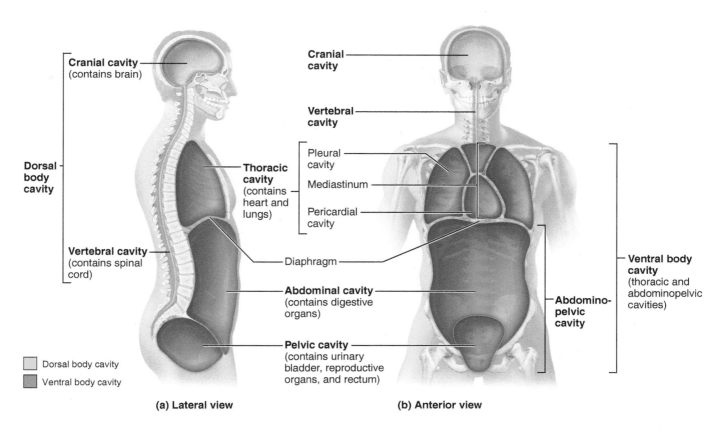

Cranial cavity
(contains brain)

Cranial cavity

Vertebral cavity

Dorsal body cavity

Pleural cavity

Mediastinum

Pericardial cavity

Thoracic cavity
(contains heart and lungs)

Vertebral cavity
(contains spinal cord)

Diaphragm

Abdominal cavity
(contains digestive organs)

Pelvic cavity
(contains urinary bladder, reproductive organs, and rectum)

Ventral body cavity
(thoracic and abdominopelvic cavities)

Abdomino-pelvic cavity

Dorsal body cavity

Ventral body cavity

(a) Lateral view   **(b) Anterior view**

Figure 1.6 Dorsal and ventral body cavities and their subdivisions.

Dorsal Body Cavity

The dorsal body cavity can be subdivided into two cavities. The **cranial cavity,** within the rigid skull, contains the brain. The **vertebral (or spinal) cavity,** which is within the bony vertebral column, protects the delicate spinal cord.

Ventral Body Cavity

Like the dorsal cavity, the ventral body cavity is subdivided. The superior **thoracic cavity** is separated from the rest of the ventral cavity by the dome-shaped diaphragm. The heart and lungs, located in the thoracic cavity, are protected by the bony rib cage. The cavity inferior to the diaphragm is often referred to as the **abdominopelvic cavity.** Although there is no further physical separation of the ventral cavity, some describe the abdominopelvic cavity as two areas, a superior **abdominal cavity** (the area that houses the stomach, intestines, liver, and other organs) and an inferior **pelvic cavity** (the region that is partially enclosed by the bony pelvis and contains the reproductive organs, bladder, and rectum).

Serous Membranes of the Ventral Body Cavity

The walls of the ventral body cavity and the outer surfaces of the organs it contains are covered with a very thin, double-layered membrane called the **serosa,** or **serous membrane.** The part of the membrane lining the cavity walls is referred to as the **parietal serosa,** and it is continuous with a similar membrane, the **visceral serosa,** covering the external surface of the organs within the cavity. These membranes produce a thin lubricating fluid that allows the visceral organs to slide over one another or to rub against the body wall without friction. Serous membranes also compartmentalize the various organs. This helps prevent infection in one organ from spreading to others.

The specific names of the serous membranes depend on the structures they surround. The serosa lining the abdominal cavity and covering its organs is the **peritoneum,** that enclosing the lungs is the **pleura,** and that around the heart is the **pericardium (Figure 1.7).** A fist pushed into a limp balloon demonstrates the relationship between the visceral and parietal serosae (Figure 1.7d).

Abdominopelvic Quadrants and Regions

Because the abdominopelvic cavity is quite large and contains many organs, it is helpful to divide it up into smaller areas for discussion or study.

Most physicians and nurses use a scheme that divides the abdominal surface and the abdominopelvic cavity into four approximately equal regions called **quadrants.** These quadrants are named according to their relative position—that is, *right upper quadrant, right lower quadrant, left upper quadrant,* and *left lower quadrant* **(Figure 1.8).** (Note that the terms *left* and *right* refer to the left and right side of the body in Figure 1.8, not the left and right side of the art on the page).

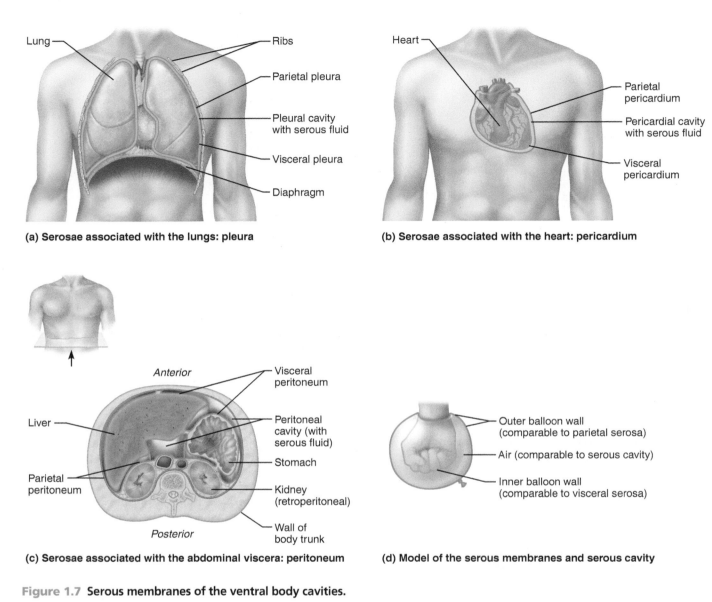

(a) Serosae associated with the lungs: pleura

Lung
Ribs
Parietal pleura
Pleural cavity with serous fluid
Visceral pleura
Diaphragm

(b) Serosae associated with the heart: pericardium

Heart
Parietal pericardium
Pericardial cavity with serous fluid
Visceral pericardium

(c) Serosae associated with the abdominal viscera: peritoneum

Anterior
Visceral peritoneum
Liver
Peritoneal cavity (with serous fluid)
Stomach
Parietal peritoneum
Kidney (retroperitoneal)
Posterior
Wall of body trunk

(d) Model of the serous membranes and serous cavity

Outer balloon wall (comparable to parietal serosa)
Air (comparable to serous cavity)
Inner balloon wall (comparable to visceral serosa)

Figure 1.7 Serous membranes of the ventral body cavities.

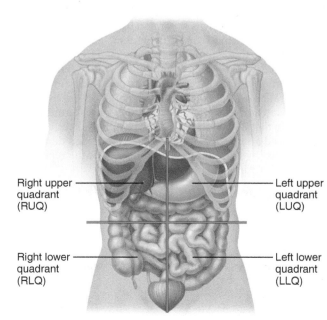

Right upper quadrant (RUQ)
Left upper quadrant (LUQ)
Right lower quadrant (RLQ)
Left lower quadrant (LLQ)

Figure 1.8 Abdominopelvic quadrants. Superficial organs are shown in each quadrant.

Identifying Organs in the Abdominopelvic Cavity

Examine the torso model to respond to the following directions and questions.

Name two organs found in the left upper quadrant.

_____ and _____

Name two organs found in the right lower quadrant.

_____ and _____

Which organ (Figure 1.8) is divided into identical halves by

the median plane? _____ ◼

A different scheme commonly used by anatomists divides the abdominal surface and abdominopelvic cavity into nine separate regions by four planes (Figure 1.9a). As you read through the descriptions of these nine regions and locate them, also note the organs the regions contain (Figure 1.9b).

Umbilical region: The centermost region, which includes the umbilicus (navel).

Epigastric region: Immediately superior to the umbilical region; overlies most of the stomach.

Pubic (hypogastric) region: Immediately inferior to the umbilical region; encompasses the pubic area.

Inguinal, or **iliac, regions:** Lateral to the hypogastric region and overlying the superior parts of the hip bones.

Lateral (lumbar) regions: Between the ribs and the flaring portions of the hip bones; lateral to the umbilical region.

Hypochondriac regions: Flanking the epigastric region laterally and overlying the lower ribs.

Locating Abdominopelvic Surface Regions

Locate the regions of the abdominopelvic surface on a human torso model. ◼

Other Body Cavities

Besides the large, closed body cavities, there are several types of smaller body cavities (Figure 1.10). Many of these are in the head, and most open to the body exterior.

Oral cavity: The oral cavity, commonly called the *mouth,* contains the tongue and teeth. It is continuous with the rest of the digestive tube, which opens to the exterior at the anus.

Nasal cavity: Located within and posterior to the nose, the nasal cavity is part of the passages of the respiratory system.

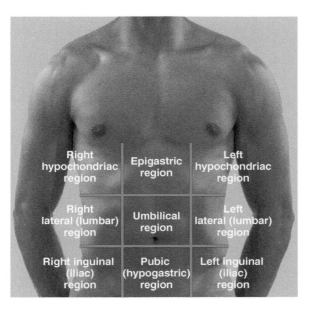

(a)

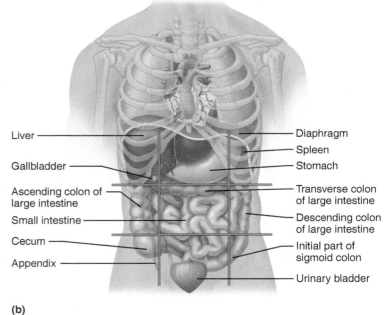

Liver
Gallbladder
Ascending colon of large intestine
Small intestine
Cecum
Appendix

Diaphragm
Spleen
Stomach
Transverse colon of large intestine
Descending colon of large intestine
Initial part of sigmoid colon
Urinary bladder

(b)

Figure 1.9 Abdominopelvic regions. Nine regions are delineated by four planes. **(a)** The superior horizontal plane is just inferior to the ribs; the inferior horizontal plane is at the superior aspect of the hip bones. The vertical planes are just medial to the nipples. **(b)** Superficial organs are shown in each region.

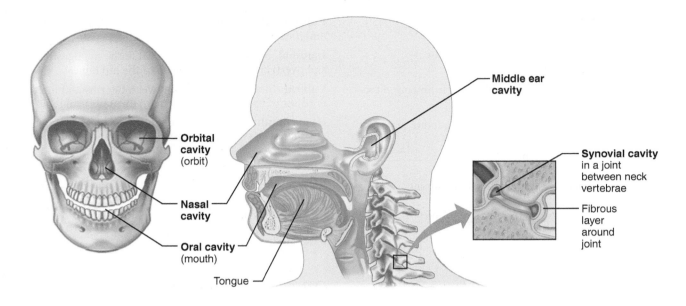

Figure 1.10 Other body cavities. The oral, nasal, orbital, and middle ear cavities are located in the head and open to the body exterior. Synovial cavities are found in joints between bones, such as the vertebrae of the spine, and at the knee, shoulder, and hip.

Orbital cavities: The orbital cavities (orbits) in the skull house the eyes and present them in an anterior position.

Middle ear cavities: Each middle ear cavity lies just medial to an eardrum and is carved into the bony skull. These cavities contain tiny bones that transmit sound vibrations to the hearing receptor in the inner ears.

Synovial cavities: Synovial cavities are joint cavities—they are enclosed within fibrous capsules that surround the freely movable joints of the body, such as those between the vertebrae and the knee and hip joints. Like the serous membranes of the ventral body cavity, membranes lining the synovial cavities secrete a lubricating fluid that reduces friction as the enclosed structures move across one another.

Name _____

Lab Time/Date _____

The Language of Anatomy

Regional Terms

1. Describe completely the standard human anatomical position. _____

_____

2. Use the regional terms to correctly label the body regions indicated on the figures below.

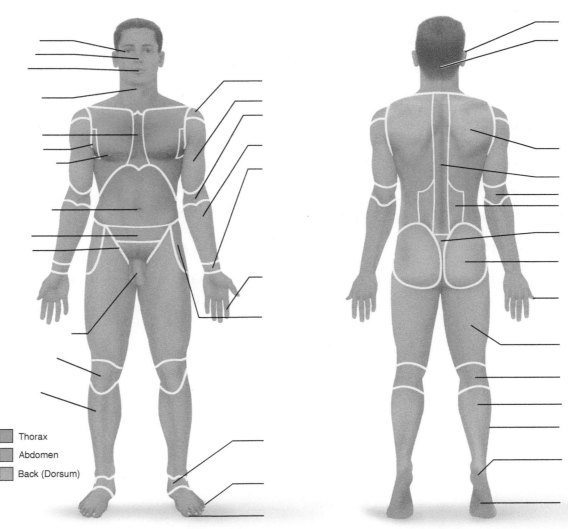

Thorax
Abdomen
Back (Dorsum)

(a) Anterior/Ventral   **(b) Posterior/Dorsal**

Directional Terms, Planes, and Sections

3. Define *plane*. _____

4. Several incomplete statements appear below. Correctly complete each statement by choosing the appropriate anatomical term from the choices. Use each term only once.

anterior	inferior	posterior	superior
distal	lateral	proximal	transverse
frontal	medial	sagittal	

    1. The thoracic cavity is _____ to the abdominopelvic cavity.

    2. The trachea (windpipe) is _____ to the vertebral column.

    3. The wrist is _____ to the hand.

    4. If an incision cuts the heart into left and right parts, a _____ plane of section was used.

    5. The nose is _____ to the cheekbones.

    6. The thumb is _____ to the ring finger.

    7. The vertebral cavity is _____ to the cranial cavity.

    8. The knee is _____ to the thigh.

    9. The plane that separates the head from the neck is the _____ plane.

   10. The popliteal region is _____ to the patellar region.

   11. The plane that separates the anterior body surface from the posterior body surface is the _____ plane.

5. Correctly identify each of the body planes by writing the appropriate term on the answer line below the drawing.

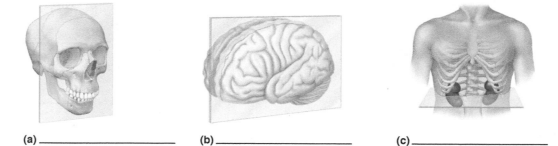

(a) _____   (b) _____   (c) _____

Body Cavities

6. Name the muscle that subdivides the ventral body cavity. _____

7. Which body cavity provides the least protection to its internal structures? _____

8. For the body cavities listed, name one organ located in each cavity.

    1. cranial cavity _____

    2. vertebral cavity _____

    3. thoracic cavity _____

4. abdominal cavity _____

5. pelvic cavity _____

6. mediastinum _____

9. Name the abdominopelvic region where each of the listed organs is located.

1. spleen _____

2. urinary bladder _____

3. stomach (largest portion) _____

4. cecum _____

10. Explain how serous membranes protect organs from infection. _____

_____

11. Which serous membrane(s) is/are found in the thoracic cavity? _____

_____

12. Which serous membrane(s) is/are found in the abdominopelvic cavity? _____

_____

13. Using the key choices, identify the small body cavities described below.

Key: a. middle ear cavity   c. oral cavity   e. synovial cavity
        b. nasal cavity   d. orbital cavity

_____ 1. holds the eyes in an anterior-facing   _____ 3. contained within the nose
             position
                                                              _____ 4. contains the tongue
_____ 2. houses three tiny bones involved in
             hearing   _____ 5. surrounds a joint

14. ✚ Name the body region that blood is usually drawn from. _____

15. ✚ A patient has been diagnosed with appendicitis. Use anatomical terminology to describe the location of the person's

pain. Assume that the pain is referred to the surface of the body above the organ. _____

16. ✚ Which body cavity would be opened to perform a hysterectomy? _____

_____

17. ✚ Which smaller body cavity would be opened to perform a total knee joint replacement? _____

_____

18. ✚ An abdominal hernia results when weakened muscles allow the protrusion of abdominal structures. In the case of an
umbilical hernia, parts of a serous membrane and the small intestine form the bulge. Which serous membrane is involved?

_____

Organ Systems Overview

MATERIALS

- Freshly killed or preserved rat (predissected by instructor as a demonstration; or for student dissection, one rat for every two to four students) or predissected human cadaver
- Dissection trays
- Twine or large dissecting pins
- Scissors
- Probes
- Forceps
- Disposable gloves
- Human torso model (dissectible)

LEARNING OUTCOMES

☐ Name the human organ systems, and indicate the major functions of each system.

☐ List several major organs of each system, and identify them in a dissected rat, human cadaver or cadaver image, or dissectible human torso model.

☐ Name the correct organ system for each organ.

PRE-LAB QUIZ

1. Name the structural and functional unit of all living things. _____

2. The small intestine is an example of a(n) _____, because it is composed of two or more tissue types that perform a particular function for the body.
   a. epithelial tissue
   b. muscle tissue
   c. organ
   d. organ system

3. The _____ system is responsible for maintaining homeostasis of the body via rapid communication.

4. The kidneys are part of the _____ system.

5. The thin muscle that separates the thoracic and abdominal cavities is the _____.

The basic unit of life is the **cell.** Cells fall into four different categories according to their structures and functions. These categories correspond to the four primary tissue types: epithelial, muscular, nervous, and connective. A **tissue** is a group of cells that are similar in structure and function. An **organ** is a structure composed of two or more tissue types that performs a specific function for the body.

An **organ system** is a group of organs that act together to perform a particular body function. For example, the organs of the digestive system work together to break down foods and absorb the end products into the bloodstream to provide nutrients and fuel for all the body's cells. In all, there are 11 organ systems **(Table 2.1)**.

Read through this summary of the body's organ systems before beginning your rat dissection or examination of the predissected human cadaver. If a human cadaver is not available, Figures 2.3 through 2.6 will serve as a partial replacement.

Table 2.1   Overview of Organ Systems of the Body

Organ system	Major component organs	Function
Integumentary	Skin, hair, and nails; cutaneous sense organs and glands	• Protects deeper organs from mechanical, chemical, and bacterial injury, and from drying out • Excretes salts and urea • Aids in regulation of body temperature • Produces vitamin D
Skeletal	Bones, cartilages, tendons, ligaments, and joints	• Supports the body and protects internal organs • Provides levers for muscular action • Cavities provide a site for blood cell formation • Bones store minerals
Muscular	Muscles attached to the skeleton	• Primary function is to contract or shorten; in doing so, skeletal muscles allow locomotion (running, walking, etc.), grasping and manipulation of the environment, and facial expression • Generates heat
Nervous	Brain, spinal cord, nerves, and sensory receptors	• Allows body to detect changes in its internal and external environment and to respond to such information by activating appropriate muscles or glands • Helps maintain homeostasis of the body via rapid communication
Endocrine	Pituitary, thymus, thyroid, parathyroid, adrenal, and pineal glands; ovaries, testes, and pancreas	• Helps maintain body homeostasis, promotes growth and development; produces chemical messengers called hormones that travel in the blood to exert their effect(s) on various target organs
Cardiovascular	Heart and blood vessels	• Primarily a transport system that carries blood containing oxygen, carbon dioxide, nutrients, wastes, ions, hormones, and other substances to and from the tissue cells where exchanges are made; blood is propelled through the blood vessels by the pumping action of the heart • Antibodies and other protein molecules in the blood protect the body
Lymphatic	Lymphatic vessels, lymph nodes, spleen, and thymus	• Picks up fluid leaked from the blood vessels and returns it to the blood • Cleanses blood of pathogens and other debris • Houses lymphocytes that act via the immune response to protect the body from foreign substances
Respiratory	Nasal cavity, pharynx, larynx, trachea, bronchi, and lungs	• Keeps the blood continuously supplied with oxygen while removing carbon dioxide • Contributes to the acid-base balance of the blood via its carbonic acid–bicarbonate buffer system
Digestive	Oral cavity, pharynx, esophagus, stomach, small and large intestines, and accessory structures including teeth, salivary glands, liver, and pancreas	• Breaks down ingested foods to smaller particles, which can be absorbed into the blood for delivery to the body cells • Undigested residue removed from the body as feces
Urinary	Kidneys, ureters, bladder, and urethra	• Filters the blood and rids the body of nitrogen-containing wastes, including urea, uric acid, and ammonia, which result from the breakdown of proteins and nucleic acids • Maintains water, electrolyte, and acid-base balance of blood
Reproductive	Male: testes, prostate, scrotum, penis, and duct system, which carries sperm to the body exterior	• Provides gametes called sperm for producing offspring
	Female: ovaries, uterine tubes, uterus, mammary glands, and vagina	• Provides gametes called eggs; the female uterus houses the developing fetus until birth; mammary glands provide nutrition for the infant

DISSECTION AND IDENTIFICATION
The Organ Systems of the Rat

Many of the external and internal structures of the rat are quite similar in structure and function to those of the human, so a study of the gross anatomy of the rat should help you understand our own anatomy. The following instructions include directions for dissecting and observing a rat. In addition, instructions for observing organs (Activity 4, "Examining the Ventral Body Cavity," page 18) also apply to superficial observations of a previously dissected human cadaver. The general instructions for observing external structures also apply to human cadaver observations. The photographs in Figures 2.3 through 2.6 will provide visual aids.

(a)

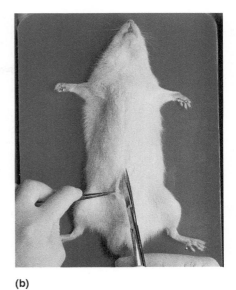

(b)

Figure 2.1 Rat dissection: Securing for dissection and the initial incision. (a) Securing the rat to the dissection tray with dissecting pins. **(b)** Using scissors to make the incision on the median line of the abdominal region.

Note that four of the organ systems listed in Table 2.1 (integumentary, skeletal, muscular, and nervous) will not be studied at this time because they require microscopic study or more detailed dissection. ■

ACTIVITY 1

Observing External Structures

1. If your instructor has provided a predissected rat, go to the demonstration area to make your observations. Alternatively, if you and/or members of your group will be dissecting the specimen, obtain a preserved or freshly killed rat, a dissecting tray, dissecting pins or twine, scissors, probe, forceps, and disposable gloves. Bring these items to your laboratory bench.

If a predissected human cadaver is available, obtain a probe, forceps, and disposable gloves before going to the demonstration area.

⚠ 2. Don the gloves before beginning your observations. This precaution is particularly important when handling freshly killed animals, which may harbor pathogens.

3. Observe the major divisions of the body—head, trunk, and extremities. If you are examining a rat, compare these divisions to those of humans. ■

ACTIVITY 2

Examining the Oral Cavity

Examine the structures of the oral cavity. Identify the teeth and tongue. Observe the extent of the hard palate (the portion underlain by bone) and the soft palate (immediately posterior to the hard palate, with no bony support). Notice that the posterior end of the oral cavity leads into the throat, or

pharynx, a passageway used by both the digestive and respiratory systems. ■

ACTIVITY 3

Opening the Ventral Body Cavity

1. Pin the animal to the wax of the dissecting tray by placing its dorsal side down and securing its extremities to the wax with large dissecting pins as shown in **Figure 2.1a**.

2. Lift the abdominal skin with a forceps, and cut through it with the scissors (Figure 2.1b). Close the scissor blades, and insert them flat under the cut skin. Moving in a cephalad direction, open and close the blades to loosen the skin from the underlying connective tissue and muscle. Now cut the skin along the body midline, from the pubic region to the lower jaw (Figure 2.1c, page 18). Finally, make a lateral cut about halfway down the ventral surface of each limb. Complete the job of freeing the skin with the scissor tips, and pin the flaps to the tray (Figure 2.1d). The underlying tissue that is now exposed is the skeletal musculature of the body wall and limbs. Notice that the muscles are packaged in sheets of pearly white connective tissue (fascia), which protect the muscles and bind them together.

3. Carefully cut through the muscles of the abdominal wall in the pubic region, avoiding the underlying organs. Remember, to *dissect* means "to separate". Now, hold and lift the muscle layer with a forceps and cut through the muscle layer from the pubic region to the bottom of the rib cage. Make two lateral cuts at the base of the rib cage **(Figure 2.2)**. A thin membrane attached to the inferior boundary of the rib cage should be obvious; this is the **diaphragm,** which separates the thoracic and abdominopelvic cavities. Cut the diaphragm where it attaches to the ventral ribs to loosen the rib cage. Cut through the rib cage on either side. You can now lift the ribs to view the contents of the thoracic cavity. Cut across the flap at the level of the neck, and remove the rib cage. ■

(c)

(d)

Figure 2.1 (*continued*) **Rat dissection: Securing for dissection and the initial incision. (c)** Completed incision from the pelvic region to the lower jaw. **(d)** Reflection (folding back) of the skin to expose the underlying muscles.

ACTIVITY 4

Examining the Ventral Body Cavity

1. Starting with the most superficial structures and working deeper, examine the structures of the thoracic cavity. Refer to **Figure 2.3** as you work. Choose the appropriate view depending on whether you are examining a rat (a) or a human cadaver (b).

Thymus: An irregular mass of glandular tissue overlying the heart (not illustrated in the human cadaver photograph).

With the probe, push the thymus to the side to view the heart.

Heart: Medial oval structure enclosed within the pericardium (serous membrane).

Lungs: Lateral to the heart on either side.

Now observe the throat region to identify the trachea.

Trachea: Tubelike "windpipe" running medially down the throat; part of the respiratory system.

Follow the trachea into the thoracic cavity; notice where it divides into two branches. These are the bronchi.

Bronchi: Two passageways that plunge laterally into the tissue of the two lungs.

To expose the esophagus, push the trachea to one side.

Esophagus: A food chute; the part of the digestive system that transports food from the pharynx (throat) to the stomach.

Diaphragm: A thin muscle attached to the inferior boundary of the rib cage.

Follow the esophagus through the diaphragm to its junction with the stomach.

Stomach: A curved organ important in food digestion and temporary food storage.

2. Examine the superficial structures of the abdominopelvic cavity. Lift the **greater omentum,** an extension of the peritoneum (serous membrane) that covers the abdominal viscera. Continuing from the stomach, trace the rest of the digestive tract **(Figure 2.4)**.

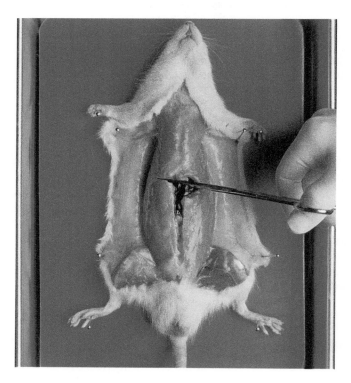

Figure 2.2 Rat dissection: Making lateral cuts at the base of the rib cage.

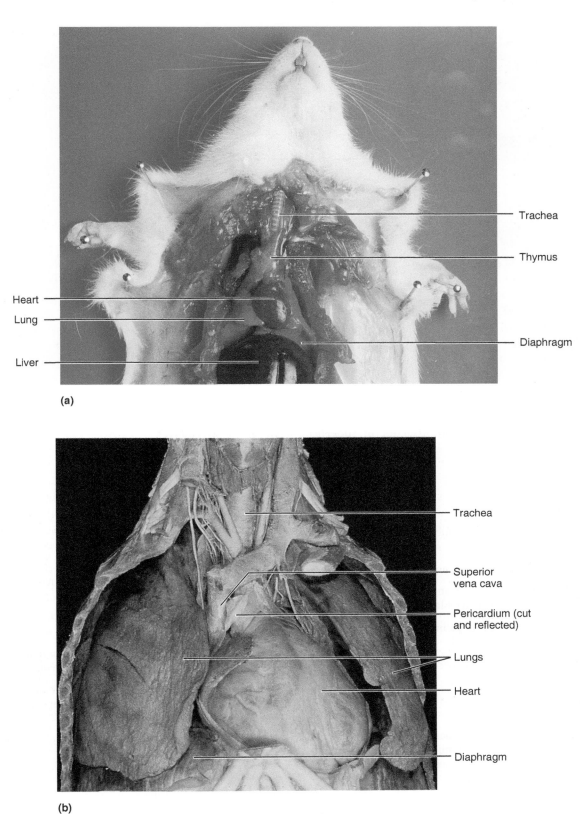

(a)

(b)

Figure 2.3 Superficial organs of the thoracic cavity. (a) Dissected rat.
(b) Human cadaver.

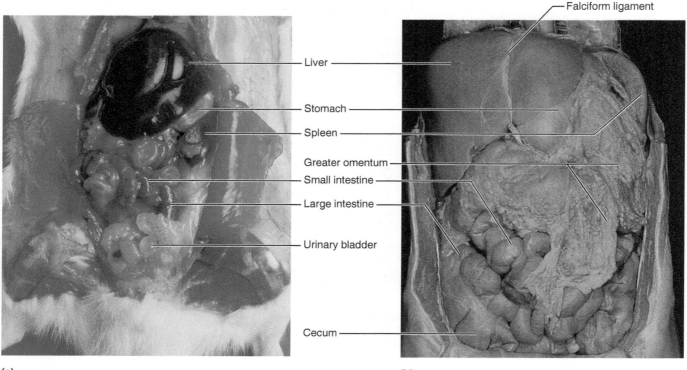

Figure 2.4 Abdominal organs. (a) Dissected rat, superficial view. **(b)** Human cadaver, superficial view.

Small intestine: Connected to the stomach and ending just before the saclike cecum.

Large intestine: A large muscular tube connected to the small intestine and ending at the anus.

Cecum: The initial portion of the large intestine.

Follow the course of the large intestine to the rectum, which is partially covered by the urinary bladder.

Rectum: Terminal part of the large intestine; continuous with the anal canal (not visible in this dissection).

Anus: The opening of the digestive tract (through the anal canal) to the exterior.

Now lift the small intestine with the forceps to view the mesentery.

Mesentery: An apronlike serous membrane; suspends many of the digestive organs in the abdominal cavity. Notice that it is heavily invested with blood vessels and, more likely than not, riddled with large fat deposits.

Locate the remaining abdominal structures.

Pancreas: A diffuse gland; rests dorsal to and in the mesentery between the first portion of the small intestine and the stomach. You will need to lift the stomach to view the pancreas.

Spleen: A dark red organ curving around the left lateral side of the stomach; an organ of the lymphatic system, and it is often called the red blood cell "graveyard."

Liver: Large and brownish red; the most superior organ in the abdominal cavity, directly beneath the diaphragm.

3.   To locate the deeper structures of the abdominopelvic cavity, move the stomach and the intestines to one side with the probe.

Examine the posterior wall of the abdominal cavity to locate the two kidneys **(Figure 2.5)**.

Kidneys: Bean-shaped organs; retroperitoneal (behind the peritoneum).

Adrenal glands: Large endocrine glands that sit on top of each kidney; considered part of the endocrine system.

Carefully strip away part of the peritoneum with forceps, and attempt to follow the course of one of the ureters to the bladder.

Ureter: Tube running from the indented region of a kidney to the urinary bladder.

Urinary bladder: The sac that serves as a reservoir for urine.

4.   In the midline of the body cavity lying between the kidneys are the two principal abdominal blood vessels:

Inferior vena cava: The large vein that returns blood to the heart from the lower body regions.

Descending aorta: Deep to the inferior vena cava; the largest artery of the body; carries blood away from the heart.

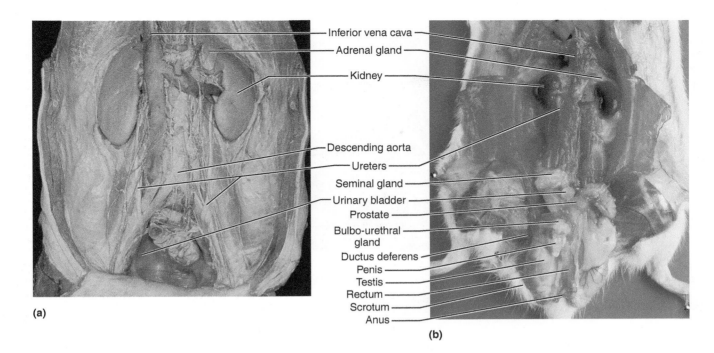

Inferior vena cava
Adrenal gland
Kidney

Descending aorta
Ureters
Seminal gland
Urinary bladder
Prostate
Bulbo-urethral gland
Ductus deferens
Penis
Testis
Rectum
Scrotum
Anus

(a)

(b)

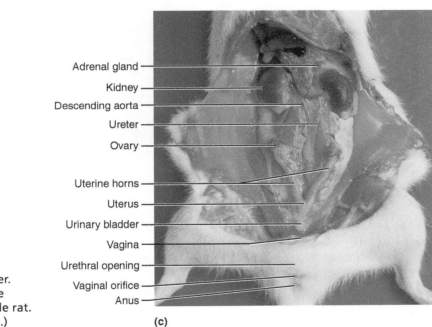

Adrenal gland
Kidney
Descending aorta
Ureter
Ovary
Uterine horns
Uterus
Urinary bladder
Vagina
Urethral opening
Vaginal orifice
Anus

(c)

Figure 2.5 Deep structures of the abdominopelvic cavity. (a) Human cadaver. **(b)** Dissected male rat. (Some reproductive structures also shown.) **(c)** Dissected female rat. (Some reproductive structures also shown.)

5.   You will perform only a brief examination of reproductive organs. If you are working with a rat, first determine whether the animal is a male or female. Observe the ventral body surface beneath the tail. If a saclike scrotum and an opening for the anus are visible, the animal is a male. If three body openings—urethral, vaginal, and anal—are present, it is a female.

Male Rat

Make a shallow incision into the **scrotum.** Loosen and lift out one oval **testis.** Exert a gentle pull on the testis to identify the slender **ductus deferens,** or **vas deferens,** which carries sperm from the testis superiorly into the abdominal cavity and joins with the urethra. The urethra runs through the penis of the male and carries both urine and sperm out of the body. Identify the **penis,** extending from the bladder to the ventral body wall. You may see other glands of the male rat's reproductive system (Figure 2.5b), but you don't need to identify them at this time.

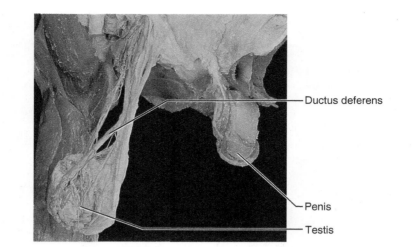

(a)

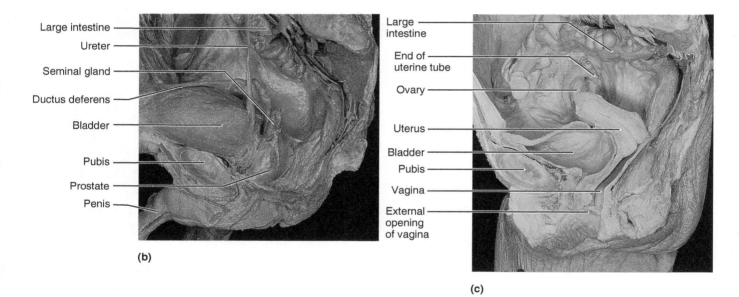

(b)

(c)

Figure 2.6 Human reproductive organs. (a) Male external genitalia. **(b)** Sagittal section of the male pelvis. **(c)** Sagittal section of the female pelvis.

Female Rat

Inspect the pelvic cavity to identify the Y-shaped **uterus** lying against the dorsal body wall and beneath the bladder (Figure 2.5c). Follow one of the uterine horns superiorly to identify an **ovary,** a small oval structure at the end of the uterine horn. (The rat uterus is quite different from the uterus of a human female, which is a single-chambered organ about the size and shape of a pear.) The inferior undivided part of the rat uterus is continuous with the **vagina,** which leads to the body exterior. Identify the **vaginal orifice** (external vaginal opening).

Male Cadaver

Make a shallow incision into the **scrotum (Figure 2.6a)**. Loosen and lift out the oval **testis.** Exert a gentle pull on the testis to identify the slender **ductus (vas) deferens,** which carries sperm from the testis superiorly into the abdominopelvic cavity (Figure 2.6b) and joins with the urethra. The urethra

runs through the penis of the male and carries both urine and sperm out of the body. Identify the **penis,** extending from the bladder to the ventral body wall.

Female Cadaver

Inspect the pelvic cavity to identify the pear-shaped **uterus** lying against the dorsal body wall and superior to the bladder. Follow one of the **uterine tubes** superiorly to identify an **ovary,** a small oval structure at the end of the uterine tube (Figure 2.6c). The inferior part of the uterus is continuous with the **vagina,** which leads to the body exterior. Identify the **vaginal orifice** (external vaginal opening).

6.   When you have finished your observations, rewrap or store the dissection animal or cadaver according to your instructor's directions. Wash the dissecting tools and equipment with laboratory detergent. Dispose of the gloves as instructed.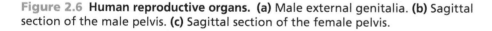

ACTIVITY 5

Examining the Human Torso Model

1. Examine a human torso model to identify the organs listed next to the photograph of the human torso model (Figure 2.7). If a torso model is not available, Figure 2.7 may be used for this part of the exercise. Some model organs will have to be removed to see the deeper organs.

2. Using the terms to the right of the figure (Figure 2.7), label each organ supplied with a leader line in the figure (Figure 2.7).

3. List each of the organs with the correct body cavity or cavities.

Dorsal body cavity_____

Thoracic cavity_____

_____

Abdominopelvic cavity _____

_____

_____

_____

_____

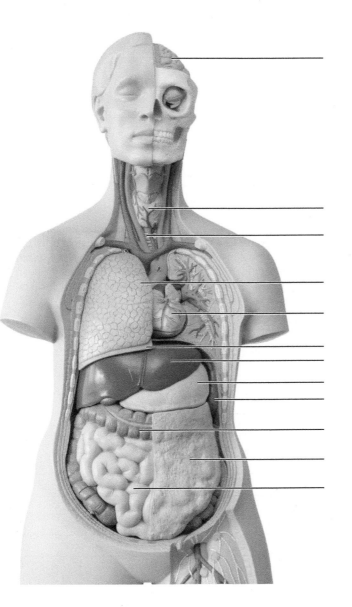

Adrenal gland
Aortic arch
Brain
Bronchi
Descending aorta
Diaphragm
Esophagus
Greater omentum
Heart
Inferior vena cava
Kidneys
Large intestine
Liver
Lungs
Pancreas
Rectum
Small intestine
Spinal cord
Spleen
Stomach
Thyroid gland
Trachea
Ureters
Urinary bladder

Figure 2.7 Human torso model.

4. Determine which organs are found in each abdominopelvic region, and record below.

Umbilical region: _____

Epigastric region: _____

Pubic (hypogastric): _____

Right inguinal (iliac) region: _____

Left inguinal (iliac) region: _____

Right lateral (lumbar) region: _____

Left lateral (lumbar) region: _____

Right hypochondriac region: _____

Left hypochondriac region: _____

Now, assign each of the organs just identified to one of the organ system categories listed below.

Digestive: _____

_____

Urinary: _____

_____

Cardiovascular: _____

_____

Endocrine: _____

_____

Reproductive: _____

_____

Respiratory: _____

_____

Lymphatic/immune: _____

_____

Nervous: _____

_____

GROUP CHALLENGE

Odd Organ Out

Each box below contains four organs. One of the listed organs in each case does *not* share a characteristic that the other three do. Work in groups of three, and discuss the characteristics of the four organs in each set. On a separate piece of paper, one student will record the characteristics of each organ in the set. For each set of four organs, discuss the possible candidates for the "odd organ" and which characteristic it lacks, based on your recorded notes. Once you have come to a consensus among your group, circle the organ that doesn't belong with the others, and explain why it is singled out. Include as many reasons as you can think of, but make sure the organ does not have the key characteristic(s). Use Table 2.1 and the pictures in your lab manual to help you select and justify your answer.

1. Which is the "odd organ"?	Why is it the odd one out?
Stomach Small intestine Teeth Oral cavity	
2. Which is the "odd organ"?	**Why is it the odd one out?**
Thyroid gland Spleen Thymus Lymph nodes	
3. Which is the "odd organ"?	**Why is it the odd one out?**
Ovaries Uterus Prostate gland Uterine tubes	
4. Which is the "odd organ"?	**Why is it the odd one out?**
Stomach Esophagus Small intestine Large intestine	

Organ Systems Overview

1. Use the key below to indicate which body systems perform the following functions. (Some responses are used more than once.) Then, circle the organ systems (in the key) that are present in all subdivisions of the ventral body cavity.

   Key: a. cardiovascular   d. integumentary   g. nervous   j. skeletal
          b. digestive   e. lymphatic/immune   h. reproductive   k. urinary
          c. endocrine   f. muscular   i. respiratory

   _____ 1. rids the body of nitrogen-containing wastes

   _____ 2. is affected by removal of the thyroid gland

   _____ 3. provides support and the levers on which the muscular system acts

   _____ 4. includes the heart

   _____ 5. protects underlying organs from drying out and from mechanical damage

   _____ 6. protects the body; destroys bacteria and tumor cells

   _____ 7. breaks down ingested food into its building blocks

   _____ 8. removes carbon dioxide from the blood

   _____ 9. delivers oxygen and nutrients to the tissues

   _____ 10. moves the limbs; facilitates facial expression

   _____ 11. regulates water balance and removes nitrogen-containing wastes from the body

   _____ and _____ 12. facilitate conception and childbearing

   _____ 13. controls the body by means of chemical molecules called hormones

   _____ 14. is damaged when you cut your finger or get a severe sunburn

2. Using the key above, choose the *organ system* to which each of the following sets of organs or body structures belongs.

   _____ 1. thymus, spleen,
                     lymphatic vessels

   _____ 2. bones, cartilages,
                     tendons

   _____ 3. pancreas, pituitary
                     gland

   _____ 4. trachea, bronchi,
                     lungs

   _____ 5. epidermis, dermis,
                     cutaneous sense organs

   _____ 6. testis, prostate

   _____ 7. liver, large intestine,
                     rectum

   _____ 8. kidneys, ureter, urethra

3. Name the cells that are produced by the testes and ovaries. _____

   _____

4. List the four primary tissue types. _____

   _____

5. Explain why an artery is an organ. _____

   _____

6. Name the two main organ systems that communicate within the body to maintain homeostasis. Briefly explain their different

   control mechanisms. _____

   _____

   _____

7. Explain the role that the skeletal system plays in facilitating cardiovascular system function. _____

   _____

8. ➕ Untreated diabetes mellitus can lead to a condition in which the blood is more acidic than normal. Name two organ

   systems that play the largest role in compensating for acid-base imbalances. _____

   _____

   _____

9. ➕ The mother of a child scheduled to receive a thymectomy (removal of the thymus gland) asks you whether there will
   be any side effects from the removal of the gland. Which two organ systems would you mention in your explanation?

   _____

   _____

10. ➕ Individuals with asplenia are missing their spleen or have a spleen that doesn't function well. It is recommended that
    these patients talk to their doctor about vaccines that are indicated for their health condition. Explain how this recommendation

    correlates to their chronic health condition. _____

    _____

    _____

The Microscope

MATERIALS

- Compound microscope
- Millimeter ruler
- Prepared slides of the letter *e* or newsprint
- Immersion oil
- Lens paper
- Prepared slide of grid ruled in millimeters
- Prepared slide of three crossed colored threads
- Clean microscope slide and coverslip
- Toothpicks (flat-tipped)
- Physiological saline in a dropper bottle
- Iodine or dilute methylene blue stain in a dropper bottle
- Filter paper or paper towels
- Beaker containing fresh 10% household bleach solution for wet mount disposal
- Disposable autoclave bag
- Prepared slide of cheek epithelial cells

Note to the Instructor: The slides and coverslips used for viewing cheek cells are to be soaked for 2 hours (or longer) in 10% bleach solution and then drained. The slides and disposable autoclave bag containing coverslips, lens paper, and used toothpicks are to be autoclaved for 15 min at 121°C and 15 pounds pressure to ensure sterility. After autoclaving, the disposable autoclave bag may be discarded in any disposal facility and the slides and glassware washed with laboratory detergent and prepared for use. These instructions also apply to any bloodstained glassware or disposable items used in other experimental procedures.

LEARNING OUTCOMES

- ☐ Identify the parts of the microscope, and list the function of each part.
- ☐ Describe and demonstrate the proper techniques for care of the microscope.
- ☐ Demonstrate proper focusing technique.
- ☐ Define *total magnification, resolution, parfocal, field, depth of field,* and *working distance*.
- ☐ Measure the field diameter for one objective lens, calculate it for all the other objective lenses, and estimate the size of objects in each field.
- ☐ Discuss the general relationships among magnification, working distance, and field diameter.

PRE-LAB QUIZ

1. The microscope slide rests on the _____ while being viewed.
   a. base   b. condenser   c. iris   d. stage
2. Your lab microscope is *parfocal*. What does this mean?
   a. The specimen is clearly in focus at this depth.
   b. The slide should be almost in focus when you change to higher magnifications.
   c. You can easily discriminate two close objects as separate.
3. If the ocular lens magnifies a specimen 10× and the objective lens magnifies the specimen 35× what is the total magnification being used to observe the specimen?_____
4. How do you clean the lenses of your microscope?
   a. with a paper towel   b. with soap and water
   c. with special lens paper and cleaner
5. Circle True or False. You should always start your observation of specimens with the oil-immersion lens.

With the invention of the microscope, biologists gained a valuable tool to observe and study structures, such as cells, that are too small to be seen by the unaided eye. The knowledge they acquired helped establish many of the theories basic to the biological sciences. This exercise will familiarize you with the workhorse of microscopes—the compound microscope—and provide you with the necessary instructions for its proper use.

Care and Structure of the Compound Microscope

The **compound microscope** is a precision instrument and should always be handled with care. *At all times you must observe the following rules for its transport, cleaning, use, and storage:*

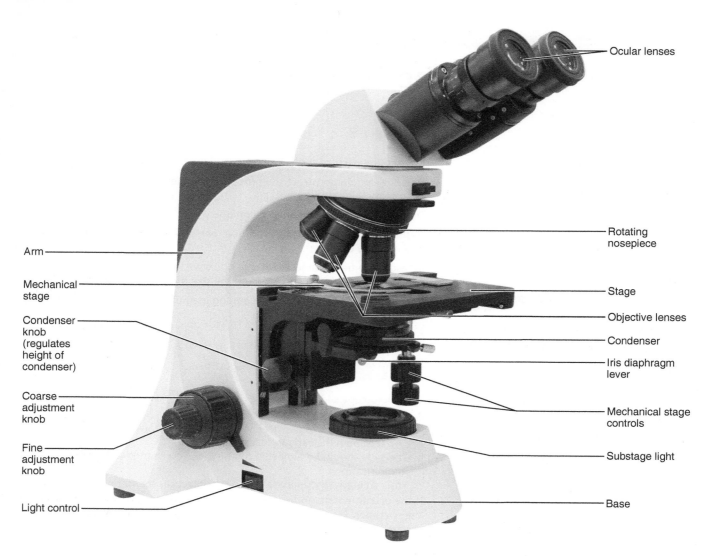

Arm

Mechanical stage

Condenser knob (regulates height of condenser)

Coarse adjustment knob

Fine adjustment knob

Light control

Ocular lenses

Rotating nosepiece

Stage

Objective lenses

Condenser

Iris diaphragm lever

Mechanical stage controls

Substage light

Base

Figure 3.1 Compound microscope and its parts.

• When transporting the microscope, hold it in an upright position with one hand on its arm and the other supporting its base. Do not swing the instrument during its transport or jar the instrument when setting it down.

• Use only special grit-free lens paper to clean the lenses. Use a circular motion to wipe the lenses, and clean all lenses before and after use.

• Always begin the focusing process with the scanning objective lens in position, changing to the higher-power lenses as necessary.

• Use the coarse adjustment knob only to adjust the scanning lens.

• Always use a coverslip with wet mount preparations.

• Before putting the microscope in the storage cabinet, remove the slide from the stage, rotate the scanning objective lens into position, wrap the cord as directed, and replace the dust cover or return the microscope to the appropriate storage area.

• Never remove any parts from the microscope; inform your instructor of any mechanical problems that arise.

ACTIVITY 1

Identifying the Parts of a Microscope

1. Using the proper transport technique, obtain a microscope and bring it to the laboratory bench.

• Record the number of your microscope in the **Summary Chart** (on page 31).

Compare your microscope with the illustration **(Figure 3.1)**, and identify the microscope parts described in **Table 3.1**.

Table 3.1   Parts of the Microscope

Microscope part	Description and function
Base	The bottom of the microscope. Provides a sturdy flat surface to support and steady the microscope.
Substage light	Located in the base. The light from the lamp passes directly upward through the microscope.
Light control	Located on the base or arm. This dial allows you to adjust the intensity of the light passing through the specimen.
Stage	The platform that the slide rests on while being viewed. The stage has a hole in it to allow light to pass through the stage and through the specimen.
Mechanical stage	Holds the slide in position for viewing and has two adjustable knobs that control the precise movement of the slide.
Condenser	Small nonmagnifying lens located beneath the stage that concentrates the light on the specimen. The condenser may have a knob that raises and lowers the condenser to vary the light delivery. Generally, the best position is close to the inferior surface of the stage.
Iris diaphragm lever	The iris diaphragm is a shutter within the condenser that can be controlled by a lever to adjust the amount of light passing through the condenser. The lever can be moved to close the diaphragm and improve contrast. If your field of view is too dark, you can open the diaphragm to let in more light.
Coarse adjustment knob	This knob allows you to make large adjustments to the height of the stage to initially focus your specimen.
Fine adjustment knob	This knob is used for precise focusing once the initial coarse focusing has been completed.
Head	Attaches to the nosepiece to support the objective lens system. It also provides for attachment of the eyepieces which house the ocular lenses.
Arm	Vertical portion of the microscope that connects the base and the head.
Nosepiece	Rotating mechanism connected to the head. Generally, it carries three or four objective lenses and permits positioning of these lenses over the hole in the stage.
Objective lenses	These lenses are attached to the nosepiece. Usually, a compound microscope has four objective lenses: scanning ($4\times$), low-power ($10\times$), high-power ($40\times$), and oil immersion ($100\times$) lenses. Typical magnifying powers for the objectives are listed in parentheses.
Ocular lens(es)	Binocular microscopes will have two lenses located in the eyepieces at the superior end of the head. Most ocular lenses have a magnification power of $10\times$. Some microscopes will have a pointer and/or reticle (micrometer), which can be positioned by rotating the ocular lens.

2.   Examine the objective lenses carefully; note their relative lengths and the numbers inscribed on their sides. On many microscopes, the scanning lens, with a magnification of $4\times$, is the shortest lens. The low-power objective lens typically has a magnification of $10\times$. The high-power objective lens is of intermediate length and has a magnification range from $40\times$ to $50\times$, depending on the microscope. The oil immersion objective lens is usually the longest of the objective lenses and has a magnifying power of $95\times$ to $100\times$. Some microscopes lack the oil immersion lens.

•   Record the magnification of each objective lens of your microscope in the first row of the Summary Chart (page 31). Also, cross out any column relating to a lens that your microscope does not have. Plan on using the same microscope for all microscopic studies. ▪

Magnification and Resolution

The microscope is an instrument of magnification. With the compound microscope, magnification is achieved through the interplay of two lenses—the ocular lens and the objective lens. The objective lens magnifies the specimen to produce a **real image** that is projected to the ocular. This real image is magnified by the ocular lens to produce the **virtual image** that your eye sees (Figure 3.2).

The **total magnification (TM)** of any specimen being viewed is equal to the power of the ocular lens multiplied by the power of the objective lens being used. For example, if the ocular lens magnifies $10\times$ and the objective lens magnifies $45\times$, the total magnification is $450\times$ (or 10×45).

•   Determine the total magnification for each of the objectives on your microscope, and record the figures on the third row of the Summary Chart.

The compound light microscope has certain limitations. Although the level of magnification is almost limitless, the **resolution** (or resolving power), that is, the ability to discriminate two close objects as separate, is not. The human eye can resolve objects about 100 μm apart, but the compound microscope has a resolution of 0.2 μm under ideal conditions. Objects closer than 0.2 μm are seen as a single fused image.

Resolution is determined by the amount and physical properties of the visible light that enters the microscope. In general, the more light delivered to the objective lens, the

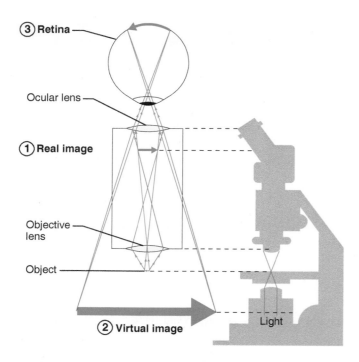

Figure 3.2 Image formation in light microscopy.
Step 1: The objective lens magnifies the object, forming the real image. **Step 2:** The ocular lens magnifies the real image, forming the virtual image. **Step 3:** The virtual image passes through the lens of the eye and is focused on the retina.

greater the resolution. The size of the objective lens aperture (opening) decreases with increasing magnification, allowing less light to enter the objective. Thus, you will probably find it necessary to increase the light intensity at the higher magnifications.

ACTIVITY 2

Viewing Objects Through the Microscope

1. Obtain a millimeter ruler, a prepared slide of the letter *e* or newsprint, a dropper bottle of immersion oil, and some lens paper. Adjust the condenser to its highest position, and switch on the light source of your microscope.

2. Secure the slide on the stage so that you can read the slide label and the letter *e* is centered over the light beam passing through the stage. On the mechanical stage, open the jaws of the slide holder by using the control lever, typically located at the rear left corner of the mechanical stage. Insert the slide squarely within the confines of the slide holder.

3. With your scanning objective lens in position over the stage, use the coarse adjustment knob to bring the objective lens and stage as close together as possible.

4. Look through the ocular lens, and using the iris diaphragm lever, adjust the light for comfort. Now use the coarse adjustment knob to focus slowly away from the *e* until it is as clearly focused as possible. Complete the focusing with the fine adjustment knob.

5. Sketch the letter *e* in the circle on the Summary Chart just as it appears in the **field** (the area you see through the microscope).

How far is the bottom of the objective lens from the surface of the slide? In other words, what is the **working distance**? Use a millimeter ruler to measure the working distance.

Record the working distance in the Summary Chart.

How has the apparent orientation of the *e* changed top to bottom, right to left, and so on?

_____

_____

6. Move the slide slowly away from you on the stage as you view it through the ocular lens. In what direction does the image move?

_____

Move the slide slowly to the left. In what direction does the image move?

_____

7. Today, most good laboratory microscopes are **parfocal;** that is, the slide should be in focus (or nearly so) at the higher magnifications once you have properly focused at the lower magnification. *Without touching the focusing knobs,* increase the magnification by rotating the next higher magnification lens into position over the stage. Make sure it clicks into position. Using the fine adjustment only, sharpen the focus. If you are unable to focus with a new lens, your microscope is not parfocal. Do not try to force the lens into position. Consult your instructor. Note the decrease in working distance. As you can see, focusing with the coarse adjustment knob could drive the objective lens through the slide, breaking the slide and possibly damaging the lens. Sketch the letter *e* in the Summary Chart. What new details become clear?

_____

_____

As best you can, measure the distance between the objective and the slide.

Record the working distance in the Summary Chart.

Is the image larger or smaller? _____

Approximately how much of the letter *e* is visible now?

_____

Is the field diameter larger or smaller? _____

Summary Chart for Microscope # _____	Scanning	Low power	High power	Oil immersion
Magnification of objective lens	_____ ×	_____ ×	_____ ×	_____ ×
Magnification of ocular lens	__10__ ×	__10__ ×	__10__ ×	__10__ ×
Total magnification	_____ ×	_____ ×	_____ ×	_____ ×
Working distance	_____ mm	_____ mm	_____ mm	_____ mm
Detail observed Letter e	◯	◯	◯	◯
Field diameter	___mm___μm	___mm___μm	___mm___μm	___mm___μm

Why is it necessary to center your object (or the portion of the slide you wish to view) before changing to a higher power?

_____

Move the iris diaphragm lever while observing the field. What happens?

_____

Is it more desirable to increase *or* to decrease the light when changing to a higher magnification?

_____Why? _____

_____

8. If you have just been using the low-power objective, repeat the steps given in direction 7 using the high-power objective lens. What new details become clear?

_____

_____

Record the working distance in the Summary Chart.

9. Without touching the focusing knob, rotate the high-power lens out of position so that the area of the slide over the opening in the stage is unobstructed. Place a drop of immersion oil over the *e* on the slide and rotate the oil immersion lens into position. Set the condenser at its highest point (closest to the stage), and open the diaphragm fully. Adjust the fine focus and fine-tune the light for the best possible resolution.

Note: If for some reason the specimen does not come into view after you adjust the fine focus, do not go back to the 40×

lens to recenter. You do not want oil from the oil immersion lens to cloud the 40× lens. Turn the revolving nosepiece in the other direction to the low-power lens, and recenter and refocus the object. Then move the immersion lens back into position, again avoiding the 40× lens. Sketch the letter *e* in the Summary Chart. What new details become clear?

_____

_____

Is the field diameter again decreased in size? _____

As best you can, estimate the working distance, and record it in the Summary Chart. Is the working distance less *or* greater than it was when the high-power lens was focused?

_____

Compare your observations on the relative working distances of the objective lenses with the illustration in **Figure 3.3**.

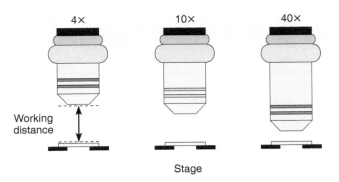

Figure 3.3 Relative working distances of the 4×,10×, and 40× objectives.

Explain why it is desirable to begin the focusing process in the lowest power.

_____

_____

_____

10. Rotate the oil immersion lens slightly to the side and remove the slide. Clean the oil immersion lens carefully with lens paper, and then clean the slide in the same manner with a fresh piece of lens paper. ▬

The Microscope Field

The microscope field decreases with increasing magnification. Measuring the diameter of each of the microscope fields will allow you to estimate of the size of the objects you view in any field. For example, if you have calculated the field diameter to be 4 mm and the object being observed extends across half this diameter, you can estimate the length of the object to be approximately 2 mm.

Microscopic specimens are usually measured in micrometers and millimeters, both units of the metric system. You can get an idea of the relationship and meaning of these units from **Table 3.2**. (A more detailed treatment appears in the Appendix.)

ACTIVITY 3

Estimating the Diameter of the Microscope Field

1. Obtain a grid slide (a slide prepared with graph paper ruled in millimeters). Each of the squares in the grid is 1 mm on each side. Use your scanning objective to bring the grid lines into focus.

2. Move the slide so that one grid line touches the edge of the field on one side, and then count the number of squares you can see across the diameter of the field. If you can see only

Table 3.2	Comparison of Metric Units of Length	
Metric unit	**Abbreviation**	**Equivalent**
Meter	m	(about 39.3 in.)
Centimeter	cm	10^{-2} m
Millimeter	mm	10^{-3} m
Micrometer (or micron)	μm (μ)	10^{-6} m
Nanometer (or millimicrometer or millimicron)	nm(mμ)	10^{-9} m
Ångstrom	Å	10^{-10} m

part of a square, as in the accompanying diagram, estimate the part of a millimeter that the partial square represents.

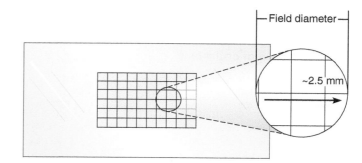

Record this figure in the appropriate space marked "field diameter" on the Summary Chart (page 31). (If you have been using the scanning lens, repeat the procedure with the low-power objective lens.)

Complete the chart by computing the approximate diameter of the high-power and oil immersion fields. The general formula for calculating the unknown field diameter is as follows:

Diameter of field A × total magnification of field A =

diameter of field B × total magnification of field B

where A represents the known or measured field and B represents the unknown field. This can be simplified to

Diameter of field B =

$$\frac{\text{diameter of field } A \times \text{total magnification of field } A}{\text{total magnification of field B}}$$

For example, if the diameter of the low-power field (field A) is 2 mm and the total magnification is 50×, you would compute the diameter of the high-power field (field B) with a total magnification of 100× as follows:

Field diameter B = (2 mm × 50)/100

Field diameter B = 1 mm

3. Estimate the length (longest dimension) of the following drawings of microscopic objects. _Base your calculations on the field diameter you have determined for your microscope and the approximate percentage of the diameter that the object occupies._

The first one is done for you.

a. Fat cell seen in 400× (total magnification, TM) field:

   Field diameter = 0.4 mm = 400 μm

   Portion of the field diameter occupied by the object = 1/3

   Approximate length = 133 μm

   400× (TM)

b. Smooth muscle seen in 400× (TM) field:

   Approximate length:

   _____ mm

   or _____ μm

c. Cheek cell seen in oil immersion field:

Approximate length:

_____ μm

Perceiving Depth

Any microscopic specimen has depth as well as length and width; it is rare indeed to view a tissue slide with just one layer of cells. Normally you can see two or three cell thicknesses. Therefore, it is important to learn how to determine relative depth with your microscope. In microscope work, the **depth of field** (the thickness of the plane that is clearly in focus) is greater at lower magnifications. As magnification increases, the depth of field decreases.

ACTIVITY 4

Perceiving Depth

1. Obtain a slide with colored crossed threads. Focusing at low magnification, locate the point where the three threads cross each other.

2. Use the iris diaphragm lever to greatly reduce the light, thus increasing the contrast. Focus down with the coarse adjustment until the threads are out of focus, then slowly focus upward again, noting which thread comes into clear focus first. (You will see two or even all three threads, so you must be very careful in determining which one first comes into clear focus.) Observe: As you rotate the adjustment knob forward (away from you), does the stage rise or fall? If the stage rises, then the first clearly focused thread is the top one; the last clearly focused thread is the bottom one.

If the stage descends, how is the order affected? _____

_____

Record your observations as to which color of thread is uppermost, in the middle, or lowest:

Top thread _____

Middle thread _____

Bottom thread _____

Viewing Cells Under the Microscope

There are various ways to prepare cells for viewing under a microscope. One method of preparation is to mix the cells in physiological saline (called a wet mount) and stain them with methylene blue stain.

If you are not instructed to prepare your own wet mount, obtain a prepared slide of epithelial cells to make the observations in step 10 of Activity 5.

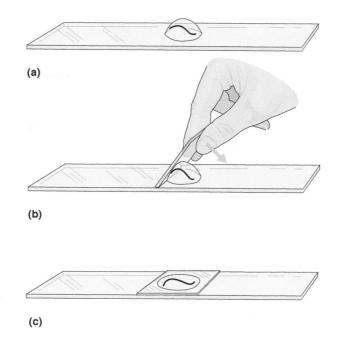

(a)

(b)

(c)

Figure 3.4 Procedure for preparing a wet mount.
(a) Place the object in a drop of water (or saline) on a clean slide, **(b)** hold a coverslip at a 45° angle with the fingertips, and **(c)** lower the coverslip slowly.

ACTIVITY 5

Preparing and Observing a Wet Mount

1. Obtain the following: a clean microscope slide and coverslip, two flat-tipped toothpicks, a dropper bottle of physiological saline, a dropper bottle of iodine or methylene blue stain, and filter paper (or paper towels). Handle only your own slides throughout the procedure.

2. Place a drop of physiological saline in the center of the slide. Using the flat end of the toothpick, *gently* scrape the inner lining of your cheek. Transfer your cheek scrapings to the slide by agitating the end of the toothpick in the drop of saline **(Figure 3.4a)**.

⚠ *Immediately* discard the used toothpick in the disposable autoclave bag provided at the supplies area.

3. Add a tiny drop of the iodine or methylene blue stain to the preparation. (These epithelial cells are nearly transparent and thus difficult to see without the stain, which colors the nuclei of the cells and makes them look much darker than the cytoplasm.) Stir with a clean toothpick.

⚠ *Immediately* discard the used toothpick in the disposable autoclave bag provided at the supplies area.

4. Hold the coverslip with your fingertips so that its bottom edge touches one side of the fluid drop (Figure 3.4b), then *slowly* lower the coverslip onto the preparation (Figure 3.4c). *Do not just drop the coverslip,* or you will trap large air bubbles under it, which will obscure the cells. *Always use a coverslip with a wet mount* to prevent soiling the lens if you should misfocus.

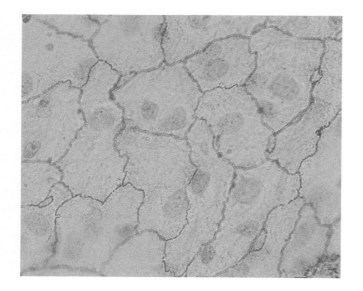

Figure 3.5 Epithelial cells of the cheek cavity (surface view, 630×).

5. Examine your preparation carefully. The coverslip should be tight against the slide. If there is excess fluid around its edges, you will need to remove it. Obtain a piece of filter paper, fold it in half, and use the folded edge to absorb the excess fluid. You may use a twist of paper towel as an alternative.

 Before continuing, discard the filter paper or paper towel in the disposable autoclave bag.

6. Place the slide on the stage, and locate the cells in low power. You will probably want to dim the light with the iris diaphragm lever to provide more contrast for viewing the lightly stained cells.

7. Cheek epithelial cells are very thin, flat cells. In the cheek, they provide a smooth, tilelike lining (Figure 3.5). Move to high power to examine the cells more closely.

8. Make a sketch of the epithelial cells that you observe.

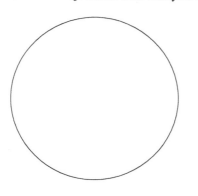

Use information on your Summary Chart (page 31) to estimate the diameter of cheek epithelial cells. Record the total magnification (TM) used.

_____ μm   _____ ×(TM)

Why do *your* cheek cells look different from those illustrated in Figure 3.5? (*Hint:* What did you have to *do* to your cheek to obtain them?)

_____

_____

⚠ 9. When you complete your observations of the wet mount, dispose of your wet mount preparation in the beaker of bleach solution, and put the coverslips in an autoclave bag.

10. Obtain a prepared slide of cheek epithelial cells, and view them under the microscope.

Estimate the diameter of one of these cheek epithelial cells using information from the Summary Chart (page 31).

_____ μm   _____ ×(TM)

Why are these cells more similar to those seen in Figure 3.5 and easier to measure than those of the wet mount?

_____

_____

11. Before leaving the laboratory, make sure all other materials are properly discarded or returned to the appropriate laboratory station. Clean the microscope lenses, and put the dust cover on the microscope before you return it to the storage cabinet. ▬

The Microscope

Care and Structure of the Compound Microscope

1. Label all indicated parts of the microscope

2. Determine whether each of the following statements is true or false. If it is true, write *T* on the answer blank. If it is false, correct the statement by writing on the blank the proper word or phrase to replace the one that is underlined.

_____ 1.   The microscope lens may be cleaned <u>with any soft tissue</u>.

_____ 2.   The microscope should be stored with the <u>oil immersion</u> lens in position over the stage.

_____ 3.   When beginning to focus, you should use the <u>scanning objective</u> lens.

_____ 4.   When focusing with the high-power lens, always use the <u>coarse</u> adjustment knob to focus.

_____ 5.   A coverslip should always be used <u>with wet mounts</u>.

3. Match the microscope structures given in column B with the statements in column A that identify or describe them.

Column A

_____ 1.   platform on which the slide rests for viewing

_____ 2.   lens located at the superior end of the body tube

_____ 3.   controls the movement of the slide on the stage

_____ 4.   delivers a concentrated beam of light to the specimen

_____ 5.   used for precise focusing once initial focusing has been done

_____ 6.   carries the objective lenses; rotates so that the different objective lenses can be brought into position over the specimen

_____ 7.   used to adjust the amount of light passing through the specimen

Column B

a.   coarse adjustment knob

b.   condenser

c.   fine adjustment knob

d.   iris diaphragm lever

e.   mechanical stage

f.   movable nosepiece

g.   objective lenses

h.   ocular lenses

i.   stage

4. Explain the proper technique for transporting the microscope.

_____

_____

5. Define the following terms.

total magnification: _____

_____

resolution: _____

_____

Viewing Objects Through the Microscope

6. Complete, or respond to, the following statements:

_____ 1. The distance from the bottom of the objective lens in use to the surface of the slide is called the_____.

_____ 2. Assume there is an object on the left side of the field that you want to bring to the center (that is, toward the apparent right). In what direction would you move your slide?

_____ 3. The area of the specimen seen when looking through the microscope is the _____.

_____ 4. If a microscope has a 10× ocular lens and the total magnification is 950×, the objective lens in use at that time is _____ ×.

_____ 5. Why should the light be dimmed when you are looking at living (nearly transparent) cells?

_____ 6. After focusing in low power, you find that you need to use only the fine adjustment to focus the specimen at the higher powers. The microscope is therefore said to be _____.

_____ 7. You are using a 10× ocular and a 15× objective. If the field diameter is 1.5 mm, the approximate field size with a 30× objective is _____ mm.

_____ 8. If the diameter of the low-power field is 1.5 mm, an object that occupies approximately a third of that field has an estimated diameter of _____ mm.

7. You have been asked to prepare a slide with the letter *k* on it (as shown below). In the circle below, draw the *k* as seen in the low-power field.

8. Estimate the length (longest dimension) of the object in μm:

Total magnification = 100×

Field diameter = 1.6 mm

Length of object = _____ μm

9. Say you are observing an object in the low-power field. When you switch to high power, it is no longer in your field of view.

Why might this occur? _____

_____

What should you have done initially to prevent this from happening? _____

_____

10. Do the following factors increase or decrease as one moves to higher magnifications with the microscope?

resolution: _____   amount of light needed: _____

working distance: _____   depth of field: _____

11. A student has the high-power lens in position and appears to be intently observing the specimen. The instructor, noting a working distance of about 1 cm, knows the student isn't actually seeing the specimen.

How so? _____

12. Describe the proper procedure for preparing a wet mount.

_____

_____

_____

13. Indicate the probable cause of the following situations during use of a microscope.

a. Only half of the field is illuminated: _____

_____

b. The field does not change as mechanical stage is moved: _____

_____

14. ➕ A blood smear is used to diagnose malaria. In patients with malaria, the protozoa can be found near and inside red blood cells. Explain why a microscope capable of high magnification and high resolution would be needed to diagnose malaria.

_____

15. ➕ Histopathology is the use of microscopes to view tissues to diagnose and track the progression of diseases. Why are

thin slices of tissue ideal for this procedure? _____

_____

The Cell: Anatomy and Division

MATERIALS

- Three-dimensional model of the "composite" animal cell or laboratory chart of cell anatomy
- Compound microscope
- Prepared slides of simple squamous epithelium (AgNO₃ stain), teased smooth muscle (l.s.), human blood cell smear, and sperm
- Video of mitosis
- Three-dimensional models of mitotic stages
- Prepared slides of whitefish blastulas
- Chenille sticks (pipe cleaners), two different colors, cut into 3" pieces, 8 pieces per group

Note to the Instructor: See directions for handling wet mount preparations and disposable supplies (pages 33–34, Exercise 3). For suggestions on video of mitosis, see Instructor's Guide.

LEARNING OUTCOMES

☐ Define *cell, organelle,* and *inclusion.*

☐ Identify on a cell model or diagram the following cellular regions, and list the major function of each region: nucleus, cytoplasm, and plasma membrane.

☐ Identify the cytoplasmic organelles, and discuss their structure and function.

☐ Compare and contrast specialized cells with the concept of the "generalized cell."

☐ Define *interphase, mitosis,* and *cytokinesis.*

☐ List the stages of mitosis, and describe the key events of each stage.

☐ Identify the mitotic phases on slides or appropriate diagrams.

☐ Explain the importance of mitotic cell division, and describe its product.

PRE-LAB QUIZ

1. Define *cell.* _____

   _____

2. When a cell is not dividing, the DNA is loosely spread throughout the nucleus in a threadlike form called:
   a. chromatin   c. cytosol
   b. chromosomes   d. ribosomes

3. The plasma membrane not only provides a protective boundary for the cell but also determines which substances enter or exit the cell. We call this characteristic:
   a. diffusion   c. osmosis
   b. membrane potential   d. selective permeability

4. Proteins are assembled on these organelles. _____

5. Because these organelles are responsible for providing most of the ATP that the cell needs, they are often referred to as the "powerhouses" of the cell. They are the:
   a. centrioles   c. mitochondria
   b. lysosomes   d. ribosomes

6. Circle the correct underlined term. During <u>cytokinesis</u> / <u>interphase</u> the cell grows and performs its usual activities.

7. Circle True or False. The end product of mitosis is four genetically identical daughter nuclei.

8. How many stages of mitosis are there? _____

9. DNA replication occurs during:
   a. cytokinesis   c. metaphase
   b. interphase   d. prophase

10. Circle True or False. All animal cells have a cell wall.

The **cell** is the structural and functional unit of all living things. The cells of the human body are highly diverse, and their differences in size, shape, and internal composition reflect their specific roles in the body. Nonetheless, cells do have many common anatomical features, and all cells must carry out certain functions to sustain life. For example, all cells can maintain their boundaries, metabolize, digest nutrients and dispose of wastes, grow and reproduce, move, and respond to a stimulus. Most of these functions are considered in detail in later exercises. This exercise begins by describing structural similarities found in many cells and illustrated by a "composite," or "generalized," cell. The only function considered here is cell reproduction (cell division).

Anatomy of the Composite Cell

In general, all animal cells have three major regions, or parts, that can readily be identified with a light microscope: the **nucleus,** the **plasma membrane,** and the **cytoplasm.** The nucleus is near the center of the cell. It is surrounded by cytoplasm, which in turn is enclosed by the plasma membrane. A diagrammatic representation of the composite cell can help you begin to identify these five structures **(Figure 4.1a).**

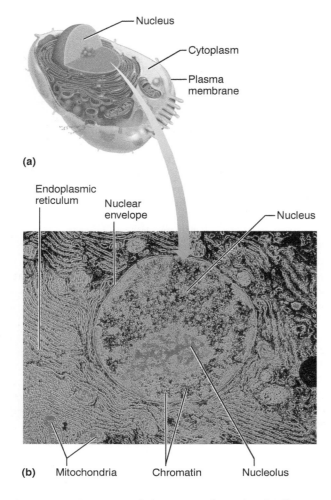

(a)

(b)   Mitochondria   Chromatin   Nucleolus

Figure 4.1 Anatomy of the composite animal cell.
(a) Diagram. **(b)** Transmission electron micrograph
(5500×).

An electron microscope can provide a detailed view of the cellular structure, particularly that of the nucleus (Figure 4.1b).

Nucleus

The nucleus contains the genetic material, DNA, sections of which are called *genes*. Often described as the control center of the cell, the nucleus is necessary for cell reproduction. A cell that has lost or ejected its nucleus is literally programmed to stop dividing.

When the cell is not dividing, the genetic material is loosely dispersed throughout the nucleus in a threadlike form called **chromatin.** When the cell is in the process of dividing to form daughter cells, the chromatin coils and condenses, forming dense, rodlike bodies called **chromosomes**—much in the way a stretched spring becomes shorter and thicker when it is released.

The nucleus also contains one or more small spherical bodies, called **nucleoli,** composed primarily of proteins and ribonucleic acid (RNA). The nucleoli are assembly sites for ribosomes that are particularly abundant in the cytoplasm.

The nucleus is bound by a double-layered porous membrane, the **nuclear envelope.** The nuclear envelope is similar in composition to other cellular membranes, but it is distinguished by its large **nuclear pores.** Nuclear pores are spanned by protein complexes that regulate what passes through, and they permit easy passage of protein and RNA molecules.

ACTIVITY 1

Identifying Parts of a Cell

Identify the nuclear envelope, chromatin, nucleolus, and the nuclear pores shown in Figure 4.1a and b and Figure 4.3. ■

Plasma Membrane

The plasma membrane separates cell contents from the surrounding environment, providing a protective barrier. Its main structural building blocks are phospholipids (fats) and globular protein molecules. Some of the externally facing proteins and lipids have sugar (carbohydrate) side chains attached to them that are important in cellular interactions **(Figure 4.2).** As described by the fluid mosaic model, the membrane is a bilayer of phospholipid molecules in which the protein molecules float. Occasional cholesterol molecules dispersed in the fluid phospholipid bilayer help stabilize it.

Because of its molecular composition, the plasma membrane is selective about what passes through it. It allows nutrients to enter the cell but keeps out undesirable substances. By the same token, valuable cell proteins and other substances are kept within the cell, and excreta or wastes pass to the exterior. This property is known as **selective permeability.**

Additionally, the plasma membrane maintains a resting potential that is essential to normal functioning of excitable cells, such as neurons and muscle cells, and plays a vital role in cell signaling and in cell-to-cell interactions. In some cells the membrane is thrown into tiny fingerlike projections or folds called **microvilli (Figure 4.3).** Microvilli greatly increase the surface area of the cell available for absorption or passage of materials and for the binding of signaling molecules.

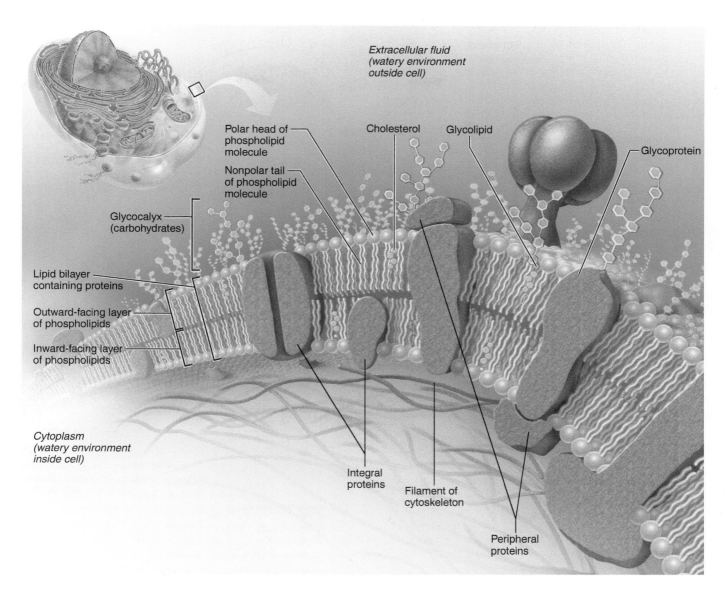

Extracellular fluid
(watery environment
outside cell)

Polar head of
phospholipid
molecule

Nonpolar tail
of phospholipid
molecule

Cholesterol

Glycolipid

Glycoprotein

Glycocalyx
(carbohydrates)

Lipid bilayer
containing proteins

Outward-facing layer
of phospholipids

Inward-facing layer
of phospholipids

Cytoplasm
(watery environment
inside cell)

Integral
proteins

Filament of
cytoskeleton

Peripheral
proteins

Figure 4.2 **Structural details of the plasma membrane.**

ACTIVITY 2

Identifying Components of a Plasma Membrane

Identify the phospholipid and protein portions of the plasma membrane shown in Figure 4.2. Also locate the (*glyco* = carbohydrate) side chains and cholesterol molecules. Identify the microvilli shown in Figure 4.3. ▬

Cytoplasm and Organelles

The cytoplasm consists of the cell contents outside the nucleus. Suspended in the **cytosol,** the fluid cytoplasmic material, are many small structures called **organelles** (literally, "small organs"). The organelles are the metabolic machinery of the cell, and they are highly organized to carry out specific functions for the cell as a whole. The cytoplasmic organelles include the ribosomes, smooth and rough endoplasmic reticulum, Golgi apparatus, lysosomes, peroxisomes, mitochondria, cytoskeletal elements, and centrioles.

ACTIVITY 3

Locating Organelles

Each organelle type is summarized in **Table 4.1**. Read through this material and then, as best you can, locate the organelles (see Figures 4.1b and 4.3). ▬

The cell cytoplasm may or may not contain **inclusions.** Examples of inclusions are stored foods (glycogen granules and lipid droplets), pigment granules, crystals of various types, water vacuoles, and ingested foreign materials.

ACTIVITY 4

Examining the Cell Model

Once you have located all of these structures in Figure 4.3, examine the cell model (or cell chart) to repeat and reinforce your identifications. ▬

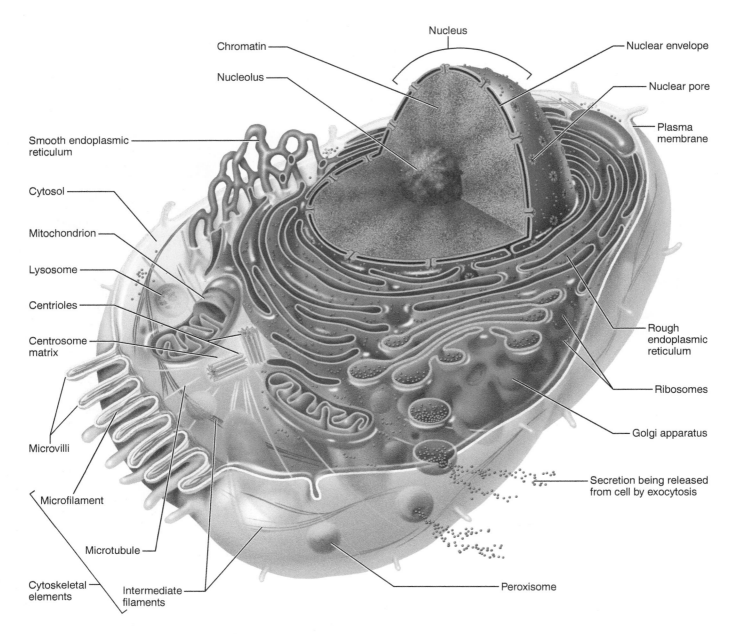

Figure 4.3 Structure of the generalized cell. No cell is exactly like this one, but this composite illustrates features common to many human cells. Note that not all organelles are drawn to the same scale in this illustration.

Differences and Similarities in Cell Structure

ACTIVITY 5

Observing Various Cell Structures

1. Obtain a compound microscope and prepared slides of simple squamous epithelium, smooth muscle cells (teased), human blood, and sperm.

2. Observe each slide under the microscope, carefully noting similarities and differences in the cells. See photomicrographs of simple squamous epithelium (Figure 3.5 in Exercise 3) and teased smooth muscle (Figure 5.7c in Exercise 5). You'll need the oil immersion lens to observe blood and sperm. Distinguish the boundaries of the individual cells, and notice the shape and position of the nucleus in each case. When you look at the human blood smear, direct your attention to the red blood cells, the pink-stained cells that are most numerous. The color photomicrographs illustrating a blood smear (Figure 22.3 in Exercise 22) and sperm (Figure 29.3 in Exercise 29) may be helpful in this cell structure study. Sketch your observations in the circles provided (page 44).

Table 4.1   Cytoplasmic Organelles

Organelle	Location and function
Ribosomes	Tiny spherical bodies composed of RNA and protein; floating free or attached to a membranous structure (the rough ER) in the cytoplasm. Actual sites of protein synthesis.
Endoplasmic reticulum (ER)	Membranous system of tubules that extends throughout the cytoplasm. Two varieties: Rough ER is studded with ribosomes (tubules of the rough ER provide an area for storage and transport of the proteins made on the ribosomes to other cell areas. Smooth ER has no function in protein synthesis; rather, it is a site of steroid and lipid synthesis, lipid metabolism, and drug detoxification.
Golgi apparatus	Stack of flattened sacs with bulbous ends and associated small vesicles; found close to the nucleus. Plays a role in packaging proteins or other substances for export from the cell or incorporation into the plasma membrane and in packaging lysosomal enzymes.
Lysosomes	Various-sized membranous sacs containing digestive enzymes (acid hydrolases). Function to digest worn-out cell organelles and foreign substances that enter the cell. They have the capacity to destroy the cell entirely if they are ruptured and are for this reason referred to as "suicide sacs."
Peroxisomes	Small lysosome-like membranous sacs containing oxidase enzymes that detoxify alcohol, free radicals, and other harmful chemicals. They are particularly abundant in liver and kidney cells.
Mitochondria	Generally rod-shaped bodies with a double-membrane wall; inner membrane is shaped into folds, or cristae. Contain enzymes that oxidize foodstuffs to produce cellular energy (ATP); often referred to as "powerhouses of the cell."
Centrioles	Paired, cylindrical bodies that lie at right angles to each other close to the nucleus. Internally, each centriole is composed of nine triplets of microtubules. As part of the centrosome, they direct the formation of the mitotic spindle during cell division; form the bases of cilia and flagella.
Cytoskeletal elements: microfilaments, intermediate filaments, and microtubules	Form an internal scaffolding called the *cytoskeleton*. Provide cellular support; function in intracellular transport. Microfilaments are formed largely of actin, a contractile protein, and thus are important in cell mobility (particularly in muscle cells). Intermediate filaments are stable elements composed of a variety of proteins and resist mechanical forces acting on cells. Microtubules form the internal structure of the centrioles and help determine cell shape.

3. Measure the length and/or diameter of each cell, and record below the appropriate sketch.

4. How do these four cell types differ in shape and size?

_____

_____

_____

How might cell shape affect cell function?

_____

_____

_____

_____

**Simple squamous
epithelium**

Diameter _____

Sperm cells

Length _____

**Human
red blood cells**

Diameter _____

**Teased smooth
muscle cells**

Length _____

Which cells have visible projections?

_____

How do these projections relate to the function of these cells?

_____

_____

Do any of these cells lack a plasma membrane? _____

A nucleus? _____

In the cells with a nucleus, can you discern nucleoli?

_____

_____

Cell Division

The cell cycle is the series of changes that a cell goes through from the time it is formed until it reproduces. The outer ring of **Figure 4.4** shows the two main periods of the cell cycle, interphase and the mitotic phase. **Interphase** is the longer period, during which the cell grows and carries out its usual activities. **Cell division,** or the **mitotic phase,** is the period when the cell reproduces itself by dividing. In an interphase cell about to divide, the genetic material (DNA) is copied exactly via DNA replication. Once this important event has occurred, cell division ensues.

Cell division is essential for growth and repair. Cell division, which is also called the **M (mitotic) phase** of the cell cycle, consists of two events called *mitosis* and *cytokinesis*.

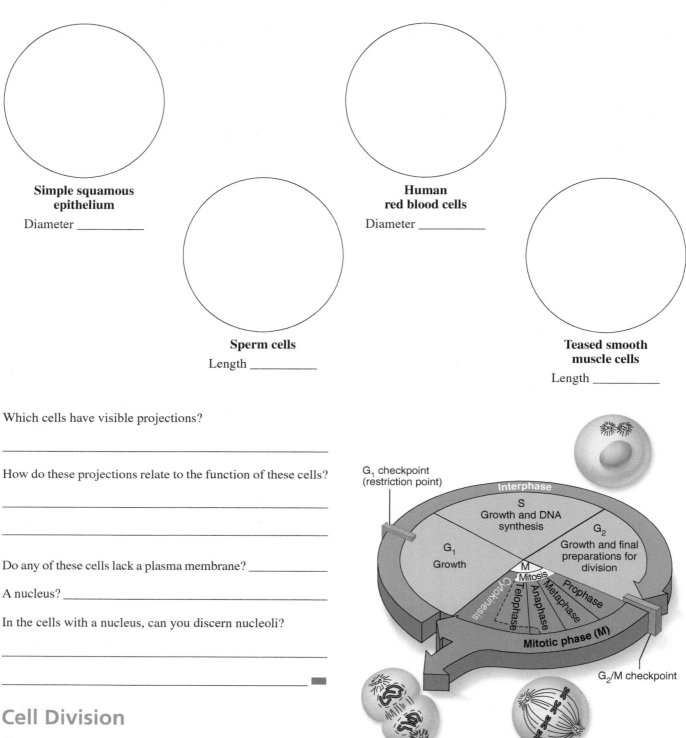

Figure 4.4 The cell cycle. The two main phases are interphase and the mitotic phase.

Mitosis is the division of the copied DNA of the mother cell to two daughter nuclei. **Cytokinesis** is the division of the cytoplasm. It begins when mitosis is nearly complete. Although mitosis is usually accompanied by cytokinesis, in some instances cytoplasmic division does not occur, leading to the formation of binucleate or multinucleate cells.

The product of **mitosis** is two daughter nuclei that are genetically identical to the mother nucleus. This distinguishes mitosis from **meiosis,** a specialized type of nuclear division that occurs only in the reproductive organs (testes or ovaries). Meiosis, which yields four daughter nuclei that differ genetically in composition from the mother nucleus, is used only for producing gametes (eggs and sperm) for sexual reproduction. The function of cell division, including mitosis and cytokinesis in the body, is to increase the number of cells for growth and repair.

The phases of mitosis include **prophase, metaphase, anaphase,** and **telophase.** The detailed events of interphase, mitosis, and cytokinesis are described and illustrated in **Figure 4.5** on pages 46-47.

Mitosis is essentially the same in all animal cells, but depending on the tissue, it takes from 5 minutes to several hours to complete. In most cells, centriole replication occurs during interphase of the next cell cycle.

At the end of cell division, two daughter cells exist—each with a smaller cytoplasmic mass than the mother cell but genetically identical to it. The daughter cells grow and carry out the normal spectrum of metabolic processes until it is their turn to divide.

Cell division is extremely important during the body's growth period. Most cells divide until puberty, when normal body size is achieved and overall body growth ceases. After this time in life, only certain cells carry out cell division routinely—for example, cells subjected to abrasion (epithelium of the skin and lining of the gut). Other cell populations—such as liver cells—stop dividing but retain this ability should some of them be removed or damaged. Skeletal muscle, cardiac muscle, and most mature neurons almost completely lose this ability to divide and, thus, are severely handicapped by injury.

ACTIVITY 6

Identifying the Mitotic Stages

1. Watch a video presentation of mitosis (if available).

2. Using the three-dimensional models of dividing cells provided, identify each of the mitotic phases described (Figure 4.5).

3. Obtain a prepared slide of whitefish blastulas to study the stages of mitosis. The cells of each *blastula* (a stage of embryonic development consisting of a hollow ball of cells) are at approximately the same mitotic stage, so it may be necessary to observe more than one blastula to view all the mitotic stages. A good analogy for a blastula is a soccer ball in which each leather piece making up the ball's surface represents an embryonic cell. The exceptionally high rate of mitosis observed in this tissue is typical of embryos, but if it occurs in specialized tissues it can indicate cancerous cells, which also have an extraordinarily high mitotic rate. Examine the slide carefully, identifying the four mitotic phases and the process of cytokinesis. Compare your observations with Figure 4.5, and verify your identifications with your instructor. ▬

"Chenille Stick" Mitosis

1. Obtain a total of eight chenille sticks, each measuring 3 inches: four of one color and four of another color (e.g., four green and four purple).

2. Assemble the chenille sticks into a total of four chromosomes (each with two sister chromatids) by twisting two sticks of the same color together at the center with a single twist.

What does the twist at the center represent? _____

3. Arrange the chromosomes as they appear in early prophase.

Name the structure that assembles during this phase.

_____

Draw early prophase in the space provided on your Review Sheet (question 8, page 52).

4. Arrange the chromosomes as they appear in late prophase.

What structure on the chromosome centromere do the

growing spindle microtubules attach to? _____

What structure is now present as fragments? _____

Draw late prophase in the space provided on your Review Sheet (question 8, page 52).

5. Arrange the chromosomes as they appear in metaphase.

What is the name of the imaginary plane that the chromo-

somes align along? _____

Draw metaphase in the space provided on your Review Sheet (question 8, page 52).

6. Arrange the chromosomes as they appear in anaphase.

What does untwisting of the chenille sticks represent?

_____

Each sister chromatid has now become a _____.

Draw anaphase in the space provided on your Review Sheet (question 8, page 52).

7. Arrange the chromosomes as they appear in telophase.

Briefly list four reasons why telophase is like the reverse of prophase.

_____

_____

_____

_____

Draw telophase in the space provided on your Review Sheet (question 8, page 52).

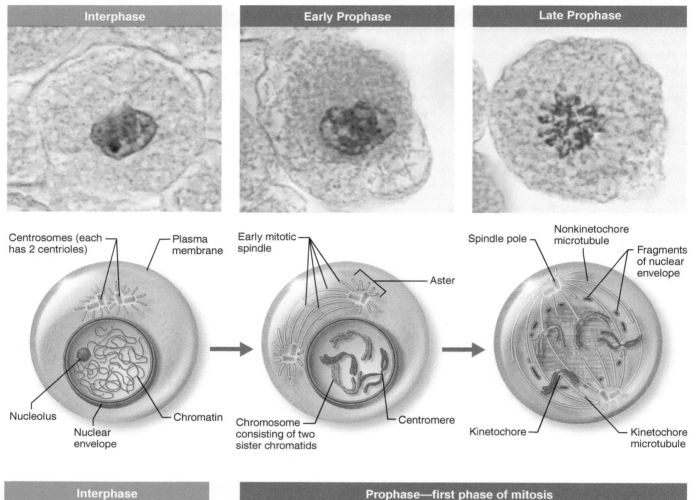

| Interphase | Early Prophase | Late Prophase |

Centrosomes (each has 2 centrioles) · Plasma membrane · Nucleolus · Nuclear envelope · Chromatin

Early mitotic spindle · Aster · Chromosome consisting of two sister chromatids · Centromere

Spindle pole · Nonkinetochore microtubule · Fragments of nuclear envelope · Kinetochore · Kinetochore microtubule

Interphase

Interphase is the period of a cell's life when it carries out its normal metabolic activities and grows. Interphase is not part of mitosis.

• During interphase, the DNA-containing material is in the form of chromatin. The nuclear envelope and one or more nucleoli are intact and visible.

• There are three distinct periods of interphase:
G_1: The centrioles begin replicating.
S: DNA is replicated.
G_2: Final preparations for mitosis are completed, and centrioles finish replicating.

Prophase—first phase of mitosis

Early Prophase
• The chromatin condenses, forming barlike chromosomes.

• Each duplicated chromosome consists of two identical threads, called **sister chromatids**, held together at the **centromere**. (Later, when the chromatids separate, each will be a new chromosome.)

• As the chromosomes appear, the nucleoli disappear, and the two centrosomes separate from one another.

• The centrosomes act as focal points for growth of a microtubule assembly called the **mitotic spindle**. As the microtubules lengthen, they propel the centrosomes toward opposite ends (poles) of the cell.

• Microtubule arrays called **asters** ("stars") extend from the centrosome matrix.

Late Prophase
• The nuclear envelope breaks up, allowing the spindle to interact with the chromosomes.

• Some of the growing spindle microtubules attach to **kinetochores**, special protein structures at each chromosome's centromere. Such microtubules are called **kinetochore microtubules**.

• The remaining spindle microtubules (not attached to any chromosomes) are called **nonkinetochore microtubules**. The microtubules slide past each other, forcing the poles apart.

• The kinetochore microtubules pull on each chromosome from both poles in a tug-of-war that ultimately draws the chromosomes to the center, or equator, of the cell.

Figure 4.5 The interphase cell and the events of cell division. The cells shown are from an early embryo of a whitefish. Photomicrographs are above; corresponding diagrams are below. (Photomicrographs approximately 1530×.)

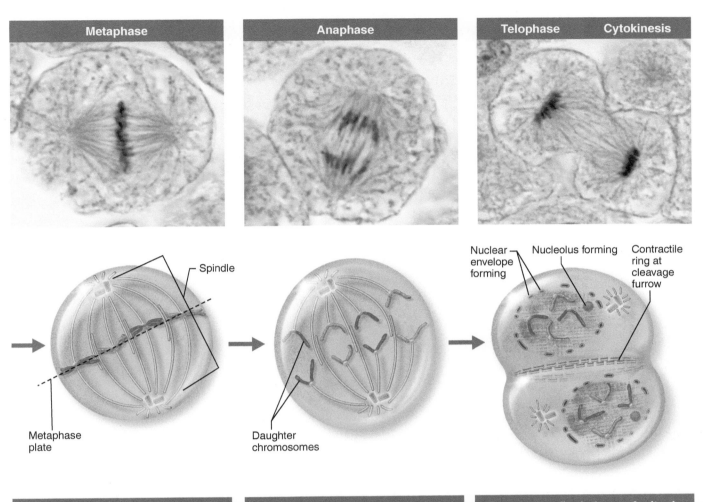

Metaphase—second phase of mitosis

- The two centrosomes are at opposite poles of the cell.
- The chromosomes cluster at the midline of the cell, with their centromeres precisely aligned at the **equator** of the spindle. This imaginary plane midway between the poles is called the **metaphase plate**.
- Enzymes act to separate the chromatids from each other.

Anaphase—third phase of mitosis

The shortest phase of mitosis, anaphase begins abruptly as the centromeres of the chromosomes split simultaneously. Each chromatid now becomes a chromosome in its own right.

- The kinetochore microtubules, moved along by motor proteins in the kinetochores, gradually pull each chromosome toward the pole it faces.
- At the same time, the nonkinetochore microtubules slide past each other, lengthen, and push the two poles of the cell apart.
- The moving chromosomes look V-shaped. The centromeres lead the way, and the chromosomal "arms" dangle behind them.
- The short, compact shape of the chromosomes makes it easier for them to move and separate. Diffuse threads of chromatin would trail, tangle, and break, resulting in imprecise "parceling out" to the daughter cells.

Telophase—final phase of mitosis

Telophase begins as soon as chromosomal movement stops. This final phase is like prophase in reverse.

- The identical sets of chromosomes at the opposite poles of the cell uncoil and resume their threadlike chromatin form.
- A new nuclear envelope forms around each chromatin mass, nucleoli reappear within the nuclei, and the spindle breaks down and disappears.
- Mitosis is now ended. The cell, for just a brief period, is binucleate (has two nuclei) and each new nucleus is identical to the original parent nucleus.

Cytokinesis—division of cytoplasm

Cytokinesis begins during late anaphase and continues through and beyond telophase. A contractile ring of actin microfilaments forms the **cleavage furrow** and pinches the cell apart.

Figure 4.5 *(continued)*

Name _____

Lab Time/Date _____

The Cell: Anatomy and Division

Anatomy of the Composite Cell

1. Define the following terms:

   organelle: _____

   _____

   cell: _____

2. Although cells have differences that reflect their specific functions in the body, what functions do they have in common?

   _____

3. Identify the following cell structures:

   _____ 1. external boundary of cell; regulates flow of materials into and out of the cell; site of cell signaling

   _____ 2. contains digestive enzymes of many varieties; "suicide sac" of the cell

   _____ 3. scattered throughout the cell; major site of ATP synthesis

   _____ 4. slender extensions of the plasma membrane that increase its surface area

   _____ 5. stored glycogen granules, crystals, pigments present in some cell types

   _____ 6. membranous system consisting of flattened sacs and vesicles; packages proteins for export

   _____ 7. control center of the cell; necessary for cell division and cell life

   _____ 8. two rod-shaped bodies near the nucleus; associated with the formation of the mitotic spindle

   _____ 9. dense nuclear body; packaging site for ribosomes

   _____ 10. contractile elements of the cytoskeleton

   _____ 11. membranous tubules covered with ribosomes; involved in intracellular transport of proteins and synthesis of membrane lipids

   _____ 12. attached to membrane systems or scattered in the cytoplasm; site of protein synthesis

   _____ 13. threadlike structures in the nucleus; contain genetic material (DNA)

   _____ 14. site of free-radical detoxification

4. Label the cell structures using the leader lines provided.

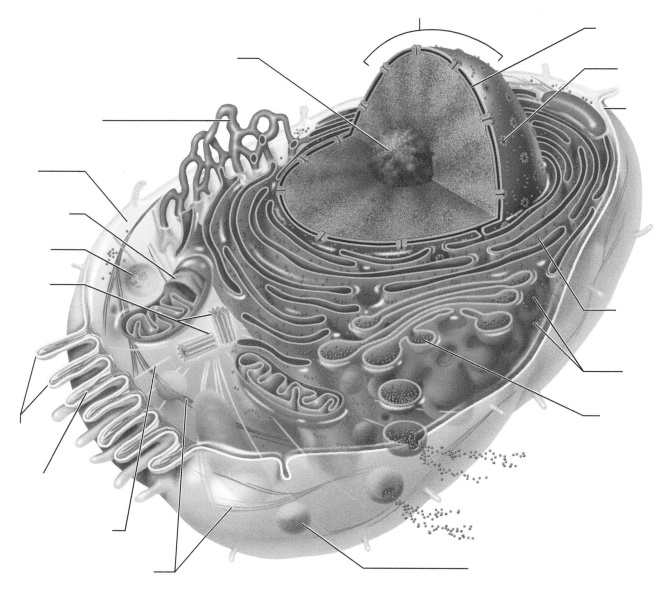

Differences and Similarities in Cell Structure

5. Choose the specimen observed in Activity 5 (squamous epithelium, sperm cells, smooth muscle, or human red blood cells) that fits the description below.

1. _____ cell has a flagellum for movement

2. _____ cells have an elongated shape (tapered at each end)

3. _____ cells are close together

4. _____ cells are circular

5. _____ cells are thin and flat with irregular borders

6. _____ cells are anucleate (without a nucleus)

Cell Division

6. Identify the four phases of mitosis shown in the following photomicrographs, and select the events from the key that correctly identify each phase. On the appropriate answer line, write the letters that correspond to these events.

Key:

a. The nuclear envelope re-forms.

b. Chromosomes line up in the center of the cell.

c. Chromatin coils and condenses, forming chromosomes.

d. Chromosomes stop moving toward the poles.

e. The chromosomes are V shaped.

f. The nuclear envelope breaks down.

g. Chromosomes attach to the spindle fibers.

h. The mitotic spindle begins to form.

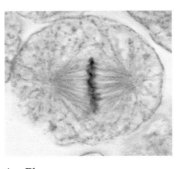

1. Phase: _____
   Events: _____

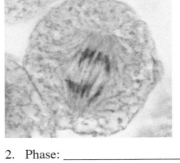

2. Phase: _____
   Events: _____

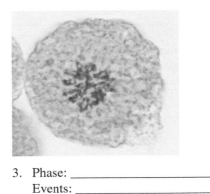

3. Phase: _____
   Events: _____

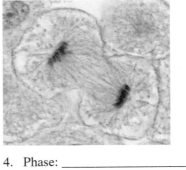

4. Phase: _____
   Events: _____

7. Complete the following statements:

Division of the __1__ is referred to as mitosis. Cytokinesis is division of the __2__. The major structural difference between chromatin and chromosomes is that the latter is __3__. Chromosomes attach to the spindle fibers by undivided structures called __4__. If a cell undergoes mitosis but not cytokinesis, the product is __5__. The structure that acts as a scaf-folding for chromosomal attachment and movement is called the __6__. __7__ is the period of cell life when the cell is not involved in division. Three cell populations in the body that do not routinely undergo cell division are __8__, __9__, and __10__.

1. _____

2. _____

3. _____

4. _____

5. _____

6. _____

7. _____

8. _____

9. _____

10. _____

8. Draw the phases of mitosis for a cell with a chromosome number of 4.

9. ➕ Plasma cells are key to the immune response because they secrete antibodies. Given that antibodies are made of protein, which

membrane-enclosed cell organelle would you expect the plasma cells to have in abundance? Why? _____

_____

10. ➕ Name which organelle you would expect to play the largest role in decomposition of the human body. Why? _____

_____

11. ➕ Some antifungal medications work by blocking DNA synthesis in the fungal cell. Describe where in the cell cycle such a

medication would halt the fungal cell and the consequences of this early termination of the cycle. _____

_____

Classification of Tissues

MATERIALS

- Compound microscope
- Immersion oil
- Prepared slides of simple squamous, simple cuboidal, simple columnar, stratified squamous (nonkeratinized), stratified cuboidal, stratified columnar, pseudostratified ciliated columnar, and transitional epithelium
- Prepared slides of mesenchyme; of adipose, areolar, reticular, and dense (regular, irregular, and elastic) connective tissues; of hyaline and elastic cartilage; of fibrocartilage; of bone (x.s.); and of blood
- Prepared slide of nervous tissue (spinal cord smear)
- Prepared slides of skeletal, cardiac, and smooth muscle (l.s.)

LEARNING OUTCOMES

☐ Name the four primary tissue types in the human body, and state a general function of each type.

☐ Name the major subcategories of the primary tissue types, and identify the tissues of each subcategory microscopically or in an appropriate image.

☐ State the locations of the various tissues in the body.

☐ List the general function and structural characteristics of each of the tissues studied.

PRE-LAB QUIZ

1. Groups of cells that are anatomically similar and share a function are called:
   a. organ systems
   b. organisms
   c. organs
   d. tissues
2. How many primary tissue types are found in the human body?_____
3. Circle True or False. Endocrine and exocrine glands are classified as epithelium because they usually develop from epithelial membranes.
4. Epithelial tissues can be classified according to cell shape. _____ epithelial cells are scalelike and flattened.
   a. Columnar
   b. Cuboidal
   c. Squamous
   d. Transitional
5. All connective tissue is derived from an embryonic tissue known as:
   a. cartilage
   b. ground substance
   c. mesenchyme
   d. reticular
6. All the following are examples of connective tissue *except:*
   a. bones
   b. ligaments
   c. neurons
   d. tendons
7. Circle True or False. Blood is a type of connective tissue.
8. Circle the correct underlined term. Of the two major cell types found in nervous tissue, neurons / neuroglial cells are highly specialized to generate and conduct electrical signals.
9. How many basic types of muscle tissue are there? _____
10. Which type of muscle tissue is found in the walls of hollow organs, has no striations, and has spindle-shaped cells? _____

ells are the building blocks of life and the all-inclusive functional units of unicellular organisms. However, in higher organisms, cells do not usually operate as isolated, independent entities. In humans and other multicellular organisms, cells depend on one another and cooperate to maintain homeostasis in the body.

With a few exceptions, even the most complex animal starts out as a single cell, the fertilized egg, which divides almost endlessly. The trillions of cells that result become specialized for a particular function; some become supportive bone, others the transparent lens of the eye, still others skin cells, and so on. Thus a division of labor exists, with certain groups of cells highly specialized to perform functions that benefit the organism as a whole.

Groups of cells that are similar in structure and function are called **tissues.** The four primary tissue types—epithelial tissue, connective tissue, nervous tissue, and muscle—have distinctive structures, patterns, and functions. The four primary tissues are further divided into subcategories, as described shortly.

To perform specific body functions, the tissues are organized into **organs** such as the heart, kidneys, and lungs. Most organs contain several representatives of the primary tissues, and the arrangement of these tissues determines the organ's structure and function. Thus **histology,** the study of tissues, complements a study of gross anatomy and provides the structural basis for a study of organ physiology.

The main objective of this exercise is to familiarize you with the major similarities and dissimilarities of the primary tissues, so that when the tissue composition of an organ is described, you will be able to more easily understand (and perhaps even predict) the organ's major function.

Epithelial Tissue

Epithelial tissue, or an **epithelium,** is a sheet of cells that covers a body surface or lines a body cavity. It occurs in the body as (1) covering and lining epithelium and (2) glandular epithelium. Covering and lining epithelium forms the outer layer of the skin and lines body cavities that open to the outside. It covers the walls and organs of the closed body cavity. Because glands almost invariably develop from epithelial sheets, glands are classified as epithelium.

Epithelial functions include protection, absorption, filtration, excretion, secretion, and sensory reception. For example, the epithelium covering the body surface protects against bacterial invasion and chemical damage. Epithelium specialized to absorb substances lines the stomach and small intestine. In the kidney tubules, the epithelium absorbs, secretes, and filters. Secretion is a specialty of the glandular epithelium.

The following characteristics distinguish epithelial tissues from other types:

- **Polarity.** The membranes always have one free surface, called the *apical surface,* and typically that surface is significantly different from the *basal surface.*

- **Specialized contacts.** Cells fit closely together to form membranes, or sheets of cells, and are bound together by specialized junctions.

- **Supported by connective tissue.** The cells are attached to and supported by an adhesive **basement membrane,** which is an acellular material secreted partly by the epithelial cells *(basal lamina)* and connective tissue cells *(reticular lamina)* that lie adjacent to each other.

- **Avascular but innervated.** Epithelial tissues are supplied by nerves but have no blood supply of their own (are avascular). Instead they depend on diffusion of nutrients from the underlying connective tissue.

- **Regeneration.** If well nourished, epithelial cells can easily regenerate themselves. This is an important characteristic because many epithelia are subjected to a good deal of abrasion.

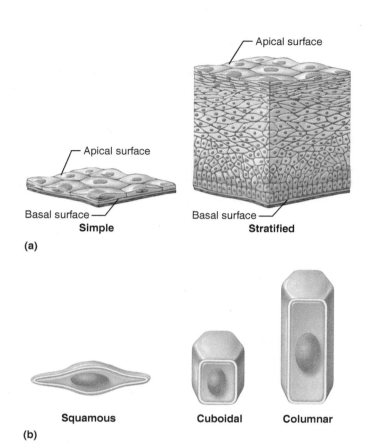

(a)

(b)

Figure 5.1 Classification of epithelia.

The covering and lining epithelia are classified according to two criteria—arrangement or relative number of layers and cell shape **(Figure 5.1)**. On the basis of arrangement, there are **simple** epithelia, consisting of one layer of cells attached to the basement membrane, and **stratified** epithelia, consisting of two or more layers of cells. The general types based on shape are **squamous** (scalelike), **cuboidal** (cubelike), and **columnar** (column-shaped) epithelial cells. The terms denoting shape and arrangement of the epithelial cells are combined to describe the epithelium fully. *Stratified epithelia are named according to the cells at the apical surface of the epithelial sheet,* not those resting on the basement membrane.

There are, in addition, two less easily categorized types of epithelia. **Pseudostratified epithelium** is actually a simple columnar epithelium (one layer of cells), but because its cells vary in height and their nuclei lie at different levels above the basement membrane, it gives the false appearance of being stratified. This epithelium is often ciliated. **Transitional epithelium** is a rather peculiar stratified squamous epithelium formed of rounded, or "plump," cells with the ability to slide over one another to allow the organ to be stretched. Transitional epithelium is found only in urinary system organs subjected to stretching, such as the bladder. The superficial cells are flattened (like true squamous cells) when the organ is full and rounded when the organ is empty.

Epithelial cells forming glands are highly specialized to remove materials from the blood and to manufacture them into new materials, which they then secrete. There are two types of glands: exocrine and endocrine glands, shown in **Figure 5.2**. **Endocrine glands** lose their surface connection (duct) as they develop; thus they are referred to as ductless glands. They secrete hormones into the extracellular fluid, and from there the hormones enter the blood or the lymphatic vessels that weave through the glands. **Exocrine glands** retain their ducts, and their secretions empty through these ducts, either to a body surface or into body cavities. The exocrine glands include the sweat and oil glands, liver, and pancreas. Glands are discussed in conjunction with the organ systems to which their products are functionally related.

The most common types of epithelia, their characteristic locations in the body, and their functions are described in **Figure 5.3**.

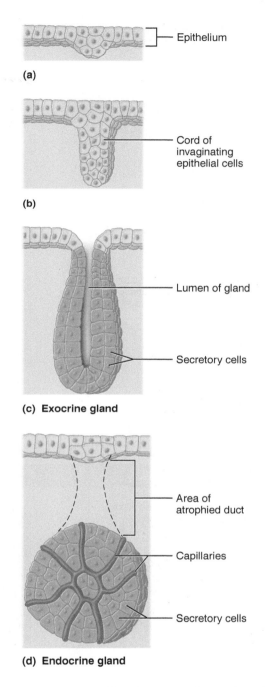

(a)

Epithelium

(b)

Cord of invaginating epithelial cells

Lumen of gland

Secretory cells

(c) Exocrine gland

Area of atrophied duct

Capillaries

Secretory cells

(d) Endocrine gland

ACTIVITY 1

Examining Epithelial Tissue Under the Microscope

Obtain slides of simple squamous, simple cuboidal, simple columnar, pseudostratified ciliated columnar, stratified squamous (nonkeratinized), stratified cuboidal, stratified columnar, and transitional epithelia. Examine each carefully, and notice how the epithelial cells fit closely together to form intact sheets of cells, a necessity for a tissue that forms linings or covering membranes. Scan each epithelial type for modifications for specific functions, such as cilia (motile cell projections that help to move substances along the cell surface), and microvilli, which increase the surface area for absorption. Also be alert for goblet cells, which secrete lubricating mucus. Compare your observations with the descriptions and photomicrographs in Figure 5.3.

While you are working, check the questions in the Review Sheet at the end of this exercise. A number of the questions there refer to some of the observations you are asked to make during your microscopic study. ■

WHY THIS
MATTERS | Buccal Swabs

A buccal, or cheek, swab is a method used to collect stratified squamous cells from the oral cavity. The cells contain DNA that can be used for DNA fingerprinting or tissue typing. DNA fingerprinting can be used in criminal investigations, and tissue typing can be used to match a recipient with a donor for organ transplant, especially a bone marrow transplant. The buccal swab procedure involves using a cotton-tipped applicator to scrape the inside of the mouth in the buccal region and remove cells at the surface. This noninvasive procedure provides an easy way to obtain the DNA profile of an individual, a unique molecular "signature." ■

Figure 5.2 Formation of endocrine and exocrine glands from epithelial sheets. (a) Epithelial cells grow and push into the underlying tissue. **(b)** A cord of epithelial cells forms. **(c)** In an exocrine gland, a lumen (cavity) forms. The inner cells form the duct, and the outer cells produce the secretion. **(d)** In a forming endocrine gland, the connecting duct cells atrophy, leaving the secretory cells with no connection to the epithelial surface. However, they do become heavily invested with blood and lymphatic vessels that receive the secretions.

Text continues on page 60.

(a) Simple squamous epithelium

Description: Single layer of flattened cells with disc-shaped central nuclei and sparse cytoplasm; the simplest of the epithelia.

Function: Allows materials to pass by diffusion and filtration in sites where protection is not important; secretes lubricating substances in serosae.

Location: Kidney glomeruli; air sacs of lungs; lining of heart, blood vessels, and lymphatic vessels; lining of ventral body cavity (serosae).

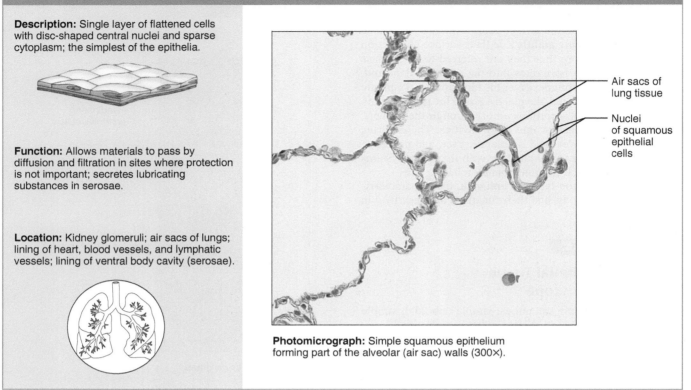

Air sacs of lung tissue

Nuclei of squamous epithelial cells

Photomicrograph: Simple squamous epithelium forming part of the alveolar (air sac) walls (300×).

(b) Simple cuboidal epithelium

Description: Single layer of cubelike cells with large, spherical central nuclei.

Function: Secretion and absorption.

Location: Kidney tubules; ducts and secretory portions of small glands; ovary surface.

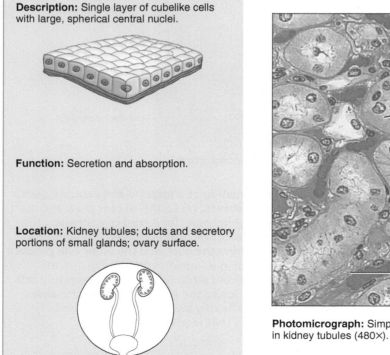

Simple cuboidal epithelial cells

Lumen

Basement membrane

Connective tissue

Photomicrograph: Simple cuboidal epithelium in kidney tubules (480×).

Figure 5.3 Epithelial tissues. Simple epithelia (**a** and **b**).

(c) Simple columnar epithelium

Description: Single layer of tall cells with *round* to *oval* nuclei; some cells bear cilia; layer may contain mucus-secreting unicellular glands (goblet cells).

Function: Absorption; secretion of mucus, enzymes, and other substances; ciliated type propels mucus (or reproductive cells) by ciliary action.

Location: Nonciliated type lines most of the digestive tract (stomach to rectum), gallbladder, and excretory ducts of some glands; ciliated variety lines small bronchi, uterine tubes, and some regions of the uterus.

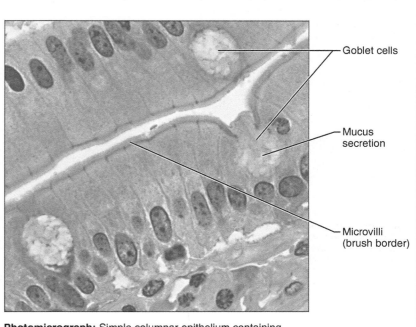

- Goblet cells
- Mucus secretion
- Microvilli (brush border)

Photomicrograph: Simple columnar epithelium containing goblet cells from the small intestine (675×).

(d) Pseudostratified columnar epithelium

Description: Single layer of cells of differing heights, some not reaching the free surface; nuclei seen at different levels; may contain mucus-secreting goblet cells and bear cilia.

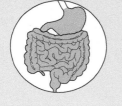

Function: Secretes substances, particularly mucus; propulsion of mucus by ciliary action.

Location: Nonciliated type in male's sperm-carrying ducts and ducts of large glands; ciliated variety lines the trachea, most of the upper respiratory tract.

Trachea

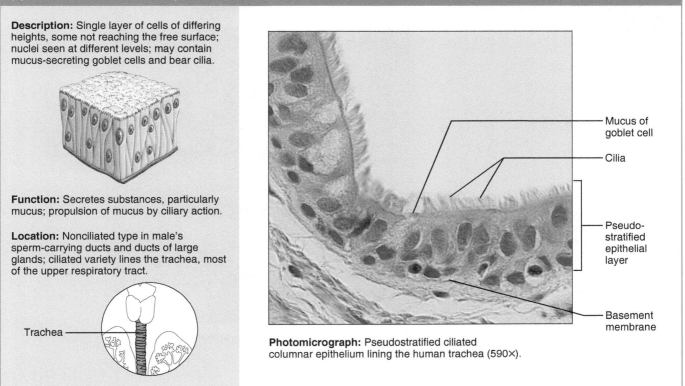

- Mucus of goblet cell
- Cilia
- Pseudo-stratified epithelial layer
- Basement membrane

Photomicrograph: Pseudostratified ciliated columnar epithelium lining the human trachea (590×).

Figure 5.3 (*continued*) (**c** and **d**).

(e) Stratified squamous epithelium

Description: Thick membrane composed of several cell layers; basal cells are cuboidal or columnar and metabolically active; surface cells are flattened (squamous); in the keratinized type, the surface cells are full of keratin and dead; basal cells are active in mitosis and produce the cells of the more superficial layers.

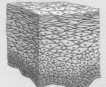

Function: Protects underlying tissues in areas subjected to abrasion.

Location: Nonkeratinized type forms the moist linings of the esophagus, mouth, and vagina; keratinized variety forms the epidermis of the skin, a dry membrane.

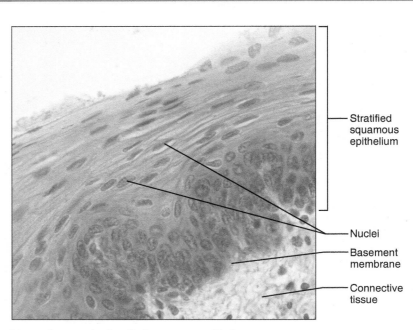

Photomicrograph: Stratified squamous epithelium lining the esophagus (590×).

(f) Stratified cuboidal epithelium

Description: Generally two layers of cubelike cells.

Function: Protection

Location: Largest ducts of sweat glands, mammary glands, and salivary glands.

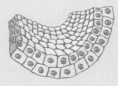

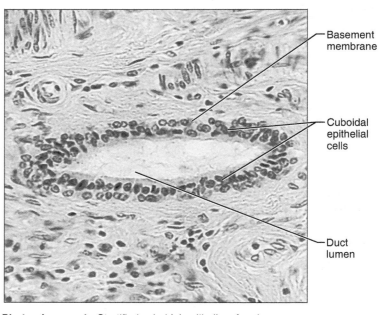

Photomicrograph: Stratified cuboidal epithelium forming a salivary gland duct (350×).

Figure 5.3 (*continued*) **Epithelial tissues.** Stratified epithelia (**e** and **f**).

(g) Stratified columnar epithelium

Description: Several cell layers; basal cells usually cuboidal; superficial cells elongated and columnar.

Function: Protection; secretion.

Location: Rare in the body; small amounts in male urethra and in large ducts of some glands.

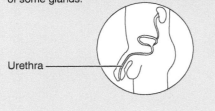

Urethra

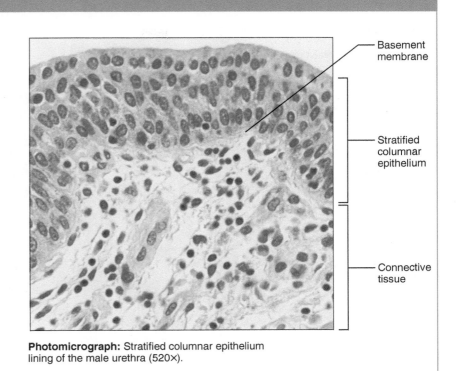

Basement membrane

Stratified columnar epithelium

Connective tissue

Photomicrograph: Stratified columnar epithelium lining of the male urethra (520×).

(h) Transitional epithelium

Description: Resembles both stratified squamous and stratified cuboidal; basal cells cuboidal or columnar; surface cells dome shaped or squamous-like, depending on degree of organ stretch.

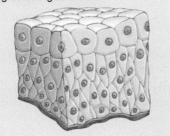

Function: Stretches readily and permits distension of urinary organ by contained urine.

Location: Lines the ureters, bladder, and part of the urethra.

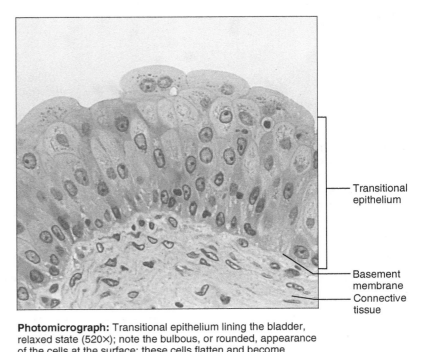

Transitional epithelium

Basement membrane

Connective tissue

Photomicrograph: Transitional epithelium lining the bladder, relaxed state (520×); note the bulbous, or rounded, appearance of the cells at the surface; these cells flatten and become elongated when the bladder is filled with urine.

Figure 5.3 (*continued*) **(g)** and Transitional epithelium **(h)**.

Cell types

Extracellular
matrix

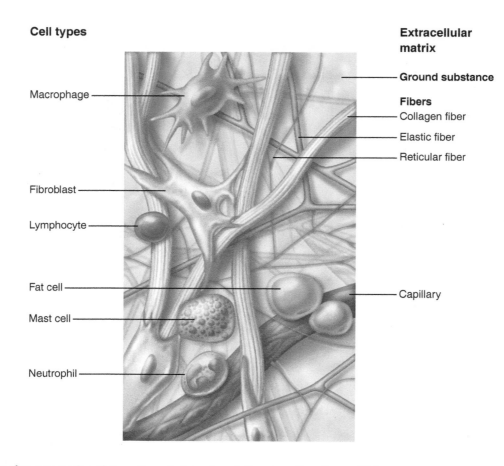

Macrophage

Ground substance

Fibers
Collagen fiber
Elastic fiber
Reticular fiber

Fibroblast

Lymphocyte

Fat cell

Capillary

Mast cell

Neutrophil

Figure 5.4 Areolar connective tissue: A prototype (model) connective tissue. This tissue underlies epithelia and surrounds capillaries. Note the various cell types and the three classes of fibers (collagen, reticular, elastic) embedded in the ground substance.

Connective Tissue

Connective tissue is found in all parts of the body as discrete structures or as part of various body organs. It is the most abundant and widely distributed of the tissue types.

There are four main types of connective tissue: **connective tissue proper, cartilage, bone,** and **blood.** Connective tissue proper has two subclasses: **loose connective tissues** (areolar, adipose, and reticular) and **dense connective tissues** (dense regular, dense irregular, and elastic). **Connective tissues** perform a variety of functions, but primarily they protect, support, and bind together other tissues of the body. For example, bones are composed of connective tissue (**bone,** or **osseous tissue),** and they protect and support other body tissues and organs. The ligaments and tendons (**dense regular connective tissue**) bind the bones together or connect skeletal muscles to bones.

Areolar connective tissue (Figure 5.4) is a soft packaging material that cushions and protects body organs. **Adipose** (fat) tissue provides insulation for the body tissues and a source of stored energy. Blood-forming (**hematopoietic**) tissue replenishes the body's supply of red blood cells. Connective tissue also serves a vital function in the repair of all body tissues; many wounds are repaired by connective tissue in the form of scar tissue.

The characteristics of connective tissue include the following:

*   **Common origin.** All connective tissues are derived from embryonic tissue *(mesenchyme).*

*   **Degrees of vascularity.** Many types of connective tissue have a rich blood supply. Exceptions include cartilage, which is avascular, and dense connective tissue, which is poorly vascularized.

*   **Extracellular matrix.** There is a great deal of noncellular, nonliving material (matrix) between the cells of connective tissues. The composition and amount of matrix vary for connective tissues.

The extracellular matrix has two components—ground substance and fibers. The **ground substance** is composed chiefly of interstitial fluid, cell adhesion proteins, and proteoglycans. Depending on its specific composition, the ground substance may be liquid, semisolid, gel-like, or very hard. When the matrix is firm, as in cartilage and bone, the connective tissue cells reside in cavities in the matrix called *lacunae*. The fibers, which provide support, include **collagen** (white) **fibers, elastic** (yellow) **fibers,** and **reticular** (fine collagen) **fibers.**

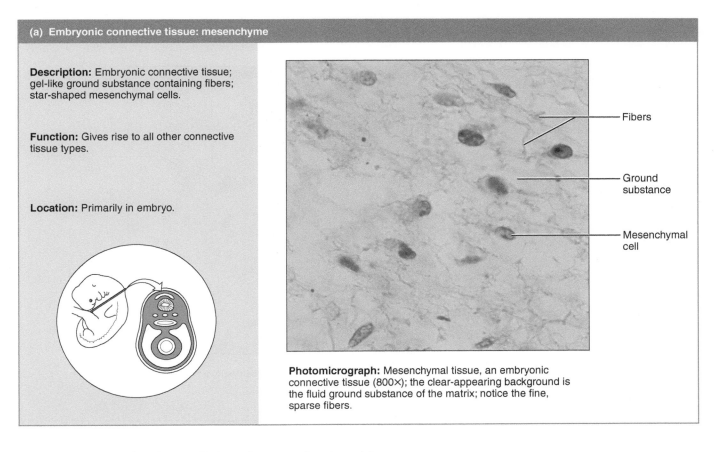

(a) Embryonic connective tissue: mesenchyme

Description: Embryonic connective tissue; gel-like ground substance containing fibers; star-shaped mesenchymal cells.

Function: Gives rise to all other connective tissue types.

Location: Primarily in embryo.

Fibers

Ground substance

Mesenchymal cell

Photomicrograph: Mesenchymal tissue, an embryonic connective tissue (800×); the clear-appearing background is the fluid ground substance of the matrix; notice the fine, sparse fibers.

Figure 5.5 Connective tissues. Embryonic connective tissue **(a).**

Of these, the collagen fibers are most abundant. The connective tissues have a common structural plan seen best in *areolar connective tissue* (Figure 5.4). Because all other connective tissues are variations of areolar, it is considered the model, or prototype, of the connective tissues. Notice that areolar tissue has all three varieties of fibers, but they are sparsely arranged in its transparent gel-like ground substance (refer to Figure 5.4). The cell type that secretes its matrix is the *fibroblast,* but a wide variety of such other cells, including phagocytic cells, such as macrophages, and certain white blood cells and mast cells that act in the inflammatory response, are present as well. The more durable connective tissues, such as bone, cartilage, and the dense connective tissues, characteristically have a firm ground substance and many more fibers.

The general characteristics, location, and function of some of the connective tissues found in the body are shown in **Figure 5.5**.

Examining Connective Tissue Under the Microscope

Obtain prepared slides of mesenchyme; of areolar, adipose, reticular, and dense irregular, dense regular, and elastic connective tissue; of hyaline and elastic cartilage and fibrocartilage; of osseous connective tissue (bone); and of blood. Compare your observations with the views illustrated in Figure 5.5.

Distinguish the living cells from the matrix, and pay particular attention to the denseness and arrangement of the matrix. For example, notice how the matrix of the dense regular connective tissues, which make up tendons and the dermis of the skin, is packed with collagen fibers. Note also that in the *regular* variety (tendon), the fibers are all running in the same direction, whereas in the *irregular* variety (dermis), they appear to be running in many directions.

While examining the areolar connective tissue, notice how much empty space there appears to be (*areol* = small empty space), and distinguish the collagen fibers from the coiled elastic fibers. Identify the starlike fibroblasts. Also, try to locate a **mast cell,** which has large, darkly staining granules in its cytoplasm (*mast* = stuffed full of granules). This cell type releases histamine which makes capillaries more permeable during inflammatory reactions and allergies.

In adipose tissue, locate a "signet ring" cell, a fat cell in which the nucleus can be seen pushed to one side by the large, fat-filled vacuole that appears to be a large empty space. Also notice how little matrix there is in adipose (fat) tissue. Distinguish the living cells from the matrix in the dense connective tissue, bone, and hyaline cartilage preparations.

Text continues on page 67.

(b) Connective tissue proper: loose connective tissue, areolar

Description: Gel-like matrix with all three fiber types; cells: fibroblasts, macrophages, mast cells, and some white blood cells.

Function: Wraps and cushions organs; its macrophages phagocytize bacteria; plays important role in inflammation; holds and conveys tissue fluid.

Location: Widely distributed under epithelia of body, e.g., forms lamina propria of mucous membranes; packages organs; surrounds capillaries.

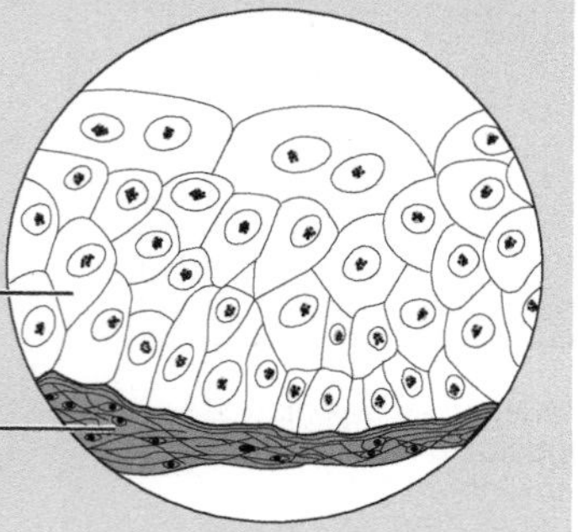

Epithelium

Lamina propria

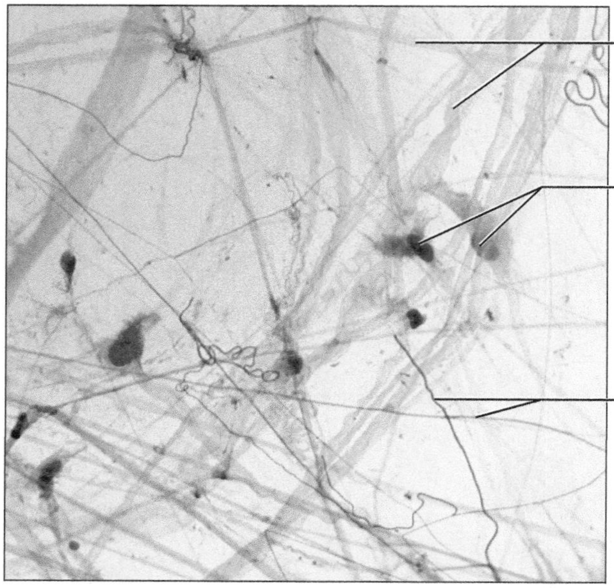

Collagen fibers

Fibroblast nuclei

Elastic fibers

Photomicrograph: Areolar connective tissue, a soft packaging tissue of the body (350×).

(c) Connective tissue proper: loose connective tissue, adipose

Description: Matrix as in areolar, but very sparse; closely packed adipocytes, or fat cells, have nucleus pushed to the side by large fat droplet.

Function: Provides reserve food fuel; insulates against heat loss; supports and protects organs.

Location: Under skin in the hypodermis; around kidneys and eyeballs; within abdomen; in breasts.

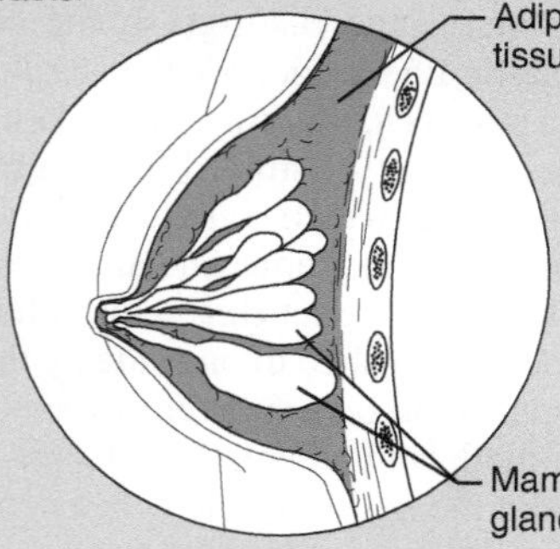

Adipose tissue

Mammary glands

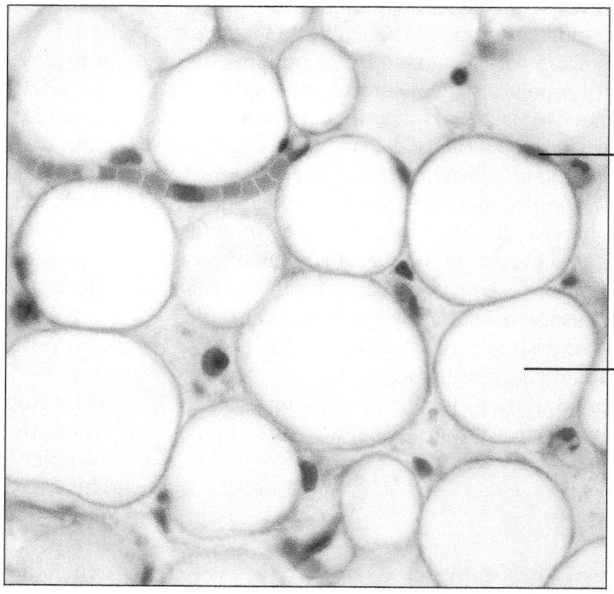

Nucleus of fat cell

Vacuole containing fat droplet

Photomicrograph: Adipose tissue from the subcutaneous layer under the skin (380×).

Figure 5.5 (*continued*) **Connective tissues.** Connective tissue proper (**b** and **c**).

(d) Connective tissue proper: loose connective tissue, reticular

Description: Network of reticular fibers in a typical loose ground substance; reticular cells lie on the network.

Function: Fibers form a soft internal skeleton (stroma) that supports other cell types including white blood cells, mast cells, and macrophages.

Location: Lymphoid organs (lymph nodes, bone marrow, and spleen).

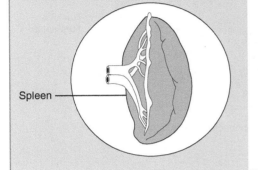

Spleen

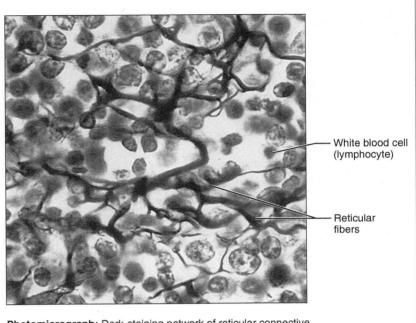

White blood cell (lymphocyte)

Reticular fibers

Photomicrograph: Dark-staining network of reticular connective tissue fibers forming the internal skeleton of the spleen (630×).

(e) Connective tissue proper: dense irregular connective tissue

Description: Primarily irregularly arranged collagen fibers; some elastic fibers; major cell type is the fibroblast.

Function: Able to withstand tension exerted in many directions; provides structural strength.

Location: Fibrous capsules of organs and of joints; dermis of the skin; submucosa of digestive tract.

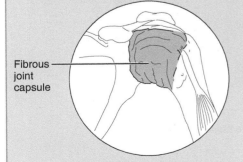

Fibrous joint capsule

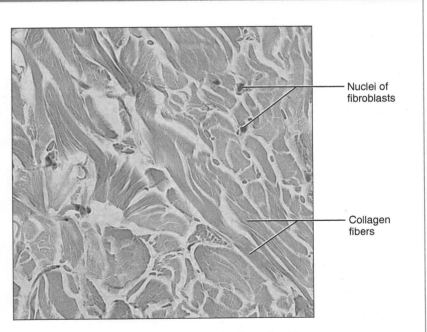

Nuclei of fibroblasts

Collagen fibers

Photomicrograph: Dense irregular connective tissue from the dermis of the skin (345×).

Figure 5.5 (*continued*) (**d** and **e**).

(f) Connective tissue proper: dense regular connective tissue

Description: Primarily parallel collagen fibers; a few elastic fibers; major cell type is the fibroblast.

Function: Attaches muscles to bones or to muscles; attaches bones to bones; withstands great tensile stress when pulling force is applied in one direction.

Location: Tendons, most ligaments, aponeuroses.

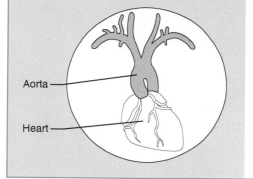

Shoulder joint

Ligament

Tendon

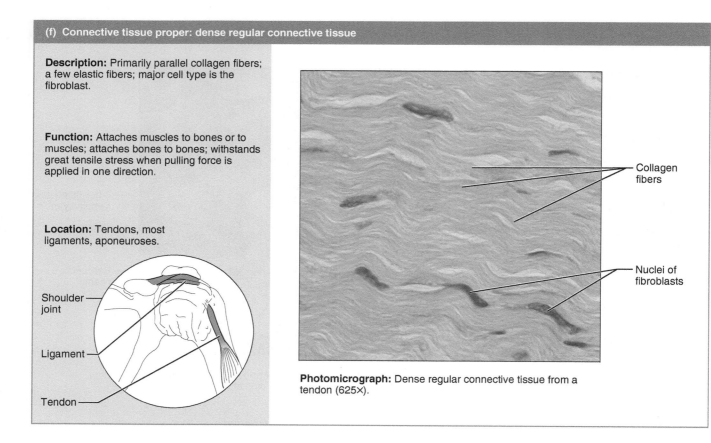

Collagen fibers

Nuclei of fibroblasts

Photomicrograph: Dense regular connective tissue from a tendon (625×).

(g) Connective tissue proper: elastic connective tissue

Description: Dense regular connective tissue containing a high proportion of elastic fibers.

Function: Allows recoil of tissue following stretching; maintains pulsatile flow of blood through arteries; aids passive recoil of lungs following inspiration.

Location: Walls of large arteries; within certain ligaments associated with the vertebral column; within the walls of the bronchial tubes.

Aorta

Heart

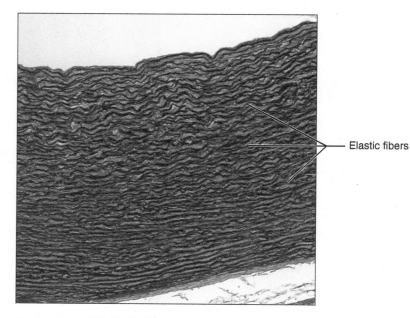

Elastic fibers

Photomicrograph: Elastic connective tissue in the wall of the aorta (145×).

Figure 5.5 (*continued*) **Connective tissues.** Connective tissue proper (**f** and **g**).

(h) Cartilage: hyaline

Description: Amorphous but firm matrix; collagen fibers form an imperceptible network; chondroblasts produce the matrix and when mature (chondrocytes) lie in lacunae.

Function: Supports and reinforces; serves as resilient cushion; resists compressive stress.

Location: Forms most of the embryonic skeleton; covers the ends of long bones in joint cavities; forms costal cartilages of the ribs; cartilages of the nose, trachea, and larynx.

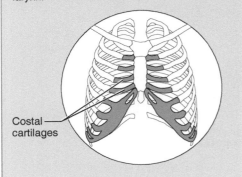

Costal cartilages

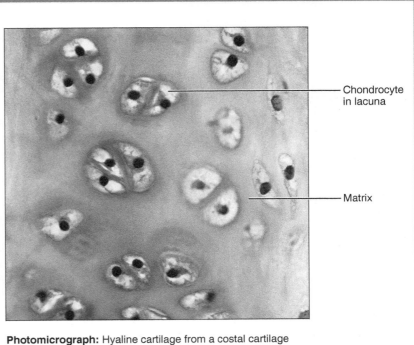

Chondrocyte in lacuna

Matrix

Photomicrograph: Hyaline cartilage from a costal cartilage of a rib (380×).

(i) Cartilage: elastic

Description: Similar to hyaline cartilage, but more elastic fibers in matrix.

Function: Maintains the shape of a structure while allowing great flexibility.

Location: Supports the external ear (auricle); epiglottis.

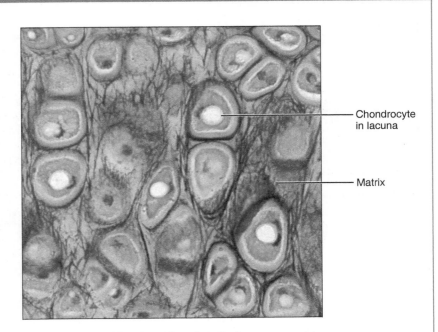

Chondrocyte in lacuna

Matrix

Photomicrograph: Elastic cartilage from the human ear auricle; forms the flexible skeleton of the ear (715×).

Figure 5.5 (*continued*) Cartilage (**h** and (**i**).

(j) Cartilage: fibrocartilage

Description: Matrix similar to but less firm than that in hyaline cartilage; thick collagen fibers predominate.

Function: Tensile strength with the ability to absorb compressive shock.

Location: Intervertebral discs; pubic symphysis; discs of knee joint.

Intervertebral discs

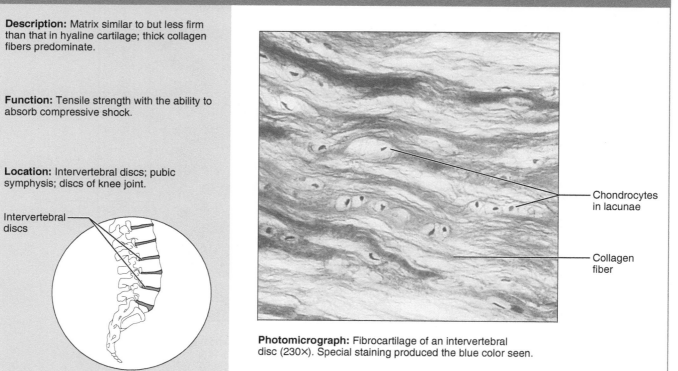

Chondrocytes in lacunae

Collagen fiber

Photomicrograph: Fibrocartilage of an intervertebral disc (230×). Special staining produced the blue color seen.

(k) Bone (osseous tissue)

Description: Hard, calcified matrix containing many collagen fibers; osteocytes lie in lacunae. Very well vascularized.

Function: Bone supports and protects (by enclosing); provides levers for the muscles to act on; stores calcium and other minerals and fat; marrow inside bones is the site for blood cell formation (hematopoiesis).

Location: Bones

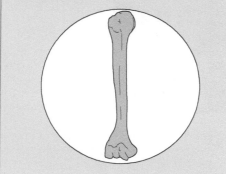

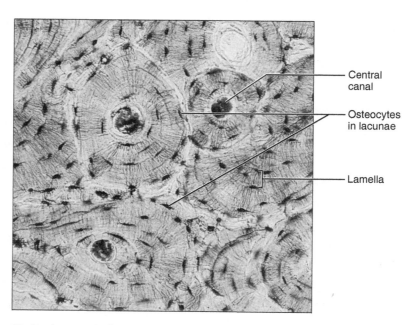

Central canal

Osteocytes in lacunae

Lamella

Photomicrograph: Cross-sectional view of bone (400×).

Figure 5.5 (*continued*) **Connective tissues.** Cartilage (j) and bone (k).

(I) Blood

Description: Red and white blood cells in a fluid matrix (plasma).

Function: Transport of respiratory gases, nutrients, wastes, and other substances.

Location: Contained within blood vessels.

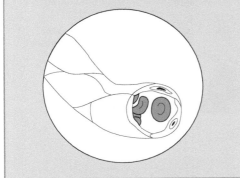

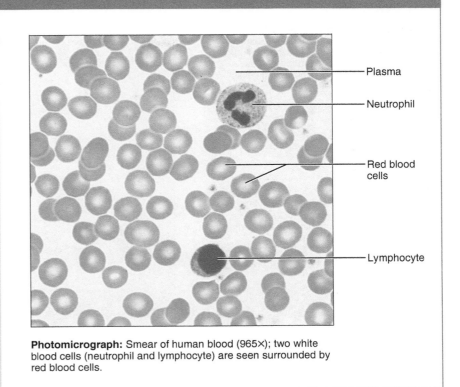

— Plasma

— Neutrophil

— Red blood cells

— Lymphocyte

Photomicrograph: Smear of human blood (965×); two white blood cells (neutrophil and lymphocyte) are seen surrounded by red blood cells.

Figure 5.5 (*continued*) Blood (I).

Scan the blood slide at low and then high power to examine the general shape of the red blood cells. Then, switch to the oil immersion lens for a closer look at the various types of white blood cells. How does the matrix of blood differ from all other connective tissues?

_____

_____

Nervous Tissue

Nervous tissue is made up of two major cell populations. The **neuroglia** are special supporting cells that protect, support, and insulate the more delicate neurons. The **neurons** are highly specialized to receive stimuli (excitability) and to generate electrical signals that may be sent to all parts of the body (conductivity).

The structure of neurons is markedly different from that of all other body cells. They have a nucleus-containing cell body, and their cytoplasm is drawn out into long extensions (cell processes)—sometimes as long as 1 m (about 3 feet), which allows a single neuron to conduct an electrical signal over relatively long distances. (More detail about the anatomy of the different classes of neurons and neuroglia appears in Exercise 13).

ACTIVITY 3

Examining Nervous Tissue Under the Microscope

Obtain a prepared slide of a spinal cord smear. Locate a neuron, and compare it to **Figure 5.6**. Keep the light dim—this will help you see the cellular extensions of the neurons. (See also Figure 13.2 in Exercise 13.) ■

Muscle Tissue

Muscle tissue (**Figure 5.7**) is highly specialized to contract and produces most types of body movement. The three basic types of muscle tissue are described briefly here.

Skeletal muscle, the flesh of the body, is attached to the skeleton. It is under conscious voluntary control, and its contraction moves the limbs and other body parts. The cells of skeletal muscles are long, cylindrical, nonbranching, and multinucleate (several nuclei per cell), with the nuclei pushed to the periphery of the cells; they have obvious *striations* (stripes).

Cardiac muscle is found only in the heart. As it contracts, the heart acts as a pump, propelling the blood into the blood vessels. Cardiac muscle, like skeletal muscle, has

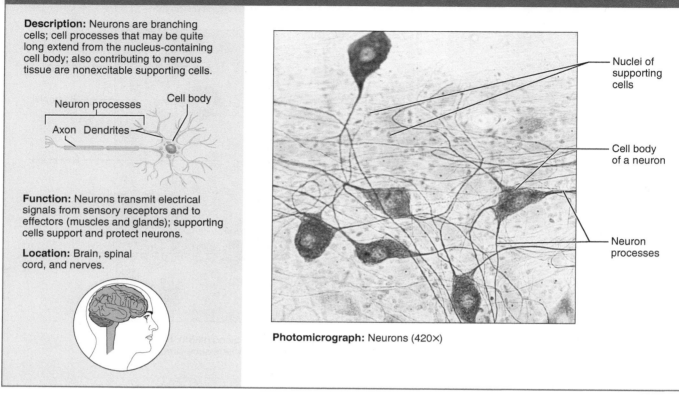

Nervous tissue

Description: Neurons are branching cells; cell processes that may be quite long extend from the nucleus-containing cell body; also contributing to nervous tissue are nonexcitable supporting cells.

Neuron processes

Axon Dendrites

Cell body

Function: Neurons transmit electrical signals from sensory receptors and to effectors (muscles and glands); supporting cells support and protect neurons.

Location: Brain, spinal cord, and nerves.

Nuclei of supporting cells

Cell body of a neuron

Neuron processes

Photomicrograph: Neurons (420×)

Figure 5.6 Nervous tissue.

striations, but cardiac cells are branching uninucleate cells that interdigitate (fit together) at junctions called **intercalated discs.** These structural modifications allow the cardiac muscle to act as a unit. Cardiac muscle is under involuntary control, which means that we cannot voluntarily or consciously control the operation of the heart.

Smooth muscle is found mainly in the walls of hollow organs (digestive and urinary tract organs, uterus, blood vessels). Typically it has two layers that run at right angles to each other; consequently its contraction can constrict or dilate the lumen (cavity) of an organ and propel substances along predetermined pathways. Smooth muscle cells are quite different in appearance from those of skeletal or cardiac muscle.

No striations are visible, and the uninucleate smooth muscle cells are spindle-shaped (tapered at the ends, like a candle). Like cardiac muscle, it is under involuntary control.

ACTIVITY 4

Examining Muscle Tissue Under the Microscope

Obtain and examine prepared slides of skeletal, cardiac, and smooth muscle. Notice their similarities and dissimilarities in your observations and in the illustrations and photomicrographs (Figure 5.7). ■

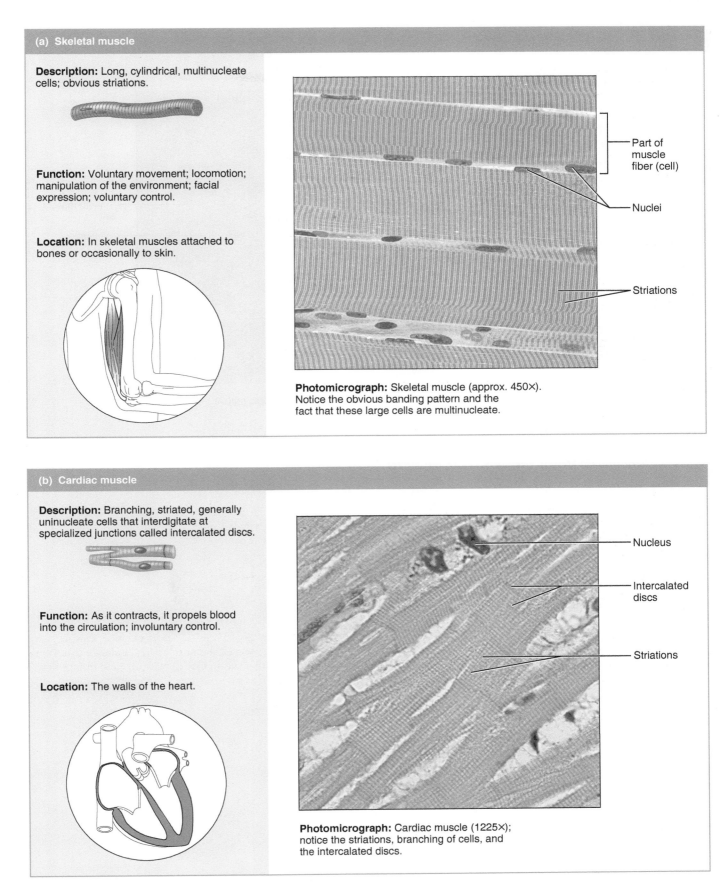

(a) Skeletal muscle

Description: Long, cylindrical, multinucleate cells; obvious striations.

Function: Voluntary movement; locomotion; manipulation of the environment; facial expression; voluntary control.

Location: In skeletal muscles attached to bones or occasionally to skin.

Part of muscle fiber (cell)

Nuclei

Striations

Photomicrograph: Skeletal muscle (approx. 450×). Notice the obvious banding pattern and the fact that these large cells are multinucleate.

(b) Cardiac muscle

Description: Branching, striated, generally uninucleate cells that interdigitate at specialized junctions called intercalated discs.

Function: As it contracts, it propels blood into the circulation; involuntary control.

Location: The walls of the heart.

Nucleus

Intercalated discs

Striations

Photomicrograph: Cardiac muscle (1225×); notice the striations, branching of cells, and the intercalated discs.

Figure 5.7 Muscle tissues. Skeletal (**a**) and cardiac (**b**) muscles.

(c) Smooth muscle

Description: Spindle-shaped cells with central nuclei; no striations; cells arranged closely to form sheets.

Function: Propels substances (foodstuffs, urine) or a baby along internal passageways; involuntary control.

Location: Mostly in the walls of hollow organs.

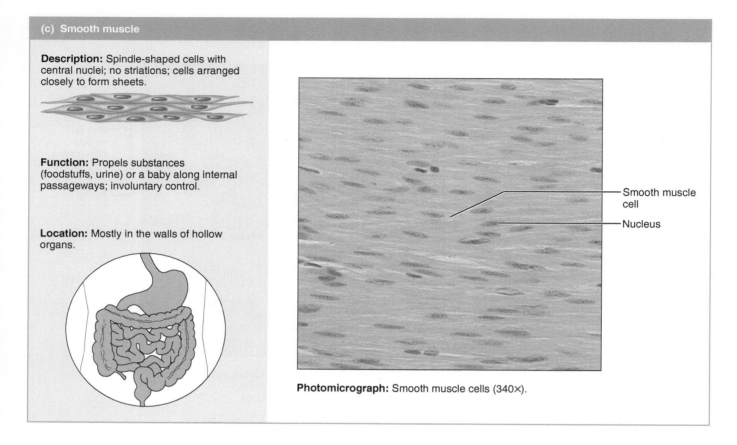

Smooth muscle cell

Nucleus

Photomicrograph: Smooth muscle cells (340×).

Figure 5.7 (*continued*) **Muscle tissues.** Smooth muscle (**c**).

Classification of Tissues

Tissue Structure and Function—General Review

1. Define *tissue.* _____

_____

2. Use the key to identify the major tissue types described below. Responses may be used more than once.

   Key: a.   connective tissue   b.   epithelium   c.   muscle   d.   nervous tissue

   _____   1. lines body cavities and covers the body's external surface

   _____   2. pumps blood, flushes urine out of the body, allows one to swing a bat

   _____   3. transmits electrical signals

   _____   4. anchors, packages, and supports body organs

   _____   5. cells may absorb, secrete, and filter

   _____   6. most involved in regulating and controlling body functions

   _____   7. major function is to contract

   _____   8. synthesizes hormones

   _____   9. includes nonliving extracellular matrix

   _____   10. most widespread tissue in the body

   _____   11. forms nerves and the brain

Epithelial Tissue

3. Describe five general characteristics of epithelial tissue. _____

_____

_____

_____

4. How are epithelial tissues classified? _____

_____

_____

WHY THIS MATTERS | 5. Which type of epithelium is removed with a buccal swab? _____

_____

6. Explain why a buccal swab procedure shouldn't cause bleeding. _____

_____

7. List five major functions of epithelium in the body, and give examples of cells or organs that provide each function.

Function 1: _____ Example: _____

Function 2: _____ Example: _____

Function 3: _____ Example: _____

Function 4: _____ Example: _____

Function 5: _____ Example: _____

8. How does the function of stratified epithelium differ from the function of simple epithelium? _____

_____

_____

9. Where is ciliated epithelium found? _____

_____

What role does it play? _____

_____

_____

10. Transitional epithelium is actually stratified squamous epithelium, but there is something special about it.

How does it differ structurally from other stratified squamous epithelia? _____

_____

How does the structural difference support its function in the body? _____

_____

11. How do the endocrine and exocrine glands differ in structure and function? _____

_____

_____

_____

12. Using the key, write the letter indicating the type of epithelial tissue that fits the description.

Key: a.   simple squamous   c.   simple columnar   e.   stratified squamous
          b.   simple cubodial   d.   pseudostratified ciliated columnar   f.   transitional

_____   1. lining of the esophagus

_____   2. lining of the stomach

_____   3. alveolar sacs of lungs

_____   4. lining of the trachea

_____   5. epidermis of the skin

_____   6. lining of bladder; peculiar cells that have the ability to slide over each other

Connective Tissue

13. What are three general characteristics of connective tissues? _____

_____

_____

14. What functions are performed by connective tissue? _____

_____

15. How are the functions of connective tissue reflected in its structure? _____

_____

_____

16. Using the key, choose the best response to identify the connective tissues described below.

_____   1. attaches bones to bones and muscles to bones

_____   2. acts as a storage depot for fat

_____   3. forms the fibrous joint capsule

_____   4. makes up the intervertebral discs

_____   5. composes basement membranes; a soft packaging tissue with a jellylike matrix

_____   6. forms the larynx, the costal cartilages of the ribs, and the embryonic skeleton

_____   7. provides a flexible framework for the external ear

Key: a.   adipose connective tissue
       b.   areolar connective tissue
       c.   dense irregular connective tissue
       d.   dense regular connective tissue
       e.   elastic cartilage
       f.   elastic connective tissue
       g.   fibrocartilage
       h.   hyaline cartilage
       i.   osseous tissue

_____   8. matrix hard owing to calcium salts; provides levers for muscles to act on

_____   9. forms the walls of large arteries

17. Why do adipose cells remind people of a signet ring (a ring with a single jewel)? _____

_____

Nervous Tissue

18. What two physiological characteristics are highly developed in neurons? _____

_____

19. In what ways are neurons similar to other cells? _____

How are they structurally different? _____

_____

20. Describe how the unique structure of a neuron relates to its function in the body. _____

_____

Muscle Tissue

21. The terms and phrases in the key relate to the muscle tissues. For each of the three muscle tissues, select the terms or phrases that characterize it, and write the corresponding letter of each term on the answer line.

Key: a. striated   e. voluntary   i. attached to bones
         b. branching cells   f. involuntary   j. intercalated discs
         c. spindle-shaped cells g. one nucleus   k. in wall of bladder and stomach
         d. cylindrical cells   h. many nuclei   l. forms heart walls

Skeletal muscle: _____ Cardiac muscle: _____ Smooth muscle: _____

22. ➕ Orthopedic surgeons are fond of saying, "It is better to break a bone than it is to tear a tendon or ligament." Based on

your understanding of these two types of connective tissue, explain why that would be true. _____

_____

23. ➕ When cardiac muscle tissue dies in adults, it is replaced with scar tissue composed of dense connective tissue. Explain

how the function of the scar tissue would differ from the function of the cardiac muscle tissue. _____

_____

24. ➕ Smoking impairs cilia because the toxins paralyze and can destroy the cilia. Based on this loss of function, explain which

types of infections smokers would be more susceptible to. _____

_____

For Review

25. Label the tissue types illustrated here and on the next page, and identify all structures with leader lines.

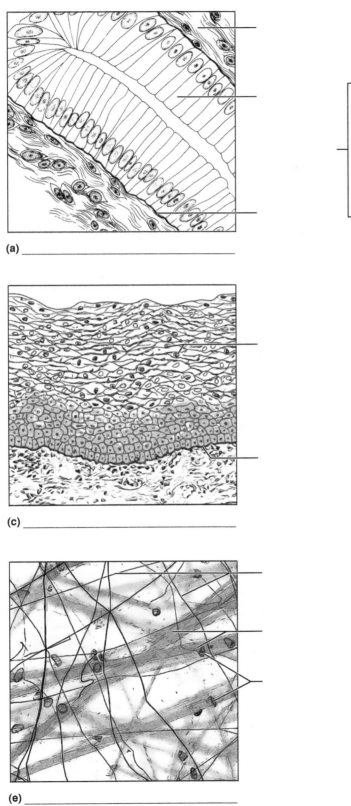

(a) _____

(c) _____

(e) _____

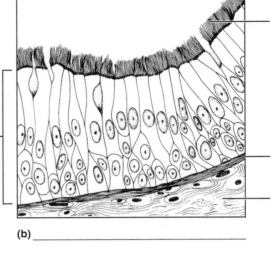

(b) _____

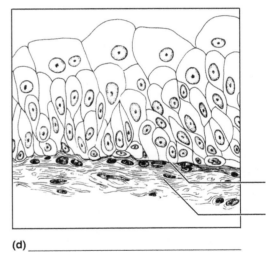

(d) _____

(f) _____

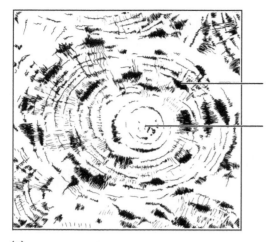

(g) _____

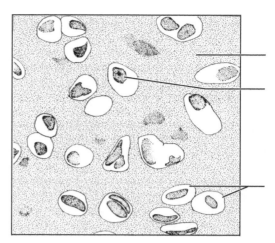

(h) _____

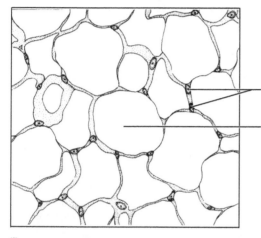

(i) _____

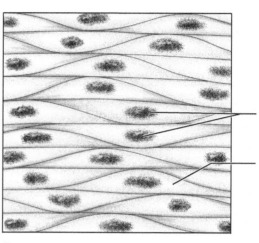

(j) _____

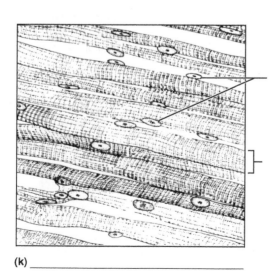

(k) _____

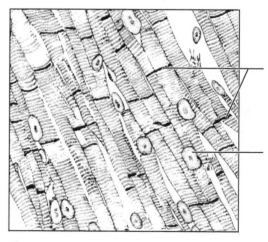

(l) _____

The Integumentary System

MATERIALS

- Skin model (three-dimensional, if available)
- Compound microscope
- Prepared slide of human scalp
- Prepared slide of skin of palm or sole
- Sheet of 20# bond paper ruled to mark off 1-cm² areas
- Scissors
- Povidone-iodine swabs, or Lugol's iodine and cotton swabs
- Adhesive tape
- Disposable gloves
- Data collection sheet for plotting distribution of sweat glands
- Porelon fingerprint pad or portable inking foils
- Ink cleaner towelettes
- Index cards (4 in. × 6 in.)
- Magnifying glasses

LEARNING OUTCOMES

- ☐ List several important functions of the skin, or integumentary system.
- ☐ Identify the following skin structures on a model, image, or microscope slide: epidermis, dermis (papillary and reticular layers), hair follicles and hair, sebaceous glands, and sweat glands.
- ☐ Name and describe the layers of the epidermis.
- ☐ List the factors that determine skin color, and describe the function of melanin.
- ☐ Identify the major regions of nails.
- ☐ Describe the distribution and function of hairs, sebaceous glands, and sweat glands.
- ☐ Discuss the difference between eccrine and apocrine sweat glands.
- ☐ Compare and contrast the structure and functions of the epidermis and the dermis.

PRE-LAB QUIZ

1. All the following are functions of the skin *except:*
   a. excretion of body wastes   c. protection from mechanical damage
   b. insulation   d. site of vitamin A synthesis
2. The skin has two distinct regions. The superficial layer is the _____, and the underlying connective tissue, the _____.
3. The most superficial layer of the epidermis is the:
   a. stratum basale   c. stratum granulosum
   b. stratum spinosum   d. stratum corneum
4. Thick skin of the epidermis contains _____ layers.
5. _____ is a yellow-orange pigment found in the stratum corneum and the hypodermis.
   a. Keratin   c. Melanin
   b. Carotene   d. Hemoglobin
6. These cells produce a brown-to-black pigment that colors the skin and protects DNA from ultraviolet radiation damage. The cells are:
   a. dendritic cells   c. melanocytes
   b. keratinocytes   d. tactile cells
7. Circle True or False. Nails are hornlike derivatives of the epidermis.
8. The portion of a hair that you see that projects from the surface of the skin is known as the:
   a. bulb   c. root
   b. matrix   d. shaft
9. Circle the correct underlined term. The ducts of <u>sebaceous</u> / <u>sweat glands</u> usually empty into a hair follicle but may also open directly onto the skin surface.
10. Circle the correct underlined term. <u>Eccrine</u> / <u>Apocrine</u> sweat glands are found primarily in the genital and axillary areas.

The **integument** is considered an organ system because it consists of multiple organs, the **skin** and its accessory organs. It is much more than an external body covering; architecturally, the skin is a marvel. It is tough yet pliable, a characteristic that enables it to withstand constant insult from outside agents.

The skin has many functions, most concerned with protection. It insulates and cushions the underlying body tissues and protects the entire body from abrasion, exposure to harmful chemicals, temperature extremes, and bacterial invasion. The hardened uppermost layer of the skin helps prevent water loss from the body surface. The skin's abundant capillary network plays an important role in temperature regulation.

The skin has other functions as well. For example, it acts as a mini excretory system; urea, salts, and water are lost through the skin pores in sweat. The skin also has important metabolic duties. For example, it is the site of vitamin D synthesis for the body. (Vitamin D plays a role in calcium absorption in the digestive system.) Finally, the sense organs for touch, pressure, pain, and temperature are located here.

Basic Structure of the Skin

The skin has two distinct regions—the superficial *epidermis* composed of epithelium and an underlying connective tissue *dermis* **(Figure 6.1)**. These layers are firmly "cemented" together along a wavy border. But friction, such as the rubbing of a poorly fitting shoe, may cause them to separate, resulting in a blister. Immediately deep to the dermis is the **subcutaneous layer,** or **hypodermis,** which is not considered part of the skin. It consists primarily of adipose tissue. The main skin areas and structures are described next.

ACTIVITY 1

Locating Structures on a Skin Model

As you read, locate the following structures in Figure 6.1 and on a skin model. ■■

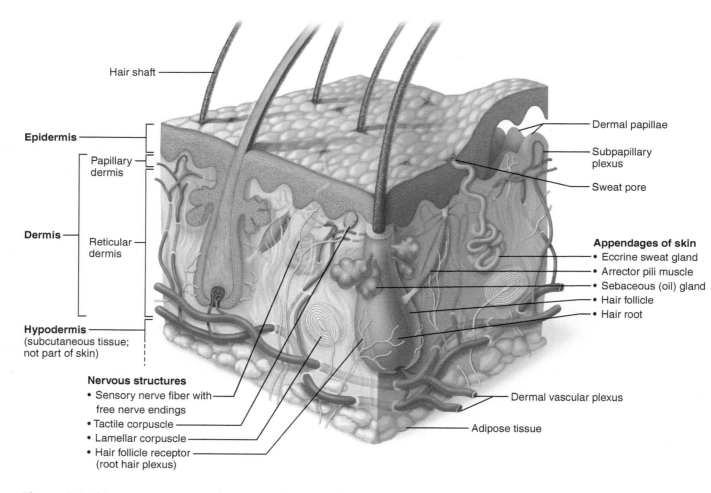

Figure 6.1 Skin structure. Three-dimensional view of the skin and the underlying tissue. The epidermis and dermis have been pulled apart at the right corner to reveal the dermal papillae. Tactile corpuscles are not common in hairy skin but they are included here for illustrative purposes.

Epidermis

Structurally, the avascular epidermis is a keratinized stratified squamous epithelium consisting of four distinct cell types and four or five distinct layers.

Cells of the Epidermis

• **Keratinocytes** (literally, keratin cells): The most abundant epidermal cells, their main function is to produce keratin fibrils. **Keratin** is a fibrous protein that gives the epidermis its durability and protective capabilities. Keratinocytes are tightly connected to each other by desmosomes.

Far less numerous are the following types of epidermal cells **(Figure 6.2)**:

• **Melanocytes:** Spidery black cells that produce the brown-to-black pigment called **melanin.** The skin tans because melanin production increases when the skin is exposed to ultraviolet radiation (UVR) in sunlight. The melanin provides a protective pigment umbrella over the nuclei of the cells in the deeper epidermal layers, thus shielding their genetic material (DNA) from the damaging effects of ultraviolet radiation. *Freckles* and *moles (nevi)* are areas where melanin is more concentrated.

• **Dendritic cells:** Also called *Langerhans cells,* these cells arise from the bone marrow and migrate to the epidermis. They ingest foreign substances and play a key role in activating the immune response.

• **Tactile epithelial cells:** Occasional spiky hemispheres that, in conjunction with disklike sensory nerve endings, form sensitive touch receptors located at the epidermal-dermal junction.

Layers of the Epidermis

The epidermis consists of four layers in thin skin, which covers most of the body. Thick skin, found on the palms of the hands and soles of the feet, contains a fifth layer, the stratum lucidum. From deep to superficial, the layers of the epidermis are the stratum basale, stratum spinosum, stratum granulosum, stratum lucidum, and stratum corneum (Figure 6.2).

The layers of the epidermis are summarized in **Table 6.1**.

Dermis

The dense irregular connective tissue making up the dermis consists of two principal regions—the papillary and reticular areas (Figure 6.1). Like the epidermis, the dermis varies in thickness. For example, it is particularly thick on the palms of the hands and soles of the feet and is quite thin on the eyelids.

• **Papillary dermis:** The more superficial dermal region composed of areolar connective tissue. It is very uneven and has fingerlike projections from its superior surface, the **dermal papillae,** which attach it to the epidermis above. These projections lie on top of the larger dermal ridges. In the palms of the hands and soles of the feet, they produce the *fingerprints,* unique patterns of *epidermal ridges* that remain unchanged throughout life. The abundant capillary networks in the papillary layer furnish nutrients for the epidermal layers and allow heat to radiate to the skin surface. The pain receptors (free nerve endings) and touch receptors (**tactile corpuscles** in hairless skin) are also found here.

• **Reticular dermis:** The deepest skin layer. It is composed of dense irregular connective tissue and contains many arteries and veins, sweat and sebaceous glands, and pressure receptors (**lamellar corpuscles**).

Both the papillary dermis and reticular dermis are abundant in collagen and elastic fibers. The elastic fibers give skin its exceptional elasticity in youth. In old age, the number of elastic fibers decreases and the subcutaneous layer loses fat, which leads to wrinkling and inelasticity of the skin. Fibroblasts, adipose cells, various types of macrophages and other cell types are found throughout the dermis.

The abundant dermal blood supply, consisting mainly of the deep *dermal vascular plexus* (between the dermis and hypodermis) and the *subpapillary plexus* (located just deep to the dermal papillae), allows the skin to play a role in regulating body temperature. When body temperature is high, the arterioles serving the skin dilate, and the capillary network of the dermis becomes engorged with the heated blood. Thus body heat is allowed to radiate from the skin surface.

Table 6.1 Layers of the Epidermis (from superficial to deep)

Epidermal layer	Description
Stratum corneum (horny layer)	The outermost layer consisting of 20–30 layers of dead, scalelike keratinocytes. They are constantly being exfoliated and replaced by the division of the deeper cells.
Stratum lucidum (clear layer)	Present only in thick skin. A very thin transparent band of flattened, dead keratinocytes with indistinct boundaries.
Stratum granulosum (granular layer)	A thin layer named for the abundant granules its cells contain. These granules are (1) *lamellar granules,* which contain a waterproofing glycolipid that is secreted into the extracellular space; and (2) *keratohyaline granules,* which help to form keratin in the more superficial layers. At the upper border of this layer, the cells are beginning to die.
Stratum spinosum (spiny layer)	Several layers of cells that contain thick, weblike bundles of intermediate filaments made of a pre-keratin protein. The cells in this layer appear spiky because when the tissue is prepared, the cells shrink, but their desmosomes hold tight to adjacent cells. Cells in this layer and the basal layer are the only ones to receive adequate nourishment from diffusion of nutrients from the dermis.
Stratum basale (basal layer)	A single row of cells immediately above the dermis. Its cells are constantly undergoing mitosis to form new cells, hence its alternate name, *stratum germinativum.* Some 10–25% of the cells in this layer are melanocytes, which thread their processes through this and adjacent layers of keratinocytes. Occasional tactile epithelial cells are also present in this layer.

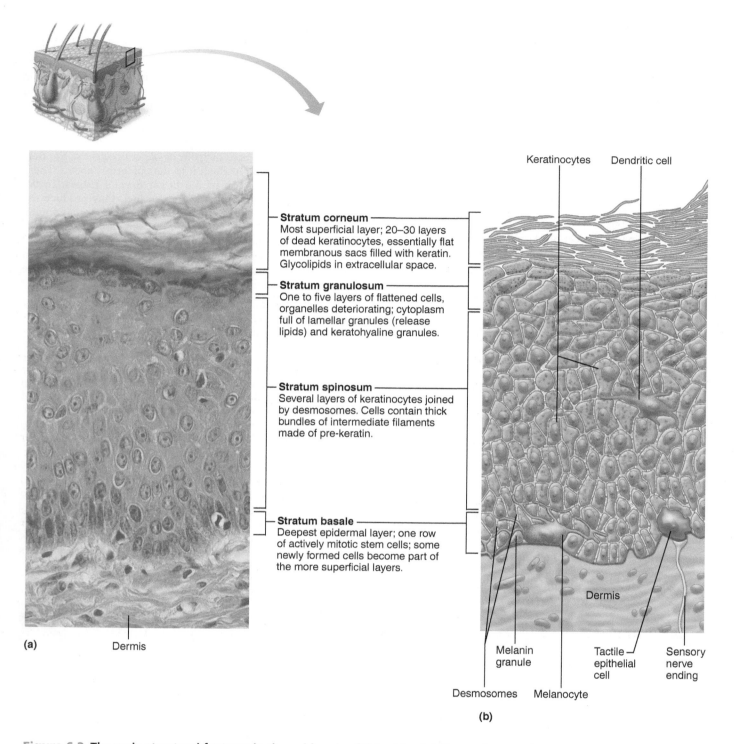

Stratum corneum
Most superficial layer; 20–30 layers of dead keratinocytes, essentially flat membranous sacs filled with keratin. Glycolipids in extracellular space.

Stratum granulosum
One to five layers of flattened cells, organelles deteriorating; cytoplasm full of lamellar granules (release lipids) and keratohyaline granules.

Stratum spinosum
Several layers of keratinocytes joined by desmosomes. Cells contain thick bundles of intermediate filaments made of pre-keratin.

Stratum basale
Deepest epidermal layer; one row of actively mitotic stem cells; some newly formed cells become part of the more superficial layers.

(a)   Dermis

Keratinocytes   Dendritic cell

Dermis

Melanin granule

Desmosomes   Melanocyte

Tactile epithelial cell

Sensory nerve ending

(b)

Figure 6.2 The main structural features in the epidermis of thin skin. (a) Photomicrograph depicting the four major epidermal layers (435×). **(b)** Diagram showing the layers and relative distribution of the different cell types. Notice that the keratinocytes are joined by numerous desmosomes. The stratum lucidum, present in thick skin, is not illustrated here.

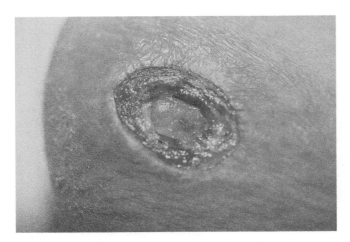

Figure 6.3 **Photograph of a deep (stage III) decubitus ulcer.**

Any restriction of the normal blood supply to the skin results in cell death and, if severe enough, skin ulcers **(Figure 6.3)**. **Bedsores (decubitus ulcers)** occur in bedridden patients who are not turned regularly enough. The weight of the body exerts pressure on the skin, especially over bony projections (hips, heels, etc.), which leads to restriction of the blood supply and tissue death. ✚

The dermis is also richly provided with lymphatic vessels and nerve fibers. Many of the nerve endings bear highly specialized receptor organs that, when stimulated by environmental changes, transmit messages to the central nervous system for interpretation. Some of these receptors include free nerve endings, lamellar corpuscles, a tactile corpuscle, and a hair follicle receptor (also called a *root hair plexus*) (Figure 6.1).

Skin Color

Skin color is a result of the relative amount of melanin in skin, the relative amount of carotene in skin, and the degree of oxygenation of the blood. People who produce large amounts of melanin have brown-toned skin. In light-skinned people, who have less melanin pigment, the dermal blood supply flushes through the rather transparent cell layers above, giving the skin a rosy glow. *Carotene* is a yellow-orange pigment present primarily in the stratum corneum and in the adipose tissue of the hypodermis. Its presence is most noticeable when large amounts of carotene-rich foods (carrots, for instance) are eaten.

Skin color can be an important diagnostic tool. For example, flushed skin may indicate hypertension or fever, whereas pale skin is typically seen in anemic individuals. When the blood is inadequately oxygenated, as during asphyxiation and serious lung disease, both the blood and the skin take on a bluish, cast, a condition called *cyanosis*.

Jaundice, in which the tissues become yellowed, is almost always diagnostic for liver disease, whereas a bronzing of the skin hints that a person's adrenal cortex is hypoactive (**Addison's disease**). ✚

Accessory Organs of the Skin

The accessory organs of the skin—cutaneous glands, hair, and nails—all originate in the epidermis, but they reside primarily in the dermis. They originate from the stratum basale and extend into the dermis.

Cutaneous Glands

The cutaneous glands fall primarily into two categories: the sebaceous glands and the sweat glands (Figure 6.1 and Figure 6.4).

Sebaceous (Oil) Glands

The sebaceous glands are found nearly all over the skin, except for the palms of the hands and the soles of the feet. Their ducts usually empty into a hair follicle, but some open directly on the skin surface.

Sebum is the product of sebaceous glands. It is a mixture of oily substances and fragmented cells that acts as a lubricant to keep the skin soft and moist (a natural skin cream) and keeps the hair from becoming brittle. The sebaceous glands become particularly active during puberty, thus the skin tends to become oilier during this period of life.

Blackheads are accumulations of dried sebum, bacteria, and melanin from epithelial cells in the oil duct. **Acne** is an active infection of the sebaceous glands. ✚

Sweat (Sudoriferous) Glands

Sweat, or sudoriferous, glands are exocrine glands and are widely distributed all over the skin. Outlets for the glands are epithelial openings called *pores*. Sweat glands are categorized by the composition of their secretions.

• **Eccrine sweat glands:** Also called **merocrine sweat glands,** these glands are distributed all over the body. They produce a clear secretion consisting primarily of water, salts (mostly NaCl), and urea. Eccrine sweat glands, under the control of the nervous system, are an important part of the body's heat-regulating apparatus. They secrete sweat when the external temperature or body temperature is high. When sweat evaporates, it carries excess body heat with it.

• **Apocrine sweat glands:** Found predominantly in the axillary and genital areas, these glands secrete the basic components of eccrine sweat plus proteins and fat-rich substances. Apocrine sweat is an excellent nutrient medium for the microorganisms typically found on the skin. This sweat is odorless, but when bacteria break down its organic components, it begins to smell unpleasant.

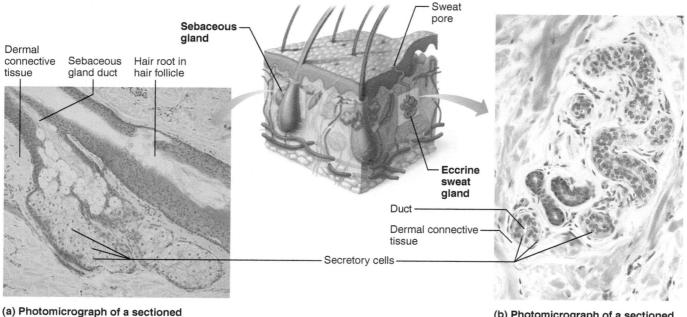

(a) Photomicrograph of a sectioned
   sebaceous gland (100×)

(b) Photomicrograph of a sectioned
   eccrine sweat gland (145×)

Figure 6.4 **Cutaneous glands.**

ACTIVITY 2

Differentiating Sebaceous and Sweat Glands Microscopically

Using the slide *thin skin with hairs,* identify sebaceous and eccrine sweat glands (refer to Figure 6.4). What characteristics relating to location or gland structure allow you to differentiate these glands?

_____

_____

_____

ACTIVITY 3

Plotting the Distribution of Sweat Glands

1. Form a hypothesis about the relative distribution of sweat glands on the palm and forearm. Justify your hypothesis.

2. The bond paper for this simple experiment has been pre-ruled in cm^2—put on disposable gloves, and cut along the lines to obtain the required squares. You will need two squares of bond paper (each 1 cm × 1 cm), adhesive tape, and a povidone-iodine swab *or* Lugol's iodine and a cotton-tipped swab.

3. Paint an area of the medial aspect of your left palm (avoid the deep crease lines) and a region of your left forearm with the iodine solution, and allow it to dry thoroughly. The painted area in each case should be slightly larger than the paper squares to be used.

4. Have your lab partner *securely* tape a square of bond paper over each iodine-painted area, and leave the paper squares in place for 20 minutes. (If it is very warm in the laboratory

while this test is being conducted, you can obtain good results within 10 to 15 minutes.)

5. After 20 minutes, remove the paper squares, and count the number of blue-black dots on each square. The presence of a blue-black dot on the paper indicates an active sweat gland. The iodine in the pore is dissolved in the sweat and reacts chemically with the starch in the bond paper to produce the blue-black color. Thus "sweat maps" have been produced for the two skin areas.

6. Which skin area tested has the greater density of sweat glands?

_____

_____

7. Tape your results (bond paper squares) to a data collection sheet labeled "palm" and "forearm" at the front of the lab. Be sure to put your paper squares in the correct columns on the data sheet.

8. Once all the data has been collected, review the class results. ■

Nails

Nails are hornlike derivatives of the epidermis (**Figure 6.5**). They consist of the following parts:

- **Nail plate:** The visible attached portion.

- **Free edge:** The portion of the nail that grows out away from the body.

- **Hyponychium:** The region beneath the free edge of the nail.

- **Nail root:** The part that is embedded in the skin and adheres to an epithelial nail bed.

- **Nail folds:** Skin folds that overlap the borders of the nail.

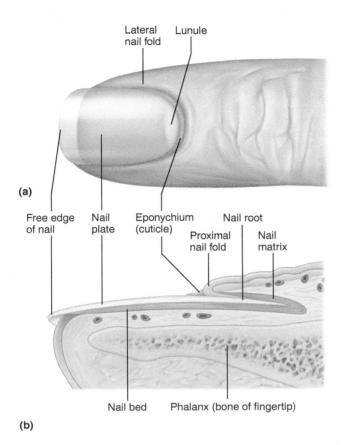

(a)

(b)

Figure 6.5 **Structure of a nail.** **(a)** Surface view of the distal part of a finger. The nail matrix that forms the nail lies beneath the lunule; the epidermis of the nail bed underlies the nail. **(b)** Sagittal section of the fingertip.

* **Eponychium:** The thick proximal nail fold, commonly called the cuticle.

* **Nail bed:** Extension of the stratum basale beneath the nail.

* **Nail matrix:** The thickened proximal part of the nail bed containing germinal cells responsible for nail growth. As the matrix produces the nail cells, they become heavily keratinized and die. Thus nails, like hairs, are mostly nonliving material.

* **Lunule:** The proximal region of the thickened nail matrix, which appears as a white crescent. Everywhere else, nails are transparent and nearly colorless, but they appear pink because of the blood supply in the underlying dermis. When someone is cyanotic because of a lack of oxygen in the blood, the nail beds take on a blue cast.

ACTIVITY 4

Identifying Nail Structures

Identify the nail structures shown in Figure 6.5 on yourself or your lab partner. ▬

Hairs and Associated Structures

Hairs, enclosed in hair follicles, are found all over the entire body surface, except for thick-skinned areas (the palms of the hands and the soles of the feet), parts of the external genitalia, the nipples, and the lips.

* **Hair:** Hair consists of two primary regions: the **hair shaft,** the region projecting from the surface of the skin, and the **hair root,** which is beneath the surface of the skin and is embedded within the **hair follicle.** The **hair bulb** is a collection of well-nourished epithelial cells at the base of the hair follicle **(Figure 6.6).** The hair shaft and the hair root have

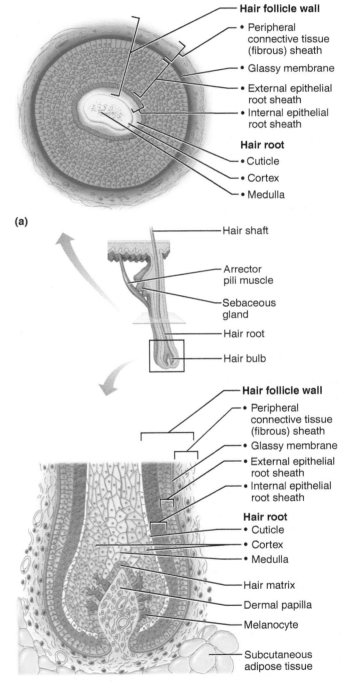

(a)

(b)

Figure 6.6 **Structure of a hair and hair follicle.**
(a) Diagram of a cross section of a hair within its follicle.
(b) Diagram of a longitudinal view of the expanded hair bulb of the follicle, which encloses the hair matrix.

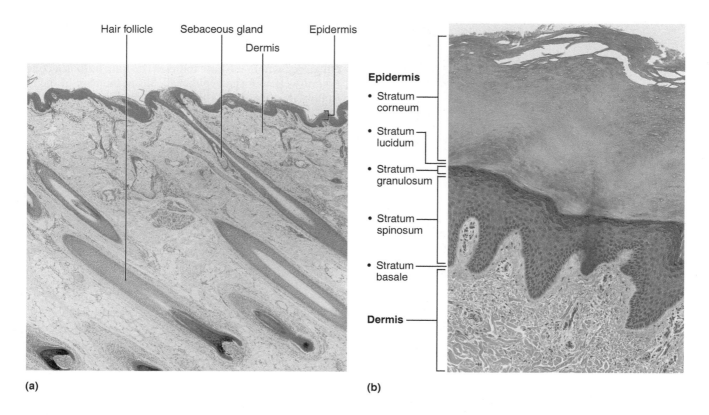

Figure 6.7 Photomicrographs of skin. (a) Thin skin with hairs (40×). **(b)** Thick hairless skin (75×).

three layers of keratinized cells: the *medulla* in the center, surrounded by the *cortex,* and the protective *cuticle.* Abrasion of the cuticle at the tip of the hair shaft results in split ends. Hair color depends on the amount and type of melanin pigment found in the hair cortex.

•   **Hair follicle:** A structure formed from both epidermal and dermal cells (see Figure 6.6). Its **epithelial root sheath,** with two parts (internal and external), is enclosed by a thickened basement membrane, the glassy membrane, and a **peripheral connective tissue** (or fibrous) **sheath,** which is essentially dermal tissue. A small nipple of dermal tissue protrudes into the hair bulb from the peripheral connective tissue sheath and provides nutrition to the growing hair. It is called the **dermal papilla.** A layer of actively dividing epithelial cells called the **hair matrix** is located on top of the dermal papilla.

•   **Arrector pili muscle:** Small bands of smooth muscle cells connecting each hair follicle to the papillary layer of the dermis (Figures 6.1 and 6.6). When these muscles contract (during cold or fright), the slanted hair follicle is pulled upright, dimpling the skin surface with goose bumps. This phenomenon is especially dramatic in a scared cat, whose fur actually stands on end to increase its apparent size.

ACTIVITY 5

Comparing Hairy and Relatively Hair-Free Skin Microscopically

Whereas thick skin has no hair follicles or sebaceous (oil) glands, thin skin usually has both. The scalp, of course, has the highest density of hair follicles.

1. Obtain a prepared slide of the human scalp, and study it carefully under the microscope. Compare your tissue slide to **Figure 6.7a,** and identify as many of the structures as possible (refer to Figure 6.1).

How is this stratified squamous epithelium different from that observed in the esophagus (Exercise 5)?

_____

_____

How do these differences relate to the functions of these two similar epithelia?

_____

_____

_____

2. Obtain a prepared slide of hairless skin of the palm or sole (Figure 6.7b). Compare the slide to the photomicrograph (Figure 6.7a). In what ways does the thick skin of the palm or sole differ from the thin skin of the scalp?

_____

_____

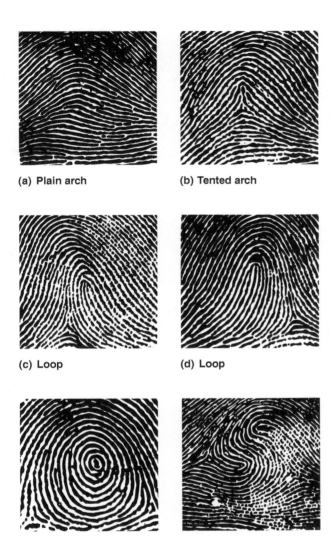

(a) Plain arch   (b) Tented arch

(c) Loop   (d) Loop

(e) Plain whorl   (f) Double loop whorl

Figure 6.8 Main types of fingerprint patterns. (a–b) Arches. **(c–d)** Loops. **(e–f)** Whorls.

Dermography: Fingerprinting

Each of us has a unique, genetically determined set of fingerprints. Because fingerprinting is useful for identifying and apprehending criminals, most people associate this craft solely with criminal investigations. However, fingerprints are also invaluable in quickly identifying amnesia victims, missing persons, and unknown deceased, such as people killed in major disasters.

The friction ridges responsible for fingerprints appear in several patterns, which are clearest when the fingertips are inked and then pressed against white paper. Impressions are also made when perspiration or any foreign material such as blood, dirt, or grease adheres to the ridges and the fingers are then pressed against a smooth, nonabsorbent surface. The three most common patterns are *arches, loops,* and *whorls* (Figure 6.8).

ACTIVITY 6

Taking and Identifying Inked Fingerprints

For this activity, you will be working as a group with your lab partners. Though the equipment for professional fingerprinting is fairly basic, consisting of a glass or metal inking plate, printer's ink (a heavy black paste), ink roller, and standard 8 in. × 8 in. cards, you will be using supplies that are even easier to handle. Each student will prepare two index cards, each bearing the thumbprint and index fingerprint of the <u>right hand</u>.

1. Obtain the following supplies and bring them to your bench: two 4 in. × 6 in. index cards per student, Porelon fingerprint pad or portable inking foils, ink cleaner towelettes, and a magnifying glass.

2. The subject should wash and dry the hands. Open the ink pad or peel back the covering over the ink foil, and position it close to the edge of the laboratory bench. The subject should position himself or herself at arm's length from the bench edge and inking object.

3. A second student, called the *operator,* will stand to the left of the subject and with two hands will hold and direct the movement of the subject's fingertip. During this process, the subject should look away, try to relax, and refrain from trying to help the operator.

4. The thumbprint is to be placed on the left side of the index card, the index fingerprint on the right. The operator should position the subject's right thumb or index finger on the side of the bulb of the finger in such a way that the area to be inked spans the distance from the fingertip to just beyond the first joint, and then roll the finger lightly across the inked surface until its bulb faces in the opposite direction. To prevent smearing, the thumb is rolled away from the body midline (from left to right as the subject sees it; see **Figure 6.9**) and the index finger is rolled toward the body midline (from right to left). The same ink foil can be reused for all the students at the bench; the ink pad is good for thousands of prints. Repeat the procedure (still using the subject's <u>right hand</u>) on the second index card.

5. If the prints are too light, too dark, or smeary, repeat the procedure.

6. While other students in the group are making clear prints of their thumb and index finger, those who have completed that activity should clean their inked fingers with a towelette

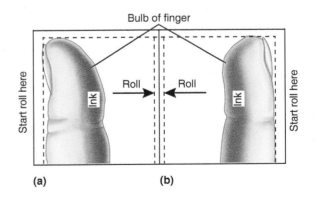

Figure 6.9 Fingerprinting. Method of inking **(a)** the thumb and **(b)** the index finger.

and attempt to classify their own prints as arches, loops, or whorls. Use the magnifying glass as necessary to see ridge details.

7.   When all group members have completed the above steps, they are to write their names on the backs of their index cards. The students combine their cards and shuffle them before transferring them to another group. Finally, students classify the patterns and identify which prints were made by the same individuals.

How difficult was it to classify the prints into one of the three

categories given?_____

Why do you think this was so?_____

Was it easy or difficult to identify the prints made by the same

individual?_____

Why do you think this was so? _____

_____

The Integumentary System

Basic Structure of the Skin

1. Complete the following statements by writing the appropriate word or phrase on the blank line:

    1.   The superficial region of the skin is the _____, composed of _____

         _____ _____ (3 words) tissue.

    2.   The deeper region tissue is the _____, composed of connective tissue.

    3.   The most numerous cell of the epidermis is the _____.

    4.   The two primary layers of the dermis are the _____ dermis, composed of areolar connective

         tissue, and the _____ dermis, composed of dense irregular connective tissue.

2. Name four protective functions of the skin:

    a. _____   c. _____

    b. _____   d. _____

3. Using the key choices, choose all responses that apply to the following descriptions. Some terms are used more than once.

    Key: a. stratum basale   d. stratum lucidum   g. reticular dermis
             b. stratum corneum   e. stratum spinosum
             c. stratum granulosum   f. papillary dermis

    _____ 1. layer of translucent cells containing dead keratinocytes

    _____ 2. two layers of dead cells

    _____ 3. dermal layer responsible for fingerprints

    _____ 4. epidermal layer exhibiting the most rapid cell division

    _____ 5. layer including scalelike dead cells, full of keratin, that constantly slough off

    _____ 6. layer named for the numerous granules present

    _____ 7. location of melanocytes and tactile epithelial cells

    _____ 8. area where weblike pre-keratin filaments first appear

    _____ 9. deep layer of the dermis

    _____ 10. layer that secretes a glycolipid that prevents water loss from the skin

4. Label the integumentary structures and areas indicated by leader lines in the figure below.

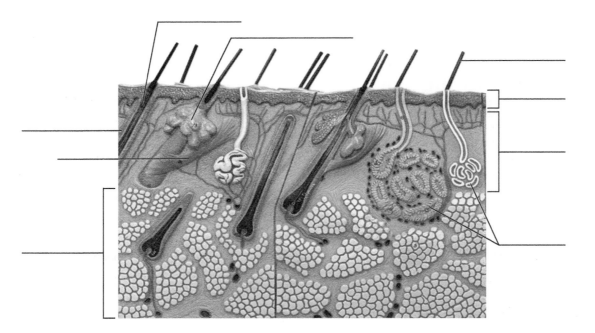

5. Label the layers of the epidermis in thick skin. Then, complete the statements that follow.

a. Glands that respond to rising androgen levels are the _____ glands.

b. _____ _____ are epidermal cells that play a role

in the immune response.

c. Tactile corpuscles are located in the _____ _____.

d. _____ corpuscles are located deep in the dermis.

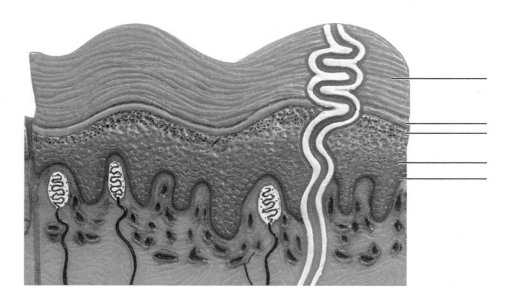

6. List the sensory receptors found in the dermis of the skin. _____

_____

_____

Accessory Organs of the Skin

7. Using the key, match the terms with the appropriate descriptions. Some terms are used more than once.

Key: a. arrector pili d. hair follicle g. sweat gland—apocrine
b. cutaneous receptors e. nail h. sweat gland—eccrine
c. hair f. sebaceous gland

_____ 1. tiny muscles, attached to hair follicles, that pull the hair upright during fright or cold

_____ 2. sweat gland with a role in temperature control

_____ 3. sheath formed of both epithelial and connective tissues

_____ 4. less numerous type of sweat-producing gland; found mainly in the pubic and axillary regions

_____ 5. primarily dead/keratinized cells (two responses from key)

_____ 6. specialized nerve endings that respond to temperature, touch, etc.

_____ 7. its secretion is a lubricant for hair and skin

_____ 8. "sports" a lunule and a cuticle

8. Describe two integumentary system mechanisms that help regulate body temperature. _____

_____

_____

_____

9. Several structures of the hair are listed below. Identify each by matching its letter with the appropriate area on the photomicrograph.

a. cortex
b. cuticle
c. hair matrix
d. hair papilla
e. medulla

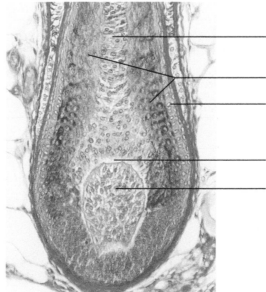

Plotting the Distribution of Sweat Glands

10. With what substance in the bond paper does the iodine painted on the skin react? _____

11. On the basis of class data, which skin area—the forearm or palm of hand—has more sweat glands?

_____

Was this an expected result? _____ Explain. _____

_____

Which other body areas would, if tested, prove to have a high density of sweat glands? _____

_____

12. What organ system controls the activity of the eccrine sweat glands? _____

Dermography: Fingerprinting

13. Why can fingerprints be used to identify individuals? _____

_____

14. Name the three common fingerprint patterns.

_____, _____, and _____

15. Henna tattoos are temporary tattoos that last about 2 weeks. Hypothesize why henna tattoos do not last as long as permanent

tattoos. _____

_____

_____

16. ➕ Vitiligo is a disorder in which the pigmentation of the skin is uneven, resulting in white patches. Recent research suggests that vitiligo might be an autoimmune disorder. Which cells would you expect to be most affected, and why?

_____

_____

17. ➕ Keratinase is an enzyme produced by dermatophytes. Which organs in the body would these pathogenic fungi tend to

proliferate in, and why? _____

_____

Overview of the Skeleton: Classification and Structure of Bones and Cartilages

MATERIALS

- Disarticulated bones (identified by number) that demonstrate classic examples of the four bone classifications (long, short, flat, and irregular)
- Long bone sawed longitudinally (beef bone from a slaughterhouse, if possible, or prepared laboratory specimen)
- Disposable gloves
- Long bone soaked in 10% hydrochloric acid (HCl) or vinegar until flexible
- Long bone baked at 250°F for more than 2 hours
- Compound microscope
- Prepared slide of ground bone (x.s.)
- Three-dimensional model of microscopic structure of compact bone
- Prepared slide of a developing long bone undergoing endochondral ossification
- Articulated skeleton

LEARNING OUTCOMES

- ☐ Name the two primary tissue types that form the skeleton.
- ☐ List the functions of the skeletal system.
- ☐ Name the four main groups of bones based on shape.
- ☐ Identify surface bone markings, and list their functions.
- ☐ Identify the major anatomical areas on a longitudinally cut long bone or on an appropriate image.
- ☐ Explain the role of inorganic salts and organic matrix in providing flexibility and hardness to bone.
- ☐ Locate and identify the major parts of an osteon microscopically or on a histological model or appropriate image of compact bone.
- ☐ Locate and identify the three major types of skeletal cartilages.

PRE-LAB QUIZ

1. All the following are functions of the skeleton *except:*
   a. attachment for muscles
   b. production of melanin
   c. site of red blood cell formation
   d. storage of lipids
2. Circle the correct underlined term. The axial / appendicular skeleton consists of bones that surround the body's center of gravity.
3. Circle the correct underlined term. Compact / Spongy bone looks smooth and homogeneous on the outer surface.
4. _____ bones are generally thin and have a layer of spongy bone between two layers of compact bone.
   a. Flat   b. Irregular   c. Long   d. Short
5. The femur is an example of a(n) _____ bone.
   a. flat   c. long
   b. irregular   d. short
6. Circle the correct underlined term. The shaft of a long bone is known as the epiphysis / diaphysis.
7. The structural unit of compact bone is the:
   a. osteon   b. canaliculus   c. lacuna
8. Circle True or False. Embryonic skeletons consist primarily of elastic cartilage, which is gradually replaced by bone during development and growth.
9. The type of cartilage that has the greatest strength and is found in the knee joint and intervertebral discs is:
   a. elastic   b. fibrocartilage   c. hyaline
10. Circle True or False. Cartilage has a covering made of dense connective tissue called a periosteum.

The **skeleton,** the body's framework, is constructed of two of the most supportive tissues found in the human body—cartilage and bone. In embryos, the skeleton is predominantly made up of hyaline cartilage, but in the adult, most of the cartilage is replaced by more rigid bone. Cartilage remains only in such isolated areas as the external ear, bridge of the nose, larynx, trachea, joints, and parts of the rib cage (see Figure 7.5, page 99).

Besides supporting and protecting the body as an internal framework, the skeleton provides a system of levers with which the skeletal muscles work to move the body. In addition, the bones store lipids and many minerals (the most

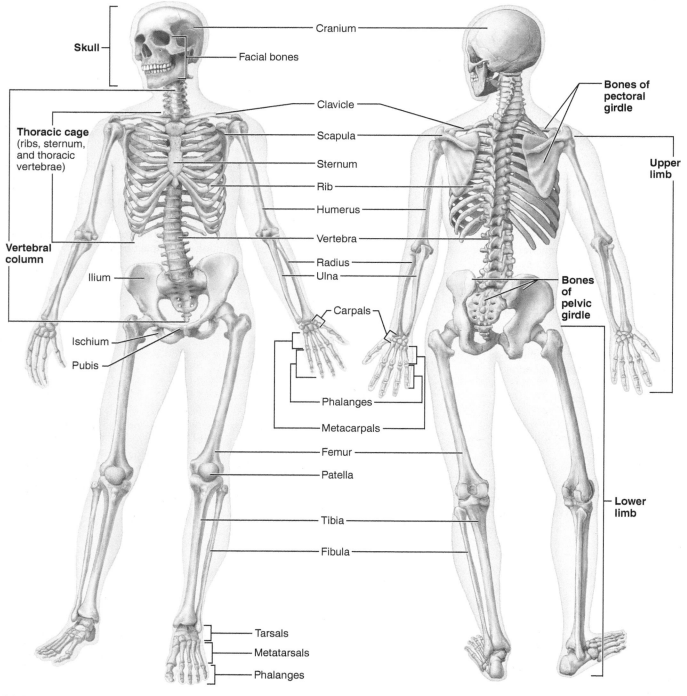

(a) **Anterior view**

(b) **Posterior view**

Figure 7.1 The human skeleton. The bones of the axial skeleton are colored green to distinguish them from the bones of the appendicular skeleton.

important of which is calcium). Finally, the red marrow of bones provides a site for blood cell formation.

The skeleton is made up of bones that are connected at *joints,* or *articulations.* The skeleton is subdivided into two divisions: the **axial skeleton** (those bones that lie around the body's center of gravity) and the **appendicular skeleton** (bones of the limbs, or appendages) **(Figure 7.1)**.

Classification of Bones

The 206 bones of the adult skeleton are composed of two basic kinds of osseous tissue that differ in their texture. **Compact bone** is dense and made up of organizational units called *osteons.* **Spongy** (or *cancellous*) **bone** is composed of small *trabeculae* (columns) of bone and lots of open space.

Bones may be classified further on the basis of their relative gross anatomy into four groups: long, short, flat, and irregular bones.

Long bones, such as the femur and bones of the fingers (phalanges) (Figure 7.1), are much longer than they are wide, generally consisting of a shaft with heads at either end. Long bones are composed mostly of compact bone. **Short bones** are typically cube-shaped, and they contain more spongy bone than compact bone. The tarsals and carpals (Figure 7.1) are examples.

Flat bones are generally thin, with two waferlike layers of compact bone sandwiching a thicker layer of spongy bone between them. Although the name "flat bone" implies a structure that is level or horizontal, many flat bones are curved (for example, the bones of the cranium). Bones that do not fall into one of the preceding categories are classified as **irregular bones.** The vertebrae are irregular bones (see Figure 7.1).

Some anatomists also recognize two other subcategories of bones. **Sesamoid bones** are special types of short bones formed within tendons. The patellas (kneecaps) are sesamoid bones. **Sutural bones** are tiny bones between cranial bones. Except for the patellas, the sesamoid and sutural bones are not included in the bone count of 206 because they vary in number and location in different individuals.

Bone Markings

Even a casual observation of the bones will reveal that bone surfaces are not featureless smooth areas but are scarred with an array of bumps, holes, and ridges. These **bone markings** reveal where bones form joints with other bones, where muscles, tendons, and ligaments were attached, and where blood vessels and nerves passed. Bone markings fall into two main categories: projections, or processes, that grow out from the bone and serve as sites of muscle attachment or help form joints; and depressions or cavities, indentations, or openings in the bone that often serve as conduits for nerves and blood vessels. Refer to the summary of bone markings **(Table 7.1)**.

ACTIVITY 1

Examining and Classifying Bones

Examine the isolated (disarticulated) bones on display to find specific examples of the bone markings described in Table 7.1. Then classify each of the bones into one of the four anatomical groups by recording its number in the accompanying chart.

Long	Short	Flat	Irregular

Gross Anatomy of the Typical Long Bone

ACTIVITY 2

Examining a Long Bone

1. Obtain a long bone that has been sawed along its longitudinal axis. If a cleaned dry bone is provided, no special preparations need be made.

> Note: If the bone supplied is a fresh beef bone, don disposable gloves before beginning your observations.

Identify the **diaphysis,** or shaft **(Figure 7.2)**. Observe its smooth surface, which is composed of compact bone. If you are using a fresh specimen, carefully pull away the **periosteum,** a fibrous membrane covering made up of dense irregular connective tissue, to view the bone surface. Notice that many fibers of the periosteum penetrate into the bone. These fibers are called **perforating collagen fiber bundles** (*Sharpey's fibers*). Blood vessels and nerves travel through the periosteum and invade the bone. *Osteoprogenitor cells* are found on the inner, or osteogenic, layer of the periosteum.

WHY THIS
MATTERS | Shin Splints

Runners, gymnasts, basketball players, and other athletes often complain of a dull ache in the shin. "Shin splints" is a term for any pain in the leg. Most cases of shin pain are classified as medial tibial stress syndrome (MTSS), a more scientific term than "shin splints" but still not very explanatory. Assuming there is no stress fracture in the tibia, the "stress" can be on muscles, tendons, the periosteum, perforating fibers, or any combination of these structures. Most commonly, though, the pain is caused by inflammation of the periosteum (periostitis) and perforating collagen fiber bundles. A variety of mechanical factors, including flat feet and turning the feet inward with impact exercise, can contribute to MTSS. ■

2. Now inspect the **epiphysis,** the end of the long bone. Notice that it is composed of a thin layer of compact bone that encloses spongy bone.

3. Identify the **articular cartilage,** which covers the epiphyseal surface in place of the periosteum. The glassy hyaline cartilage provides a smooth surface to minimize friction at joints.

4. If the animal was still young and growing, you will be able to see the **epiphyseal plate,** a thin area of hyaline cartilage that provides for longitudinal growth of the bone during youth. Once the long bone has stopped growing, these areas are replaced with bone and appear as thin, barely discernible remnants—the **epiphyseal lines.**

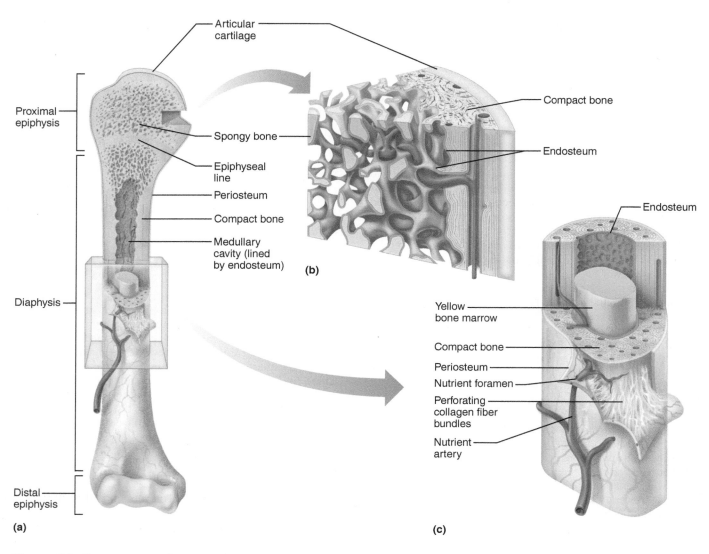

Figure 7.2 The structure of a long bone (humerus of the arm). (a) Anterior view with longitudinal section cut away at the proximal end. **(b)** Pie-shaped, three-dimensional view of spongy bone and compact bone of the epiphysis. **(c)** Cross section of diaphysis (shaft). Note that the external surface of the diaphysis is covered by a periosteum but that the articular surface of the epiphysis is covered with hyaline cartilage.

5.   In an adult animal, the central cavity of the shaft (*medullary cavity*) is essentially a storage region for adipose tissue, or **yellow bone marrow.** In the infant, this area is involved in forming blood cells, and so **red bone marrow** is found in the marrow cavities. In adult bones, the red bone marrow is confined to the interior of the epiphyses, where it occupies the spaces between the trabeculae of spongy bone.

6.   If you are examining a fresh bone, look carefully to see if you can distinguish the delicate **endosteum** lining the shaft. The endosteum also covers the trabeculae of spongy bone and lines the central and perforating canals of compact bone. Like the periosteum, the endosteum contains osteoprogenitor cells that differentiate into osteoblasts. As the bone grows in diameter on its external surface, it is constantly being broken down

on its inner surface. Thus, the thickness of the compact bone layer composing the shaft remains relatively constant.

7.   If you have been working with a fresh bone specimen, return it to the appropriate area, and properly dispose of your gloves. Wash your hands before continuing on to the microscope study. ■

Longitudinal bone growth at epiphyseal plates (growth plates) follows a predictable sequence and provides a reliable indicator of the age of children exhibiting normal growth. If problems of long-bone growth are suspected (for example, pituitary dwarfism), X-ray films are taken to view the width of the growth plates. An abnormally thin epiphyseal plate indicates growth retardation. ✚

Table 7.1   Bone Markings

Name of bone marking	Description	Illustration
Projections That Are Sites of Muscle and Ligament Attachment		
Tuberosity (too″be-ros′ĭ-te)	Large rounded projection; may be roughened	
Crest	Narrow ridge of bone; usually prominent	
Trochanter (tro-kan′ter)	Very large, blunt, irregularly shaped process (the only examples are on the femur)	
Line	Narrow ridge of bone; less prominent than a crest	
Tubercle (too′ber-kl)	Small rounded projection or process	
Epicondyle (ep″ĭ-kon′dīl)	Raised area on or above a condyle	
Spine	Sharp, slender, often pointed projection	
Process	Any bony prominence	
Surfaces That Form Joints		
Head	Bony expansion carried on a narrow neck	
Facet	Smooth, nearly flat articular surface	
Condyle (kon′dīl)	Rounded articular projection, often articulates with a corresponding fossa	
Ramus (ra′mus)	Armlike bar of bone	
Depressions and Openings		
For passage of vessels and nerves		
Foramen (fo-ra′men)	Round or oval opening through a bone	
Groove	Furrow	
Fissure	Narrow, slitlike opening	
Notch	Indentation at the edge of a structure	
Others		
Fossa (fos′ah)	Shallow basinlike depression in a bone, often serving as an articular surface	
Meatus (me-a′tus)	Canal-like passageway	
Sinus	Bone cavity, filled with air and lined with mucous membrane	

Illustration labels: Iliac crest, Trochanters, Intertrochanteric line, Ischial spine, Hip bone, Ischial tuberosity, Vertebra, Adductor tubercle, Femur of thigh, Medial epicondyle, Condyle, Spinous process. Head, Facets, Rib, Condyle, Ramus, Mandible. Meatus, Sinus, Fossa, Notch, Groove, Inferior orbital fissure, Foramen, Skull.

Chemical Composition of Bone

Bone is one of the hardest materials in the body. Although relatively light, bone has a remarkable ability to resist tension and shear forces that continually act on it. An engineer would tell you that a cylinder (like a long bone) is one of the strongest structures for its mass.

The hardness of bone is due to the inorganic calcium salts deposited in its ground substance. Its flexibility comes from the organic elements of the matrix, particularly the collagen fibers.

Examining the Effects of Heat and Hydrochloric Acid on Bones

Obtain a bone sample that has been soaked in hydrochloric acid (HCl) (or in vinegar) and one that has been baked. Heating removes the organic part of bone, whereas acid dissolves out the minerals. Do the treated bones retain the shape of untreated specimens?

_____

Gently apply pressure to each bone sample. What happens to the heated bone?

_____

_____

What happens to the bone treated with acid?

_____

What does the acid appear to remove from the bone?

_____

What does baking appear to do to the bone?

_____

_____

Microscopic Structure of Compact Bone

As you have seen, spongy bone has a spiky, open-work appearance, resulting from the arrangement of the **trabeculae** that compose it, whereas compact bone appears to be dense and homogeneous on the outer surface. However, microscopic examination of compact bone reveals that it is riddled with passageways carrying blood vessels, nerves, and lymphatic vessels that provide the living bone cells with needed substances and a way to eliminate wastes **(Figure 7.3)**. Indeed, bone histology is much easier to understand when you recognize that bone tissue is organized around its blood supply.

Examining the Microscopic Structure of Compact Bone

1. Obtain a prepared slide of ground bone and examine it under low power. Using the photomicrograph (Figure 7.3c) as a guide, focus on a central canal. The **central (Haversian)** canal runs parallel to the long axis of the bone and carries blood vessels, nerves, and lymphatic vessels through the bony matrix. Identify the **osteocytes** (mature bone cells) in **lacunae** (chambers), which are arranged in concentric circles called **concentric lamellae** around the central canal. Because bone remodeling is going on all the time, you will also see some _interstitial lamellae,_ remnants of osteons that have been broken down (Figure 7.3c).

A central canal and all the concentric lamellae surrounding it are referred to as an **osteon,** or **Haversian system.** Also identify **canaliculi,** tiny canals radiating outward from a central canal to the lacunae of the first lamella and then from lamella to lamella. The canaliculi form a dense transportation network through the hard bone matrix, connecting all the living cells of the osteon to the nutrient supply. You may need a higher-power magnification to see the fine canaliculi.

2. Also note the **perforating canals** _(Volkmann's canals)_ (Figure 7.3). These canals run at right angles to the shaft and connect the blood and nerve supply of the medullary cavity to the central canals.

3. If a model of bone histology is available, identify the same structures on the model. ■

Ossification: Bone Formation and Growth in Length

Except for the collarbones (clavicles), all bones of the body inferior to the skull form in the embryo by the process of **endochondral ossification,** which uses hyaline cartilage "bones" as a model for bone formation. The major events of this process, which begins in the (primary ossification) center of the shaft of a developing long bone, are as follows:

•   Blood vessels invade the perichondrium covering the hyaline cartilage model and convert it to a periosteum.

•   Osteoblasts at the inner surface of the periosteum secrete bone matrix around the hyaline cartilage model, forming a bone collar.

•   Cartilage in the shaft center calcifies and then hollows out, forming an internal cavity.

•   A _periosteal bud_ (blood vessels, nerves, red marrow elements, osteoblasts, and osteoclasts) invades the cavity and forms spongy bone, which is removed by osteoclasts, thus, producing the medullary cavity. This process proceeds in both directions from the _primary ossification center._ As bones grow longer, the medullary cavity gets larger and longer. Chondroblasts lay down new cartilage matrix on the epiphyseal face of the epiphyseal plate, and it is eroded away and replaced by bony spicules on the diaphyseal face **(Figure 7.4)**. This process continues until late adolescence, when the entire epiphyseal plate is replaced by bone.

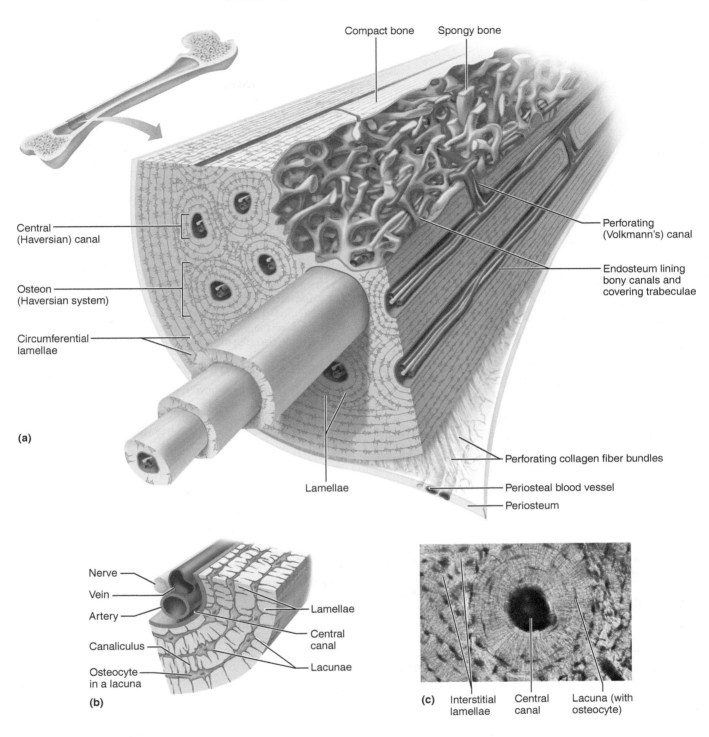

(a)

Compact bone

Spongy bone

Central (Haversian) canal

Osteon (Haversian system)

Circumferential lamellae

Lamellae

Perforating (Volkmann's) canal

Endosteum lining bony canals and covering trabeculae

Perforating collagen fiber bundles

Periosteal blood vessel

Periosteum

(b)

Nerve

Vein

Artery

Canaliculus

Osteocyte in a lacuna

Lamellae

Central canal

Lacunae

(c)   Interstitial lamellae   Central canal   Lacuna (with osteocyte)

Figure 7.3 Microscopic structure of compact bone. (a) Diagram of a pie-shaped segment of compact bone, illustrating its structural units (osteons). **(b)** Higher magnification view of a portion of one osteon. Note the position of osteocytes in lacunae. **(c)** Photomicrograph of a cross-sectional view of an osteon (320×).

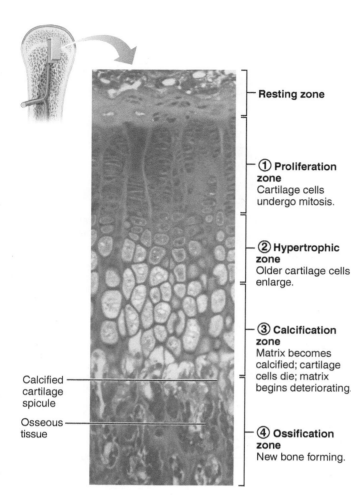

- Resting zone
- ① **Proliferation zone**
  Cartilage cells undergo mitosis.
- ② **Hypertrophic zone**
  Older cartilage cells enlarge.
- ③ **Calcification zone**
  Matrix becomes calcified; cartilage cells die; matrix begins deteriorating.
- ④ **Ossification zone**
  New bone forming.

Calcified cartilage spicule

Osseous tissue

Figure 7.4 Growth in length of a long bone occurs at the epiphyseal plate. The side of the epiphyseal plate facing the epiphysis contains resting cartilage cells. Cells proximal to the resting zone are arranged in four zones: proliferation, hypertrophic, calcification, and ossification from the earliest stage of growth① to the region where bone replaces cartilage④ (300×).

ACTIVITY 5

Examining the Osteogenic Epiphyseal Plate

Obtain a slide depicting endochondral ossification (cartilage bone formation) and bring it to your bench to examine under the microscope. Identify the proliferation, hypertrophic, calcification, and ossification zones of the epiphyseal plate (refer to Figure 7.4). Then, identify the area of resting cartilage cells distal to the growth zone, some hypertrophied chondrocytes, bony spicules, the periosteal bone collar, and the medullary cavity. ■

Cartilages of the Skeleton

Location and Basic Structure

As mentioned earlier, cartilaginous regions of the skeleton have a fairly limited distribution in adults **(Figure 7.5)**. The most important of these skeletal cartilages are (1) **articular cartilages,** which cover the bone ends at movable joints; (2) **costal cartilages,** found connecting the ribs to the sternum (breastbone); (3) **laryngeal cartilages,** which largely construct the larynx (voice box); (4) **tracheal** and **bronchial cartilages,** which reinforce other passageways of the respiratory system; (5) **nasal cartilages,** which support the external nose; (6) **intervertebral discs,** which separate and cushion bones of the spine (vertebrae); and (7) the cartilage supporting the external ear.

Cartilage tissues are distinguished by the fact that they contain no nerves and very few blood vessels. Like bones, each cartilage is surrounded by a covering of dense irregular connective tissue, called a *perichondrium* (rather than a periosteum). The perichondrium acts like a girdle to resist distortion of the cartilage when the cartilage is subjected to pressure. It also plays a role in cartilage growth and repair.

Classification of Cartilage

The skeletal cartilages have representatives from each of the three cartilage tissue types—hyaline, elastic, and fibrocartilage. Because cartilage tissues are covered in another lab (Exercise 5), we will only briefly discuss that information here.

Hyaline Cartilage

Hyaline cartilage looks like frosted glass when viewed by the unaided eye. Most skeletal cartilages are composed of hyaline cartilage (Figure 7.5). Hyaline cartilage provides sturdy support with some resilience, or "give." (Review Figure 5.5h, page 65.)

Elastic Cartilage

Elastic cartilage can be envisioned as "hyaline cartilage with more elastic fibers." Consequently, it is much more flexible than hyaline cartilage, and it tolerates repeated bending better. Essentially, only the cartilages of the external ear and the epiglottis (which flops over and covers the larynx when we swallow) are made of elastic cartilage. (Review Figure 5.5i, page 65.)

Fibrocartilage

Fibrocartilage consists of rows of chondrocytes alternating with rows of thick collagen fibers. This tissue looks like a cartilage-dense regular connective tissue hybrid. Fibrocartilage has great tensile strength and can withstand heavy compression. Hence, its use to construct the intervertebral discs and the cartilages within the knee joint makes a lot of sense. (See Figure 7.5 and Figure 5.5j, page 66.)

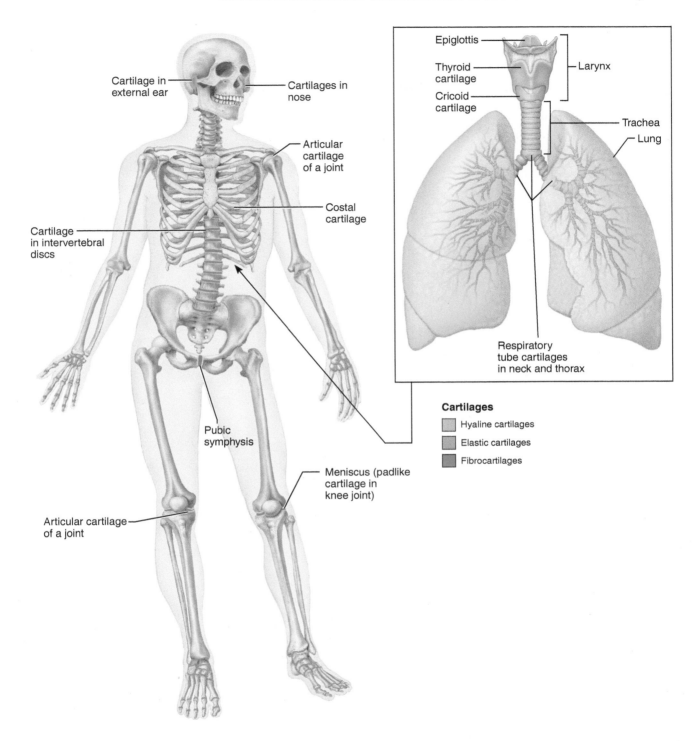

Figure 7.5 Cartilages in the adult skeleton and body. The cartilages that support the respiratory tubes and larynx are shown separately at the upper right.

Name _____

Lab Time/Date _____

Overview of the Skeleton: Classification and Structure of Bones and Cartilages

Classification of Bones

1. Classify each of the bones in the chart below as either long, short, flat, or irregular by placing a check mark in the appropriate column. Also use a check mark to indicate whether the bone is a part of the axial or the appendicular skeleton. Use Figure 7.1 as a guide

	Long	Short	Flat	Irregular	Axial skeleton	Appendicular skeleton
Sternum						
Radius						
Calcaneus (tarsal bone)						
Parietal bone (cranial bone)						
Phalanx (single bone of a digit)						
Vertebra						

Bone Markings

2. Match the terms in column B with the appropriate description in column A.

Column A

_____ 1. sharp, slender process*

_____ 2. small rounded projection*

_____ 3. narrow ridge of bone*

_____ 4. large rounded projection*

_____ 5. structure supported on neck†

_____ 6. armlike projection†

_____ 7. rounded, convex projection†

_____ 8. narrow slitlike opening‡

_____ 9. canal-like structure

_____ 10. round or oval opening through a bone‡

_____ 11. shallow depression

Column B

a. condyle

b. crest

c. epicondyle

d. facet

e. fissure

f. foramen

g. fossa

h. head

i. meatus

j. process

k. ramus

_____ 12. air-filled cavity

_____ 13. large, irregularly shaped projection*

_____ 14. raised area on or above a condyle*

_____ 15. bony projection

_____ 16. smooth, nearly flat articular surface†

l. sinus

m. spine

n. trochanter

o. tubercle

p. tuberosity

*a site of muscle and ligament attachment
†takes part in joint formation
‡a passageway for nerves or blood vessels

Gross Anatomy of the Typical Long Bone

3. Use the terms in the key below to identify the structures marked by leader lines and braces in the diagrams. Some terms are used more than once.

Key:
a. articular cartilage
b. compact bone
c. diaphysis
d. endosteum

e. epiphyseal line
f. epiphysis
g. medullary cavity
h. nutrient artery

i. periosteum
j. spongy bone
k. yellow bone marrow

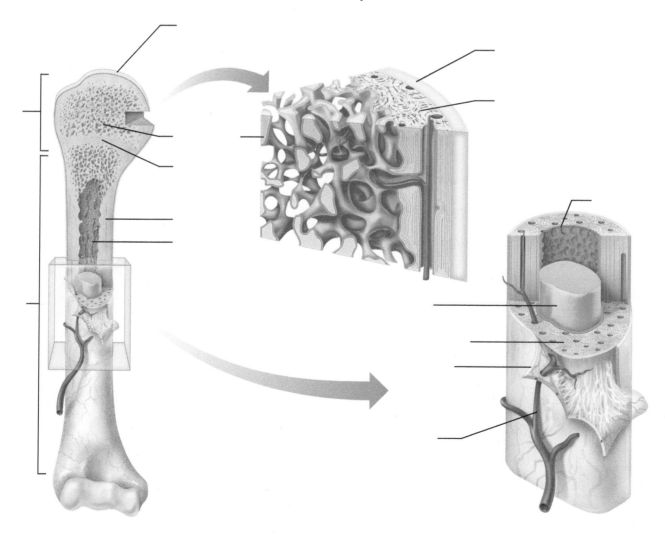

4. Match the key terms with the descriptions.

Key: a. articular cartilage   d. epiphyseal line   g. periosteum
        b. diaphysis   e. epiphysis   h. red bone marrow
        c. endosteum   f. medullary cavity

_____ 1. end portion of a long bone

_____ 2. helps reduce friction at joints

_____ 3. site of blood cell formation

_____, _____ 4. two major submembranous sites of osteoprogenitor cells

_____ 5. scientific term for bone shaft

_____ 6. contains yellow bone marrow in adult bones

_____ 7. growth plate remnant

WHY THIS MATTERS | **5.** Which type of connective tissue is inflamed in cases of periostitis? _____

      6. In cases of periostitis, the cells of the periosteum attempt to repair the microtears to the periosteum and perforating collagen fiber bundles. Which cells are most likely to be involved in this repair process?

_____

Chemical Composition of Bone

7. What is the function of the organic matrix in bone? _____

8. Name the important organic bone components. _____

9. Calcium salts form the bulk of the inorganic material in bone. What is the function of the calcium salts?

_____

10. Baking removes _____ from bone. Soaking bone in acid removes _____.

Microscopic Structure of Compact Bone

11. Several descriptions of bone structure are given below. Identify the structure involved by choosing the appropriate term from the key and placing its letter in the blank. Then, on the photomicrograph of bone on the right, identify all structures that are named in the key, and draw a bracket enclosing a single osteon.

Key: a. canaliculi   c. lacuna
        b. central canal   d. lamella

_____ 1. layer of bony matrix around a central canal

_____ 2. site of osteocytes

_____ 3. longitudinal canal carrying blood vessels, lymphatics, and nerves

_____ 4. tiny canals connecting osteocytes of an osteon

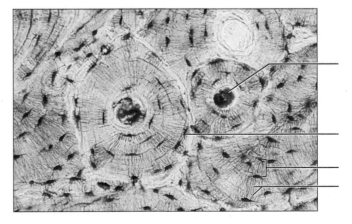

Ossification: Bone Formation and Growth in Length

12. Compare and contrast events occurring on the epiphyseal and diaphyseal faces of the epiphyseal plate.

epiphyseal face: _____

diaphyseal face: _____

_____

Cartilages of the Skeleton

13. Using the key choices, identify each type of cartilage described (in terms of its body location or function) below. Terms may be used more than once.

Key: a. elastic   b. fibrocartilage   c. hyaline

_____ 1. supports the external ear

_____ 2. between the vertebrae

_____ 3. forms the walls of the
                        voice box (larynx)

_____ 4. the epiglottis

_____ 5. articular cartilages

_____ 6. meniscus in a knee joint

_____ 7. connects the ribs to the
                        sternum

_____ 8. most effective at resisting
                        compression

_____ 9. most springy and flexible

_____ 10. most abundant

14. ✚ In a child with rickets, the bones are not properly calcified. Which treated bone in Activity 3 most closely resembles

the bones of a child with rickets? Why?_____

_____

_____

15. ✚ Achondroplasia is a type of dwarfism in which the long bones stop growing during childhood, resulting in limbs that are disproportionately shorter than the torso. This genetic disorder is characterized by deficiencies in the epiphyseal plate that include a low number of chondrocytes and inability of chondrocytes to enlarge. Which zones do you think would be

most affected by this disorder, and why?_____

_____

_____

The Axial Skeleton

MATERIALS

- Intact skull and Beauchene skull
- X-ray images of individuals with scoliosis, lordosis, and kyphosis (if available)
- Articulated skeleton, articulated vertebral column, removable intervertebral discs
- Isolated cervical, thoracic, and lumbar vertebrae, sacrum, and coccyx
- Isolated fetal skull

LEARNING OUTCOMES

- ☐ Name the three parts of the axial skeleton.
- ☐ Identify the bones of the axial skeleton, either by examining disarticulated bones or by pointing them out on an articulated skeleton or skull, and name the important bone markings on each bone.
- ☐ Name and describe the different types of vertebrae.
- ☐ Discuss the importance of intervertebral discs and spinal curvatures.
- ☐ Identify three abnormal spinal curvatures.
- ☐ List the components of the thoracic cage.
- ☐ Identify the bones of the fetal skull by examining an articulated skull or image.
- ☐ Define *fontanelle,* and discuss the function and fate of fontanelles in the fetus.
- ☐ Discuss important differences between the fetal and adult skulls.

PRE-LAB QUIZ

1. The axial skeleton can be divided into the skull, the vertebral column, and the:
   a. thoracic cage
   b. femur
   c. hip bones
   d. humerus
2. Eight bones make up the _____, which encloses and protects the brain.
   a. cranium
   b. face
   c. skull
3. How many bones of the skull are considered facial bones? _____
4. Circle the correct underlined term. The lower jawbone, or <u>maxilla</u> / <u>mandible</u>, articulates with the temporal bones in the only freely movable joints in the skull.
5. Circle the correct underlined term. The <u>body</u> / <u>spinous process</u> of a typical vertebra forms the rounded, central portion that faces anteriorly in the human vertebral column.
6. The seven bones of the neck are called _____ vertebrae.
7. The _____ vertebrae articulate with the corresponding ribs.
8. The _____ is a flat bone formed by the fusion of three bones: the manubrium, the body, and the xiphoid process.
   a. coccyx
   b. sacrum
   c. sternum
9. Circle True or False. The first seven pairs of ribs are called floating ribs because they have only indirect cartilage attachments to the sternum.
10. A fontanelle:
    a. is found only in the fetal skull
    b. is a fibrous membrane
    c. allows for compression of the skull during birth
    d. all of the above

The **axial skeleton** (the green portion of Figure 7.1 on page 92) can be divided into three parts: the skull, the vertebral column, and the thoracic cage.

The Skull

The **skull** is composed of two sets of bones. The bones of the **cranium** (8 bones) enclose and protect the fragile brain tissue. The **facial bones** (14 bones) support the eyes and position them anteriorly. They also provide attachment sites for facial muscles, which make it possible for us to present our feelings to the world. All but one of the bones of the skull are joined by interlocking fibrous joints called *sutures*. The mandible, or lower jawbone, is attached to the rest of the skull by a freely movable joint.

ACTIVITY 1

Identifying the Bones of the Skull

The bones of the skull (**Figure 8.1** through **Figure 8.10**) are described in **Tables 8.1** and **8.2** on page 111. As you read through this material, identify each bone on an intact and/or Beauchene skull (see Figure 8.10).

Note: Important bone markings are listed in the tables for the bones on which they appear, and each bone name is colored to correspond to the bone color in the figures.

Text continues on page 110.

Table 8.1A   The Axial Skeleton: Cranial Bones and Important Bone Markings

Cranial bone	Important markings	Description
Frontal (1) Figures 8.1, 8.3, 8.7, 8.9, and 8.10	N/A	Forms the forehead, superior part of the orbit, and the floor of the anterior cranial fossa.
	Supraorbital margin	Thick margin of the eye socket that lies beneath the eyebrows.
	Supraorbital foramen (notch)	Opening above each orbit allowing blood vessels and nerves to pass.
	Glabella	Smooth area between the eyes.
Parietal (2) Figures 8.1, 8.3, 8.6, 8.7, and 8.10	N/A	Form the superior and lateral aspects of the skull.
Temporal (2) Figures 8.1, 8.2, 8.3, 8.6, 8.7, and 8.10	N/A	Form the inferolateral aspects of the skull and contribute to the middle cranial fossa; each has squamous, tympanic, and petrous parts.
	Squamous part	Located inferior to the squamous suture. The next two markings are located in this part.
	Zygomatic process	A bridgelike projection that articulates with the zygomatic bone to form the zygomatic arch.
	Mandibular fossa	Located on the inferior surface of the zygomatic process; receives the condylar process of the mandible to form the temporomandibular joint.
	Tympanic part	Surrounds the external ear opening. The next two markings are located in this part.
	External acoustic meatus	Canal leading to the middle ear and eardrum.
	Styloid process	Needlelike projection that serves as an attachment point for ligaments and muscles of the neck. (This process is often missing from demonstration skulls because it has broken off.)
	Petrous part	Forms a bony wedge between the sphenoid and occipital bones and contributes to the cranial base. The remaining temporal markings are located in this part.
	Jugular foramen	Located where the petrous part of the temporal bone joins the occipital bone. Forms an opening which the internal jugular vein and cranial nerves IX, X, and XI pass.
	Carotid canal	Opening through which the internal carotid artery passes into the cranial cavity.
	Foramen lacerum	Almost completely closed by cartilage in the living person but forms a jagged opening in dried skulls.
	Stylomastoid foramen	Tiny opening between the mastoid and styloid processes through which cranial nerve VII leaves the cranium.
	Mastoid process	Located posterior to the external acoustic meatus; serves as an attachment point for neck muscles.
Occipital (1) Figures 8.1, 8.2, 8.3, and 8.6	N/A	Forms the posterior aspect and most of the base of the skull.
	Foramen magnum	Large opening in the base of the bone, which allows the spinal cord to join with the brain stem.
	Occipital condyles	Rounded projections lateral to the foramen magnum that articulate with the first cervical vertebra (atlas).
	Hypoglossal canal	Opening medial and superior to the occipital condyle through which cranial nerve XII (the hypoglossal nerve) passes.
	External occipital protuberance	Midline prominence posterior to the foramen magnum.

The number in parentheses ( ) following the bone name indicates the total number of such bones in the body.

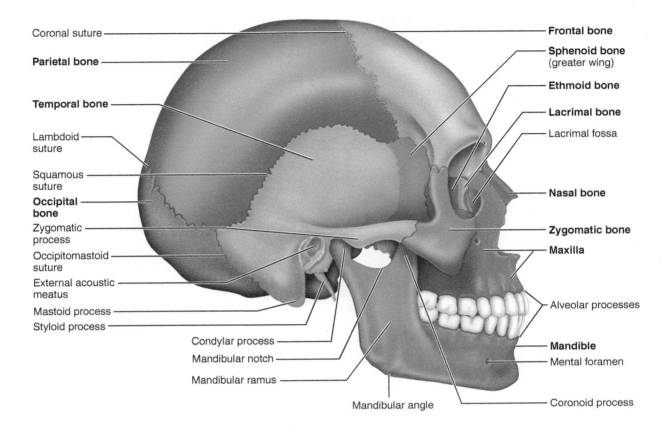

Coronal suture

Parietal bone

Temporal bone

Lambdoid suture

Squamous suture

Occipital bone

Zygomatic process

Occipitomastoid suture

External acoustic meatus

Mastoid process

Styloid process

Condylar process

Mandibular notch

Mandibular ramus

Mandibular angle

Frontal bone

Sphenoid bone (greater wing)

Ethmoid bone

Lacrimal bone

Lacrimal fossa

Nasal bone

Zygomatic bone

Maxilla

Alveolar processes

Mandible

Mental foramen

Coronoid process

Figure 8.1 **External anatomy of the right lateral aspect of the skull.**

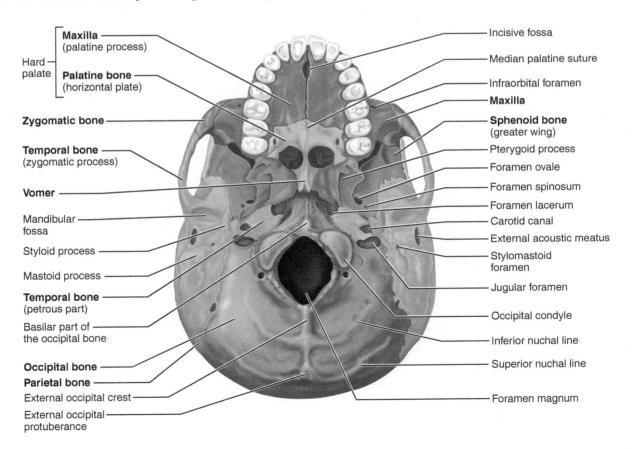

Maxilla (palatine process)

Hard palate

Palatine bone (horizontal plate)

Zygomatic bone

Temporal bone (zygomatic process)

Vomer

Mandibular fossa

Styloid process

Mastoid process

Temporal bone (petrous part)

Basilar part of the occipital bone

Occipital bone

Parietal bone

External occipital crest

External occipital protuberance

Incisive fossa

Median palatine suture

Infraorbital foramen

Maxilla

Sphenoid bone (greater wing)

Pterygoid process

Foramen ovale

Foramen spinosum

Foramen lacerum

Carotid canal

External acoustic meatus

Stylomastoid foramen

Jugular foramen

Occipital condyle

Inferior nuchal line

Superior nuchal line

Foramen magnum

Figure 8.2 **Inferior view of the skull, mandible removed.**

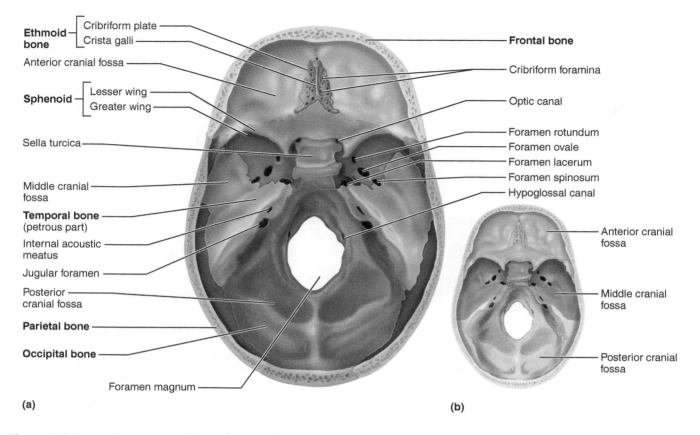

Figure 8.3 Internal anatomy of the inferior portion of the skull. (a) Superior view of the cranial cavity, calvaria removed. **(b)** Diagram of the cranial base showing the extent of its major fossae.

Table 8.1B The Axial Skeleton: Cranial Bones and Important Bone Markings		
Cranial bone	**Important markings**	**Description**
Sphenoid bone (1) Figures 8.1, 8.2, 8.3, 8.4, 8.7, and 8.10	N/A	Bat-shaped bone that is described as the keystone bone of the cranium because it articulates with all other cranial bones.
	Greater wings	Project laterally from the sphenoid body, forming parts of the middle cranial fossa and the orbits.
	Pterygoid processes	Project inferiorly from the greater wings; attachment site for chewing muscles (pterygoid muscles).
	Superior orbital fissures	Slits in the orbits providing passage of cranial nerves that control eye movements (III, IV, VI, and the ophthalmic division of V).
	Sella turcica	"Turkish saddle" located on the superior surface of the body; the seat of the saddle, called the *hypophyseal fossa,* holds the pituitary gland.
	Lesser wings	Form part of the floor of the anterior cranial fossa and part of the orbit.
	Optic canals	Openings in the base of the lesser wings; cranial nerve II (optic nerve) passes through to serve the eye.
	Foramen rotundum	Openings located in the medial part of the greater wing; a branch of cranial nerve V (maxillary division) passes through.
	Foramen ovale	Openings located posterolateral to the foramen rotundum; a branch of cranial nerve V (mandibular division) passes through.
	Foramen spinosum	Openings located posterolateral to the foramen ovale; provides passageway for the middle meningeal artery.

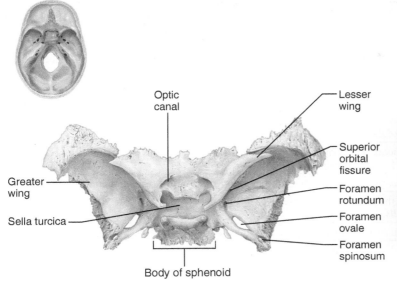

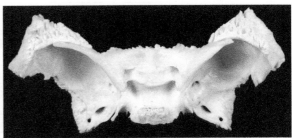

Optic
canal

Lesser
wing

Superior
orbital
fissure

Greater
wing

Foramen
rotundum

Sella turcica

Foramen
ovale

Foramen
spinosum

Body of sphenoid

(a) Superior view

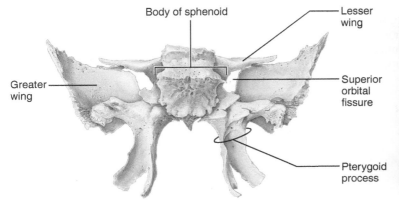

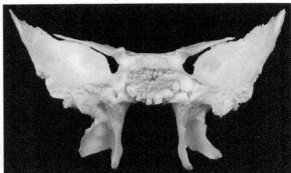

Body of sphenoid

Lesser
wing

Greater
wing

Superior
orbital
fissure

Pterygoid
process

(b) Posterior view

Figure 8.4 The sphenoid bone.

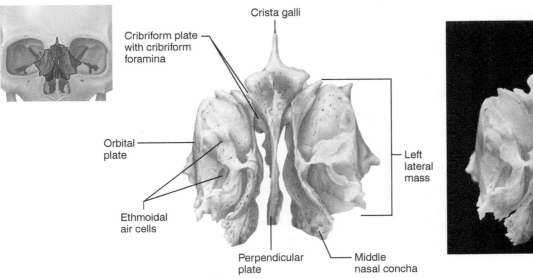

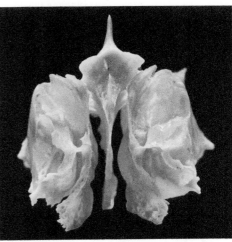

Crista galli

Cribriform plate
with cribriform
foramina

Orbital
plate

Left
lateral
mass

Ethmoidal
air cells

Perpendicular
plate

Middle
nasal concha

Figure 8.5 The ethmoid bone. Anterior view.

Table 8.1C   The Axial Skeleton: Cranial Bones and Important Bone Markings

Cranial bone	Important markings	Description
Ethmoid (1) Figures 8.1, 8.3, 8.5, 8.7, and 8.10	N/A	Contributes to the anterior cranial fossa; forms part of the nasal septum and the nasal cavity; contributes to the medial wall of the orbit.
	Crista galli	"Rooster's comb"; a superior projection that attaches to the dura mater, helping to secure the brain within the skull.
	Cribriform plates	Located lateral to the crista galli; form a portion of the roof of the nasal cavity and the floor of the anterior cranial fossa.
	Cribriform foramina	Tiny holes in the cribriform plates that allow for the passage of filaments of cranial nerve I (olfactory nerve).
	Perpendicular plate	Inferior projection that forms the superior portion of the nasal septum.
	Lateral masses	Flank the perpendicular plate on each side and are filled with sinuses called *ethmoidal air cells.*
	Orbital plates	Lateral surface of the lateral masses that contribute to the medial wall of the orbits.
	Superior and middle nasal conchae	Extend medially from the lateral masses; act as turbinates to improve airflow through the nasal cavity.

The Cranium

The cranium may be divided into two major areas for study—the **calvaria,** forming the superior, lateral, and posterior walls of the skull, and the **cranial base,** forming the skull bottom. Internally, the cranial base has three distinct depressions, the **anterior, middle,** and **posterior cranial fossae** (Figure 8.3). The brain sits in these fossae, completely enclosed by the calvaria.

Eight large bones construct the cranium. *With the exception of two paired bones (the parietals and the temporals), all are single bones.*

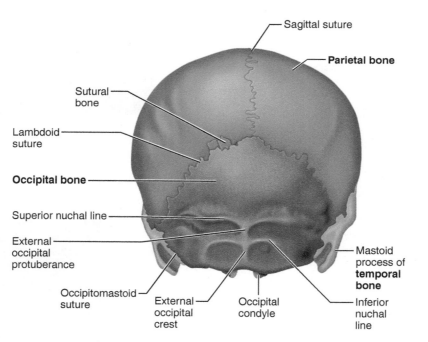

Figure 8.6 Posterior view of the skull.

Major Sutures

The four largest sutures are located where the parietal bones articulate with each other and where the parietal bones articulate with other cranial bones:

• **Sagittal suture:** Occurs where the left and right parietal bones meet superiorly in the midline of the cranium (Figure 8.6).

• **Coronal suture:** Running in the frontal plane where the parietal bones meet the frontal bone anteriorly (Figure 8.1).

• **Squamous suture:** Occurs where a parietal bone and temporal bone meet on the lateral aspect of the skull (Figure 8.1).

• **Lambdoid suture:** Occurs where the parietal bones meet the occipital bone posteriorly (Figure 8.6).

Facial Bones

Of the 14 bones composing the face, 12 are paired. *Only the mandible and vomer are single bones.* An additional bone, the hyoid bone, although not a facial bone, is considered here because of its location.

Facial bone	Important markings	Description
Nasal (2)	N/A	Small rectangular bones forming the bridge of the nose.
Lacrimal (2)	N/A	Each forms part of the medial orbit in between the maxilla and ethmoid bone.
	Lacrimal fossa	Houses the lacrimal sac, which helps to drain tears from the nasal cavity.
Zygomatic (2) (also Figure 8.2)	N/A	Commonly called the cheekbones; each forms part of the lateral orbit.
Inferior nasal concha (2)	N/A	Inferior turbinate; each forms part of the lateral walls of the nasal cavities; improves the airflow through the nasal cavity.
Palatine (2) (also Figure 8.2)	N/A	Forms the posterior hard palate, a small part of the nasal cavity, and part of the orbit.
	Horizontal plate	Forms the posterior portion of the hard palate.
	Median palatine suture	Median fusion point of the horizontal plates of the palatine bones.
Vomer (1)	N/A	Thin, blade-shaped bone that forms the inferior nasal septum.

Table 8.2 The Axial Skeleton: Facial Bones and Important Bone Markings (Figures 8.1, 8.7, 8.9, and 8.10, with additional figures listed for specific bones)

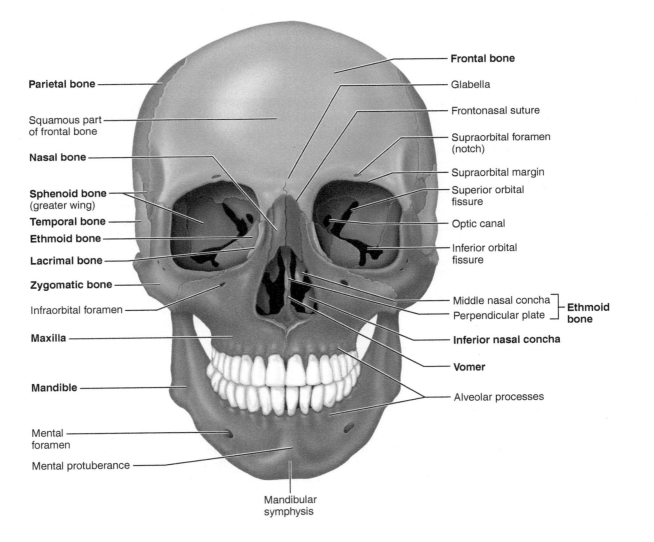

Figure 8.7 Anterior view of the skull.

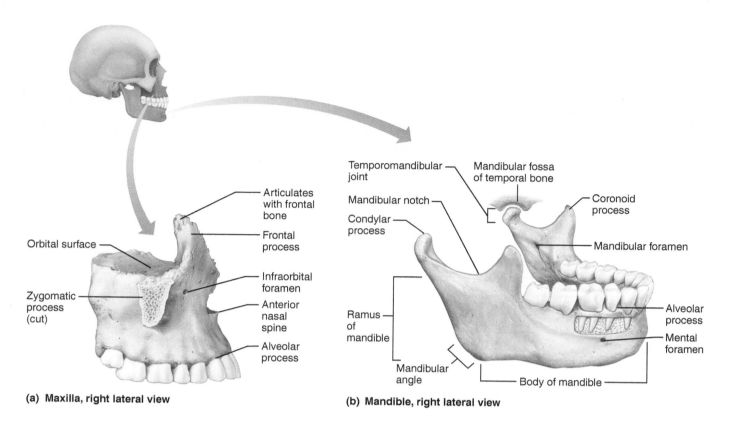

(a) Maxilla, right lateral view

(b) Mandible, right lateral view

Figure 8.8 Detailed anatomy of the maxilla and mandible.

Facial bone	Important markings	Description
Table 8.2 The Axial Skeleton: Facial Bones and Important Bone Markings (Figures 8.1, 8.7, 8.9, and 8.10, with additional figures listed for specific bones) (continued)		
Maxilla (2) (also Figures 8.2 and 8.8)	N/A	Keystone facial bones because they articulate with all other facial bones except the mandible; form the upper jaw and parts of the hard palate, orbits, and nasal cavity.
	Frontal process	Forms part of the lateral aspect of the bridge of the nose.
	Infraorbital foramen	Opening under the orbit that forms a passageway for the infraorbital artery and nerve.
	Palatine process	Forms the anterior hard palate; the two processes meet anteriorly (Note: Seen in inferior view).
	Zygomatic process	Articulation process for zygomatic bone.
	Alveolar process	Inferior margin; contains dental alveoli in which the teeth lie.
Mandible (1) (also Figures 8.2 and 8.8)	N/A	The lower jawbone, which articulates with the temporal bone to form the only freely movable joints in the skull (the temporomandibular joint).
	Condylar processes	Articulate with the mandibular fossae of the temporal bones.
	Coronoid processes	"Crown-shaped" portion of the ramus for muscle attachment.
	Mandibular notches	Separate the condylar process and the coronoid process.
	Body	Horizontal portion that forms the chin.
	Ramus	Vertical extension of the body.
	Mandibular angles	Posterior points where the ramus meets the body.
	Mental foramina	Paired openings on the body (lateral to the midline); transmit blood vessels and nerves to the lower lip and skin of the chin.
	Alveolar process	Superior margin of the mandible; contains dental alveoli in which the teeth lie.
	Mandibular foramina	Located on the medial surface of each ramus; passageway for the nerve involved in tooth sensation. (Dentists inject anesthetic into this foramen before working on the lower teeth.)

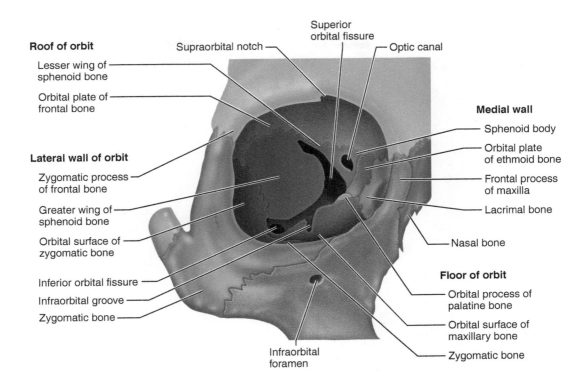

Roof of orbit
Lesser wing of sphenoid bone
Orbital plate of frontal bone

Lateral wall of orbit
Zygomatic process of frontal bone
Greater wing of sphenoid bone
Orbital surface of zygomatic bone
Inferior orbital fissure
Infraorbital groove
Zygomatic bone

Supraorbital notch
Superior orbital fissure
Optic canal

Medial wall
Sphenoid body
Orbital plate of ethmoid bone
Frontal process of maxilla
Lacrimal bone
Nasal bone

Floor of orbit
Orbital process of palatine bone
Orbital surface of maxillary bone
Zygomatic bone

Infraorbital foramen

Figure 8.9 Bones that form the orbit.

GROUP CHALLENGE

Odd Bone Out

Each box below contains four bones. One of the listed bones does *not* share a characteristic that the other three do. Work in groups of three, and discuss the characteristics of the bones in each group. On a separate piece of paper, one student will record the characteristics of each bone. For each set of bones, discuss the possible candidates for the "odd bone out" and which characteristic it lacks, based on your notes. Once your group has come to a consensus, circle the bone that doesn't belong with the others, and explain why it is singled out. What characteristic is it missing? Include as many characteristics as you can think of, but make sure the bone does not have the key characteristic(s). Use an articulated skull, disarticulated skull bones, and the pictures in your lab manual to help you select and justify your answer.

1. Which is the "odd bone"?	Why is it the odd one out?
Zygomatic bone Maxilla Vomer Nasal bone	
2. Which is the "odd bone"?	**Why is it the odd one out?**
Parietal bone Sphenoid bone Frontal bone Occipital bone	
3. Which is the "odd bone"?	**Why is it the odd one out?**
Lacrimal bone Nasal bone Zygomatic bone Maxilla	

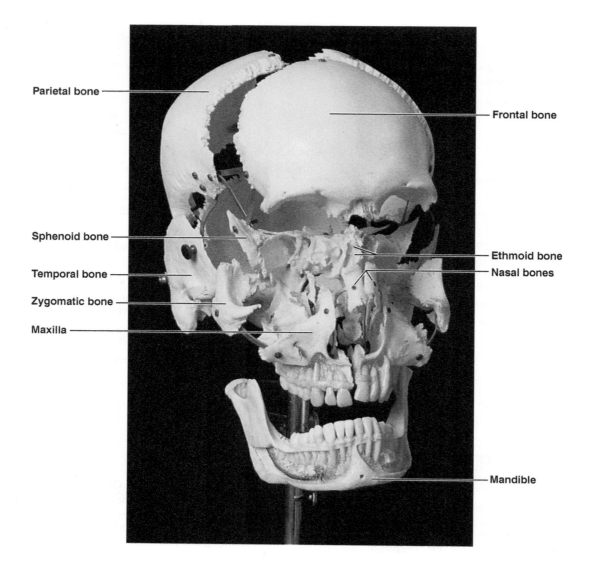

Figure 8.10 Anterior view of the Beauchene skull.

The Orbit

Seven bones of the skull form the orbit, the bony cavity that surrounds the eye: the frontal, sphenoid, ethmoid, lacrimal, maxilla, palatine, and zygomatic (Figure 8.9).

Hyoid Bone

Not really considered or counted as a skull bone, the hyoid bone is located in the throat above the larynx **(Figure 8.11)** where it serves as a point of attachment for many tongue and neck muscles. It does not articulate with any other bone and is thus unique. It is horseshoe-shaped with a body and two pairs of **horns,** or **cornua.**

Paranasal Sinuses

The bones surrounding the nasal cavity—maxillary, sphenoid, ethmoid, and frontal—contain sinuses (mucosa-lined air cavities), which lead into the nasal passages (Figure 8.5 and **Figure 8.12**). These paranasal sinuses lighten the facial bones and may act as resonance chambers for speech.

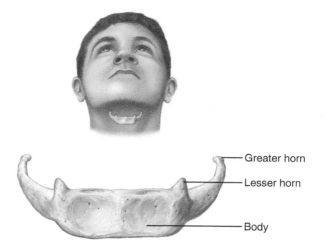

Figure 8.11 Hyoid bone.

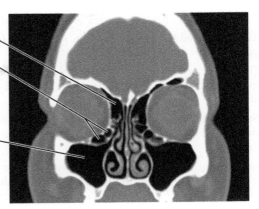

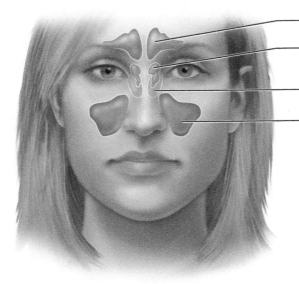

Frontal
sinus

Ethmoidal
air cells
(sinus)

Sphenoidal
sinus

Maxillary
sinus

(a) Anterior aspect

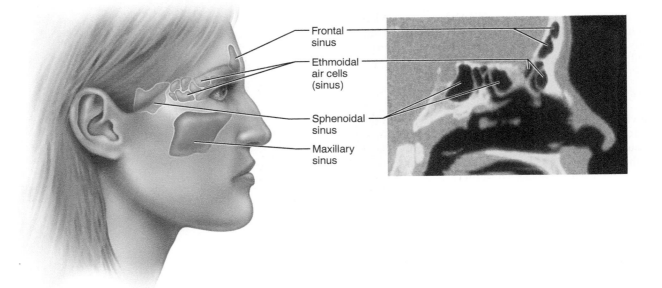

Frontal
sinus

Ethmoidal
air cells
(sinus)

Sphenoidal
sinus

Maxillary
sinus

(b) Medial aspect

Figure 8.12 Paranasal sinuses. Illustrations on left, CT scans on right.

The maxillary sinus is the largest of the sinuses found in the skull.

Sinusitis, or inflammation of the sinuses, sometimes occurs as a result of an allergy or bacterial invasion of the sinus cavities. In such cases, some of the connecting passageways between the sinuses and nasal passages may become blocked with thick mucus or infectious material. Then, as the air in the sinus cavities is absorbed, a partial vacuum forms. The result is a sinus headache localized over the inflamed sinus area. Severe sinus infections may require surgical drainage to relieve this painful condition. ✚

ACTIVITY 2

Palpating Skull Markings

Palpate the following areas on yourself. Place a check mark in the boxes as you locate the skull markings. Ask your instructor for help with any markings that you are unable to locate.

☐ Zygomatic bone and arch. (The most prominent part of your cheek is your zygomatic bone. Follow the posterior course of the zygomatic arch to its junction with your temporal bone.)

☐ Mastoid process (the rough area behind your ear).

☐ Temporomandibular joints. (Open and close your jaws to locate these.)

☐ Greater wing of sphenoid. (Find the indentation posterior to the orbit and superior to the zygomatic arch on your lateral skull.)

☐ Supraorbital foramen. (Apply firm pressure along the superior orbital margin to find the indentation resulting from this foramen.)

☐ Infraorbital foramen. (Apply firm pressure just inferior to the inferomedial border of the orbit to locate this large foramen.)

☐ Mandibular angle (most inferior and posterior aspect of the mandible).

☐ Mandibular symphysis (midline of chin).

☐ Nasal bones. (Run your index finger and thumb along opposite sides of the bridge of your nose until they "slip" medially at the inferior end of the nasal bones.)

☐ External occipital protuberance. (This midline projection is easily felt by running your fingers up the furrow at the back of your neck to the skull.)

☐ Hyoid bone. (Place a thumb and index finger beneath the chin just anterior to the mandibular angles, and squeeze gently. Exert pressure with the thumb, and feel the horn of the hyoid with the index finger.) ▬▬

The Vertebral Column

The **vertebral column,** extending from the skull to the pelvis, forms the body's major axial support. Additionally, it surrounds and protects the delicate spinal cord while allowing the spinal nerves to issue from the cord via openings between adjacent vertebrae. The term *vertebral column* might suggest a rather rigid supporting rod, but this is far from the truth. The vertebral column consists of 24 single bones called **vertebrae** and two composite, or fused, bones (the sacrum and coccyx) that are connected in such a way as to provide a flexible curved structure **(Figure 8.13)**. Of the 24 single vertebrae, the 7 bones of the neck are called *cervical vertebrae;* the next 12 are *thoracic vertebrae;* and the 5 supporting the lower back are *lumbar vertebrae.* Remembering common mealtimes for breakfast, lunch, and dinner (7 A.M., 12 noon, and 5 P.M.) may help you to remember the number of bones in each region.

The vertebrae are separated by pads of fibrocartilage, **intervertebral discs,** that cushion the vertebrae and absorb shocks. Each disc is composed of two major regions, a central gelatinous *nucleus pulposus* that behaves like a rubber ball and the *anulus fibrosus* that stabilizes the disc and contains the nucleus pulposus. The anulus fibrosus is composed of an outer ring of collagen fibers and an inner ring of fibrocartilage.

As a person ages, the water content of the discs decreases (as it does in other tissues throughout the body), and the discs become thinner and less compressible. This situation, along with other degenerative changes such as weakening of the ligaments and muscle tendons associated with the vertebral column, predisposes older people to ruptured discs. In a ruptured, or **herniated, disc,** the annulus

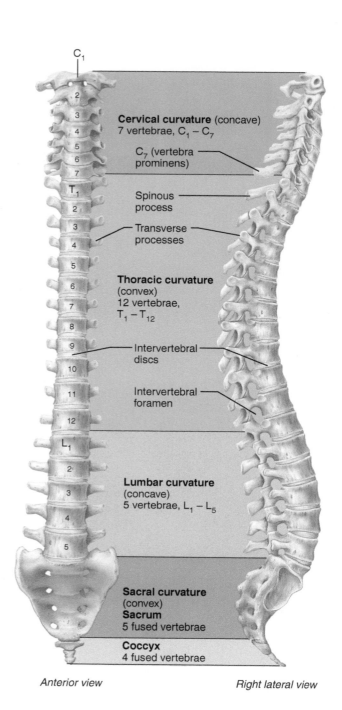

Anterior view   Right lateral view

Figure 8.13 The vertebral column. Notice the curvatures in the lateral view. (The terms *convex* and *concave* refer to the curvature of the posterior aspect of the vertebral column.)

fibrosus commonly ruptures, and the nucleus pulposus protrudes (herniates) through it. This event typically compresses adjacent nerves causing pain. ✚

The thoracic and sacral curvatures of the spine are referred to as *primary curvatures* because they are present and well developed at birth. Later the *secondary curvatures* are formed. The cervical curvature becomes prominent when the baby begins to hold its head up independently, and the lumbar curvature develops when the baby begins to walk.

Examining Spinal Curvatures

1. Observe the normal curvature of the vertebral column in the articulated vertebral column or laboratory skeleton, and compare it to Figure 8.13. Abnormal spinal curvatures, including *scoliosis, kyphosis,* and *lordosis,* may result from disease or poor posture (Figure 8.14). Also examine X-ray films, if they are available, showing these same conditions in a living patient.

2. Then, using the articulated vertebral column (or an articulated skeleton), examine the freedom of movement between two lumbar vertebrae separated by an intervertebral disc.

When the fibrous disc is properly positioned, are the spinal cord or peripheral nerves impaired in any way?

_____

Remove the disc, and put the two vertebrae back together. What happens to the nerve?

_____

What would happen to the spinal nerves in areas of malpositioned, or "slipped," discs?

_____

_____

Structure of a Typical Vertebra

Although they differ in size and specific features, all vertebrae have some features in common (Figure 8.15).

*   **Body:** Rounded central weight-bearing portion of the vertebra, which faces anteriorly in the human vertebral column.

*   **Vertebral arch:** Composed of pedicles and laminae, it represents the junction of all posterior extensions from the vertebral body.

*   **Vertebral foramen:** Opening enclosed by the body and vertebral arch; a conduit for the spinal cord.

*   **Transverse processes:** Two lateral projections from the vertebral arch.

*   **Spinous process:** Single medial and posterior projection formed at the junction of the two laminae. The transverse and spinous processes are attachment sites for muscles.

*   **Superior and inferior articular processes:** Paired projections lateral to the vertebral foramen that enable articulation with adjacent vertebrae. The superior articular processes typically face toward the spinous process (posteriorly), whereas the inferior articular processes face (anteriorly) away from the spinous process. Each process also has a corresponding facet, a smooth articular surface that is covered with hyaline cartilage.

*   **Intervertebral foramina:** The right and left pedicles have notches on their inferior and superior surfaces that create openings, the intervertebral foramina (see Figure 8.13), for spinal nerves to leave the spinal cord between adjacent vertebrae.

The following sections describe how specific vertebrae differ with respect to structure and function (refer to Figure 8.16 through 8.18 and Table 8.3).

Cervical Vertebrae

The seven cervical vertebrae (referred to as C_1 through C_7) form the neck portion of the vertebral column. The first two cervical vertebrae (atlas and axis) are highly modified to perform special functions (see Figure 8.16). The **atlas** (C_1) lacks a body, and its lateral processes contain large concave depressions on their superior surfaces that receive the occipital condyles of the skull. This joint enables you to nod "yes." The **axis** (C_2) acts as a pivot for the rotation of the atlas (and skull) above. It bears a large vertical process, the **dens,** that serves

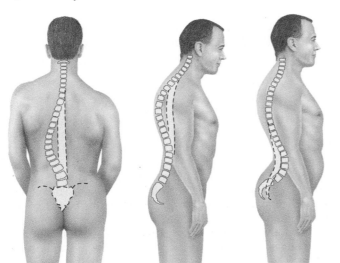

Figure 8.14 **Abnormal spinal curvatures.**

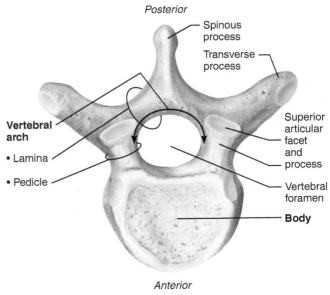

Figure 8.15 **A typical vertebra, superior view.** Inferior articulating surfaces not shown.

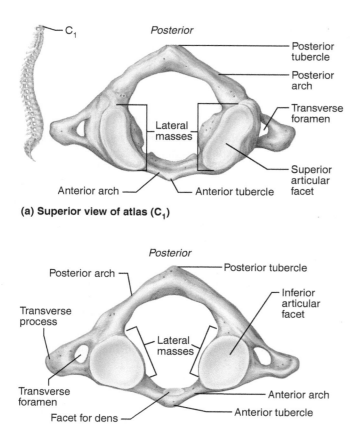

(a) Superior view of atlas (C₁)

(b) Inferior view of atlas (C₁)

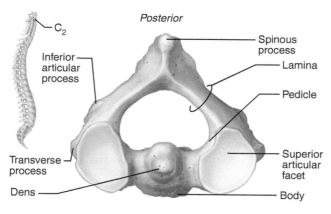

(c) Superoposterior view of axis (C₂)

Figure 8.16 The first and second cervical vertebrae.

as the pivot point. The articulation between C_1 and C_2 allows you to rotate your head from side to side to indicate "no."

The more typical cervical vertebrae (C_3 through C_7) are distinguished from the thoracic and lumbar vertebrae by several features (see Table 8.3 and Figure 8.17). They are the smallest, lightest vertebrae, and the vertebral foramen is triangular. The spinous process is short and often bifurcated (divided into two branches). The spinous process of C_7 is not branched, however, and is substantially longer than that of the

other cervical vertebrae. Because the spinous process of C_7 is visible through the skin at the base of the neck, it is called the *vertebra prominens* (Figure 8.13) and is used as a landmark for counting the vertebrae. Transverse processes of the cervical vertebrae are wide, and they contain foramina through which the vertebral arteries pass superiorly on their way to the brain. Any time you see these foramina in a vertebra, you can be sure that it is a cervical vertebra.

☐ Palpate your vertebra prominens.

Place a check mark in the box when you locate the structure.

Thoracic Vertebrae

The 12 thoracic vertebrae (referred to as T_1 through T_{12}) may be recognized by the following structural characteristics. They have a larger body than the cervical vertebrae (Figure 8.17). The body is somewhat heart-shaped, with two small articulating surfaces, or **costal facets,** on each side (one superior, the other inferior) close to the origin of the vertebral arch. These facets articulate with the heads of the corresponding ribs. The vertebral foramen is oval or round, and the spinous process is long, with a sharp downward hook. The closer the thoracic vertebra is to the lumbar region, the less sharp and shorter the spinous process. Articular facets on the transverse processes articulate with the tubercles of the ribs. Besides forming the thoracic part of the spine, these vertebrae form the posterior aspect of the thoracic cage (rib cage). Indeed, they are the only vertebrae that articulate with the ribs.

Lumbar Vertebrae

The five lumbar vertebrae (L_1 through L_5) have massive blocklike bodies and short, thick, hatchet-shaped spinous processes extending directly backward (see Table 8.3 and Figure 8.17). The superior articular facets face posteromedially; the inferior ones are directed anterolaterally. These structural features reduce the mobility of the lumbar region of the spine. Because most stress on the vertebral column occurs in the lumbar region, these are also the sturdiest of the vertebrae.

The spinal cord ends at the superior edge of L_2, but the outer covering of the cord, filled with cerebrospinal fluid, extends an appreciable distance beyond. Thus a *lumbar puncture* (for examination of the cerebrospinal fluid) or the administration of "saddle block" anesthesia for childbirth is normally done between L_3 and L_4 or L_4 and L_5, where there is little or no chance of injuring the delicate spinal cord.

The Sacrum

The **sacrum** (Figure 8.18) is a composite bone formed from the fusion of five vertebrae. Superiorly it articulates with L_5, and inferiorly it connects with the coccyx. The **median sacral crest** is a remnant of the spinous processes of the fused vertebrae. The winglike **alae,** formed by fusion of the transverse processes, articulate laterally with the hip bones. The sacrum is concave anteriorly and forms the posterior border of the pelvis. Four ridges (lines of fusion) cross the anterior part of the sacrum, and **sacral foramina** are located at either end of these ridges. These foramina allow blood vessels and nerves to pass. The vertebral canal continues inside the sacrum as the **sacral canal** and terminates near the coccyx via an enlarged opening called the **sacral hiatus.** The **sacral promontory**

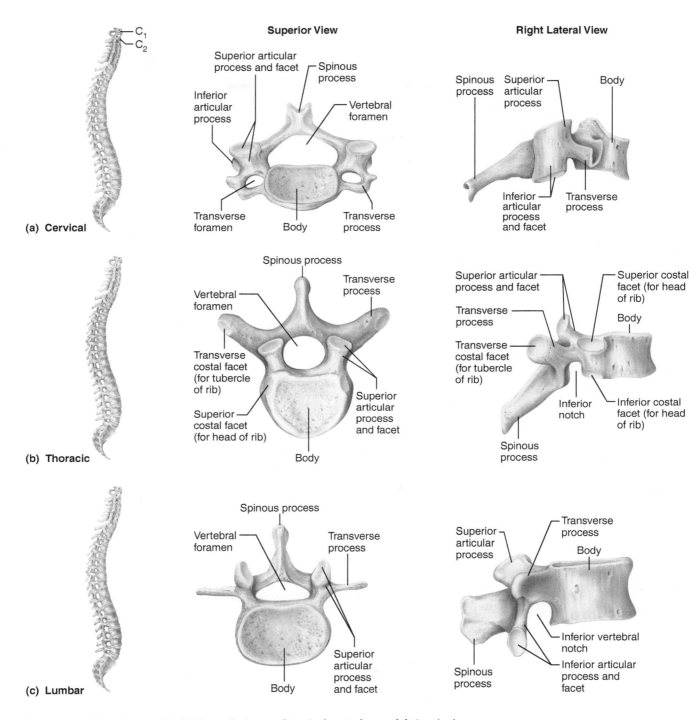

Figure 8.17 Superior and right lateral views of typical vertebrae. (a) Cervical. **(b)** Thoracic. **(c)** Lumbar.

(anterior border of the body of S_1) is an important anatomical landmark for obstetricians.

☐   Attempt to palpate the median sacral crest of your sacrum. (This is more easily done by thin people and obviously in privacy.)

Place a check mark in the box when you locate the structure.

The Coccyx

The **coccyx** (see Figure 8.18) is formed from the fusion of three to five small irregularly shaped vertebrae. It is literally the human tailbone, a vestige of the tail that other vertebrates have. The coccyx is attached to the sacrum by ligaments.

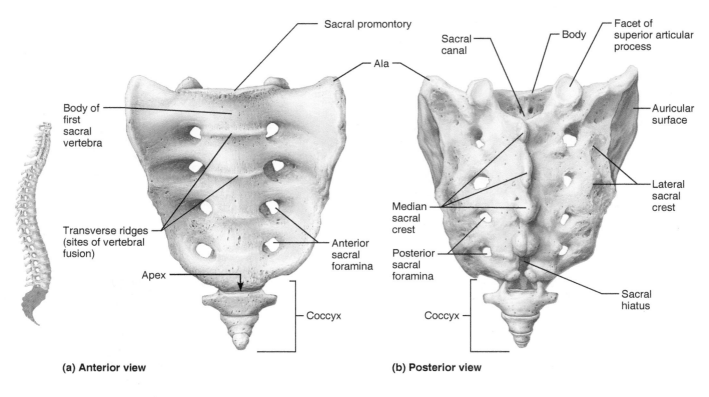

(a) Anterior view

(b) Posterior view

Figure 8.18 Sacrum and coccyx.

ACTIVITY 4

Examining Vertebral Structure

Obtain examples of each type of vertebra and examine them carefully, comparing them to each other (refer to Figures 8.16 through 8.18 and Table 8.3). ∎

The Thoracic Cage

The **thoracic cage** consists of the bony thorax, which is composed of the sternum, ribs, and thoracic vertebrae, plus costal cartilages **(Figure 8.19)**. Its cone-shaped, cagelike structure protects the organs of the thoracic cavity, including the critically important heart and lungs.

Table 8.3   Regional Characteristics of Cervical, Thoracic, and Lumbar Vertebrae

Characteristic	(a) Cervical (C_3–C_7)	(b) Thoracic	(c) Lumbar
Body	Small, wide side to side	Larger than cervical; heart-shaped; bears costal facets	Massive; kidney-shaped
Spinous process	Short; bifid; projects directly posteriorly	Long; sharp; projects inferiorly	Short; blunt; projects directly posteriorly
Vertebral foramen	Triangular	Circular	Triangular
Transverse processes	Contain foramina	Have costal facets for ribs (except T_{11} and T_{12})	Thin and tapered
Superior and inferior articulating processes	Superior facets directed superoposteriorly	Superior facets directed posteriorly	Superior facets directed posteromedially (or medially)
	Inferior facets directed inferoanteriorly	Inferior facets directed anteriorly	Inferior facets directed anterolaterally (or laterally)
Movements allowed	Flexion and extension; lateral flexion; rotation; the spine region with the greatest range of movement	Rotation; lateral flexion possible but restricted by ribs; flexion and extension limited	Flexion and extension; some lateral flexion; rotation prevented

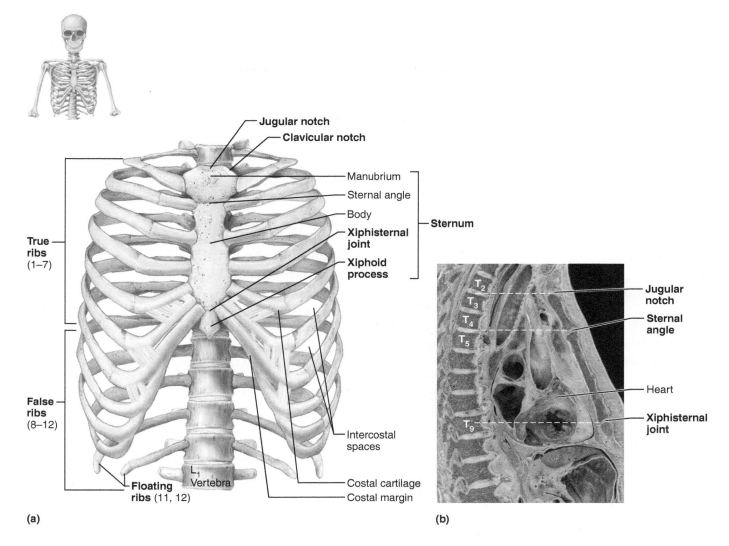

Figure 8.19 The thoracic cage. (a) Anterior view with costal cartilages shown in blue. **(b)** Median section of the thorax, illustrating the relationship of the surface anatomical landmarks of the thorax to the vertebral column (thoracic portion).

The Sternum

The **sternum** (breastbone), a typical flat bone, is a result of the fusion of three bones—the manubrium, body, and xiphoid process. It is attached to the first seven pairs of ribs. The superiormost **manubrium** looks like the knot of a tie; it articulates with the clavicle (collarbone) laterally. The **body** forms the bulk of the sternum. The **xiphoid process** constructs the inferior end of the sternum and lies at the level of the fifth intercostal space. Although it is made of hyaline cartilage in children, it is usually ossified in adults over age 40.

In some people, the xiphoid process projects dorsally. This may present a problem because underlying physical trauma to the chest can push such a xiphoid into the underlying heart or liver, causing massive hemorrhage. ✚

The sternum has three important bony landmarks—the jugular notch, the sternal angle, and the xiphisternal joint. The **jugular notch** (concave upper border of the manubrium) can be palpated easily; generally it is at the level of the disc between the second and third thoracic vertebrae. The **sternal angle** is a result of the manubrium and body meeting at a slight angle to each other, so that a transverse ridge is formed at the level of the second ribs. It provides a handy reference point for counting ribs to locate the second intercostal space for listening to certain heart valves, and is an important anatomical landmark for thoracic surgery. The **xiphisternal joint,** the point where the sternal body and xiphoid process fuse, lies at the level of the ninth thoracic vertebra.

☐ Palpate your sternal angle and jugular notch.

Place a check mark in the box when you locate the structure.

Because of its accessibility, the sternum is a favored site for obtaining samples of blood-forming (hematopoietic) tissue for the diagnosis of suspected blood diseases. A needle is inserted into the marrow of the sternum, and the sample is withdrawn (sternal puncture).

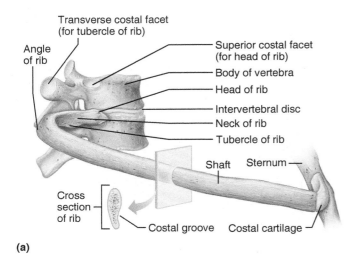

(a)

(b)

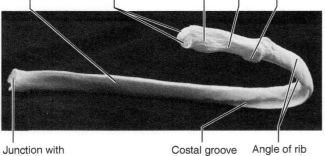

(c)

Figure 8.20 Structure of a "typical" true rib and its articulations. (a) Vertebral and sternal articulations of a typical true rib. **(b)** Superior view of the articulation between a rib and a thoracic vertebra, with costovertebral ligaments shown on left side only. **(c)** Right rib 6, posterior view.

The Ribs

The 12 pairs of **ribs** form the walls of the thoracic cage (see Figure 8.19 and **Figure 8.20**). All of the ribs articulate posteriorly with the vertebral column via their heads and tubercles and then curve downward and toward the anterior body surface. The first seven pairs, called the *true,* or *vertebrosternal, ribs,* attach directly to the sternum by their "own" costal cartilages. The next five pairs are called *false ribs;* they attach indirectly to the sternum or entirely lack a sternal attachment. Of these, rib pairs 8–10, which are also called *vertebrochondral ribs,* have indirect cartilage attachments to the sternum via the costal cartilage of rib 7. The last two pairs, called *floating,* or *vertebral, ribs,* have no sternal attachment.

ACTIVITY 5

Examining the Relationship Between Ribs and Vertebrae

First take a deep breath to expand your chest. Notice how your ribs seem to move outward and how your sternum rises. Then examine an articulated skeleton to observe the relationship between the ribs and the vertebrae. (Refer to Activity 3, "Palpating Landmarks of the Trunk," and Activity 4, "Palpating Landmarks of the Abdomen," in Exercise 30, Surface Anatomy Roundup.) ▄

The Fetal Skull

One of the most obvious differences between fetal and adult skeletons is the huge size of the fetal skull relative to the rest of the skeleton. Skull bones are incompletely formed at birth and connected by fibrous membranes called **fontanelles.** The fontanelles allow the fetal skull to be compressed slightly during birth and also allow for brain growth in the fetus and infant. They ossify (become bone) as the infant ages, completing the process by the time the child is $1\frac{1}{2}$ to 2 years old.

ACTIVITY 6

Examining a Fetal Skull

1. Obtain a fetal skull and study it carefully.

   Does it have the same bones as the adult skull?

   _____

   How does the size of the fetal face relate to the cranium?

   _____

   How does this compare to what is seen in the adult?

   _____

2. Locate the following fontanelles on the fetal skull **(Figure 8.21)**: *anterior* (or *frontal) fontanelle, mastoid fontanelle, sphenoidal fontanelle,* and *posterior* (or *occipital) fontanelle.*

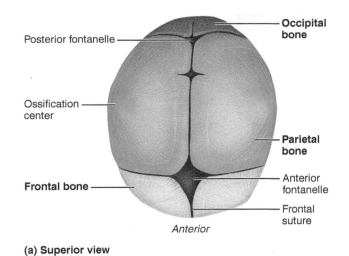

(a) **Superior view**

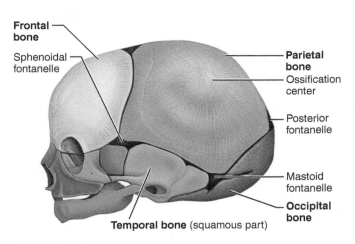

(b) **Left lateral view**

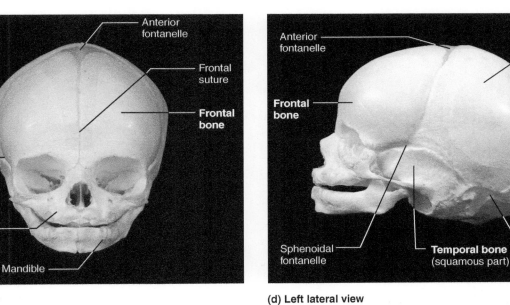

(c) **Anterior view**

(d) **Left lateral view**

Figure 8.21   Skull of a newborn.

3.   Notice that some of the cranial bones have conical protrusions. These are **ossification (growth) centers.** Notice also that the frontal bone is still bipartite and that the temporal bone is incompletely ossified—it is little more than a ring of bone.

4.   Before completing this study, check the questions on the Review Sheet at the end of this exercise to ensure that you have made all of the necessary observations. ■

Name _____

Lab Time/Date _____

The Axial Skeleton

The Skull

1. Match the bone names in column B with the descriptions in column A. Some responses in column B may be used more than once.

Column A

_____ 1. forms the anterior cranium

_____ 2. cheekbone

_____ 3. lower jaw

_____ 4. bridge of nose

_____ 5. posterior bones of the hard palate

_____ 6. much of the lateral and superior cranium

_____ 7. most posterior part of cranium

_____ 8. single, irregular, bat-shaped bone forming part of the cranial base

_____ 9. tiny bones bearing tear ducts

_____ 10. anterior part of hard palate

_____ 11. superior and middle nasal conchae form from its projections

_____ 12. site of mastoid process

_____ 13. site of sella turcica

_____ 14. site of cribriform plate

_____ 15. site of mental foramen

_____ 16. site of styloid processes

_____, _____, _____, and _____ 17. four bones containing paranasal sinuses

_____ 18. has condyles that articulate with the atlas

_____ 19. foramen magnum contained here

_____ 20. small U-shaped bone in neck, where many tongue muscles attach

_____ 21. organ of hearing found here

_____, _____ 22. two bones that form the nasal septum

_____ 23. bears an upward protrusion, the "rooster's comb," or crista galli

_____, _____ 24. contain sockets that bear teeth

_____ 25. forms the most inferior turbinate

Column B

a. ethmoid

b. frontal

c. hyoid

d. inferior nasal concha

e. lacrimal

f. mandible

g. maxilla

h. nasal

i. occipital

j. palatine

k. parietal

l. sphenoid

m. temporal

n. vomer

o. zygomatic

2. Using choices from the numbered key to the right, identify all bones and bone markings provided with various leader lines in the two following photographs. A colored dot at the end of a leader line indicates a bone. Leader lines without a colored dot indicate bone markings. Note that vomer, sphenoid bone, and zygomatic bone will each be labeled twice.

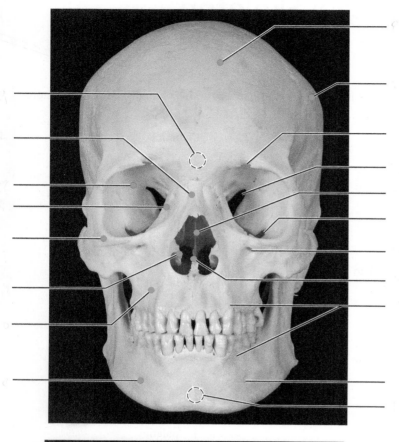

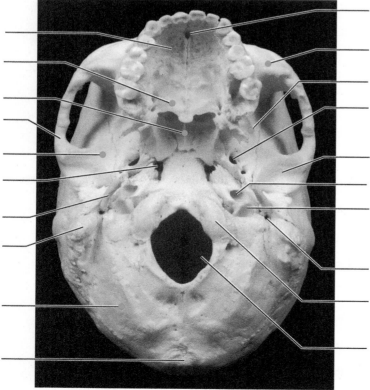

Key:

1. alveolar processes
2. carotid canal
3. ethmoid bone (perpendicular plate)
4. external occipital protuberance
5. foramen lacerum
6. foramen magnum
7. foramen ovale
8. frontal bone
9. glabella
10. incisive fossa
11. inferior nasal concha
12. inferior orbital fissure
13. infraorbital foramen
14. jugular foramen
15. lacrimal bone
16. mandible
17. mandibular fossa
18. mandibular symphysis
19. mastoid process
20. maxilla
21. mental foramen
22. nasal bone
23. occipital bone
24. occipital condyle
25. palatine bone
26. palatine process of maxilla
27. parietal bone
28. sphenoid bone
29. styloid process
30. stylomastoid foramen
31. superior orbital fissure
32. supraorbital foramen
33. temporal bone
34. vomer bone
35. zygomatic bone
36. zygomatic process

3. Name the eight bones of the cranium. (Remember to include left and right).

_____   _____   _____   _____

_____   _____   _____   _____

4. List the bones that have sinuses and give two possible functions of the sinuses. _____

_____

_____

5. What is the bony orbit? _____

What bones contribute to the formation of the orbit? _____

_____

6. Why can the sphenoid bone be called the keystone bone of the cranium? _____

_____

The Vertebral Column

7. The distinguishing characteristics of the vertebrae that make up the vertebral column are noted below. Correctly identify each described structure/region by choosing a response from the key.

Key: a. atlas   d. coccyx   f. sacrum
         b. axis   e. lumbar vertebra   g. thoracic vertebra
         c. cervical vertebra—typical

_____   1. vertebra type containing foramina in the transverse processes, through which the vertebral arteries ascend to reach the brain

_____   2. dens here provides a pivot for rotation of the first cervical vertebra (C_1)

_____   3. transverse processes faceted for articulation with ribs; spinous process pointing sharply downward

_____   4. composite bone; articulates with the hip bone laterally

_____   5. massive vertebra; weight-sustaining

_____   6. "tailbone"; vestigial fused vertebrae

_____   7. supports the head; allows a rocking motion in conjunction with the occipital condyles

8. Using the key, correctly identify the vertebral parts/areas described below. (More than one choice may apply in some cases.) Also use the key letters to correctly identify the vertebral areas in the diagram.

Key: a.   body
       b.   intervertebral foramina
       c.   lamina

       d.   pedicle
       e.   spinous process
       f.   superior articular facet

       g.   transverse process
       h.   vertebral arch
       i.   vertebral foramen

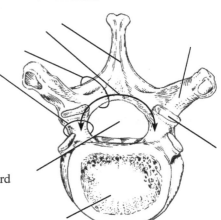

_____   1.   cavity enclosing the spinal cord

_____   2.   weight-bearing portion of the vertebra

_____   3.   provide levers against which muscles pull

_____   4.   provide an articulation point for the ribs

_____   5.   openings providing for exit of spinal nerves

_____   6.   structures that form an enclosure for the spinal cord

9. Describe how a spinal nerve exits from the vertebral column. _____

_____

10. Name two factors/structures that permit flexibility of the vertebral column. _____

_____ and _____

11. What kind of tissue makes up the intervertebral discs? _____

12. What is a herniated disc? _____

What problems might it cause? _____

_____

13. Which two spinal curvatures are obvious at birth? _____ and _____

Under what conditions do the secondary curvatures develop?

_____

_____

14. Use the key to label the structures on the thoracic region of the vertebral column.

> *Key:* a.   intervertebral discs
>
>   b.   intervertebral foramina
>
>   c.   spinous processes
>
>   d.   thoracic vertebrae
>
>   e.   transverse processes

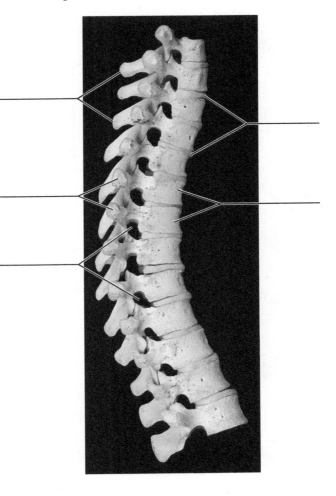

The Thoracic Cage

15. The major components of the thorax (excluding the vertebral column) are the _____

and the _____.

16. Differentiate a true rib from a false rib. _____

_____

_____

Is a floating rib a true or a false rib? _____

17. What is the general shape of the thoracic cage? _____

18. Using the terms in the key, identify the regions and landmarks of the thoracic cage.

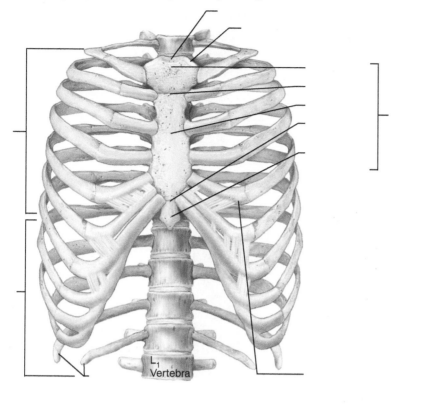

L₁
Vertebra

Key: a. body

b. clavicular notch

c. costal cartilage

d. false ribs

e. floating ribs

f. jugular notch

g. manubrium

h. sternal angle

i. sternum

j. true ribs

k. xiphisternal joint

l. xiphoid process

The Fetal Skull

19. Are the same skull bones seen in the adult also found in the fetal skull? _____

20. How does the size of the fetal face compare to the fetal cranium? _____

_____

How does this compare to the adult skull? _____

_____

_____

21. What are the outward conical projections on some of the fetal cranial bones? _____

22. What is a fontanelle? _____

What is its fate? _____

What is the function of the fontanelles in the fetal skull? _____

_____

23. Using the terms listed, identify each of the fontanelles shown on the fetal skull below.

 Key: a.   anterior fontanelle   b.   mastoid fontanelle   c.   posterior fontanelle   d.   sphenoidal fontanelle

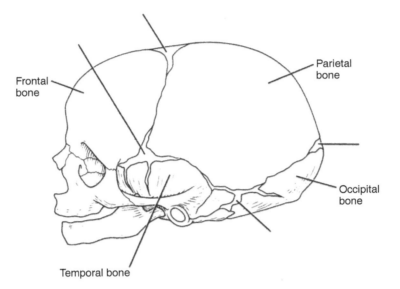

Frontal bone

Parietal bone

Occipital bone

Temporal bone

24. ✚ Craniosynostosis is a condition in which one or more of the fontanelles is replaced by bone prematurely. Discuss the ramifications of this early closure.

_____

_____

25. ✚ As we age, we often become shorter. Explain why this might occur. _____

_____

_____

26. ✚ The xiphoid process is often missing from the sternum in bone collections. Hypothesize why it might be missing. ——

_____

The Appendicular Skeleton

MATERIALS

- Articulated skeletons
- Disarticulated skeletons (complete)
- Articulated pelves (male and female for comparative study)
- X-ray images of bones of the appendicular skeleton

LEARNING OUTCOMES

☐ Identify the bones of the pectoral and pelvic girdles and their attached limbs by examining isolated bones or an articulated skeleton, and name the important bone markings on each.

☐ Describe the differences between a male and a female pelvis, and explain why these differences are important.

☐ Compare the features of the human pectoral and pelvic girdles, and discuss how their structures relate to their specialized functions.

☐ Arrange unmarked, disarticulated bones in their proper places to form an entire skeleton.

PRE-LAB QUIZ

1. The _____ skeleton is made up of 126 bones of the limbs and girdles.

2. Circle the correct underlined term. The <u>pectoral</u> / <u>pelvic</u> girdle attaches the upper limb to the axial skeleton.

3. The _____, on the posterior thorax, are roughly triangular in shape. They have no direct attachment to the axial skeleton but are held in place by trunk muscles.

4. The arm consists of one long bone, the:
   a. femur
   b. humerus
   c. tibia
   d. ulna

5. The hand consists of three groups of bones. The carpals make up the wrist. The _____ make up the palm, and the phalanges make up the fingers.

6. You are studying a pelvis that is wide and shallow. The acetabula are small and far apart. The pubic arch is rounded and greater than 90°. It appears to be tilted forward, with a wide, short sacrum. Is this a male or a female pelvis? _____

7. The strongest, heaviest bone of the body is in the thigh. It is the:
   a. femur
   b. fibula
   c. tibia

8. The _____, or "kneecap," is a sesamoid bone that is found within the quadriceps tendon. It guards the knee joint anteriorly and improves leverage of the thigh muscles acting across the knee joint.

9. Circle True or False. The fingers of the hand and the toes of the foot—with the exception of the great toe and the thumb—each have three phalanges.

10. Each foot has a total of _____ bones.

The **appendicular skeleton** (the gold-colored portion of Figure 7.1 on page 92) is composed of the 126 bones of the appendages and the pectoral and pelvic girdles, which attach the limbs to the axial skeleton. Although the upper and lower limbs differ in their functions and mobility, they have the same fundamental plan, with each limb made up of three major segments connected together by freely movable joints.

Examining and Identifying Bones of the Appendicular Skeleton

Carefully examine each of the bones described in this exercise, and identify the characteristic bone markings of each (See Tables 9.1 to 9.5). The markings help you determine whether a bone is the right or left member of its pair; for example, the glenoid cavity is on the lateral aspect of the scapula, and the spine is on its posterior aspect. *This is a very important instruction because you will be constructing your own skeleton to finish this laboratory exercise.* Additionally, when corresponding X-ray films are available, compare the actual bone specimen to its X-ray image. ■

Bones of the Pectoral Girdle and Upper Limb

The Pectoral (Shoulder) Girdle

The paired **pectoral**, or **shoulder, girdles** (**Figure 9.1** and **9.2** and **Table 9.1**) each consist of two bones—the anterior clavicle and the posterior scapula. The clavicle articulates with the axial skeleton via the sternum. The scapula does not articulate with the axial skeleton. The shoulder girdles attach the upper limbs to the axial skeleton, and they serve as attachment points for many trunk and neck muscles.

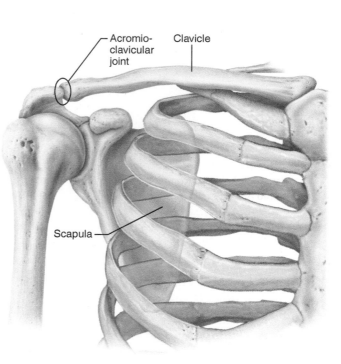

Figure 9.1 Articulated bones of the pectoral (shoulder) girdle. The right pectoral girdle is articulated to show the relationship of the girdle to the bones of the thorax and arm.

Table 9.1	The Appendicular Skeleton: The Pectoral (Shoulder) Girdle (Figures 9.1 and 9.2)	
Bone	**Important markings**	**Description**
Clavicle ("collarbone") (Figures 9.1 and 9.2)	Acromial (lateral) end	Flattened lateral end that articulates with the acromion of the scapula to form the acromioclavicular (AC) joint
	Sternal (medial) end	Oval or triangular medial end that articulates with the sternum to form the lateral walls of the jugular notch (see Figure 8.19, p. 121)
	Conoid tubercle	A small, cone-shaped projection located on the lateral, inferior end of the bone; serves to anchor ligaments
Scapula ("shoulder blade") (Figures 9.1 and 9.2)	Superior border	Short, sharp border located superiorly
	Medial (vertebral) border	Thin, long border that runs roughly parallel to the vertebral column
	Lateral (axillary) border	Thick border that is closest to the armpit and ends superiorly with the glenoid cavity
	Glenoid cavity	A shallow socket that articulates with the head of the humerus
	Spine	A ridge of bone on the posterior surface that is easily felt through the skin
	Acromion	The lateral end of the spine of the scapula that articulates with the clavicle to form the AC joint
	Coracoid process	Projects above the glenoid cavity as a hooklike process; helps attach the biceps brachii muscle
	Suprascapular notch	Small notch located medial to the coracoid process that allows for the passage of blood vessels and a nerve
	Subscapular fossa	A large shallow depression that forms the anterior surface of the scapula
	Supraspinous fossa	A depression located superior to the spine of the scapula
	Infraspinous fossa	A broad depression located inferior to the spine of the scapula

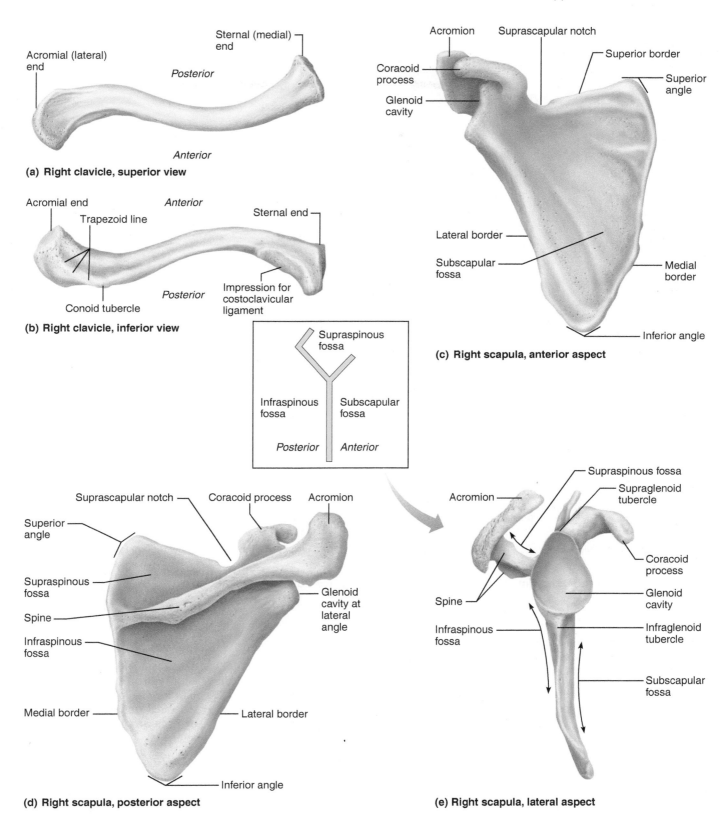

(a) **Right clavicle, superior view**

(b) **Right clavicle, inferior view**

(c) **Right scapula, anterior aspect**

(d) **Right scapula, posterior aspect**

(e) **Right scapula, lateral aspect**

Figure 9.2 Individual bones of the pectoral (shoulder) girdle. View (e) is accompanied by a schematic representation of its orientation.

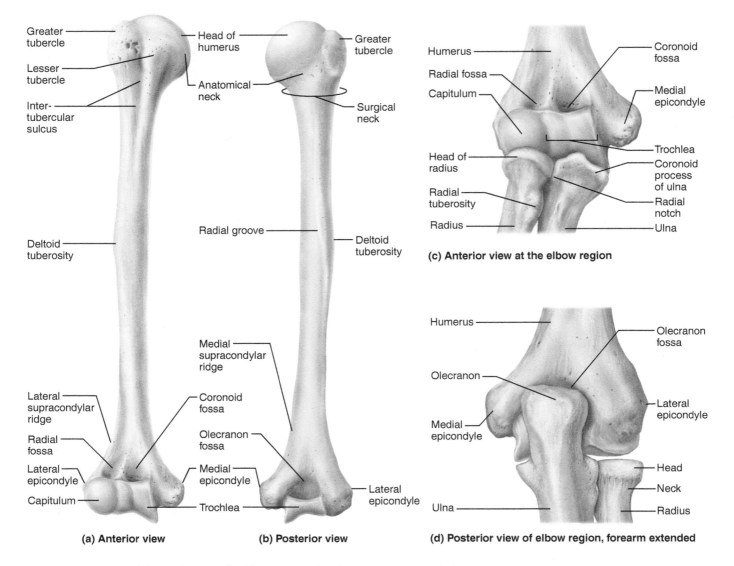

Figure 9.3 Bone of the right arm. (a, b) Humerus. **(c, d)** Detailed views of elbow region.

The shoulder girdle is exceptionally light and allows the upper limb a degree of mobility not seen anywhere else in the body. This is due to the following factors:

•   The sternoclavicular joints are the *only* site where the shoulder girdles attach to the axial skeleton.

•   The relative looseness of the scapular attachment allows it to slide back and forth against the thorax with muscular activity.

•   The glenoid cavity is shallow and does little to stabilize the shoulder joint.

However, this exceptional flexibility exacts a price: The arm bone (humerus) is very susceptible to dislocation, and fracture of the clavicle disables the entire upper limb. This is because the clavicle acts as a brace to hold the scapula and arm away from the top of the thoracic cage.

The Arm

The arm, or brachium (**Figure 9.3** and **Table 9.2**), contains a single bone—the **humerus,** a typical long bone. Proximally, its rounded *head* fits into the shallow glenoid cavity of the scapula. The head is separated from the shaft by the **anatomical neck** and by the more constricted **surgical neck,** which is a common site of fracture.

The Forearm

Two bones, the radius and the ulna, compose the skeleton of the forearm, or antebrachium **(Figure 9.4)**. When the body is in the anatomical position, the **radius** is lateral, the ulna is medial, and the radius and ulna are parallel.

The Hand

The skeleton of the hand, or manus **(Figure 9.5)**, includes three groups of bones, those of the carpus, the metacarpals, and the phalanges.

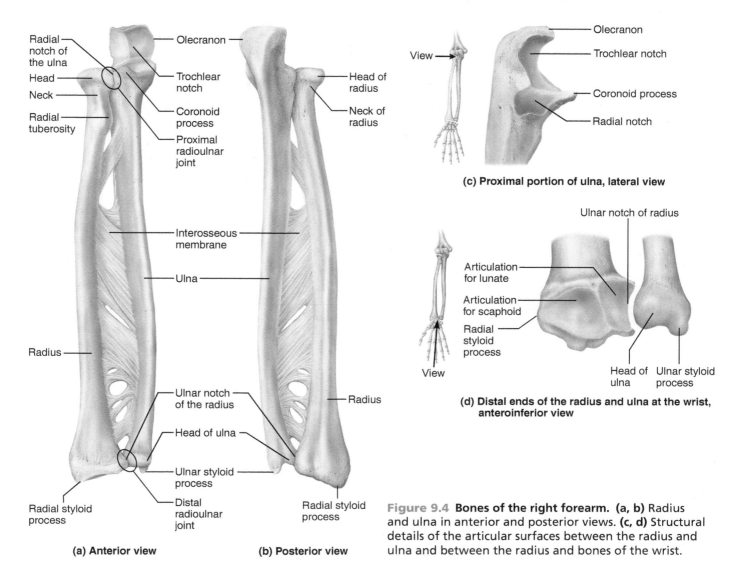

Figure 9.4 Bones of the right forearm. (a, b) Radius and ulna in anterior and posterior views. **(c, d)** Structural details of the articular surfaces between the radius and ulna and between the radius and bones of the wrist.

Table 9.2A	The Appendicular Skeleton: The Upper Limb (Figure 9.3)	
Bone	**Important markings**	**Description**
Humerus (Figure 9.3) (only bone of the arm)	Greater tubercle	Large lateral prominence; site of the attachment of rotator cuff muscles
	Lesser tubercle	Small medial prominence; site of attachment of rotator cuff muscles
	Intertubercular sulcus	A groove separating the greater and lesser tubercles; the tendon of the biceps brachii lies in this groove
	Deltoid tuberosity	A roughened area about midway down the shaft of the lateral humerus; site of attachment of the deltoid muscle
	Radial fossa	Small lateral depression; receives the head of the radius when the forearm is flexed
	Coronoid fossa	Small medial anterior depression; receives the coronoid process of the ulna when the forearm is flexed
	Capitulum	A rounded lateral condyle that articulates with the radius
	Trochlea	A flared medial condyle that articulates with the ulna
	Lateral epicondyle	Small condyle proximal to the capitulum
	Medial epicondyle	Rough condyle proximal to the trochlea
	Radial groove	Small posterior groove, marks the course of the radial nerve
	Olecranon fossa	Large distal posterior depression that accommodates the olecranon of the ulna

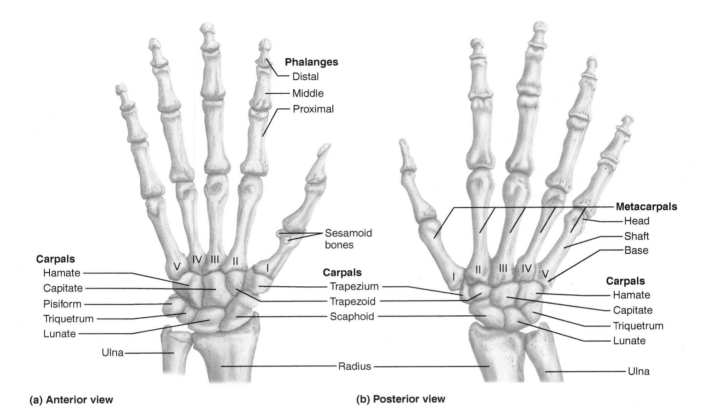

(a) Anterior view

(b) Posterior view

Figure 9.5 Bones of the right hand. (a) Anterior view and **(b)** posterior view, showing relationship of carpals, metacarpals, and phalanges.

The wrist is the proximal portion of the hand. It is referred to anatomically as the **carpus;** the eight marble-sized bones composing it are the **carpals.** The carpals are arranged in two irregular rows of four bones each (illustrated in Figure 9.5). In the proximal row (lateral to medial) are the *scaphoid, lunate, triquetrum,* and *pisiform bones;* the scaphoid and lunate articulate with the distal end of the radius. In the distal row (lateral to medial) are the *trapezium, trapezoid, capitate,* and

Table 9.2B	The Appendicular Skeleton: The Upper Limb (Figures 9.3 and 9.4)	
Bone	**Important markings**	**Description**
Radius (lateral bone of the forearm in the anatomical position) (Figures 9.4 and 9.3c, d)	Head	Proximal end of the radius that forms part of the proximal radioulnar joint and articulates with the capitulum of the humerus
	Radial tuberosity	Medial prominence just below the head of the radius; site of attachment of the biceps brachii
	Radial styloid process	Distal prominence; site of attachment for ligaments that travel to the wrist
	Ulnar notch	Small distal depression that accommodates the head of the ulna, forming the distal radioulnar joint
Ulna (medial bone of the forearm in the anatomical position) (Figures 9.4 and 9.3c, d)	Olecranon	Prominent process on the posterior proximal ulna; articulates with the olecranon fossa of the humerus when the forearm is extended
	Trochlear notch	Deep notch that separates the olecranon and the coronoid process; articulates with the trochlea of the humerus
	Coronoid process	Shaped like a point on a crown; articulates with the trochlea of the humerus
	Radial notch	Small proximal lateral notch that articulates with the head of the radius; forms part of the proximal radioulnar joint
	Head	Slim distal end of the ulna; forms part of the distal radioulnar joint
	Ulnar styloid process	Distal pointed projection; located medial to the head of the ulna

hamate. The carpals are bound closely together by ligaments, which restrict movements between them to allow only gliding movements.

The **metacarpals,** numbered I to V from the thumb side of the hand toward the little finger, radiate out from the wrist like spokes to form the palm of the hand. The *bases* of the metacarpals articulate with the carpals of the wrist; their more bulbous *heads* articulate with the phalanges of the fingers distally. When the fist is clenched, the heads of the metacarpals become prominent as the knuckles.

The 14 bones of the fingers, or digits, are miniature long bones, called **phalanges** (singular, *phalanx*) as noted above. Each finger contains three phalanges (proximal, middle, and distal) except the thumb *(pollex),* which has only two (proximal and distal).

ACTIVITY 2

Palpating the Surface Anatomy of the Pectoral Girdle and Upper Limb

Identify the following bone markings on the skin surface of the upper limb. It is usually preferable to palpate the bone markings on your lab partner because many of these markings can be seen only from the posterior aspect.

Place a check mark in the boxes as you locate the bone markings. Ask your instructor for help with any markings that you are unable to locate.

☐ Clavicle: Palpate the clavicle along its entire length from sternum to shoulder.

☐ Acromioclavicular (AC) joint: The high point of the shoulder, which represents the junction point between the clavicle and the acromion of the scapula.

☐ Spine of the scapula: Extend your arm at the shoulder so that your scapula moves posteriorly. As you do this, your scapular spine will be seen as a winglike protrusion on your posterior thorax and can be easily palpated by your lab partner.

☐ Lateral epicondyle of the humerus: After you have located the epicondyle, run your finger posteriorly into the hollow immediately dorsal to the epicondyle. This is the site where the extensor muscles of the forearm are attached and is a common site of the excruciating pain of tennis elbow.

☐ Medial epicondyle of the humerus: Feel this medial projection at the distal end of the humerus. The ulnar nerve, which runs behind the medial epicondyle, is responsible for the tingling, painful sensation felt when you hit your "funny bone."

☐ Olecranon of the ulna: Work your elbow—flexing and extending—as you palpate its posterior aspect to feel the olecranon moving into and out of the olecranon fossa on the posterior aspect of the humerus.

☐ Ulnar styloid process: With the hand in the anatomical position, feel out this small inferior projection on the medial aspect of the distal end of the ulna.

☐ Radial styloid process: Find this projection at the distal end of the radius (lateral aspect). It is most easily located by moving the hand medially at the wrist. Next, move your fingers just medially onto the anterior wrist. Press firmly, and then let up slightly on the pressure. You should be able to feel your pulse at this pressure point, which lies over the radial artery (radial pulse).

☐ Pisiform: Just distal to the ulnar styloid process, feel the rounded pealike pisiform bone.

☐ Metacarpophalangeal joints (knuckles): Clench your fist, and locate your knuckles—these are your metacarpophalangeal joints. ■

Bones of the Pelvic Girdle and Lower Limb

The Pelvic (Hip) Girdle

As with the bones of the pectoral girdle and upper limb, pay particular attention to bone markings needed to identify right and left bones.

The **pelvic girdle,** or **hip girdle** (**Figure 9.6** and **Table 9.3**), is formed by the two **hip bones** (also called the **ossa coxae,** or coxal bones) and the sacrum (part of the axial skeleton).* The deep structure formed by the hip bones, sacrum, and coccyx is called the **pelvis** or *bony pelvis.* In contrast to the bones of the shoulder girdle, those of the pelvic girdle are heavy and massive, and they attach securely to the axial skeleton. The sockets for the heads of the femurs (thigh bones) are deep and heavily reinforced by ligaments to ensure a stable, strong limb attachment. The ability to bear weight is more important here than mobility and flexibility. The combined weight of the upper body rests on the pelvic girdle.

Each hip bone is a result of the fusion of three bones—the **ilium, ischium,** and **pubis**—which are distinguishable in the young child.

The ilium, ischium, and pubis fuse at the deep socket called the **acetabulum** (literally, "vinegar cup"), which receives the head of the thigh bone. The rami of the pubis and ischium form a bar of bone enclosing the **obturator foramen** through which a few blood vessels pass.

ACTIVITY 3

Observing Pelvic Articulations

Examine an articulated pelvis. Notice how each hip bone articulates with the sacrum posteriorly and how the two hip bones join at the pubic symphysis. The sacroiliac joint is a common site of lower back problems because of the pressure it must bear. ■

Comparison of the Male and Female Pelves

The female pelvis reflects differences for childbearing—it is wider, shallower, lighter, and rounder than that of the male. Her pelvis must not only support the increasing size of a fetus, but also be large enough to allow the infant's head (its largest dimension) to descend through the birth canal at birth.

The **pelvic brim** is a continuous oval ridge of bone that runs along the pubic symphysis, pubic crests, arcuate lines, sacral alae, and sacral promontory. The **false pelvis** is that portion superior to the pelvic brim; it is bounded by the alae of the ilia laterally and the sacral promontory and lumbar vertebrae posteriorly. The false pelvis supports the abdominal viscera; it does not restrict childbirth in any way. The **true pelvis** is the region inferior to the pelvic brim that is almost

*Some authorities do not consider the sacrum part of the pelvic girdle, but here we follow the convention in *Terminologia Anatomica.*

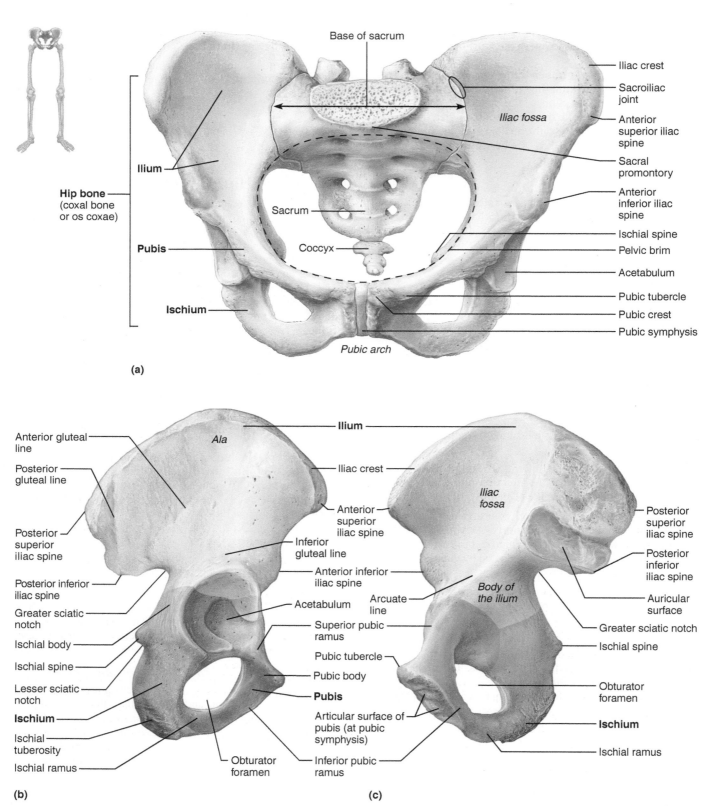

Figure 9.6 Bones of the pelvic girdle. (a) Articulated pelvis, showing the two hip bones, which together with the sacrum comprise the pelvic girdle, and the coccyx. **(b)** Right hip bone, lateral view, showing the point of fusion of the ilium, ischium, and pubis. **(c)** Right hip bone, medial view.

Table 9.3 The Appendicular Skeleton: The Pelvic (Hip) Girdle (Figure 9.6)

Bone	Important markings	Description
Ilium (Figure 9.6)	Iliac crest	Thick superior margin of bone
	Anterior superior iliac spine	The blunt anterior end of the iliac crest
	Posterior superior iliac spine	The sharp posterior end of the iliac crest
	Anterior inferior iliac spine	Small projection located just below the anterior superior iliac spine
	Posterior inferior iliac spine	Small projection located just below the posterior superior iliac spine
	Greater sciatic notch	Deep notch located inferior to the posterior inferior iliac spine; allows the sciatic nerve to enter the thigh
	Iliac fossa	Shallow depression below the iliac crest; forms the internal surface of the ilium
	Auricular surface	Rough medial surface that articulates with the auricular surface of the sacrum, forming the sacroiliac joint
	Arcuate line	A ridge of bone that runs inferiorly and anteriorly from the auricular surface
Ischium ("sit-down" bone) (Figure 9.6)	Ischial tuberosity	Rough projection that receives the weight of our body when we are sitting
	Ischial spine	Located superior to the ischial tuberosity and projects medially into the pelvic cavity
	Lesser sciatic notch	A small notch located inferior to the ischial spine
	Ischial ramus	Narrow portion of the bone that articulates with the pubis
Pubis (Figure 9.6)	Superior pubic ramus	Superior extension of the body of the pubis
	Inferior pubic ramus	Inferior extension of the body of the pubis; articulates with the ischium
	Pubic crest	Thick anterior border
	Pubic tubercle	Lateral end of the pubic crest; lateral attachment for the inguinal ligament
	Articular surface	Surface of each pubis that combines with fibrocartilage to form the pubic symphysis

entirely surrounded by bone. Its posterior boundary is formed by the sacrum. The ilia, ischia, and pubic bones define its limits laterally and anteriorly.

The **pelvic inlet** is the opening delineated by the pelvic brim. The widest dimension of the pelvic inlet is from left to right, that is, along the frontal plane. The **pelvic outlet** is the inferior margin of the true pelvis. It is bounded anteriorly by the pubic arch, laterally by the ischia, and posteriorly by the sacrum and coccyx. Because both the coccyx and the ischial spines protrude into the outlet opening, a sharply angled coccyx or large, sharp ischial spines can dramatically narrow the outlet. The largest dimension of the outlet is the anterior-posterior diameter.

The major differences between the male and female pelves are summarized in (Table 9.4).

ACTIVITY 4

Comparing Male and Female Pelves

Examine male and female pelves for the following differences:

• The female inlet is larger and more circular.

• The female pelvis as a whole is shallower, and the bones are lighter and thinner.

• The female sacrum is broader and less curved, and the pubic arch is more rounded.

• The female acetabula are smaller and farther apart, and the ilia flare more laterally.

• The female ischial spines are shorter, farther apart, and everted, thus enlarging the pelvic outlet. ▬

The Thigh

The **femur,** or thigh bone (Figure 9.7a and Table 9.5A), is the only bone of the thigh. It is the heaviest, strongest bone in the body. The ball-like head of the femur articulates with the hip bone via the deep, secure socket of the acetabulum.

The femur angles medially as it runs downward to the leg bones; this brings the knees in line with the body's center of gravity. The medial course of the femur is more noticeable in females because of the wider female pelvis.

The **patella** (Figure 9.7a) is a triangular sesamoid bone enclosed in the (quadriceps) tendon that secures the anterior thigh muscles to the tibia.

The Leg

Two bones, the tibia and the fibula, form the skeleton of the leg (Figure 9.8 and Table 9.5B). The **tibia,** or *shinbone,* is the larger, medial, weight-bearing bone of the leg.

The **fibula,** which lies parallel to the tibia, takes no part in forming the knee joint. Its proximal head articulates with the lateral condyle of the tibia.

Text continues on page 145.

Table 9.4   Comparison of the Male and Female Pelves

Characteristic	Female	Male
General structure and functional modifications	Tilted forward; adapted for childbearing; true pelvis defines the birth canal; cavity of the true pelvis is broad, shallow, and has a greater capacity	Tilted less forward; adapted for support of a male's heavier build and stronger muscles; cavity of the true pelvis is narrow and deep
Bone thickness	Bones lighter, thinner, and smoother	Bones heavier and thicker, and markings are more prominent
Acetabula	Smaller; farther apart	Larger; closer together
Pubic arch	Broader angle (80°–90°); more rounded	Angle is more acute (50°–60°)
Anterior view		
Sacrum	Wider; shorter; sacrum is less curved	Narrow; longer; sacral promontory projects anteriorly
Coccyx	More movable; straighter, projects inferiorly	Less movable; curves and projects anteriorly
Left lateral view		
Pelvic inlet	Wider; oval from side to side	Narrow; basically heart-shaped
Pelvic outlet	Wider; ischial spines shorter, farther apart, and everted	Narrower; ischial spines longer, sharper, and point more medially
Posteroinferior view		

Pelvic brim

Pubic arch

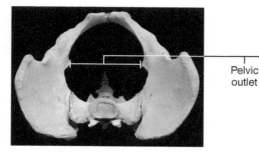
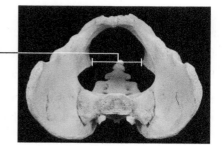

Pelvic outlet

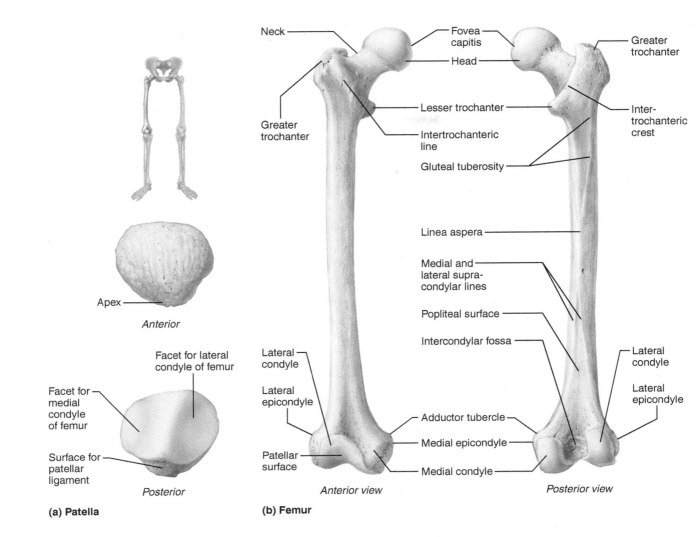

Figure 9.7 **Bones of the right knee and thigh. (a)** The patella (kneecap). **(b)** The femur (thigh bone).

Table 9.5A	The Appendicular Skeleton: The Lower Limb (Figure 9.7)	
Bone	**Important markings**	**Description**
Femur (thigh bone) (Figure 9.7)	Fovea capitis	A small pit in the head of the femur for the attachment of a short ligament that runs to the acetabulum
	Neck	Weakest part of the femur, the usual fracture site of a "broken hip"
	Greater trochanter	Large lateral projection; serves as site for muscle attachment on the proximal femur
	Lesser trochanter	Large posteromedial projection; serves as site for muscle attachment on the proximal femur
	Intertrochanteric line	Thin ridge of bone that connects the two trochanters anteriorly
	Intertrochanteric crest	Prominent ridge of bone that connects the two trochanters posteriorly
	Gluteal tuberosity	Thin ridge of bone located posteriorly; serves as a site for muscle attachment on the proximal femur
	Linea aspera	Long vertical ridge of bone on the posterior shaft of the femur
	Medial and lateral supracondylar lines	Two lines that diverge from the linea aspera and travel to their respective condyles
	Medial and lateral condyles	Distal "wheel-shaped" projections that articulate with the tibia, each condyle has a corresponding epicondyle
	Intercondylar fossa	Deep depression located between the condyles and beneath the popliteal surface
	Adductor tubercle	A small bump on the superior portion of the medial epicondyle; attachment site for the large adductor magnus muscle
	Patellar surface	Smooth distal anterior surface between the condyles; articulates with the patella

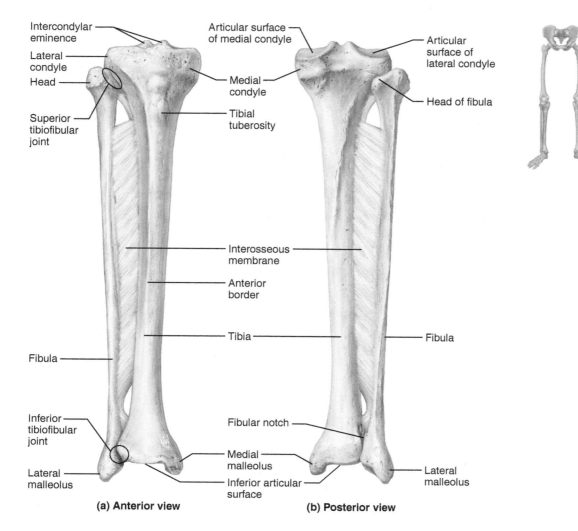

Figure 9.8 Bones of the right leg. Tibia and fibula, **(a)** anterior view, **(b)** posterior view.

Table 9.5B	The Appendicular Skeleton: The Lower Limb (Figure 9.8)	
Bone	**Important markings**	**Description**
Tibia (shin bone, medial bone of the leg) (Figure 9.8)	Lateral condyle	Slightly concave surface that articulates with the lateral condyle of the femur; the inferior region of this condyle articulates with the fibula to form the superior tibiofibular joint
	Medial condyle	Slightly concave surface that articulates with the medial condyle of the femur
	Intercondylar eminence	Irregular projection located between the two condyles
	Tibial tuberosity	Roughened anterior surface; site of patellar ligament attachment
	Anterior border	Sharp ridge of bone easily palpated because it is close to the surface
	Medial malleolus	Forms the medial bulge of the ankle
	Inferior articular surface	Distal surface of the tibia that articulates with the talus
	Fibular notch	Smooth lateral surface that articulates with the fibula to form the inferior tibiofibular joint
Fibula (lateral bone of the leg) (Figure 9.8)	Head	Proximal end of the fibula that articulates with the tibia to form the superior tibiofibular joint
	Lateral malleolus	Forms the lateral bulge of the ankle and articulates with the talus

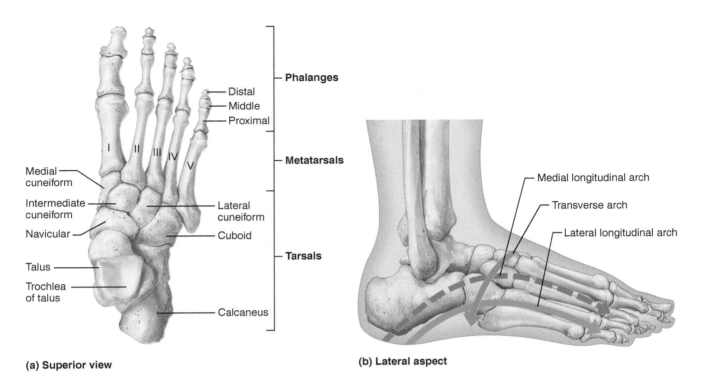

Figure 9.9 Bones of the right foot. Arches of the right foot are diagrammed in (b).

The Foot

The bones of the foot include the 7 **tarsal** bones, 5 **metatarsals,** which form the instep, and 14 **phalanges,** which form the toes **(Figure 9.9)**. Body weight is concentrated on the two largest tarsals, which form the posterior aspect of the foot, the larger *calcaneus* (heel bone) and the *talus,* which lies between the tibia and the calcaneus. The other tarsals are named and identified in the figure (Figure 9.9). The metatarsals are numbered I through V, medial to lateral. Like the fingers of the hand, each toe has three phalanges except the great toe, which has two.

The bones in the foot are arranged to produce three strong arches—two longitudinal arches (medial and lateral) and one transverse arch (Figure 9.9b). Ligaments, binding the foot bones together, and tendons of the foot muscles hold the bones firmly in the arched position but still allow a certain degree of give. Weakened tendons and ligaments supporting the arches are referred to as *fallen arches* or *flat feet.*

ACTIVITY 5

Palpating the Surface Anatomy of the Pelvic Girdle and Lower Limb

Locate and palpate the following bone markings on yourself and/or your lab partner.

Place a check mark in the boxes as you locate the bone markings. Ask your instructor for help with any markings that you are unable to locate.

☐ Iliac crest and anterior superior iliac spine: Rest your hands on your hips—they will be overlying the iliac crests. Trace the crest as far posteriorly as you can, and then follow it anteriorly to the anterior superior iliac spine. This latter bone marking is clearly visible through the skin of very slim people.

☐ Greater trochanter of the femur: This is easier to locate in females than in males because of the wider female pelvis. Try to locate it on yourself as the most lateral point of the proximal femur. It typically lies about 6 to 8 inches below the iliac crest.

☐ Patella and tibial tuberosity: Feel your kneecap, and palpate the ligaments attached to its borders. Follow the inferior patellar ligament to the tibial tuberosity.

☐ Medial and lateral condyles of the femur and tibia: As you move from the patella inferiorly on the medial (and then the lateral) knee surface, you will feel first the femoral and then the tibial condyle.

☐ Medial malleolus: Feel the medial protrusion of your ankle, the medial malleolus of the distal tibia.

☐ Lateral malleolus: Feel the bulge of the lateral aspect of your ankle, the lateral malleolus of the fibula.

☐ Calcaneus: Attempt to follow the extent of your calcaneus, or heel bone. ■

ACTIVITY 6

Constructing a Skeleton

1. When you finish examining yourself and the disarticulated bones of the appendicular skeleton, work with your lab partner to arrange the unlabeled disarticulated bones on the laboratory bench in their proper relative positions to form an entire skeleton. Careful observation of bone markings should help you distinguish between right and left members of bone pairs.

2. When you believe that you have accomplished this task correctly, ask the instructor to check your arrangement. If it is not correct, go to the articulated skeleton and check your bone arrangements. Also review the descriptions of the bone markings as necessary to correct your bone arrangement. ■

The Appendicular Skeleton

Bones of the Pectoral Girdle and Upper Limb

1. Fill in the blanks to complete the statements below:

   a. The bones that form the pectoral girdle are the _____ and _____.

   b. The upper limb is formed by the arm bone, the _____, and the two bones of the forearm, the

       _____ and _____.

   c. The _____ are the wrist bones. List the proximal row of wrist bones from lateral to medial: _____

       _____.

   List the distal row of wrist bones from lateral to medial: _____

   _____.

   d. The _____ form the palm of the hand, and the heads of these bones form the knuckles.

   e. A single finger bone is called a _____. Each hand has _____ finger bones, called _____.

2. Match the bone markings in Column B with the descriptions in Column A.

   Column A   **Column B**

   _____ 1. depression in the scapula that articulates with the humerus   a. acromion

   _____ 2. surface on the radius that receives the head of the ulna   b. capitulum

   _____ 3. lateral rounded knob on the distal humerus   c. coracoid process

   _____ 4. posterior depression on the distal humerus   d. coronoid fossa

   _____ 5. a roughened area on the lateral humerus: deltoid attachment site   e. deltoid tuberosity

   _____ 6. hooklike process; biceps brachii attachment site   f. glenoid cavity

   _____ 7. surface on the ulna that receives the head of the radius   g. medial epicondyle

   _____ 8. medial condyle of the humerus that articulates with the ulna   h. olecranon fossa

   _____ 9. lateral end of the spine of the scapula; clavicle articulation site   i. radial notch

   _____ 10. small bump on the humerus, often called the "funny bone"   j. trochlea

   _____ 11. anterior depression, superior to the trochlea, that receives part of   k. ulnar notch
                             the ulna when bending at the elbow

3. Using terms from the list at the right, identify the anatomical landmarks and regions of the scapula.

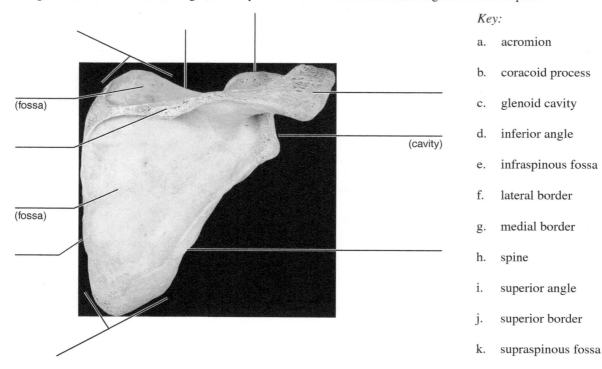

(fossa)

(fossa)

(cavity)

Key:

a. acromion

b. coracoid process

c. glenoid cavity

d. inferior angle

e. infraspinous fossa

f. lateral border

g. medial border

h. spine

i. superior angle

j. superior border

k. supraspinous fossa

4. Match the terms in the key with the appropriate leader lines on the photograph of the humerus.

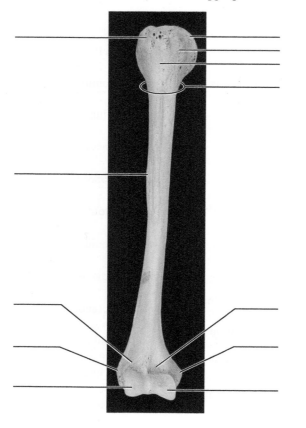

Key:

a. capitulum

b. coronoid fossa

c. deltoid tuberosity

d. greater tubercle

e. head

f. intertubercular sulcus

g. lateral epicondyle

h. lesser tubercle

i. medial epicondyle

j. radial fossa

k. surgical neck

l. trochlea

5. Match the terms in the key with the appropriate leader lines on the photographs of the posterior view of the radius on the left and the lateral view of the ulna on the right.

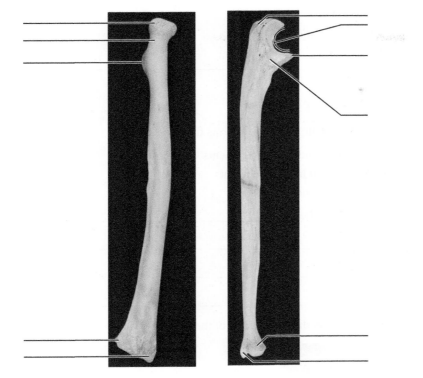

Key:

a. coronoid process

b. head of the radius

c. head of the ulna

d. neck of the radius

e. olecranon

f. radial notch of the ulna

g. radial styloid process

h. radial tuberosity

i. trochlear notch

j. ulnar notch of the radius

k. ulnar styloid process

6. Match the terms in the key with the appropriate leader lines on the photograph of the anterior view of the hand.

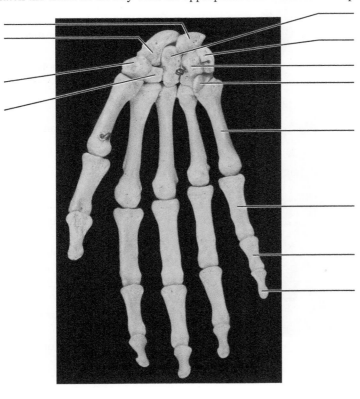

Key:

a. capitate

b. distal phalanx

c. hamate

d. lunate

e. metacarpal

f. middle phalanx

g. pisiform

h. proximal phalanx

i. scaphoid

j. trapezium

k. trapezoid

l. triquetrum

7. Name the two bone markings that form the proximal radioulnar joint.

_____

8. Name the two bone markings that form the distal radioulnar joint.

_____

Bones of the Pelvic Girdle and Lower Limb

9. Compare the pectoral and pelvic girdles by choosing appropriate descriptive terms from the key.

Key: a. flexibility most important   d. insecure axial and limb attachments

  b. massive   e. secure axial and limb attachments

  c. lightweight   f. weight-bearing most important

Pectoral: _____, _____, _____ Pelvic: _____, _____, _____

10. Distinguish between the true pelvis and the false pelvis. _____

_____

_____

11. Match the terms in the key with the appropriate leader lines on the photograph of the lateral view of the hip bone.

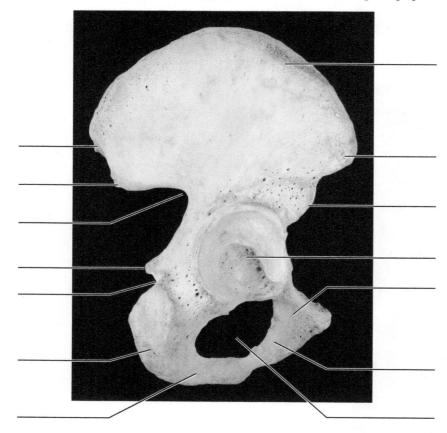

Key:

a.   acetabulum

b.   anterior inferior iliac spine

c.   anterior superior iliac spine

d.   greater sciatic notch

e.   iliac crest

f.   inferior pubic ramus

g.   ischial ramus

h.   ischial spine

i.   ischial tuberosity

j.   lesser sciatic notch

k.   obturator foramen

l.   posterior inferior iliac spine

m. posterior superior iliac spine

n.   superior pubic ramus

12. Match the bone names and markings in column B with the descriptions in column A. The items in column B may be used more than once.

Column A

Column B

_____, _____, and

a. acetabulum

_____ 1. fuse to form the hip bone

b. calcaneus

_____ 2. rough projection that supports body weight when sitting

c. femur

_____ 3. point where the hip bones join anteriorly

d. fibula

_____ 4. superiormost margin of the hip bone

e. gluteal tuberosity

_____ 5. deep socket in the hip bone that receives the head of the thigh bone

f. greater and lesser trochanters

g. greater sciatic notch

_____ 6. joint between axial skeleton and pelvic girdle

h. iliac crest

_____ 7. longest, strongest bone in body

i. ilium

_____ 8. thin lateral leg bone

_____ 9. permits passage of the sciatic nerve

j. ischial tuberosity

_____ 10. notch located inferior to the ischial spine

k. ischium

_____ 11. point where the patellar ligament attaches

l. lateral malleolus

_____ 12. kneecap

m. lesser sciatic notch

_____ 13. shinbone

n. medial malleolus

_____ 14. medial ankle projection

o. metatarsals

_____ 15. lateral ankle projection

p. obturator foramen

_____ 16. largest tarsal bone

q. patella

_____ 17. ankle bones

r. pubic symphysis

_____ 18. bones forming the instep of the foot

s. pubis

_____ 19. opening in hip bone formed by the pubic and ischial rami

t. sacroiliac joint

u. talus

_____ and _____ 20. sites of muscle attachment on the proximal femur

v. tarsals

_____ 21. tarsal bone that "sits" on the calcaneus

w. tibia

_____ 22. weight-bearing bone of the leg

x. tibial tuberosity

_____ 23. tarsal bone that articulates with the tibia

13. Match the terms in the key with the appropriate leader lines on the photograph of the anterior view of the femur.

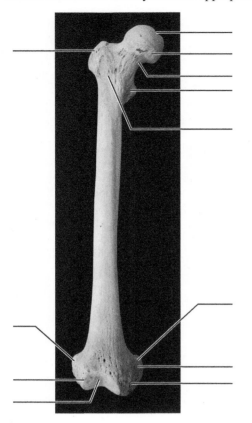

Key:

a. adductor tubercle

b. fovea capitis

c. greater trochanter

d. head

e. intertrochanteric line

f. lateral condyle

g. lateral epicondyle

h. lesser trochanter

i. medial condyle

j. medial epicondyle

k. neck

l. patellar surface

14. Match the terms in the key with the appropriate leader lines on the photograph of the anterior view of the tibia.

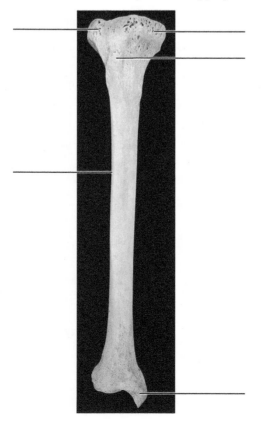

Key:

a. anterior border

b. lateral condyle

c. medial condyle

d. medial malleolus

e. tibial tuberosity

15. Match the terms in the key with the appropriate leader lines on the photograph of the posterior view of the articulated tibia and fibula.

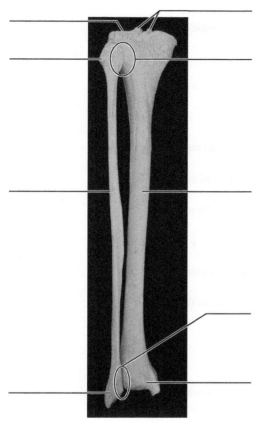

Key:

a. articular surface of the lateral condyle

b. head of the fibula

c. inferior tibiofibular joint

d. intercondylar eminence

e. lateral malleolus

f. medial malleolus

g. shaft of the fibula

h. shaft of the tibia

i. superior tibiofibular joint

16. Are the bones of the leg shown above from the left or from the right leg? _____

Explain how you can tell which side of the body they are from. _____

_____

17. Match the terms in the key with the appropriate leader lines on the photograph of the superior view of the articulated foot.

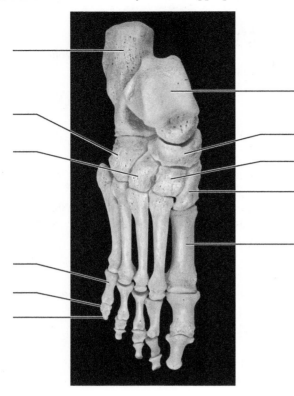

Key:

a. calcaneus

b. cuboid

c. distal phalanx

d. intermediate cuneiform

e. lateral cuneiform

f. medial cuneiform

g. metatarsal

h. middle phalanx

i. navicular

j. proximal phalanx

k. talus

18. ✚ FOOSH is an acronym that stands for Fall on Outstretched Hand. Discuss possible fractures and dislocations that might occur with an injury of this type.

_____

_____

19. ✚ Describe some of the features of the female pelvis that provide for compatibility with vaginal birth. _____

_____

_____

20. ✚ Your X-ray exam reveals that you have fractured your fibula. Your physician remarks, "Well, it's better than breaking

your tibia." Explain why a fracture of the tibia would be worse than a fracture of the fibula. _____

_____

Articulations and Body Movements

MATERIALS

- Skull
- Articulated skeleton
- Anatomical chart of joint types (if available)
- Diarthrotic joint (fresh or preserved), preferably a beef knee joint sectioned sagittally; alternatively, pig's feet with phalanges sectioned frontally
- Disposable gloves
- Water balloons and clamps
- Functional models of hip, knee, and shoulder joints (if available)
- X-ray images of normal and arthritic joints (if available)

LEARNING OUTCOMES

- ☐ Name and describe the three functional categories of joints.
- ☐ Name and describe the three structural categories of joints, and discuss how their structure is related to mobility.
- ☐ Identify the types of synovial joints; indicate whether they are nonaxial, uniaxial, biaxial, or multiaxial; and describe the movements each type makes.
- ☐ Define *origin* and *insertion* of muscles.
- ☐ Demonstrate or describe the various body movements.
- ☐ Compare and contrast the structure and function of the shoulder and hip joints.
- ☐ Describe the structure and function of the knee and temporomandibular joints.

PRE-LAB QUIZ

1. Name one of the two functions of an articulation, or joint. _____
   _____
2. The functional classification of joints is based on:
   a. a joint cavity
   b. amount of connective tissue
   c. amount of movement allowed by the joint
3. Structural classification of joints includes fibrous, cartilaginous, and _____, which have a fluid-filled cavity between articulating bones.
4. Circle the correct underlined term. Sutures, whose irregular edges of bone are joined by short fibers of connective tissue, are an example of <u>fibrous</u> / <u>cartilaginous</u> joints.
5. Circle True or False. All synovial joints are diarthroses, or freely movable joints.
6. Circle the correct underlined term. Every muscle of the body is attached to a bone or other connective tissue structure at two points. The origin / <u>insertion</u> is the more movable attachment.
7. The hip joint is an example of a _____ synovial joint.
   a. ball-and-socket   c. pivot
   b. hinge   d. plane
8. Movement of a limb away from the midline or median plane of the body in the frontal plane is known as:
   a. abduction   c. extension
   b. eversion   d. rotation
9. Circle the correct underlined term. This type of movement is common in ball-and-socket joints and can be described as the movement of a bone around its longitudinal axis. It is <u>rotation</u> / <u>flexion</u>.
10. Circle True or False. The knee joint is the most freely movable joint in the body.

With rare exceptions, every bone in the body is connected to, or forms a joint with, at least one other bone. **Articulations,** or joints, perform two functions for the body. They (1) hold the bones together and (2) allow the rigid skeletal system some flexibility so that gross body movements can occur.

Classification of Joints

Joints are classified structurally and functionally. The structural classification is based on the type of connective tissue between the bones and the presence or absence of a joint cavity between the articulating bones. Structurally, there are *fibrous, cartilaginous,* and *synovial joints.*

The functional classification focuses on the amount of movement allowed at the joint. On this basis, there are **synarthroses,** or immovable joints; **amphiarthroses,** or slightly movable joints; and **diarthroses,** or freely movable joints.

The structural and functional classifications of joints are summarized in **Table 10.1.** A sample of the types of joints is shown in **Figure 10.1.**

ACTIVITY 1

Identifying Fibrous Joints

Examine a human skull again. Notice that adjacent bone surfaces do not actually touch but are separated by fibrous connective tissue. ▬▬

ACTIVITY 2

Identifying Cartilaginous Joints

Identify the cartilaginous joints on a human skeleton, in Table 10.2, and on an anatomical chart of joint types. View an X-ray image of the cartilaginous growth plate (epiphyseal plate) of a child's bone, if one is available. If an X-ray image is not available, refer to **Figure 10.2.** ▬▬

Synovial Joints

Synovial joints are joints in which the articulating bone ends are separated by a joint cavity containing synovial fluid (see Figure 10.1f–h). All synovial joints are diarthroses, or freely movable joints. Most joints in the body are synovial joints.

Synovial joints typically have the following structural characteristics (**Figure 10.3**):

• **Joint (articular) cavity:** A space between the articulating bones that contains a small amount of synovial fluid.

• **Articular cartilage:** Hyaline cartilage that covers the surfaces of the bones forming the joint.

• **Articular capsule:** Two layers that enclose the joint cavity. The tough external layer is the *fibrous layer* composed of dense irregular connective tissue. The inner layer is the *synovial membrane* composed of loose connective tissue.

• **Synovial fluid:** A viscous fluid, the consistency of egg whites, located in the joint cavity and the articular cartilage. This fluid acts as a lubricant, reducing friction.

Table 10.1   Summary of Joint Classifications (Figure 10.1)

Structural class	Structural characteristics	Structural types	Examples	Functional classification
Fibrous	Adjoining bones connected by dense fibrous connective tissue; no joint cavity	Suture (short fibers)	Squamous suture between the parietal and temporal bones	Synarthrosis (immovable)
		Syndesmosis (longer fibers)	Between the tibia and fibula	Amphiarthrosis (slightly movable and immovable)
		Gomphosis (periodontal ligament)	Tooth in a bony socket (Figure 27.12, page 472)	Synarthrosis (immovable)
Cartilaginous	Adjoining bones united by cartilage; no joint cavity	Synchondrosis (hyaline cartilage)	Between the costal cartilage of rib 1 and the sternum and the epiphyseal plate in growing long bones (Figure 7.4, page 98)	Synarthrosis (immovable)
		Symphysis (fibrocartilage)	Intervertebral discs between adjacent vertebrae and the anterior connection between the pubic bones	Amphiarthrosis (slightly movable)
Synovial	Adjoining bones covered in articular cartilage; separated by a joint cavity and enclosed in an articular capsule lined with a synovial membrane	Plane joint	Between the carpals of the wrist	Diarthrosis (freely movable)
		Hinge joint	Elbow joint	
		Pivot joint	Proximal radioulnar joint	
		Condylar joint	Between the metacarpals and the proximal phalanx	
		Saddle joint	Between the trapezium (carpal) and metatarsal I	
		Ball-and-socket joint	Shoulder joint	

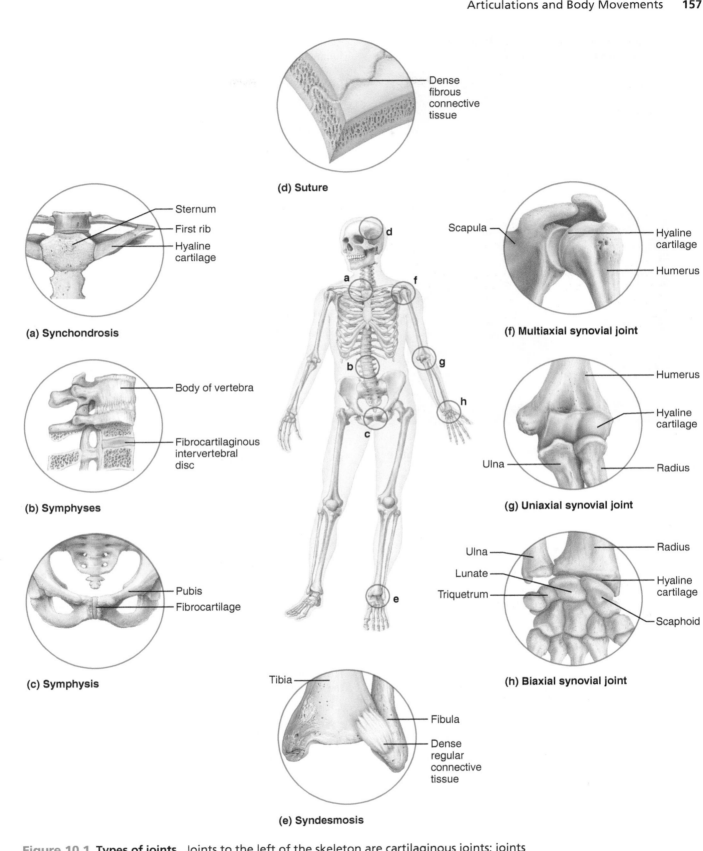

Figure 10.1 **Types of joints.** Joints to the left of the skeleton are cartilaginous joints; joints above and below the skeleton are fibrous joints; joints to the right of the skeleton are synovial joints. **(a)** Joint between costal cartilage of rib 1 and the sternum. **(b)** Intervertebral discs of fibrocartilage connecting adjacent vertebrae. **(c)** Fibrocartilaginous pubic symphysis connecting the pubic bones anteriorly. **(d)** Dense fibrous connective tissue connecting interlocking skull bones. **(e)** Ligament connecting the inferior ends of the tibia and fibula. **(f)** Shoulder joint. **(g)** Elbow joint. **(h)** Wrist joint.

158

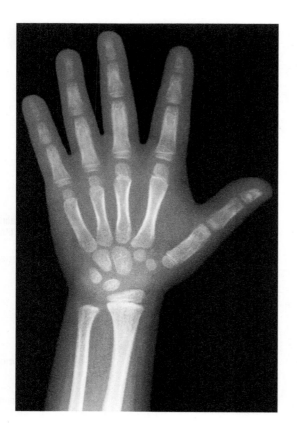

Figure 10.2 X-ray image of the hand of a child. Notice the cartilaginous epiphyseal plates, examples of temporary synchondroses.

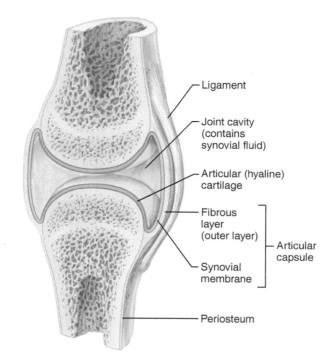

Figure 10.3 Major structural features of a synovial joint.

- **Reinforcing ligaments:** Some joints are reinforced by ligaments. Most ligaments are **capsular ligaments,** which are thickenings of the fibrous layer of the articular capsule. Some ligaments are **extracapsular,** found outside the articular capsule. **Intracapsular ligaments** are found within the articular capsule.

- **Nerves and blood vessels:** Sensory nerve fibers detect pain and joint stretching. Most of the blood vessels supply the synovial membrane.

- **Articular discs:** Pads composed of fibrocartilage (menisci) may also be present to minimize wear and tear on the bone surfaces.

- **Bursa and tendon sheath:** Sac filled with synovial fluid that reduces friction where ligaments, muscles, skin, tendons, and bones rub together. A tendon sheath is an elongated bursa that wraps around a tendon like a bun around a hot dog.

ACTIVITY 3

Examining Synovial Joint Structure

Examine a beef or pig joint to identify the general structural features of synovial joints in the previous list.

⚠ If the joint is fresh and you will be handling it, don disposable gloves before beginning your observations. ▪

ACTIVITY 4

Demonstrating the Importance of Friction-Reducing Structures

1. Obtain a small water balloon and clamp. Partially fill the balloon with water (it should still be flaccid), and clamp it closed.

2. Position the balloon atop one of your fists, and press down on its top surface with the other fist. Push on the balloon until your two fists touch, and move your fists back and forth over one another. Assess the amount of friction generated.

3. Unclamp the balloon, and add more water. The goal is to get just enough water in the balloon so that your fists cannot come into contact with one another, but instead remain separated by a thin water layer when pressure is applied to the balloon.

4. Repeat the movements in step 2 to assess the amount of friction generated.

How does the presence of a cavity containing fluid influence the amount of friction generated?

_____

What anatomical structure(s) does the water-containing balloon mimic?

_____ ▪

Types of Synovial Joints

The many types of synovial joints can be subdivided according to their function and structure. The shape of the articular surfaces determines the types of movements that can occur at the joint, and it also determines the structural classification of the joints (**Table 10.2** and **Table 10.3**, pages 159, 167–168 and **Figure 10.4**).

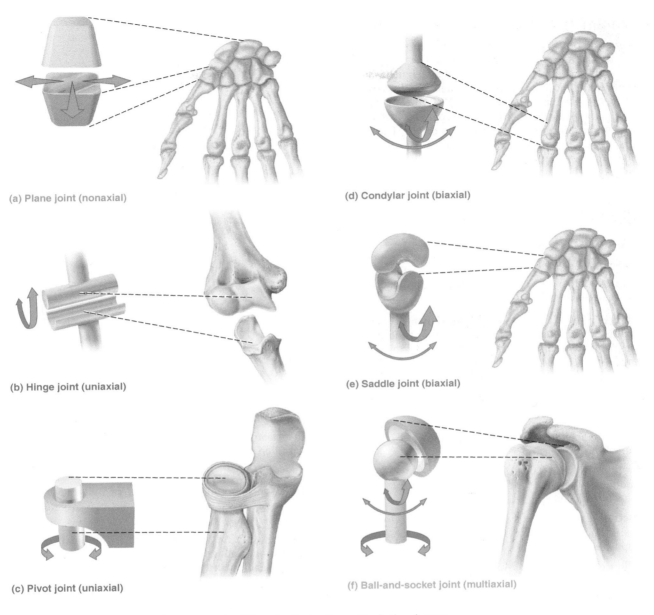

(a) Plane joint (nonaxial)

(b) Hinge joint (uniaxial)

(c) Pivot joint (uniaxial)

(d) Condylar joint (biaxial)

(e) Saddle joint (biaxial)

(f) Ball-and-socket joint (multiaxial)

Figure 10.4 Types of synovial joints. Dashed lines indicate the articulating bones.

Table 10.2	Types of Synovial Joints (Figure 10.4)		
Synovial joint	**Description of articulating surfaces**	**Movement**	**Examples**
Plane	Flat or slightly curved bones	Nonaxial: gliding	Intertarsal, intercarpal joints
Hinge	A rounded or cylindrical bone fits into a concave surface on the other bone	Uniaxial: flexion and extension	Elbow, interphalangeal joints
Pivot	A rounded bone fits into a sleeve (a concave bone plus a ligament)	Uniaxial: rotation	Proximal radioulnar, atlantoaxial joints
Condylar	An oval condyle fits into an oval depression on the other bone	Biaxial: flexion, extension, adduction, and abduction	Metacarpophalangeal (knuckle) and radiocarpal joints
Saddle	Articulating surfaces are saddle shaped; one surface is concave, the other surface is convex	Biaxial: flexion, extension, adduction, and abduction	Carpometacarpal joint of the thumb
Ball-and-socket	The ball-shaped head of one bone fits into the cuplike depression of the other bone	Multiaxial: flexion, extension, adduction, abduction, and rotation	Shoulder, hip joints

Movements Allowed by Synovial Joints

Every muscle of the body is attached to bone (or other connective tissue structures) at two points. The **origin** is the stationary, immovable, or less movable attachment, and the **insertion** is the more movable attachment. Body movement occurs when muscles contract across diarthrotic synovial joints (**Figure 10.5**). When the muscle contracts and its fibers shorten, the insertion moves toward the origin. The type of movement depends on the construction of the joint and on the placement of the muscle relative to the joint. The most common types of body movements are described below and illustrated in **Figure 10.6**.

ACTIVITY 5

Demonstrating Movements of Synovial Joints

Attempt to demonstrate each movement as you read through the following material:

Flexion (Figure 10.6a, b, and c): A movement, generally in the sagittal plane, that decreases the angle of the joint and reduces the distance between the two bones. Flexion is typical of hinge joints (bending the knee or elbow) but is also common at ball-and-socket joints (bending forward at the hip).

Extension (Figure 10.6a, b, and c): A movement that increases the angle of a joint and the distance between two bones or parts of the body; the opposite of flexion. If extension proceeds beyond anatomical position (bends the trunk backward), it is called *hyperextension.*

Abduction (Figure 10.6d): Movement of a limb away from the midline of the body, or the fanning movement of fingers or toes when they are spread apart.

Adduction (Figure 10.6d): Movement of a limb toward the midline of the body or drawing the fingers or toes together; the opposite of abduction.

Rotation (Figure 10.6e): Movement of a bone around its longitudinal axis without lateral or medial displacement.

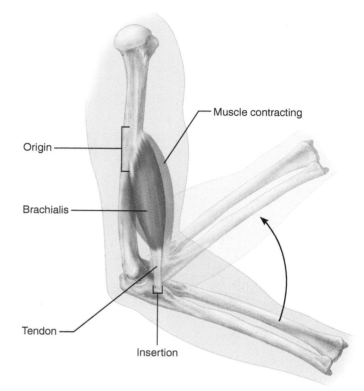

Figure 10.5 Muscle attachments (origin and insertion). When a skeletal muscle contracts, its insertion is pulled toward its origin.

Rotation, a common movement of ball-and-socket joints, also describes the movement of the atlas around the dens of the axis.

Circumduction (Figure 10.6d): A combination of flexion, extension, abduction, and adduction, commonly observed in ball-and-socket joints such as the shoulder. The limb as a whole outlines a cone.

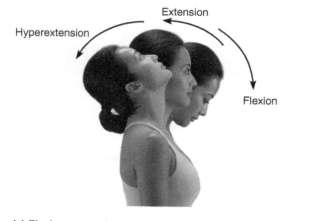

(a) Flexion, extension, and hyperextension of the neck

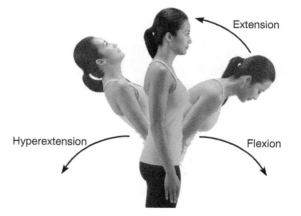

(b) Flexion, extension, and hyperextension of the vertebral column

Figure 10.6 Movements occurring at synovial joints of the body. The axis around which each movement occurs is indicated by a red dot in (c) and (d) and a red bar in (e).

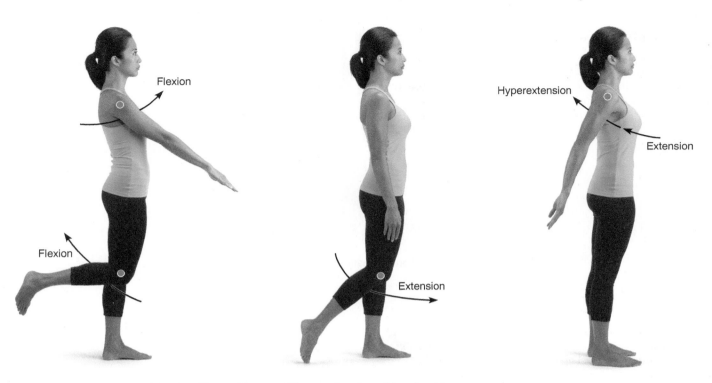

(c) Flexion and extension at the shoulder and knee, and hyperextension of the shoulder

(d) Abduction, adduction, and circumduction of the upper limb at the shoulder

(e) Rotation of the head and lower limb

Figure 10.6 *(continued)*

Pronation (Figure 10.6f): Movement of the palm of the hand from an anterior or upward-facing position to a posterior or downward-facing position. The distal end of the radius rotates over the ulna so that the bones form an **X** with pronation of the forearm.

Supination (Figure 10.6f): Movement of the palm from a posterior position to an anterior position (the anatomical position); the opposite of pronation. During supination, the radius and ulna are parallel.

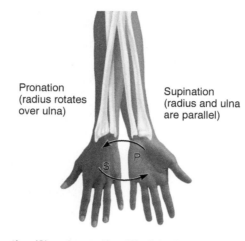

Pronation
(radius rotates
over ulna)

Supination
(radius and ulna
are parallel)

(f) Supination (S) and pronation (P) of the forearm

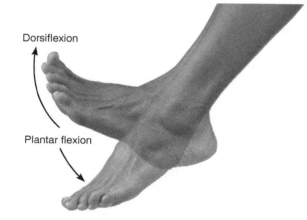

Dorsiflexion

Plantar flexion

(g) Dorsiflexion and plantar flexion of the foot

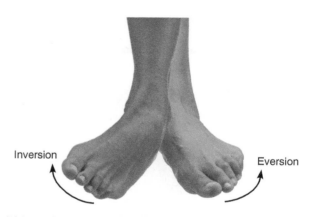

Inversion

Eversion

(h) Inversion and eversion of the foot

Figure 10.6 *(continued)* **Movements occurring at synovial joints of the body.**

The last four terms refer to movements of the foot:

Dorsiflexion (Figure 10.6g): A movement of the ankle joint that lifts the foot so that its superior surface approaches the shin.

Plantar flexion (Figure 10.6g): A movement of the ankle joint in which the foot is flexed downward as in standing on one's toes or pointing the toes.

Inversion (Figure 10.6h): A movement that turns the sole of the foot medially.

Eversion (Figure 10.6h): A movement that turns the sole of the foot laterally; the opposite of inversion. ■

Selected Synovial Joints

The Hip and Knee Joints

Both the hip and knee joints are large weight-bearing joints of the lower limb, but they differ substantially in their security. Read through the brief descriptive material that follows, and

look at the "Selected Synovial Joints" section of the Review Sheet for the questions that pertain to these joints before beginning your comparison.

The Hip Joint The hip joint is a ball-and-socket joint, so movements can occur in all possible planes. However, its movements are definitely limited by its deep socket and strong reinforcing ligaments, the two factors that account for its exceptional stability (Figure 10.7).

The deeply cupped acetabulum that receives the head of the femur is enhanced by a circular rim of fibrocartilage called the **acetabular labrum.** Because the diameter of the labrum is smaller than that of the femur's head, dislocations of the hip are rare. A short ligament, the **ligament of the head of the femur** *(ligamentum teres),* runs from the pitlike **fovea capitis** on the femur head to the acetabulum, where it helps to secure the femur. Several strong ligaments, including the **iliofemoral** and **pubofemoral** anteriorly and the **ischiofemoral** that spirals posteriorly (not shown), are arranged so that they "screw" the femur head into the socket when a person stands upright.

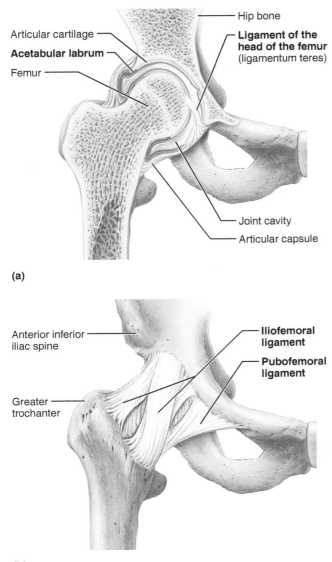

(a)

(b)

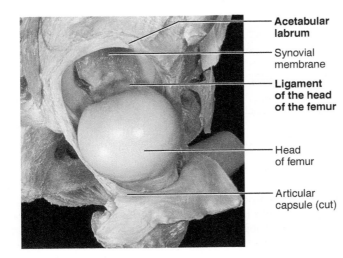

(c)

Figure 10.7 **Hip joint relationships. (a)** Frontal section through the right hip joint. **(b)** Anterior superficial view of the right hip joint. **(c)** Photograph of the interior of the hip joint, lateral view.

ACTIVITY 6

Demonstrating Actions at the Hip Joint

If a functional hip joint model is available, identify the joint parts, and manipulate it to demonstrate the following movements that can occur at this joint: flexion, extension, abduction, and medial and lateral rotation.

Reread the information on what movements the associated ligaments restrict, and verify that information during your joint manipulations. ▪▪

The Knee Joint The knee is the largest and most complex joint in the body. Three joints in one **(Figure 10.8)**, it allows extension, flexion, and a little rotation. The **tibiofemoral joint,** actually a bicondyloid joint between the femoral condyles above and the **menisci** (semilunar cartilages) of the tibia below, is functionally a hinge joint, a very unstable one made slightly more secure by the menisci (Figure 10.8b and d). Some rotation occurs when the knee is partly flexed, but during extension, the menisci and ligaments counteract rotation and side-to-side movements. The other joint is the **femoropatellar joint,** the intermediate joint anteriorly (Figure 10.8a and c).

The knee is unique in that it is only partly enclosed by an articular capsule. Anteriorly, where the capsule is absent, are three broad ligaments, the **patellar ligament** and the **medial** and **lateral patellar retinacula,** which run from the patella to the tibia below and merge with the capsule on either side.

Extracapsular ligaments, including the **fibular** and **tibial collateral ligaments** (which prevent rotation during extension) and the **oblique popliteal** and **arcuate popliteal ligaments,** are crucial in reinforcing the knee. The knees have a built-in locking device that must be "unlocked" by the popliteus muscles (Figure 10.8e) before the knees can be flexed again. The **anterior** and **posterior cruciate ligaments** are intracapsular ligaments that cross (*cruci* = cross) in the notch between the femoral condyles. They help to prevent anterior-posterior displacement of the joint and overflexion and hyperextension of the joint.

ACTIVITY 7

Demonstrating Actions at the Knee Joint

If a functional model of a knee joint is available, identify the joint parts, and manipulate it to illustrate the following movements: flexion, extension, and medial and lateral rotation.

Reread the information on what movements the various associated ligaments restrict, and verify that information during your joint manipulations. ▪▪

The Temporomandibular Joint

The **temporomandibular joint (TMJ)** lies just anterior to the ear **(Figure 10.9)**, where the egg-shaped condylar process of the mandible articulates with the inferior surface of the

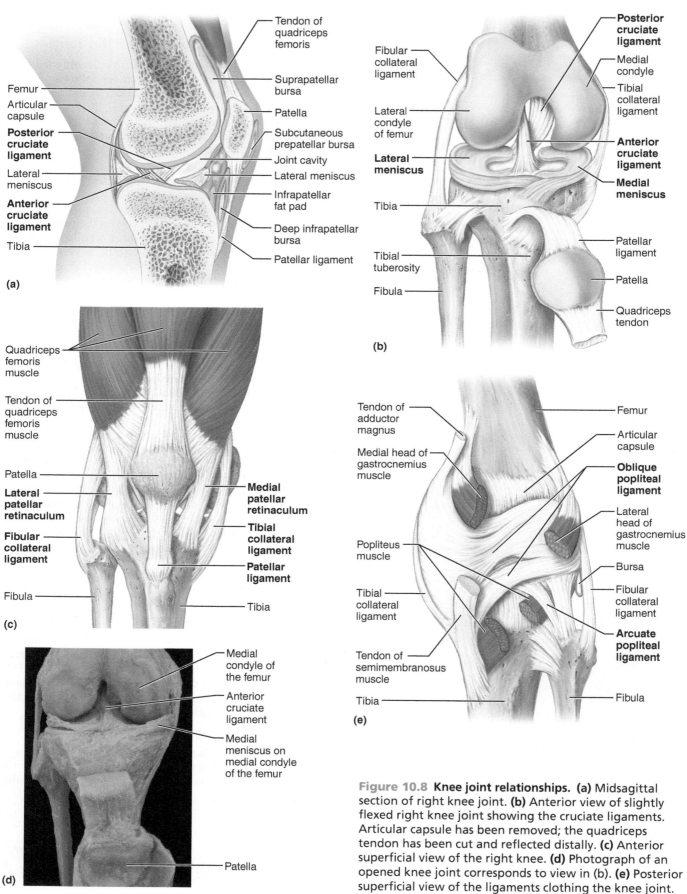

(a)

Femur
Articular capsule
Posterior cruciate ligament
Lateral meniscus
Anterior cruciate ligament
Tibia

Tendon of quadriceps femoris
Suprapatellar bursa
Patella
Subcutaneous prepatellar bursa
Joint cavity
Lateral meniscus
Infrapatellar fat pad
Deep infrapatellar bursa
Patellar ligament

(b)

Fibular collateral ligament
Lateral condyle of femur
Lateral meniscus
Tibia
Tibial tuberosity
Fibula

Posterior cruciate ligament
Medial condyle
Tibial collateral ligament
Anterior cruciate ligament
Medial meniscus
Patellar ligament
Patella
Quadriceps tendon

(c)

Quadriceps femoris muscle
Tendon of quadriceps femoris muscle
Patella
Lateral patellar retinaculum
Fibular collateral ligament
Fibula

Medial patellar retinaculum
Tibial collateral ligament
Patellar ligament
Tibia

(d)

Medial condyle of the femur
Anterior cruciate ligament
Medial meniscus on medial condyle of the femur
Patella

(e)

Tendon of adductor magnus
Medial head of gastrocnemius muscle
Popliteus muscle
Tibial collateral ligament
Tendon of semimembranosus muscle
Tibia

Femur
Articular capsule
Oblique popliteal ligament
Lateral head of gastrocnemius muscle
Bursa
Fibular collateral ligament
Arcuate popliteal ligament
Fibula

Figure 10.8 Knee joint relationships. (a) Midsagittal section of right knee joint. **(b)** Anterior view of slightly flexed right knee joint showing the cruciate ligaments. Articular capsule has been removed; the quadriceps tendon has been cut and reflected distally. **(c)** Anterior superficial view of the right knee. **(d)** Photograph of an opened knee joint corresponds to view in (b). **(e)** Posterior superficial view of the ligaments clothing the knee joint.

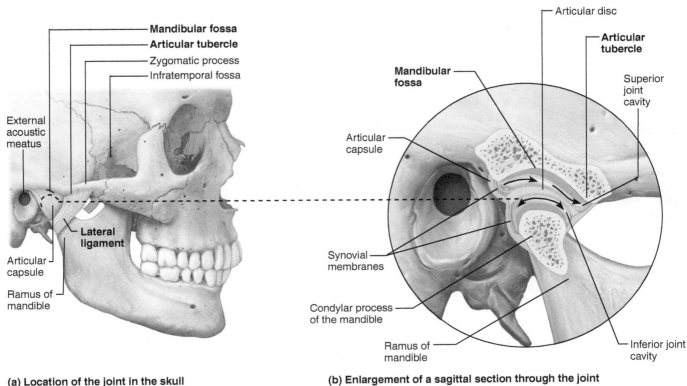

(a) Location of the joint in the skull

(b) Enlargement of a sagittal section through the joint

Figure 10.9 The temporomandibular (jaw) joint relationships. Note that the superior and inferior compartments of the joint cavity allow different movements indicated by arrows.

squamous part of the temporal bone. The temporal bone joint surface has a complicated shape: posteriorly is the **mandibular fossa,** and anteriorly is a bony knob called the **articular tubercle.** The joint's articular capsule, though strengthened by the **lateral ligament,** is loose; an articular disc divides the joint cavity into superior and inferior compartments. Typically, the condylar process–mandibular fossa connection allows the familiar hingelike movements of elevating and depressing the mandible to open and close the mouth. However, when the mouth is opened wide, the condylar process glides anteriorly and is braced against the dense bone of the articular tubercle so that the mandible is not forced superiorly through the thin mandibular fossa when we bite hard foods.

ACTIVITY 8

Examining the Action at the Temporomandibular Joint

While placing your fingers over the area just anterior to the ear, open and close your mouth to feel the hinge action at the TMJ. Then, keeping your fingers on the TMJ, yawn to demonstrate the anterior gliding of the condylar process of the mandible. ■

The Shoulder Joint

The shoulder joint, or **glenohumeral joint,** is the most freely moving joint of the body. The rounded head of the humerus fits the shallow glenoid cavity of the scapula **(Figure 10.10)**. A rim of fibrocartilage, the **glenoid labrum,** deepens the cavity slightly.

The articular capsule enclosing the joint is thin and loose, contributing to ease of movement. The **coracohumeral ligament** helps support the weight of the upper limb; three weak **glenohumeral ligaments** strengthen the front of the capsule. Muscle tendons from the biceps brachii and **rotator cuff** muscles contribute most to shoulder stability.

ACTIVITY 9

Demonstrating Actions at the Shoulder Joint

If a functional shoulder joint model is available, identify the joint parts and manipulate it to demonstrate the following movements: flexion, extension, abduction, adduction, circumduction, and medial and lateral rotation.

Note where the joint is weakest, and verify the most common direction of a dislocated humerus. ■

Joint Disorders

Joint pains and malfunctions have a variety of causes. For example, a hard blow to the knee can cause a painful bursitis, known as "water on the knee," due to damage to, or inflammation of, the patellar bursa.

Sprains and dislocations are other types of joint problems. In a **sprain,** the ligaments reinforcing a joint are damaged by overstretching or are torn away from the bony attachment. Because both ligaments and tendons are cords of dense regular connective tissue with a poor blood supply, sprains heal slowly and are quite painful.

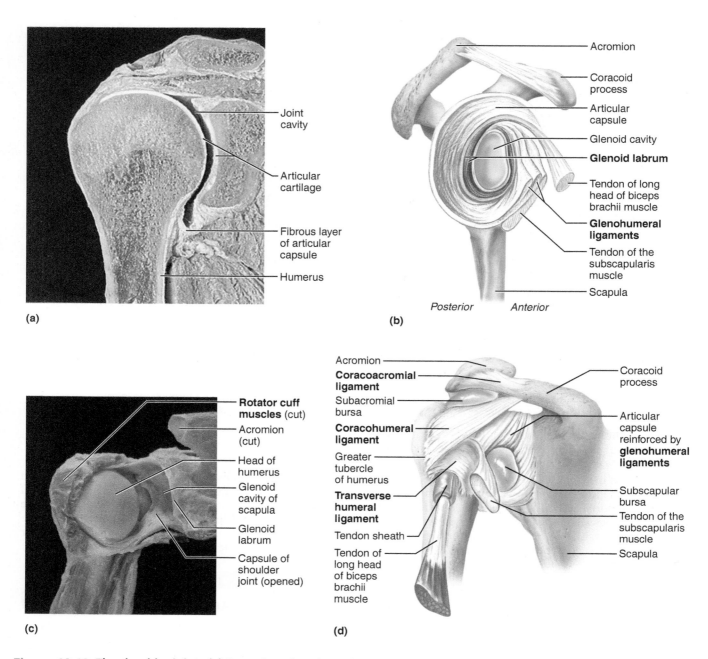

(a)

(b)

Posterior   Anterior

(c)

(d)

Figure 10.10 The shoulder joint. (a) Frontal section through the shoulder. **(b)** The right shoulder joint, cut open and viewed from the lateral aspect; the humerus has been removed. **(c)** Photo of an opened shoulder joint, anterior view. **(d)** Anterior superficial view of the right shoulder joint.

Dislocations occur when bones are forced out of their normal position in the joint cavity. They are normally accompanied by torn or stressed ligaments and considerable inflammation. The process of returning the bone to its proper position, called *reduction,* should only be done by a physician. Attempts by an untrained person to "snap the bone back into its socket" are often more harmful than helpful.

Advancing years also take their toll on joints. Weight-bearing joints in particular eventually begin to degenerate.

Adhesions (fibrous bands) may form between the surfaces where bones join, and extraneous bone tissue *(spurs)* may grow along the joint edges. Such degenerative changes lead to the complaint so often heard from the elderly: "My joints are getting so stiff. . . . "

• If possible, compare an X-ray image of an arthritic joint to one of a normal joint. ✚

Table 10.3   Structural and Functional Characteristics of Body Joints

Illustration	Joint	Articulating bones	Structural type*	Functional type; movements allowed
	Skull	Cranial and facial bones	Fibrous; suture	Synarthrotic; no movement
	Temporo-mandibular	Temporal bone of skull and mandible	Synovial; modified hinge† (contains articular discs)	Diarthrotic; gliding and uniaxial rotation; slight lateral movement, elevation, depression, protraction, and retraction of mandible
	Atlanto-occipital	Occipital bone of skull and atlas	Synovial; condylar	Diarthrotic; biaxial; flexion, extension, lateral flexion, circumduction of head on neck
	Atlantoaxial	Atlas (C_1) and axis (C_2)	Synovial; pivot	Diarthrotic; uniaxial; rotation of the head
	Intervertebral	Between adjacent vertebral bodies	Cartilaginous; symphysis	Amphiarthrotic; slight movement
	Intervertebral	Between articular processes	Synovial; plane	Diarthrotic; gliding
	Costovertebral	Vertebrae (transverse process or bodies) and ribs	Synovial; plane	Diarthrotic; gliding of ribs
	Sternoclavicular	Sternum and clavicle	Synovial; shallow saddle (contains articular disc)	Diarthrotic; multiaxial (allows clavicle to move in all axes)
	Sternocostal (first)	Sternum and rib 1	Cartilaginous; synchondrosis	Synarthrotic; no movement
	Sternocostal	Sternum and ribs 2–7	Synovial; double plane	Diarthrotic; gliding
	Acromio-clavicular	Acromion of scapula and clavicle	Synovial; plane (contains articular disc)	Diarthrotic; gliding and rotation of scapula on clavicle
	Shoulder (glenohumeral)	Scapula and humerus	Synovial; ball and socket	Diarthrotic; multiaxial; flexion, extension, abduction, adduction, circumduction, rotation of humerus
	Elbow	Ulna (and radius) with humerus	Synovial; hinge	Diarthrotic; uniaxial; flexion, extension of forearm
	Proximal radioulnar	Radius and ulna	Synovial; pivot	Diarthrotic; uniaxial; pivot (head of radius rotates in radial notch of ulna)
	Distal radioulnar	Radius and ulna	Synovial; pivot (contains articular disc)	Diarthrotic; uniaxial; rotation of radius around long axis of forearm to allow pronation and supination
	Wrist	Radius and proximal carpals	Synovial; condylar	Diarthrotic; biaxial; flexion, extension, abduction, adduction, circumduction of hand
	Intercarpal	Adjacent carpals	Synovial; plane	Diarthrotic; gliding
	Carpometacarpal of digit I (thumb)	Carpal (trapezium) and metacarpal I	Synovial; saddle	Diarthrotic; biaxial; flexion, extension, abduction, adduction, circumduction, opposition of metacarpal I
	Carpometacarpal of digits II–V	Carpal(s) and metacarpal(s)	Synovial; plane	Diarthrotic; gliding of metacarpals
	Metacarpo-phalangeal (knuckle)	Metacarpal and proximal phalanx	Synovial; condylar	Diarthrotic; biaxial; flexion, extension, abduction, adduction, circumduction of fingers
	Interphalangeal (finger)	Adjacent phalanges	Synovial; hinge	Diarthrotic; uniaxial; flexion, extension of fingers

Table 10.3   Structural and Functional Characteristics of Body Joints *(continued)*

Illustration	Joint	Articulating bones	Structural type*	Functional type; movements allowed
	Sacroiliac	Sacrum and hip bone	Synovial; plane	Diarthrotic; little movement, slight gliding possible (more during pregnancy)
	Pubic symphysis	Pubic bones	Cartilaginous; symphysis	Amphiarthrotic; slight movement (enhanced during pregnancy)
	Hip (coxal)	Hip bone and femur	Synovial; ball and socket	Diarthrotic; multiaxial; flexion, extension, abduction, adduction, rotation, circumduction of femur
	Knee (tibiofemoral)	Femur and tibia	Synovial; modified hinge† (contains articular disc)	Diarthrotic; biaxial; flexion, extension of leg, some rotation allowed
	Knee (femoropatellar)	Femur and patella	Synovial; plane	Diarthrotic; gliding of patella
	Superior tibiofibular	Tibia and fibula (proximally)	Synovial; plane	Diarthrotic; gliding of fibula
	Inferior tibiofibular	Tibia and fibula (distally)	Fibrous; syndesmosis	Synarthrotic; slight "give" during dorsiflexion of foot
	Ankle	Tibia and fibula with talus	Synovial; hinge	Diarthrotic; uniaxial; dorsiflexion and plantar flexion of foot
	Intertarsal	Adjacent tarsals	Synovial; plane	Diarthrotic; gliding; inversion and eversion of foot
	Tarsometatarsal	Tarsal(s) and metatarsal(s)	Synovial; plane	Diarthrotic; gliding of metatarsals
	Metatarsophalangeal	Metatarsal and proximal phalanx	Synovial; condylar	Diarthrotic; biaxial; flexion, extension, abduction, adduction, circumduction of great toe
	Interphalangeal (toe)	Adjacent phalanges	Synovial; hinge	Diarthrotic; uniaxial; flexion, extension of toes

*__Fibrous joints__ are indicated by orange circles; __cartilaginous joints__ by blue circles; __synovial joints__ by purple circles.
†These modified hinge joints are structurally bicondylar.

Articulations and Body Movements

Fibrous, Cartilaginous, and Synovial Joints

1. Use the key to identify the joint types described below. Some responses may be used more than once.

   Key: a. cartilaginous   b. fibrous   c. synovial

   ——————————— 1. includes shoulder, elbow, and wrist joints

   ——————————— 2. includes joints between the vertebral bodies and the pubic symphysis

   ——————————— 3. sutures are memorable examples

   ——————————— 4. found in the epiphyseal plate

   ——————————— 5. found in a gomphosis

   ——————————— 6. all are freely movable or diarthrotic

2. Label the diagram of a typical synovial joint using the terms provided in the key and the appropriate leader lines.

   Key:

   a. articular capsule

   b. articular cartilage

   c. fibrous layer

   d. joint cavity

   e. ligament

   f. periosteum

   g. synovial membrane

3. How does a tendon sheath differ from a bursa? ———————————————————————

   ——

4. Which structure in the synovial joint produces synovial fluid? ———————————————

5. Match the joint subcategories in column B with their descriptions in column A, and place an asterisk (*) beside all choices that are examples of synovial joints. Some responses may be used more than once.

Column A

Column B

_____ 1. joint between skull bones

a. ball-and-socket

_____ 2. joint between the axis and atlas

b. condylar

_____ 3. hip joint

c. hinge

_____ 4. intervertebral joints (between articular processes)

d. pivot

_____ 5. joint between forearm bones and wrist

e. plane

_____ 6. elbow

f. saddle

_____ 7. interphalangeal joints

g. suture

_____ 8. intercarpal joints

h. symphysis

_____ 9. joint between talus and tibia/fibula

i. synchondrosis

_____ 10. joint between skull and vertebral column

j. syndesmosis

_____ 11. joint between jaw and skull

_____ 12. joints between proximal phalanges and metacarpal bones

_____ 13. epiphyseal plate of a child's long bone

_____ 14. a multiaxial joint

_____, _____ 15. biaxial joints

_____, _____ 16. uniaxial joints

Selected Synovial Joints

6. Which joint, the hip or the knee, is more stable? _____

Name two important factors that contribute to the stability of the hip joint.

_____ and _____

Name two important factors that contribute to the stability of the knee.

_____ and _____

7. Label the photograph of a knee joint model using the terms provided in the key and the appropriate leader lines.

Key:

   a. anterior cruciate ligament

   b. fibula

   c. fibular collateral ligament

   d. lateral condyle of the femur

   e. lateral meniscus

   f. medial meniscus

   g. patella

   h. patellar ligament

   i. tibia

   j. tibial collateral ligament

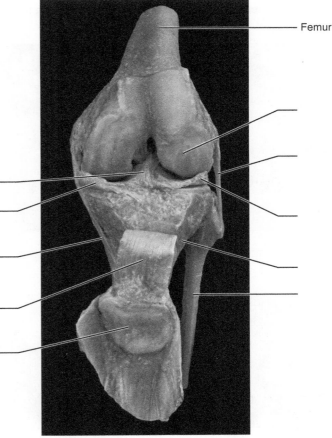

Femur

8. The shoulder joint is built for mobility. List three factors that contribute to the large range of motion at the shoulder:

   1. _____

   2. _____

   3. _____

9. In which direction does the shoulder usually dislocate? _____

Movements Allowed by Synovial Joints

10. Complete the descriptions below the diagrams by inserting the type of movement in each answer blank.

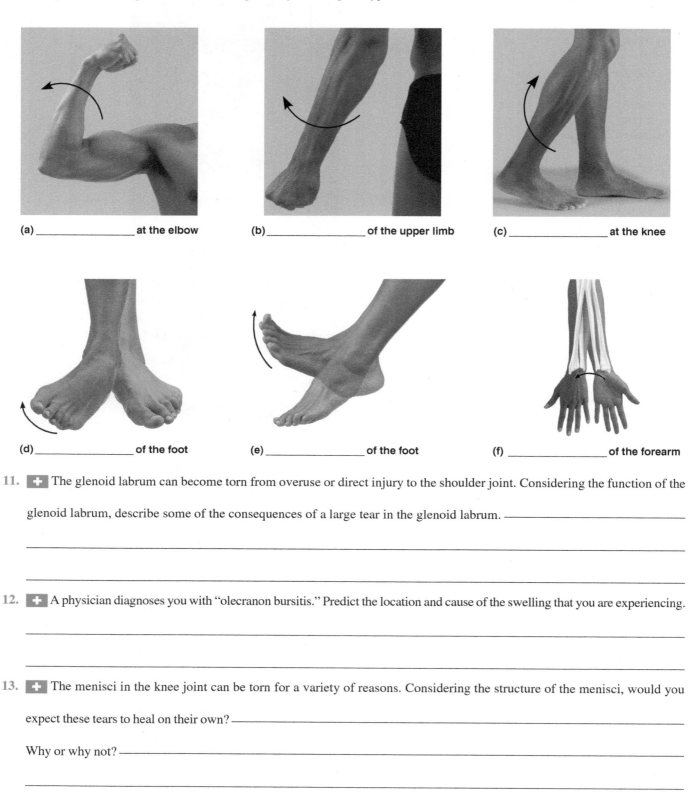

(a) _____ at the elbow

(b) _____ of the upper limb

(c) _____ at the knee

(d) _____ of the foot

(e) _____ of the foot

(f) _____ of the forearm

11. ➕ The glenoid labrum can become torn from overuse or direct injury to the shoulder joint. Considering the function of the

glenoid labrum, describe some of the consequences of a large tear in the glenoid labrum. _____

_____

_____

12. ➕ A physician diagnoses you with "olecranon bursitis." Predict the location and cause of the swelling that you are experiencing.

_____

_____

13. ➕ The menisci in the knee joint can be torn for a variety of reasons. Considering the structure of the menisci, would you

expect these tears to heal on their own? _____

Why or why not? _____

_____

Microscopic Anatomy and Organization of Skeletal Muscle

MATERIALS

- Three-dimensional model of skeletal muscle fibers (if available)
- Forceps
- Dissecting needles
- Clean microscope slides and coverslips
- 0.9% saline solution in dropper bottles
- Chicken breast or thigh muscle (freshly obtained)
- Compound microscope
- Prepared slides of skeletal muscle (l.s. and x.s. views) and skeletal muscle showing neuromuscular junctions
- Three-dimensional model of skeletal muscle showing neuromuscular junction (if available)

LEARNING OUTCOMES

☐ Define *muscle fiber, myofibril,* and *myofilament,* and describe the structural relationships among them.

☐ Describe thick (myosin) and thin (actin) filaments and their relationship to the sarcomere.

☐ Discuss the structure and location of T tubules and terminal cisterns.

☐ Define *endomysium, perimysium,* and *epimysium,* and relate them to muscle fibers, fascicles, and entire muscles.

☐ Define *tendon* and *aponeurosis,* and describe the difference between them.

☐ Describe the structure of skeletal muscle from gross to microscopic levels.

☐ Explain the connection between motor neurons and skeletal muscle, and discuss the structure and function of the neuromuscular junction.

PRE-LAB QUIZ

1. Which is *false* of skeletal muscle?
   a. It enables you to manipulate your environment.
   b. It influences the body's contours and shape.
   c. It is one of the major components of hollow organs.
   d. It provides a means of locomotion.
2. Circle the correct underlined term. Because the cells of skeletal muscle are relatively large and cylindrical in shape, they are also known as <u>fibers</u> / <u>tubules</u>.
3. Circle True or False. Skeletal muscle cells have more than one nucleus.
4. The two contractile proteins that make up the myofilaments of skeletal muscle are _____ and _____.
5. Each muscle fiber is surrounded by thin connective tissue called the:
   a. aponeurosis   c. epimysium
   b. endomysium   d. perimysium
6. A cordlike structure that connects a muscle to another muscle or bone is:
   a. a fascicle
   b. a tendon
   c. deep fascia
7. The junction between an axon and a muscle fiber is called a _____ _____.
8. Circle True or False. The neuron and muscle fiber membranes do not actually touch but are separated by a fluid-filled gap.
9. Circle the correct underlined term. The contractile unit of muscle is the <u>sarcolemma</u> / <u>sarcomere</u>.
10. Circle True or False. Larger, more powerful muscles have relatively less connective tissue than smaller muscles.

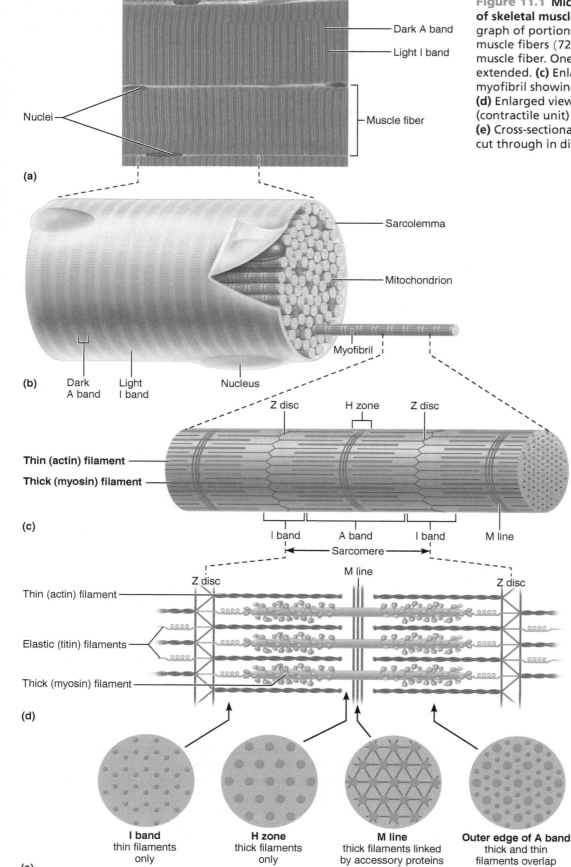

(a)

Nuclei

Dark A band

Light I band

Muscle fiber

(b)

Sarcolemma

Mitochondrion

Myofibril

Dark A band

Light I band

Nucleus

(c)

Z disc

H zone

Z disc

Thin (actin) filament

Thick (myosin) filament

I band

A band

I band

M line

Sarcomere

(d)

Z disc

M line

Z disc

Thin (actin) filament

Elastic (titin) filaments

Thick (myosin) filament

(e)

I band
thin filaments
only

H zone
thick filaments
only

M line
thick filaments linked
by accessory proteins

Outer edge of A band
thick and thin
filaments overlap

Figure 11.1 Microscopic anatomy of skeletal muscle. (a) Photomicrograph of portions of two isolated muscle fibers (725×). **(b)** Part of a muscle fiber. One myofibril has been extended. **(c)** Enlarged view of one myofibril showing its banding pattern. **(d)** Enlarged view of one sarcomere (contractile unit) of a myofibril. **(e)** Cross-sectional view of a sarcomere cut through in different areas.

Most of the muscle tissue in the body is **skeletal muscle,** which attaches to the skeleton or associated connective tissue. Skeletal muscle shapes the body and gives you the ability to move—to walk, run, jump, and dance; to draw, paint, and play a musical instrument; and to smile and frown. The remaining muscle tissue of the body consists of smooth muscle, which forms the walls of hollow organs, and cardiac muscle, which forms the walls of the heart. Smooth and cardiac muscle move materials within the body. For example, smooth muscle moves digesting food through the digestive system and moves urine from the kidneys to the exterior of the body. Cardiac muscle moves blood through the blood vessels.

Each of the three muscle types has a structure and function uniquely suited to its function in the body. However, because the term *muscular system* applies specifically to skeletal muscle, our primary objective in this exercise is to investigate the structure and function of skeletal muscle.

Skeletal muscle is also known as *voluntary muscle* (because it can be consciously controlled) and as *striated muscle* (because it has a striped appearance).

The Cells of Skeletal Muscle

Skeletal muscle is made up of relatively large, long cylindrical cells, called **muscle fibers.** These cells range from 10 to 100 μm in diameter and up to 6 cm in length.

Because hundreds of embryonic cells fuse to produce each muscle fiber, the skeletal muscle fibers (**Figure 11.1a** and **b**) are multinucleate; multiple oval nuclei can be seen just beneath the plasma membrane (called the *sarcolemma* in these cells). The nuclei are pushed peripherally by the longitudinally arranged **myofibrils,** long rod-shaped organelles that nearly fill the sarcoplasm. Alternating light (I) and dark (A) bands along the length of the perfectly aligned myofibrils give the muscle fiber as a whole its striped appearance.

Electron microscope studies have revealed that the myofibrils are made up of even smaller threadlike structures called **myofilaments** (Figure 11.1d). The myofilaments are composed largely of two varieties of contractile proteins— **actin** and **myosin**—which slide past each other during muscle activity to bring about shortening or contraction of the muscle cells. The actual contractile units of muscle, called **sarcomeres,** extend from the middle of one I band (its Z disc) to the middle of the next along the length of the myofibrils (Figure 11.1c and d). Cross sections of the sarcomere in areas where **thick filaments** and **thin filaments** overlap show that each thick filament is surrounded by six thin filaments; each thin filament is enclosed by three thick filaments (Figure 11.1e).

At each junction of the A and I bands, the sarcolemma indents into the muscle fiber, forming a **transverse tubule (T tubule).** These tubules run deep into the muscle cell between cross channels, or **terminal cisterns,** of the elaborate smooth endoplasmic reticulum called the **sarcoplasmic reticulum (SR)** (**Figure 11.2**). Regions where the SR terminal cisterns abut a T tubule on each side are called **triads.**

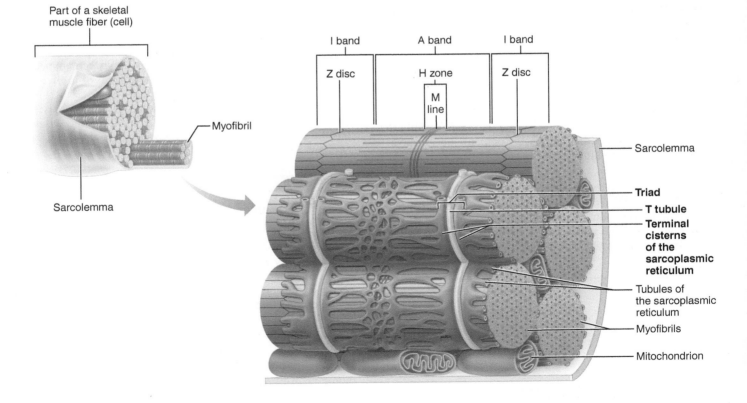

Figure 11.2 Relationship of the sarcoplasmic reticulum and T tubules to the myofibrils of skeletal muscle.

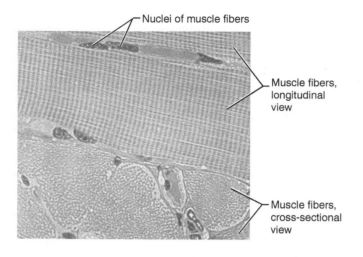

Figure 11.3 **Photomicrograph of muscle fibers, longitudinal and cross sections (800×).**

ACTIVITY 1

Examining Skeletal Muscle Fiber Anatomy

1. Look at the three-dimensional model of skeletal muscle fibers, noting the relative shape and size of the cells. Identify the nuclei, myofibrils, and light and dark bands.

2. Obtain forceps, two dissecting needles, a slide and coverslip, and a dropper bottle of saline solution. With forceps, remove a very small piece of muscle (about 1 mm diameter) from a fresh chicken breast or thigh. Place the tissue on a clean microscope slide, and add a drop of the saline solution.

3. Pull the muscle fibers apart (tease them) with the dissecting needles or forceps until you have a fluffy-looking mass of tissue. Cover the teased tissue with a coverslip, and observe under the high-power lens of a compound microscope. Look for the banding pattern by examining muscle fibers isolated at the edge of the tissue mass. Regulate the light carefully to obtain the highest possible contrast. Compare your observations with the photomicrograph (Figure 11.3).

4. Now compare your observations with what can be seen with professionally prepared muscle tissue. Obtain a slide of skeletal muscle (longitudinal section), and view it under high power. From your observations, draw a small section of a muscle fiber in the space provided below. Label the nuclei, sarcolemma, and A and I bands.

What structural details become apparent with the prepared slide?

_____

_____

Organization of Skeletal Muscle Fibers into Muscles

Muscle fibers are soft and surprisingly fragile. Thousands of muscle fibers are bundled together with connective tissue to form the organs we refer to as skeletal muscles (Figure 11.4). Each muscle fiber is enclosed in a delicate, areolar connective tissue sheath called **endomysium.** Several sheathed muscle fibers are wrapped by a collagenic membrane called **perimysium,** forming a bundle of muscle fibers called a **fascicle.** A large number of fascicles are bound together by a substantially coarser "overcoat" of dense connective tissue called an **epimysium,** which sheathes the entire muscle. All three sheaths converge to form strong cordlike **tendons** or sheetlike **aponeuroses,** which attach muscles to each other or indirectly to bones. A muscle's more movable attachment is called its *insertion,* whereas its fixed (or immovable) attachment is the *origin* (Exercise 10).

Tendons perform several functions. Two of the most important are to provide durability and to conserve space. Because tendons are tough dense regular connective tissue, they can span rough bony projections that would destroy the more delicate muscle tissues. Because of their relatively small size, more tendons than fleshy muscles can pass over a joint.

In addition to supporting and binding the muscle fibers, and providing strength to the muscle as a whole, the connective tissue wrappings provide a route for the entry and exit of nerves and blood vessels that serve the muscle fibers. The larger, more powerful muscles have relatively more connective tissue than muscles involved in fine or delicate movements.

As we age, the mass of the muscle fibers decreases, and the amount of connective tissue increases; thus the skeletal muscles gradually become more sinewy, or "stringier." +

ACTIVITY 2

Observing the Histologic Structure of a Skeletal Muscle

Obtain a slide showing a cross section of skeletal muscle tissue. Identify the muscle fibers, their peripherally located nuclei, and their connective tissue wrappings, the endomysium, perimysium, and epimysium if visible (use Figure 11.4 as a reference).

The Neuromuscular Junction

The voluntary skeletal muscle fibers must be stimulated by motor neurons via nerve impulses. The junction between an axon of a motor neuron and a muscle fiber is called a **neuromuscular junction** (Figure 11.5).

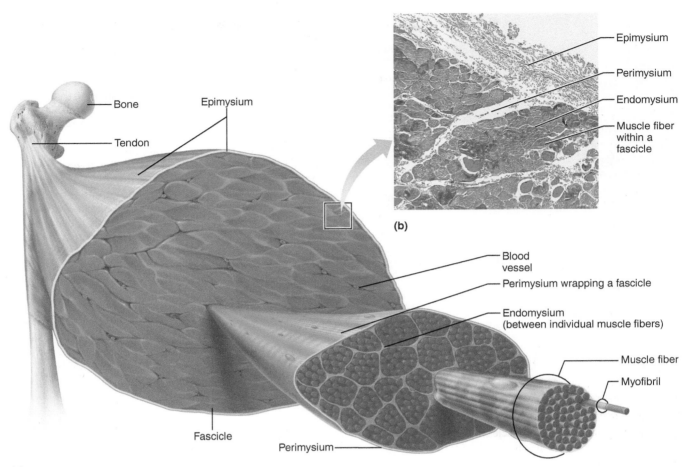

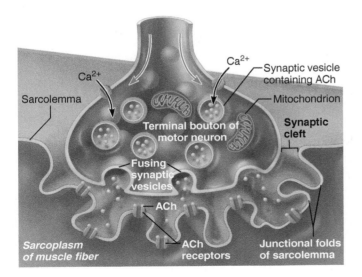

Figure 11.4 Connective tissue sheaths of skeletal muscle. (a) Diagram view.
(b) Photomicrograph of a cross section of skeletal muscle (40×).

Figure 11.5 The neuromuscular junction. Pink arrows indicate arrival of the action potential, which ultimately causes vesicles to release acetylcholine (ACh). The ACh receptor is part of an ion channel that opens briefly, causing depolarization of the sarcolemma.

Each axon of a motor neuron breaks up into many branches called **terminal boutons** or *axon terminals* as it approaches the muscle. Each of these branches participates in forming a neuromuscular junction with a single muscle fiber. Thus, a single neuron may stimulate many muscle fibers. Together, a neuron and all the muscle fibers it stimulates make up the functional structure called the **motor unit.** (Part of a motor unit is shown in **Figure 11.6.**) The neuron and muscle fiber membranes, close as they are, do not actually touch. They are separated by a small fluid-filled gap called the **synaptic cleft** (see Figure 11.5).

Within the terminal boutons are many mitochondria and vesicles containing a neurotransmitter chemical called acetylcholine (ACh). When an action potential reaches the terminal boutons, voltage-gated Ca^{2+} channels open. Ca^{2+} enters the terminal bouton and causes ACh to be released by exocytosis. The ACh diffuses across the synaptic cleft and combines with the receptors on the sarcolemma. When receptors bind ACh, the permeability of the sarcolemma changes. Ion channels open briefly, depolarizing the sarcolemma, and the muscle fiber subsequently contracts.

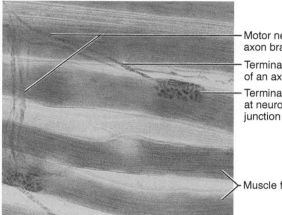

Motor neuron
axon branches

Terminal branch
of an axon

Terminal bouton
at neuromuscular
junction

Muscle fibers

Figure 11.6 Photomicrograph of neuromuscular junction (750×).

ACTIVITY 3

Studying the Structure of a Neuromuscular Junction

1. If possible, examine a three-dimensional model of skeletal muscle fibers that illustrates the neuromuscular junction. Identify the structures just described.

2. Obtain a slide of skeletal muscle stained to show a portion of a motor unit. Examine the slide under high power to identify the axon branches that extend like a leash to the muscle fibers. Follow one of the axon fibers to its terminal branch to identify the oval-shaped terminal bouton at the end of the branch. Compare your observations to the photomicrograph (Figure 11.6). Sketch a small section in the space provided below. Label the axon of the motor neuron, a terminal branch, terminal boutons, and muscle fibers. ▬▬

WHY THIS **MATTERS** | Botox—The Good, the Bad, and the Wrinkle-Free

A diluted form of the botulinum toxin, Botox, received FDA approval to treat uncontrollable blinking (blepharospasm) in 1989. Prior to this time, the botulinum toxin was better known for causing respiratory failure with the foodborne illness known as botulism. The bacterium *Clostridium botulinum* secretes botulinum toxin, which is a neurotoxin. The toxin causes paralysis by blocking the release of the neurotransmitter acetylcholine (ACh) from the terminal bouton of a motor neuron. If the neurotoxin blocks terminal boutons serving the diaphragm, death is a common result. If the ACh receptors are located in a facial muscle, such as the corrugator supercilii, a brow is unfurrowed, and a wrinkle-free glabella results. ■

Microscopic Anatomy and Organization of Skeletal Muscle

Skeletal Muscle Fibers and Their Packaging into Muscles

1. Use the terms in the key to correctly identify the structures described below. Not all terms will be used.

_____	1. connective tissue covering a bundle of muscle fibers
_____	2. bundle of muscle fibers
_____	3. contractile unit of muscle
_____	4. superficial sheath that covers the entire muscle
_____	5. thin areolar connective tissue investing each muscle fiber
_____	6. plasma membrane of the muscle fiber
_____	7. a long organelle with a banded appearance found within muscle fibers
_____	8. actin- or myosin-containing structure
_____	9. cord of collagen fibers that attaches a muscle to a bone

Key:

a. endomysium

b. epimysium

c. fascicle

d. myofibril

e. myofilament

f. perimysium

g. sarcolemma

h. sarcomere

i. tendon

2. List three reasons the connective tissue sheaths of skeletal muscle are important.

_____

_____

_____

3. Why are there more indirect—that is, tendinous—muscle attachments to bone than there are direct attachments?

_____

_____

_____

4. How does an aponeurosis differ from a tendon structurally? _____

_____

How is an aponeurosis functionally similar to a tendon? _____

_____

5. The drawing and photomicrograph below show a relaxed sarcomere. Using the terms from the key, identify each structure indicated by a leader line or bracket. The number 2 in parentheses indicates that the structure will be labeled twice.

Key:

a. actin filament

b. A band

c. H zone

d. I band (2)

e. M line

f. myosin filament

g. Z disc (2)

6. On the following figure, label the endomysium, epimysium, a fascicle, a muscle fiber, a myofibril, perimysium, and the tendon.

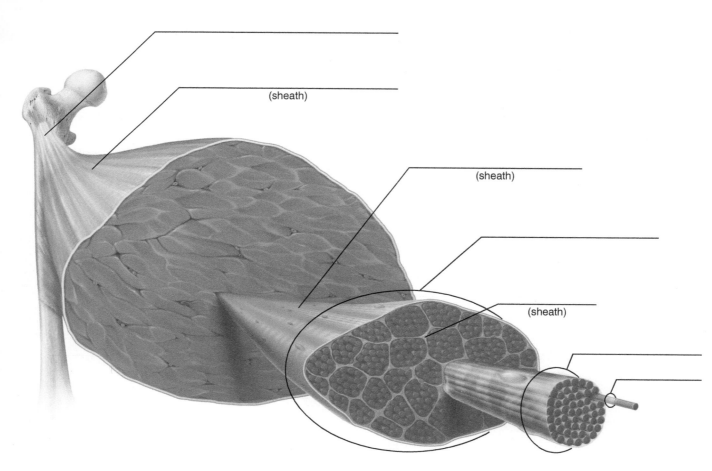

(sheath)

(sheath)

(sheath)

The Neuromuscular Junction

7. Complete the following statements:

The junction between a motor neuron's terminal bouton and the muscle fiber plasma membrane is called a __1__ junction. A motor neuron and all of the skeletal muscle fibers it stimulates is called a __2__ . The actual gap between the terminal bouton and the muscle fiber is called a __3__ . Within the terminal bouton are many small vesicles containing a neurotransmitter substance called __4__ . When the __5__ reaches the ends of the axon, the neurotransmitter is released and diffuses to the muscle cell membrane to combine with receptors there. The combining of the neurotransmitter with the muscle membrane receptors causes a change in permeability of the sarcolemma, resulting in __6__ of the membrane. Then contraction of the muscle fiber occurs.

1. _____

2. _____

3. _____

4. _____

5. _____

6. _____

8. The events that occur at a neuromuscular junction are depicted below. Identify by labeling every structure provided with a leader line.

Note: The pink arrows depict the propagation of the action potential.

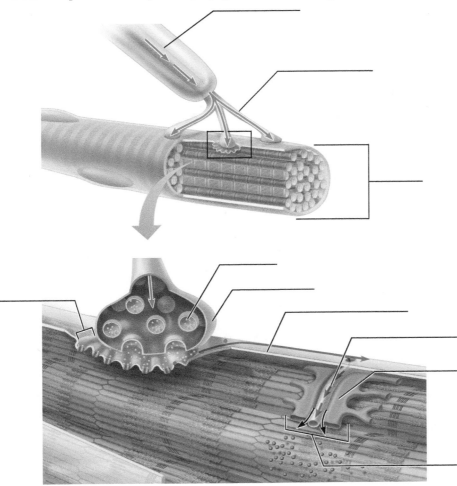

Key:

a.   motor neuron axon branch

b.   myelinated axon of motor neuron

c.   muscle fiber

d.   sarcolemma of muscle fiber

e.   synaptic cleft

f.   synaptic vesicle containing ACh

g.   terminal bouton of motor neuron

h.   terminal cistern of the SR

i.   triad

j.   T tubule

WHY THIS
MATTERS

9. Botulinum toxin binds to receptors present at the terminal bouton in order to enter the neuron. What vesicular

transport process do you think is involved in the toxin entering the neuron? _____

_____

10. Explain how botulinum toxin alters the normal sequence of events at the neuromuscular junction to result in

paralysis. Be sure to include the vesicular transport process that is inhibited. _____

_____

11. ✚ Necrotizing fasciitis is a serious bacterial infection. Necrosis is death of tissues in the body. Considering the organization of the connective tissue sheaths of skeletal muscle, explain how this infection could spread rapidly throughout the body.

_____

_____

Gross Anatomy of the Muscular System

LEARNING OUTCOMES

- ☐ Define *prime mover (agonist), antagonist, synergist,* and *fixator.*
- ☐ List the criteria used in naming skeletal muscles.
- ☐ Identify the major muscles of the human body on a torso model, a human cadaver, lab chart, or image, and state the action of each.
- ☐ Name muscle origins and insertions as required by the instructor.
- ☐ Explain how muscle actions are related to their location.
- ☐ List antagonists for the prime movers listed.
- ☐ Name and locate muscles on a dissected cat.
- ☐ Recognize similarities and differences between human and cat musculature.

PRE-LAB QUIZ

1. A prime mover, or _____, produces a particular type of movement.
   a. agonist
   b. antagonist
   c. fixator
   d. synergist
2. Skeletal muscles are named on the basis of many criteria. Name one.
   _____
3. Circle True or False. Muscles of facial expression differ from most skeletal muscles because they usually do not insert into a bone.
4. The _____ musculature includes muscles that move the vertebral column, the ribs, and the abdominal wall.
   a. head and neck
   b. lower limb
   c. trunk
5. Muscles that act on the _____ cause movement at the hip, knee, and foot joints.
   a. lower limb
   b. trunk
   c. upper limb
6. This two-headed muscle bulges when the forearm is flexed. It is the most familiar muscle of the anterior humerus. It is the:
   a. biceps brachii
   b. extensor digitorum
   c. flexor carpii radialis
   d. triceps brachii
7. These abdominal muscles are responsible for giving me my "six-pack." They also stabilize my pelvis when walking. They are the _____ muscles.
   a. internal intercostal
   b. quadriceps
   c. rectus abdominis
   d. triceps femoris
8. Circle the correct underlined term. This lower limb muscle, which attaches to the calcaneus via the calcaneal tendon and plantar flexes the foot when the leg is extended, is the <u>tibialis anterior</u> / <u>gastrocnemius</u>.
9. The _____ is the largest and most superficial of the gluteal muscles.
   a. gluteus internus
   b. gluteus maximus
   c. gluteus medius
   d. gluteus minimus
10. Circle True or False. The biceps femoris is located in the anterior compartment.

Classification of Skeletal Muscles

Types of Muscles

Most often, body movements do not result from the contraction of a single muscle but instead reflect the coordinated action of several muscles acting together. Muscles that are primarily responsible for producing a particular movement are called **prime movers,** or **agonists.**

Muscles that oppose or reverse a movement are called **antagonists.** When a prime mover is active, the fibers of the antagonist are stretched and in the relaxed state. The antagonist can also regulate the prime mover by providing some resistance, to prevent overshoot or to stop its action.

It should be noted that antagonists can be prime movers in their own right. For example, the biceps muscle of the arm (a prime mover of flexion at the elbow) is antagonized by the triceps (a prime mover of extension at the elbow).

Synergists help the action of agonists by reducing undesirable or unnecessary movement. Contraction of a muscle crossing two or more joints would cause movement at all joints spanned if the synergists were not there to stabilize them. For example, the muscles that flex the fingers cross both the wrist and finger joints, but you can make a fist without bending at the wrist because synergist muscles stabilize the wrist joint.

Fixators, or fixation muscles, are specialized synergists. They immobilize the origin of a prime mover so that all the tension is exerted at the insertion. Muscles that help maintain posture are fixators—so too are muscles of the back that stabilize, or "fix," the scapula during arm movements.

Naming Skeletal Muscles

Remembering the names of the skeletal muscles is a monumental task, but certain clues help. Muscles are named on the basis of the following criteria:

- **Direction of muscle fibers:** Some muscles are named in reference to some imaginary line, usually the midline of the body. A muscle with fibers running parallel to that imaginary line will have the term *rectus* (straight) in its name. For example, the rectus abdominis is the straight muscle of the abdomen. Likewise, the terms *transverse* and *oblique* indicate that the muscle fibers run at right angles and obliquely, respectively, to the imaginary line. Muscle structure is determined by fascicle arrangement (Figure 12.1).

- **Muscle size:** Terms such as *maximus* (largest), *minimus* (smallest), *longus* (long), and *brevis* (short) are often used in naming muscles—as in gluteus maximus and gluteus minimus.

- **Muscle location:** Some muscles are named for the bone with which they are associated. For example, the temporalis muscle overlies the temporal bone.

- **Number of origins:** When the term *biceps, triceps,* or *quadriceps* forms part of a muscle name, you can generally assume that the muscle has two, three, or four origins (respectively). For example, the biceps brachii has two heads, or origins.

- **Location of the attachments:** For example, the sternocleidomastoid muscle has its origin on the sternum *(sterno)* and clavicle *(cleido)* and inserts on the mastoid process of the temporal bone.

- **Muscle shape:** For example, the deltoid muscle is roughly triangular (*deltoid* = triangle), and the trapezius muscle resembles a trapezoid.

- **Muscle action:** For example, all the adductor muscles of the anterior thigh bring about its adduction, and all the extensor muscles of the wrist extend the hand.

Identification of Human Muscles

While reading the tables and identifying the various human muscles in the figures, try to visualize what happens when the muscle contracts. Since muscles have many actions, we have indicated the primary action of each muscle in blue type in the tables. Then, use a torso model or an anatomical chart to identify as many muscles as possible. If a human cadaver is available, your instructor will provide specific instructions. Then carry out the instructions for demonstrating and palpating muscles.

Muscles of the Head and Neck

The muscles of the head serve many specific functions. For instance, the muscles of facial expression differ from most skeletal muscles because they insert into the skin or other muscles rather than into bone. As a result, they move the facial skin, allowing the face to show a wide range of emotions. Other muscles of the head are the muscles of mastication, which move the mandible during chewing, and the six extrinsic eye muscles located within the orbit, which aim the eye. (Orbital muscles are studied in Exercise 17.)

ACTIVITY 1

Identifying Muscles of the Head and Neck

Neck muscles are concerned primarily with the movement of the head and shoulder girdle. (Figure 12.2 and Figure 12.3 are summary figures illustrating the superficial musculature of the body). Read the descriptions of specific head and neck muscles, and identify the various muscles in the figures (Table 12.1, Table 12.2, Figure 12.3, Figure 12.4, and Figure 12.5), trying to visualize their action when they contract. Then identify them on a torso model or anatomical chart.

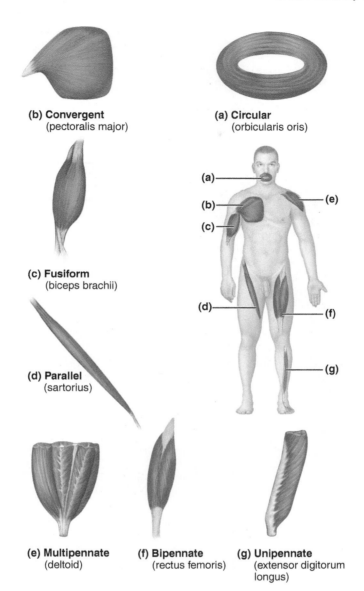

(b) Convergent
(pectoralis major)

(a) Circular
(orbicularis oris)

(c) Fusiform
(biceps brachii)

(d) Parallel
(sartorius)

(e) Multipennate
(deltoid)

(f) Bipennate
(rectus femoris)

(g) Unipennate
(extensor digitorum
longus)

Figure 12.1 Patterns of fascicle arrangement in muscles.

Demonstrating Operations of Head Muscles

1. Raise your eyebrow to wrinkle your forehead. You are using the *frontal belly* of the *epicranius* muscle.

2. Blink your eyes; wink. You are contracting *orbicularis oculi.*

3. Close your lips and pucker up. This requires contraction of *orbicularis oris.*

4. Smile. You are using *zygomaticus.*

5. To demonstrate the temporalis, place your hands on your temples, and clench your teeth. Now you can also palpate the masseter at the angle of the jaw. ▄

Muscles of the Trunk

The trunk musculature includes muscles that move the vertebral column; anterior thorax muscles that act to move ribs, head, and arms; and muscles of the abdominal wall that play a role in the movement of the vertebral column but, even more important, form the "natural girdle," or the major portion of the abdominal body wall.

Text continues on page 193.

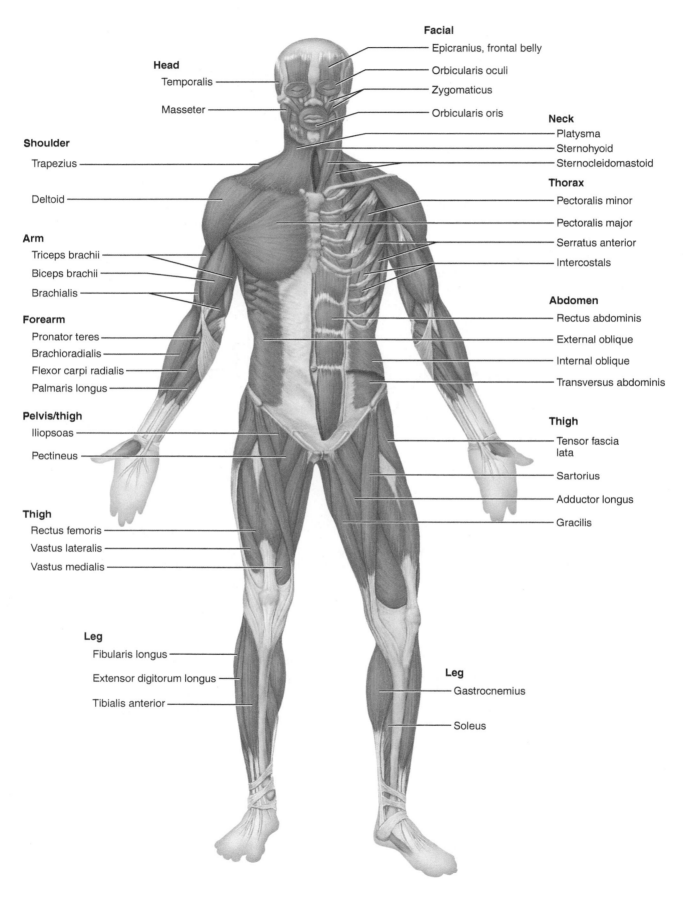

Facial
Epicranius, frontal belly
Orbicularis oculi
Zygomaticus
Orbicularis oris

Head
Temporalis
Masseter

Neck
Platysma
Sternohyoid
Sternocleidomastoid

Shoulder
Trapezius
Deltoid

Thorax
Pectoralis minor
Pectoralis major
Serratus anterior
Intercostals

Arm
Triceps brachii
Biceps brachii
Brachialis

Abdomen
Rectus abdominis
External oblique
Internal oblique
Transversus abdominis

Forearm
Pronator teres
Brachioradialis
Flexor carpi radialis
Palmaris longus

Pelvis/thigh
Iliopsoas
Pectineus

Thigh
Tensor fascia lata
Sartorius
Adductor longus
Gracilis

Thigh
Rectus femoris
Vastus lateralis
Vastus medialis

Leg
Fibularis longus
Extensor digitorum longus
Tibialis anterior

Leg
Gastrocnemius
Soleus

Figure 12.2 Anterior view of superficial muscles of the body. The abdominal surface has been partially dissected on the left side of the body to show somewhat deeper muscles.

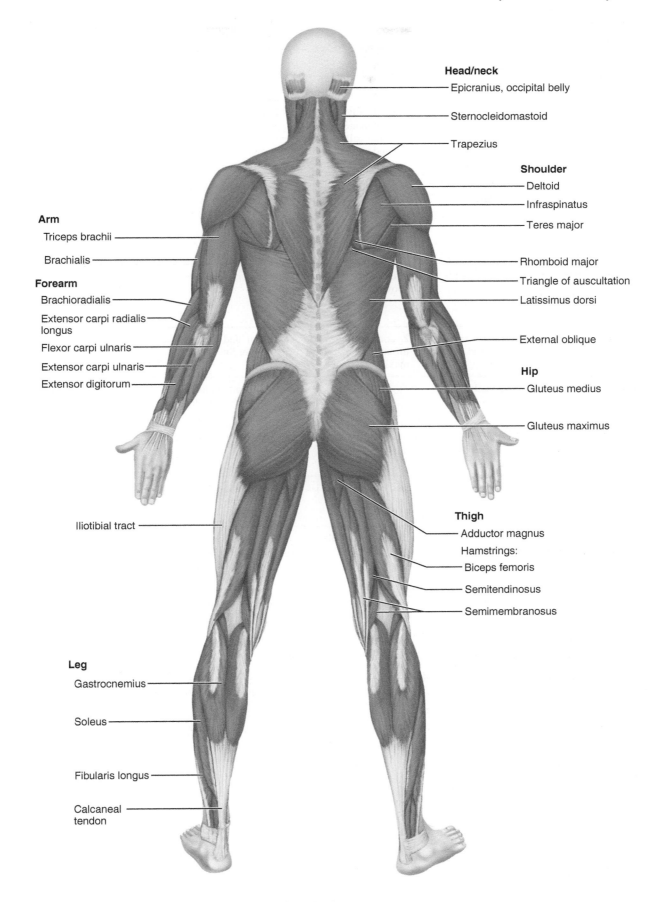

Head/neck
Epicranius, occipital belly

Sternocleidomastoid

Trapezius

Shoulder
Deltoid

Infraspinatus

Teres major

Rhomboid major

Triangle of auscultation

Latissimus dorsi

External oblique

Hip
Gluteus medius

Gluteus maximus

Arm
Triceps brachii

Brachialis

Forearm
Brachioradialis

Extensor carpi radialis longus

Flexor carpi ulnaris

Extensor carpi ulnaris

Extensor digitorum

Iliotibial tract

Thigh
Adductor magnus

Hamstrings:
Biceps femoris

Semitendinosus

Semimembranosus

Leg
Gastrocnemius

Soleus

Fibularis longus

Calcaneal tendon

Figure 12.3 Posterior view of superficial muscles of the body.

Table 12.1   Major Muscles of the Human Head (see Figure 12.4)

Muscle	Comments	Origin	Insertion	Action
Facial Expression (Figure 12.4)				
Epicranius— frontal and occipital bellies	Bipartite muscle consisting of frontal and occipital bellies, which covers dome of skull	Frontal belly—epicranial aponeurosis; occipital belly—occipital and temporal bones	Frontal belly—skin of eyebrows and root of nose; occipital belly—epicranial aponeurosis	With aponeurosis fixed, frontal belly raises eyebrows; occipital belly fixes aponeurosis and pulls scalp posteriorly
Orbicularis oculi	Tripartite sphincter muscle of eyelids	Frontal and maxillary bones and ligaments around orbit	Tissue of eyelid	Closes eye; various parts can be activated individually; produces blinking, squinting, and draws eyebrows inferiorly
Corrugator supercilii	Small muscle; acts with orbicularis oculi	Arch of frontal bone above nasal bone	Skin of eyebrow	Draws eyebrows together and inferiorly; wrinkles skin of forehead vertically
Levator labii superioris	Thin muscle between orbicularis oris and inferior eye margin	Zygomatic bone and infraorbital margin of maxilla	Skin and muscle of upper lip	Opens lips; raises and furrows upper lip
Zygomaticus— major and minor	Extends diagonally from corner of mouth to cheekbone	Zygomatic bone	Skin and muscle at corner of mouth	Raises lateral corners of mouth upward (smiling muscle)
Risorius	Slender muscle; runs inferior and lateral to zygomaticus	Fascia of masseter muscle	Skin at angle of mouth	Draws corner of lip laterally; tenses lip; zygomaticus synergist
Depressor labii inferioris	Small muscle running from lower lip to mandible	Body of mandible lateral to its midline	Skin and muscle of lower lip	Draws lower lip inferiorly
Depressor anguli oris	Small muscle lateral to depressor labii inferioris	Body of mandible below incisors	Skin and muscle at angle of mouth below insertion of zygomaticus	Draws corners of mouth downward and laterally; zygomaticus antagonist
Orbicularis oris	Multilayered muscle of lips with fibers that run in many different directions; most run circularly	Arises indirectly from maxilla and mandible; fibers blended with fibers of other muscles associated with lips	Encircles mouth; inserts into muscle and skin at angles of mouth	Closes lips; purses and protrudes lips (kissing and whistling muscle)
Mentalis	One of muscle pair forming V-shaped muscle mass on chin	Mandible below incisors	Skin of chin	Protrudes lower lip; wrinkles chin
Buccinator	Principal muscle of cheek; runs horizontally, deep to the masseter	Molar region of maxilla and mandible	Orbicularis oris	Draws corner of mouth laterally; compresses cheek (as in whistling); holds food between teeth during chewing

(Table continues on page 190.)

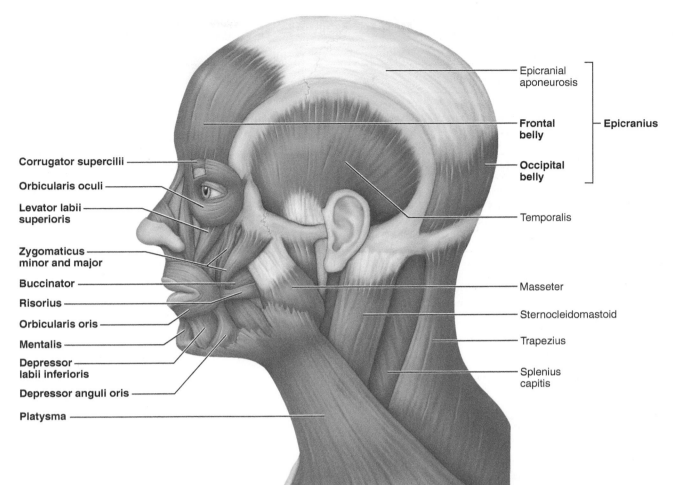

(a)

Epicranial
aponeurosis

**Frontal
belly**   — Epicranius

**Occipital
belly**

Temporalis

Corrugator supercilii

Orbicularis oculi

Levator labii
superioris

Zygomaticus
minor and major

Buccinator

Risorius

Orbicularis oris

Mentalis

Depressor
labii inferioris

Depressor anguli oris

Platysma

Masseter

Sternocleidomastoid

Trapezius

Splenius
capitis

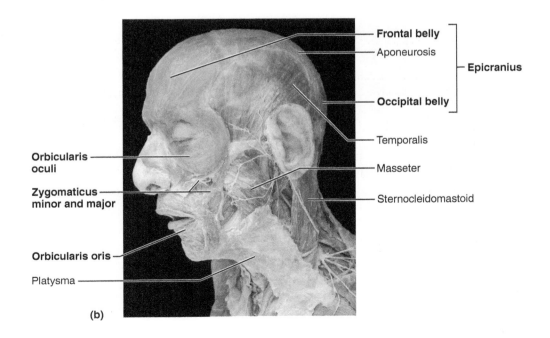

Frontal belly

Aponeurosis   — **Epicranius**

Occipital belly

Temporalis

Masseter

Sternocleidomastoid

Orbicularis
oculi

Zygomaticus
minor and major

Orbicularis oris

Platysma

(b)

Figure 12.4 Muscles of the head (left lateral view). (a) Superficial muscles.
(b) Cadaver photo of superficial structures of the head and neck.

Table 12.1 Major Muscles of the Human Head (see Figure 12.4) *(continued)*

Muscle	Comments	Origin	Insertion	Action
Muscles of Mastication (Figure 12.4c and d)				
Masseter	Covers lateral aspect of mandibular ramus; can be palpated on forcible closure of jaws	Zygomatic arch and zygomatic bone	Angle and ramus of mandible	Prime mover of jaw closure and elevates mandible
Temporalis	Fan-shaped muscle lying over parts of frontal, parietal, and temporal bones	Temporal fossa	Coronoid process of mandible	Closes jaw; elevates and retracts mandible
Buccinator	(See muscles of facial expression.)			
Medial pterygoid	Runs along internal (medial) surface of mandible (thus largely concealed by that bone)	Sphenoid, palatine, and maxillary bones	Medial surface of mandible, near its angle	Synergist of temporalis and masseter; elevates mandible; in conjunction with lateral pterygoid, aids in grinding movements of teeth
Lateral pterygoid	Superior to medial pterygoid	Greater wing of sphenoid bone	Condylar process of mandible	Protracts mandible (moves it anteriorly); in conjunction with medial pterygoid, aids in grinding movements of teeth

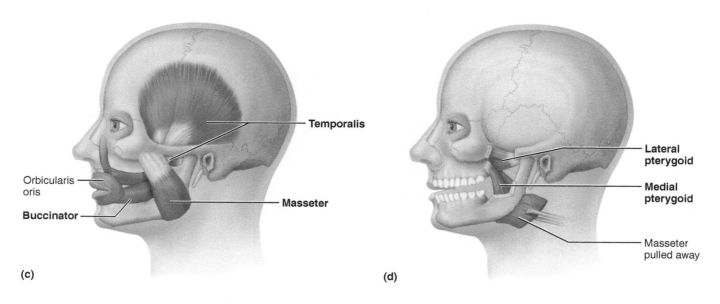

(c) (d)

Figure 12.4 *(continued)* **Muscles of the head: (mastication). (c)** Lateral view of the temporalis, masseter, and buccinator muscles. **(d)** Lateral view of the deep chewing muscles, the medial and lateral pterygoid muscles.

Table 12.2   Anterolateral Muscles of the Human Neck (see Figure 12.5)

Muscle	Comments	Origin	Insertion	Action
Superficial				
Platysma	Unpaired, thin, sheet-like superficial neck muscle, plays role in facial expression (see also Figure 12.4a)	Fascia of chest (over pectoral muscles and deltoid)	Lower margin of mandible, skin, and muscle at corner of mouth	Depresses mandible; pulls lower lip back and down (i.e., produces downward sag of the mouth); tenses skin of neck
Sternocleidomastoid	Two-headed muscle located deep to platysma on anterolateral surface of neck; fleshy parts on either side indicate limits of anterior and posterior triangles of neck	Manubrium of sternum and medial portion of clavicle	Mastoid process of temporal bone and superior nuchal line of occipital bone	Simultaneous contraction of both muscles of pair causes flexion of head; acting independently, rotate head toward shoulder on opposite side
Scalenes—anterior, middle, and posterior	Located more on lateral than anterior neck; deep to platysma and sternocleidomastoid (see Figure 12.5c)	Transverse processes of cervical vertebrae	Anterolaterally on ribs 1–2	Elevate ribs 1–2 (aid in inspiration); flex and slightly rotate neck

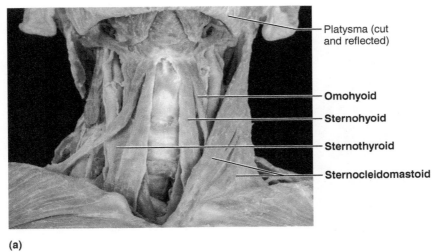

(a)

— Platysma (cut and reflected)

— Omohyoid

— Sternohyoid

— Sternothyroid

— Sternocleidomastoid

Figure 12.5 Muscles of the anterolateral neck and throat. (a) Cadaver photo of the anterior and lateral regions of the neck.

Table 12.2   Anterolateral Muscles of the Human Neck (see Figure 12.5) *(continued)*

Muscle	Comments	Origin	Insertion	Action
Deep (Figure 12.5a and b)				
Digastric	Consists of two bellies united by an intermediate tendon; assumes a V-shape under chin	Lower margin of mandible (anterior belly) and mastoid process (posterior belly)	By a connective tissue loop to hyoid bone	Open mouth and depress mandible; acting together, elevate hyoid bone
Stylohyoid	Slender muscle parallels posterior border of digastric; below angle of jaw	Styloid process of temporal bone	Hyoid bone	Elevates and retracts hyoid bone
Mylohyoid	Just deep to digastric; forms floor of mouth	Medial surface of mandible	Hyoid bone and median raphe	Elevates hyoid bone and floor of mouth during swallowing
Sternohyoid	Runs most medially along neck	Manubrium and medial end of clavicle	Lower margin of hyoid bone	Depresses larynx and hyoid bone if mandible is fixed; may also flex skull
Sternothyroid	Lateral and deep to sternohyoid	Posterior surface of manubrium	Thyroid cartilage	Depresses larynx and hyoid bone
Omohyoid	Straplike with two bellies; lateral to sternohyoid	Superior surface of scapula	Hyoid bone; inferior border	Depresses and retracts hyoid bone
Thyrohyoid	Appears as a superior continuation of sternothyroid muscle	Thyroid cartilage	Hyoid bone	Depresses hyoid bone; elevates larynx if hyoid is fixed

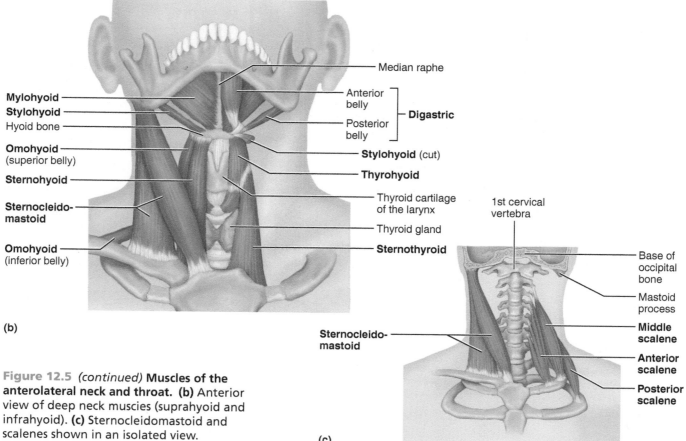

Figure 12.5 *(continued)* **Muscles of the anterolateral neck and throat. (b)** Anterior view of deep neck muscles (suprahyoid and infrahyoid). **(c)** Sternocleidomastoid and scalenes shown in an isolated view.

Identifying Muscles of the Trunk

Read the descriptions of specific trunk muscles, and identify them in the figures (**Table 12.3** and **Table 12.4** and **Figures 12.6**, **12.7**, **12.8**, and **12.9**), visualizing their action when they contract. Then identify them on a torso model or anatomical chart.

Demonstrating Operation of the Trunk Muscles

Now, work with a partner to demonstrate the operation of the following muscles. One of you can demonstrate the

movement; the following steps are addressed to this partner. The other can supply resistance and palpate the muscle being tested.

1. Fully abduct the arm and extend at the elbow. Now adduct the arm against resistance. You are using the *latissimus dorsi.*

2. To observe the action of the *deltoid,* try to abduct your arm against resistance. Now attempt to elevate your shoulder against resistance; you are contracting the upper portion of the *trapezius.*

3. You use the *pectoralis major* when you press your hands together at chest level. ■

Text continues on page 199.

Table 12.3	Anterior Muscles of the Human Thorax, Shoulder, and Abdominal Wall (see Figures 12.6, 12.7, and 12.8)			
Muscle	**Comments**	**Origin**	**Insertion**	**Action**
Thorax and Shoulder, Superficial (Figure 12.6)				
Pectoralis major	Large fan-shaped muscle covering upper portion of chest	Clavicle, sternum, cartilage of ribs 1–6 (or 7), and aponeurosis of external oblique muscle	Fibers converge to insert by short tendon into intertubercular sulcus of humerus	Prime mover of arm flexion; adducts, medially rotates arm; with arm fixed, pulls chest upward (thus also acts in forced inspiration)
Serratus anterior	Fan-shaped muscle deep to scapula; deep and inferior to pectoral muscles on lateral rib cage	Lateral aspect of ribs 1–8 (or 9)	Vertebral border of anterior surface of scapula	Prime mover to protect and hold scapula against chest wall; rotates scapula, causing inferior angle to move laterally and upward; abduction and raising of arm (called "boxer's muscle")
Deltoid (see also Figure 12.9a)	Fleshy triangular muscle forming shoulder muscle mass; intramuscular injection site	Lateral third of clavicle; acromion and spine of scapula	Deltoid tuberosity of humerus	Acting as a whole, prime mover of arm abduction; when only specific fibers are active, can act as a synergist in flexion, extension, and rotation of arm
Pectoralis minor	Flat, thin muscle directly beneath and obscured by pectoralis major	Anterior surface of ribs 3–5, near their costal cartilages	Coracoid process of scapula	With ribs fixed, draws scapula forward and inferiorly; with scapula fixed, draws rib cage superiorly
Thorax, Deep: Muscles of Respiration (Figure 12.7)				
External intercostals	11 pairs lie between ribs; fibers run obliquely downward and forward toward sternum	Inferior border of rib above (not shown in figure)	Superior border of rib below	Pull ribs toward one another to elevate rib cage; aid in inspiration
Internal intercostals	11 pairs lie between ribs; fibers run deep and at right angles to those of external intercostals	Superior border of rib below	Inferior border of rib above (not shown in figure)	Draw ribs together to depress rib cage; aid in forced expiration; antagonistic to external intercostals
Diaphragm	Broad muscle; forms floor of thoracic cavity; dome-shaped in relaxed state; fibers converge toward a central tendon	Inferior border of rib cage and sternum, costal cartilages of last six ribs and lumbar vertebrae	Central tendon	Prime mover of inspiration: flattens on contraction, increasing vertical dimensions of thorax; increases intra-abdominal pressure

Table 12.3 Anterior Muscles of the Human Thorax, Shoulder, and Abdominal Wall (see Figures 12.6, 12.7, and 12.8) *(continued)*

Muscle	Comments	Origin	Insertion	Action
Abdominal Wall (Figure 12.8)				
Rectus abdominis	Medial superficial muscle, extends from pubis to rib cage; ensheathed by aponeuroses of oblique muscles; segmented	Pubic crest and symphysis	Xiphoid process and costal cartilages of ribs 5–7	Flexes vertebral column; increases abdominal pressure; fixes and depresses ribs; stabilizes pelvis during walking; used in sit-ups and curls
External oblique	Most superficial lateral muscle; fibers run downward and medially; ensheathed by an aponeurosis	Anterior surface of lower eight ribs	Linea alba,* pubic crest and tubercles, and iliac crest	Flex vertebral column and compress abdominal wall; also aids muscles of back in trunk rotation and lateral flexion; used in oblique curls
Internal oblique	Most fibers run at right angles to those of external oblique, which it underlies	Lumbar fascia, iliac crest, and inguinal ligament	Linea alba, pubic crest, and costal cartilages of lower three ribs	As for external oblique
Transversus abdominis	Deepest muscle of abdominal wall; fibers run horizontally	Inguinal ligament, iliac crest, cartilages of last five or six ribs, and lumbar fascia	Linea alba and pubic crest	Compresses abdominal contents

*The linea alba (white line) is a narrow, tendinous sheath that runs along the middle of the abdomen from the sternum to the pubic symphysis. It is formed by the fusion of the aponeurosis of the external oblique and transversus muscles.

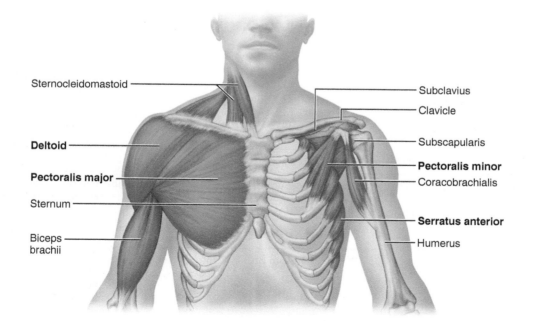

Figure 12.6 Superficial muscles of the thorax and shoulder acting on the scapula and arm (anterior view). The superficial muscles, which move the arm, are shown on the left. These muscles have been removed on the right side of the figure to show the muscles that stabilize or move the pectoral girdle.

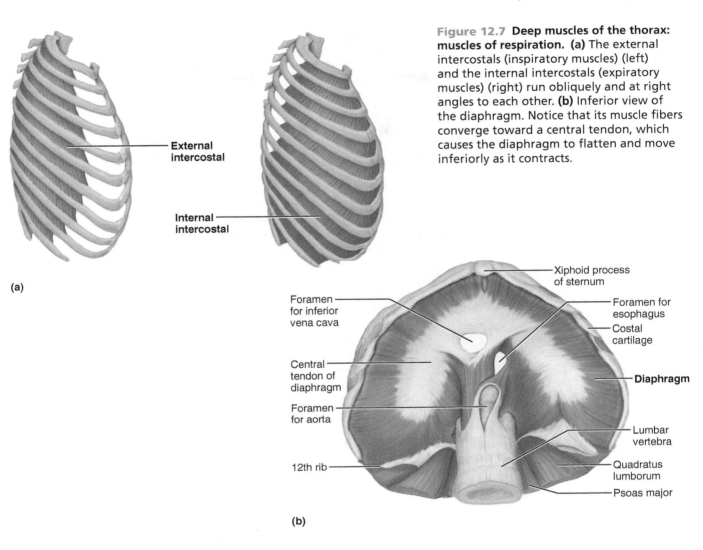

(a)

Figure 12.7 Deep muscles of the thorax: muscles of respiration. (a) The external intercostals (inspiratory muscles) (left) and the internal intercostals (expiratory muscles) (right) run obliquely and at right angles to each other. **(b)** Inferior view of the diaphragm. Notice that its muscle fibers converge toward a central tendon, which causes the diaphragm to flatten and move inferiorly as it contracts.

External intercostal

Internal intercostal

Xiphoid process of sternum

Foramen for inferior vena cava

Foramen for esophagus

Costal cartilage

Central tendon of diaphragm

Diaphragm

Foramen for aorta

Lumbar vertebra

12th rib

Quadratus lumborum

Psoas major

(b)

Figure 12.8 Anterior view of the muscles forming the anterolateral abdominal wall. (a) The superficial muscles have been partially cut away on the left side of the diagram to reveal the deeper internal oblique and transversus abdominis muscles.

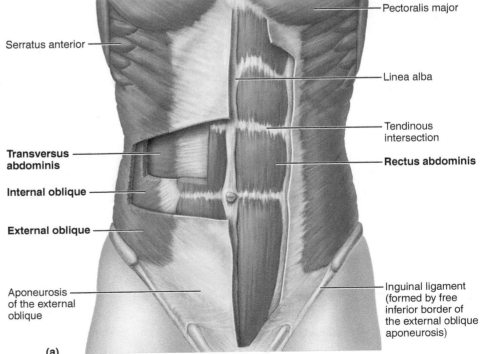

Serratus anterior

Pectoralis major

Linea alba

Tendinous intersection

Transversus abdominis

Rectus abdominis

Internal oblique

External oblique

Aponeurosis of the external oblique

Inguinal ligament (formed by free inferior border of the external oblique aponeurosis)

(a)

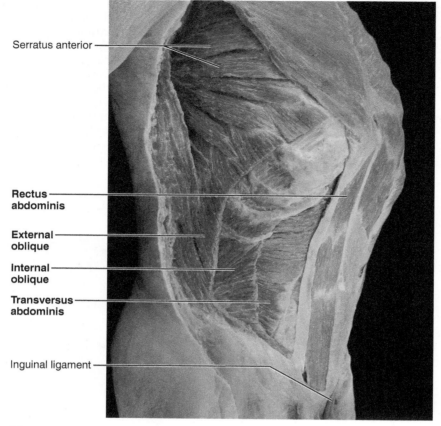

Serratus anterior

Rectus abdominis

External oblique

Internal oblique

Transversus abdominis

Inguinal ligament

(b)

Figure 12.8 *(continued)* **Anterior view of the muscles forming the anterolateral abdominal wall.** **(b)** Cadaver photo of the anterolateral abdominal wall.

Table 12.4   Posterior Muscles of the Human Trunk (see Figure 12.9)

Muscle	Comments	Origin	Insertion	Action
Muscles of the Neck, Shoulder, and Thorax (Figure 12.9a)				
Trapezius	Most superficial muscle of posterior thorax; very broad origin and insertion	Occipital bone; ligamentum nuchae; spines of C_7 and all thoracic vertebrae	Acromion and spine of scapula; lateral third of clavicle	Stabilizes, elevates, rotates, and retracts scapula; extends the head; superior fibers elevate scapula (as in shrugging the shoulders); inferior fibers depress scapula
Latissimus dorsi	Broad flat muscle of lower back (lumbar region); extensive superficial origins	Indirect attachment to spinous processes of lower six thoracic vertebrae, lumbar vertebrae, last three to four ribs, and iliac crest	Floor of intertubercular sulcus of humerus	Prime mover of arm extension; adducts and medially rotates arm; brings arm down in power stroke, as in striking a blow
Infraspinatus	A rotator cuff muscle; partially covered by deltoid and trapezius	Infraspinous fossa of scapula	Greater tubercle of humerus	Lateral rotation of humerus; helps hold head of humerus in glenoid cavity; stabilizes shoulder
Teres minor	A rotator cuff muscle; small muscle inferior to infraspinatus	Lateral margin of scapula posterior	Greater tubercle of humerus	Same as for infraspinatus

Table 12.4 *(continued)*

Muscle	Comments	Origin	Insertion	Action
Teres major	Located inferiorly to teres minor	Posterior surface at inferior angle of scapula	Intertubercular sulcus of humerus	Extends, medially rotates, and adducts humerus; synergist of latissimus dorsi
Supraspinatus	A rotator cuff muscle; obscured by trapezius	Supraspinous fossa of scapula	Greater tubercle of humerus	Initiates abduction of humerus; stabilizes shoulder joint
Levator scapulae	Located at back and side of neck, deep to trapezius	Transverse processes of C_1-C_4	Medial border of scapula superior to spine	Elevates and adducts scapula; with fixed scapula, laterally flexes neck to the same side
Rhomboids—major and minor	Beneath trapezius and inferior to levator scapulae; rhomboid minor is the more superior muscle	Spinous processes of C_7 and T_1-T_5	Medial border of scapula	Pulls scapula medially (retraction); stabilizes scapula; rotates glenoid cavity downward

Muscles Associated with the Vertebral Column (Figure 12.9b)

Muscle	Comments	Origin	Insertion	Action
Semispinalis	Deep composite muscle of the back—thoracis, cervicis, and capitis portions	Transverse processes of C_7-T_{12}	Occipital bone and spinous processes of cervical vertebrae and T_1-T_4	Acting together, extend head and vertebral column; independently cause rotation toward opposite side

(Table continues on page 199.)

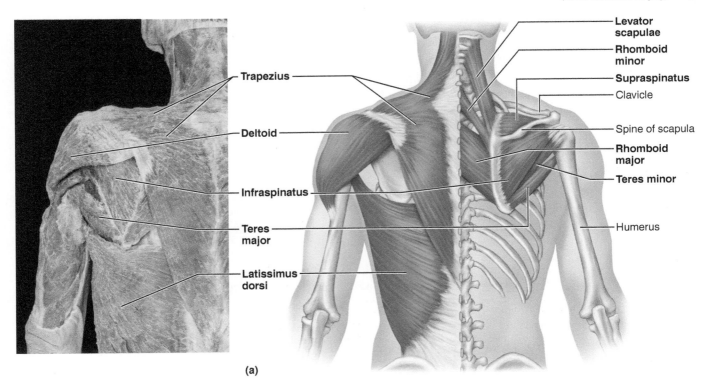

(a)

Figure 12.9 Muscles of the neck, shoulder, and thorax (posterior view). (a) The superficial muscles of the back are shown for the left side of the body, with a corresponding cadaver photograph. The superficial muscles are removed on the right side of the diagram to reveal the deeper muscles acting on the scapula and the rotator cuff muscles that help to stabilize the shoulder joint.

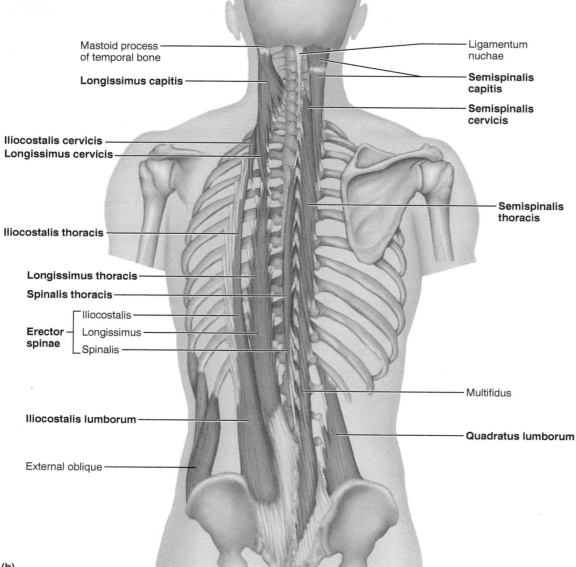

Mastoid process of temporal bone

Longissimus capitis

Ligamentum nuchae

Semispinalis capitis

Semispinalis cervicis

Iliocostalis cervicis
Longissimus cervicis

Semispinalis thoracis

Iliocostalis thoracis

Longissimus thoracis

Spinalis thoracis

Erector spinae — [Iliocostalis — Longissimus — Spinalis]

Multifidus

Iliocostalis lumborum

Quadratus lumborum

External oblique

(b)

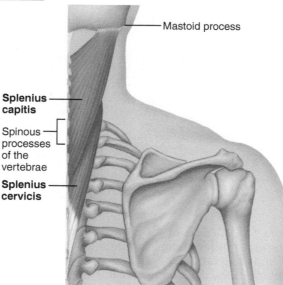

Mastoid process

Splenius capitis

Spinous processes of the vertebrae

Splenius cervicis

(c)

Figure 12.9 *(continued)* **Muscles of the neck, shoulder, and thorax (posterior view). (b)** The erector spinae and semispinalis muscles, which respectively form the intermediate and deep muscle layers of the back associated with the vertebral column. **(c)** Deep (splenius) muscles of the posterior neck. Superficial muscles have been removed.

Table 12.4   Posterior Muscles of the Human Trunk (see Figure 12.9) *(continued)*

Muscle	Comments	Origin	Insertion	Action
Muscles Associated with the Vertebral Column (Figure 12.9b), *(continued)*				
Erector spinae	A long tripartite muscle composed of iliocostalis (lateral), longissimus, and spinalis (medial) muscle columns; superficial to semispinalis muscles; extends from pelvis to head	Iliac crest, transverse processes of lumbar, thoracic, and cervical vertebrae, and/or ribs 3–6 depending on specific part	Ribs and transverse processes of vertebrae about six segments above origin; longissimus also inserts into mastoid process	Extend and laterally flex the vertebral column; fibers of the longissimus also extend and rotate head
Splenius (see Figure 12.9c)	Superficial muscle (capitis and cervicis parts) extending from upper thoracic region to skull	Ligamentum nuchae and spinous processes of C_7–T_6	Mastoid process, occipital bone, and transverse processes of C_2–C_4	As a group, extend or hyperextend head; when only one side is active, head is rotated and bent toward the same side
Quadratus lumborum	Forms greater portion of posterior abdominal wall	Iliac crest and lumbar fascia	Inferior border of rib 12; transverse processes of lumbar vertebrae	Each flexes vertebral column laterally; together extend the lumbar spine and fix rib 12; maintains upright posture

Muscles of the Upper Limb

The muscles that act on the upper limb fall into three groups: those moving the arm, those causing movement of the forearm, and those moving the hand and fingers.

The muscles that cross the shoulder joint to insert on the humerus and move the arm (subscapularis, supraspinatus and infraspinatus, deltoid, and so on) are primarily trunk muscles that originate on the axial skeleton or shoulder girdle. These muscles are included with the trunk muscles.

The second group of muscles, which cross the elbow joint and move the forearm, consists of muscles forming the musculature of the humerus. These muscles arise primarily from the humerus and insert in forearm bones. They are responsible for flexion, extension, pronation, and supination.

The third group forms the musculature of the forearm. For the most part, these muscles insert on the digits and produce movements of the hand and fingers.

ACTIVITY 3

Identifying Muscles of the Upper Limb

Study the origins, insertions, and actions of muscles that move the forearm, and identify them in the figure (**Table 12.5** and **Figure 12.10**).

Do the same for muscles acting on the wrist and hand (**Table 12.6** and **Figure 12.11**). You can identify them more easily if you locate their insertion tendons first. Muscles of the hand are described in **Table 12.7** and illustrated in **Figure 12.12**.

As before, identify these muscles on a torso model, anatomical chart, or cadaver.

Demonstrating Operations of the Upper Limb Muscles

1. To observe the *biceps brachii,* attempt to flex your forearm (hand supinated) against resistance. You can also feel the insertion tendon of this biceps muscle in the lateral aspect of the cubital fossa (where it runs toward the radius to attach).

2. If you acutely flex at your elbow and then try to extend the forearm against resistance, you can demonstrate the action of your *triceps brachii.*

3. Strongly flex your hand, and make a fist. Palpate your contracting wrist flexor muscles (which originate from the medial epicondyle of the humerus) and their insertion tendons, which you can easily feel at the anterior aspect of the wrist.

4. Flare your fingers to identify the tendons of the *extensor digitorum* muscle on the dorsum of your hand. ■

Text continues on page 206.

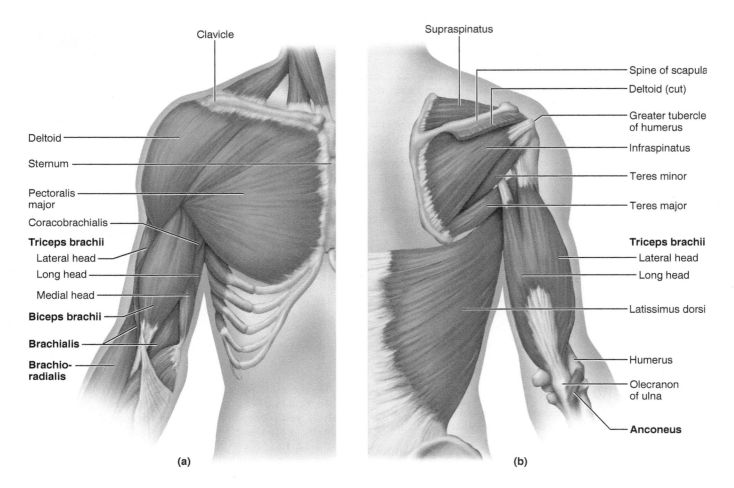

Figure 12.10 Muscles acting on the arm and forearm. (a) Superficial muscles of the anterior thorax, shoulder, and arm, anterior view. **(b)** Posterior aspect of the arm showing the lateral and long heads of the triceps brachii muscle. The supraspinatus, infraspinatus, teres minor, and subscapularis (Figure 12.6) are rotator cuff muscles.

Table 12.5	Muscles of the Human Humerus That Move the Forearm (see Figure 12.10)			
Muscle	**Comments**	**Origin**	**Insertion**	**Action**
Triceps brachii	Large fleshy muscle of posterior humerus; three-headed origin	Long head—inferior margin of glenoid cavity; lateral head—posterior humerus; medial head—distal radial groove on posterior humerus	Olecranon of ulna	Powerful forearm extensor; antagonist of forearm flexors (brachialis and biceps brachii)
Anconeus (see also Figure 12.11d and e)	Short triangular muscle blended with triceps	Lateral epicondyle of humerus	Lateral aspect of olecranon of ulna	Abducts ulna during forearm pronation; synergist of triceps brachii in forearm extension
Biceps brachii	Most familiar muscle of anterior humerus because this two-headed muscle bulges when forearm is flexed	Short head: coracoid process; long head; supraglenoid tubercle and tip of glenoid cavity; tendon of long head runs in intertubercular sulcus and within capsule of shoulder joint	Radial tuberosity	Flexion (powerful) and supination of forearm; "it turns the corkscrew and pulls the cork"; weak arm flexor
Brachioradialis (see also Figure 12.11a)	Superficial muscle of lateral forearm; forms lateral boundary of cubital fossa	Lateral ridge at distal end of humerus	Base of radial styloid process	Synergist in forearm flexion
Brachialis	Immediately deep to biceps brachii	Distal portion of anterior humerus	Coronoid process of ulna	Flexor of forearm

Table 12.6	Muscles of the Human Forearm That Move the Hand and Fingers (see Figure 12.11)			
Muscle	Comments	Origin	Insertion	Action
Anterior Compartment (Figure 12.11a, b, c)				
Superficial				
Pronator teres	Seen in a superficial view between proximal margins of brachioradialis and flexor carpi radialis	Medial epicondyle of humerus and coronoid process of ulna	Midshaft of radius	Acts synergistically with pronator quadratus to pronate forearm; weak forearm flexor
Flexor carpi radialis	Superficial; runs diagonally across forearm	Medial epicondyle of humerus	Base of metacarpals II and III	Powerful flexor and abductor of the hand
Palmaris longus	Small fleshy muscle with a long tendon; medial to flexor carpi radialis	Medial epicondyle of humerus	Palmar aponeurosis; skin and fascia of palm	Flexes hand (weak); tenses skin and fascia of palm

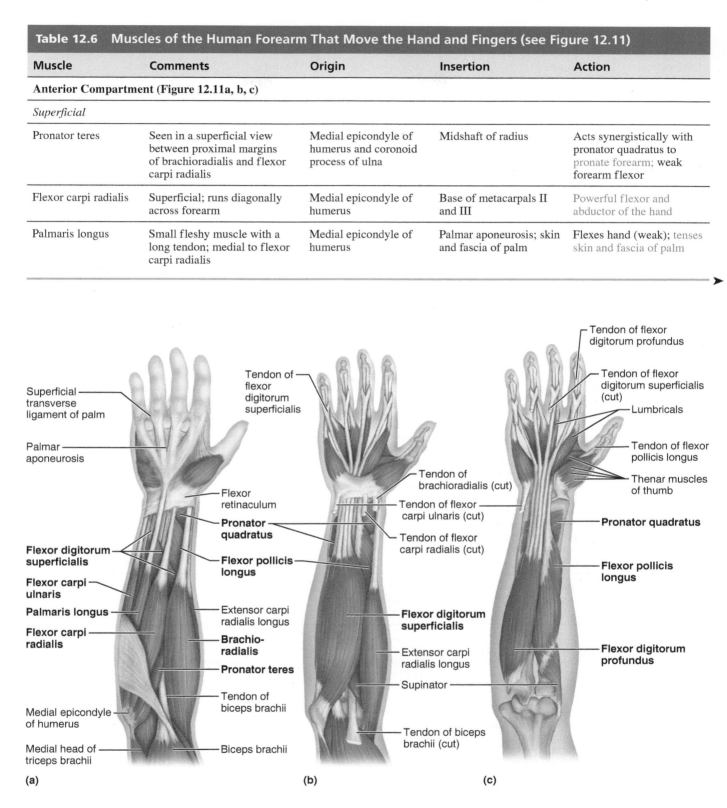

(a)   (b)   (c)

Figure 12.11 Muscles of the forearm and wrist. (a) Superficial anterior view of right forearm and hand. **(b)** The brachioradialis, flexors carpi radialis and ulnaris, and palmaris longus muscles have been removed to reveal the position of the somewhat deeper flexor digitorum superficialis. **(c)** Deep muscles of the anterior compartment. Superficial muscles have been removed. *Note:* The thenar muscles of the thumb and the lumbricals that help move the fingers are illustrated here but not described in the table (Table 12.6).

Table 12.6   Muscles of the Human Forearm That Move the Hand and Fingers (see Figure 12.11) *(continued)*

Muscle	Comments	Origin	Insertion	Action
Flexor carpi ulnaris	Superficial; medial to palmaris longus	Medial epicondyle of humerus, olecranon process and posterior surface of ulna	Base of metacarpal V; pisiform and hamate bones	Flexes and adducts hand
Flexor digitorum superficialis	Deeper muscle (deep to muscles named above); visible at distal end of forearm	Medial epicondyle of humerus, coronoid process of ulna, and shaft of radius	Middle phalanges of fingers II–V	Flexes hand and middle phalanges of fingers II–V
Deep				
Flexor pollicis longus	Deep muscle of anterior forearm; distal to and paralleling lower margin of flexor digitorum superficialis	Anterior surface of radius, and interosseous membrane	Distal phalanx of thumb	Flexes distal phalanx of thumb
Flexor digitorum profundus	Deep muscle; overlain entirely by flexor digitorum superficialis	Anteromedial surface of ulna, interosseous membrane, and coronoid process	Distal phalanges of fingers II–V	Sole muscle that flexes distal phalanges; assists in hand flexion
Pronator quadratus	Deepest muscle of distal forearm	Distal portion of anterior ulnar surface	Anterior surface of radius, distal end	Pronates forearm
Posterior Compartment (Figure 12.11d, e, f)				
Superficial				
Extensor carpi radialis longus	Superficial; parallels brachioradialis on lateral forearm	Lateral supracondylar ridge of humerus	Base of metacarpal II	Extends and abducts hand
Extensor carpi radialis brevis	Deep to extensor carpi radialis longus	Lateral epicondyle of humerus	Base of metacarpal III	Extends and abducts wrist; steadies hand during finger flexion
Extensor digitorum	Superficial; medial to extensor carpi radialis brevis	Lateral epicondyle of humerus	By four tendons into distal phalanges of fingers II–V	Prime mover of finger extension; extends hand; can abduct (flare) fingers
Extensor carpi ulnaris	Superficial; medial posterior forearm	Lateral epicondyle of humerus; posterior border of ulna	Base of metacarpal V	Extends and adducts hand
Deep				
Extensor pollicis longus and brevis	Muscle pair with a common origin and action; deep to extensor carpi ulnaris	Dorsal shaft of ulna and radius, interosseous membrane	Base of distal phalanx of thumb (longus) and proximal phalanx of thumb (brevis)	Extends thumb
Abductor pollicis longus	Deep muscle; lateral and parallel to extensor pollicis longus	Posterior surface of radius and ulna; interosseous membrane	Metacarpal I and trapezium	Abducts and extends thumb
Supinator	Deep muscle at posterior aspect of elbow	Lateral epicondyle of humerus; proximal ulna	Proximal end of radius	Synergist of biceps brachii to supinate forearm; antagonist of pronator muscles

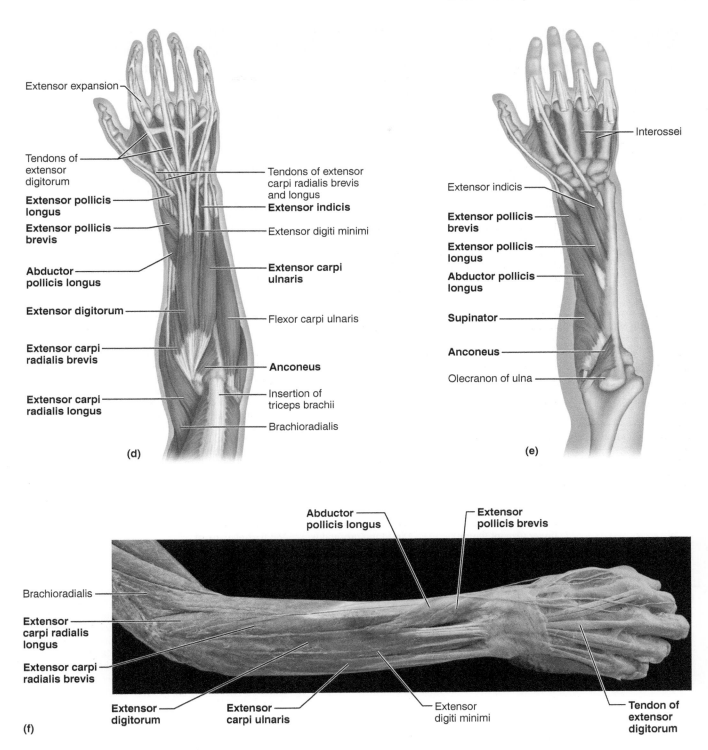

Figure 12.11 *(continued)* **Muscles of the forearm and wrist. (d)** Superficial muscles, posterior view. **(e)** Deep posterior muscles; superficial muscles have been removed. The interossei, the deepest layer of intrinsic hand muscles, are also illustrated. **(f)** Cadaver photo of posterior muscles of the right forearm.

Table 12.7　Intrinsic Muscles of the Hand: Fine Movements of the Fingers (see Figure 12.12)

Muscle	Comments	Origin	Insertion	Action
Thenar Muscles in Ball of Thumb (Figure 12.12a and b)				
Abductor pollicis brevis	Lateral muscle of thenar group; superficial	Flexor retinaculum and nearby carpals	Lateral base of thumb's proximal phalanx	Abducts thumb (at carpo-metacarpal joint)
Flexor pollicis brevis	Medial and deep muscle of thenar group	Flexor retinaculum and trapezium	Lateral side of base of proximal phalanx of thumb	Flexes thumb (at carpometa-carpal and metacarpophalan-geal joints)
Opponens pollicis	Deep to abductor pollicis brevis, on metacarpal I	Flexor retinaculum and trapezium	Whole anterior side of metacarpal I	Opposition: moves thumb to touch tip of little finger
Adductor pollicis	Fan-shaped with horizontal fibers; distal to other thenar muscles; oblique and transverse heads	Capitate bone and bases of metacarpals II–IV (oblique head); front of metacarpal III (transverse head)	Medial side of base of proximal phalanx of thumb	Adducts and helps to oppose thumb
Hypothenar Muscles in Ball of Little Finger (Figure 12.12a and b)				
Abductor digiti minimi	Medial muscle of hypothenar group; superficial	Pisiform bone	Medial side of proximal phalanx of little finger	Abducts little finger at metacarpophalangeal joint
Flexor digiti minimi brevis	Lateral deep muscle of hypothenar group	Hamate bone and flexor retinaculum	Same as abductor digiti minimi	Flexes little finger at metacarpophalangeal joint
Opponens digiti minimi	Deep to abductor digiti minimi	Same as flexor digiti minimi brevis	Most of length of medial side of metacarpal V	Helps in opposition: brings metacarpal V toward thumb to cup the hand
Midpalmar Muscles (Figure 12.12a, b, c, d)				
Lumbricals	Four worm-shaped muscles in palm, one to each finger (except thumb); odd because they originate from the tendons of another muscle	Lateral side of each tendon of flexor digitorum profundus in palm	Lateral edge of extensor expansion on first phalanx of fingers II–V	Flex fingers at metacarpophalangeal joints but extend fingers at interphalangeal joints
Palmar interossei	Four long, cone-shaped muscles; lie ventral to the dorsal interossei and between the metacarpals	The side of each metacarpal that faces the midaxis of the hand (metacarpal III) where it's absent	Extensor expansion on first phalanx of each finger (except finger III), on side facing midaxis of hand	Adduct fingers; pull fingers in toward third digit; act with lumbricals to extend fingers at interphalangeal joints and flex them at metacarpophalangeal joints
Dorsal interossei	Four bipennate muscles filling spaces between the metacarpals; deepest palm muscles, also visible on dorsal side of hand	Sides of metacarpals	Exterior expansion over first phalanx of fingers II–IV on side opposite midaxis of hand (finger III), but on *both* sides of finger III	Abduct fingers; extend fingers at interphalangeal joints and flex them at metacarpophalangeal joints

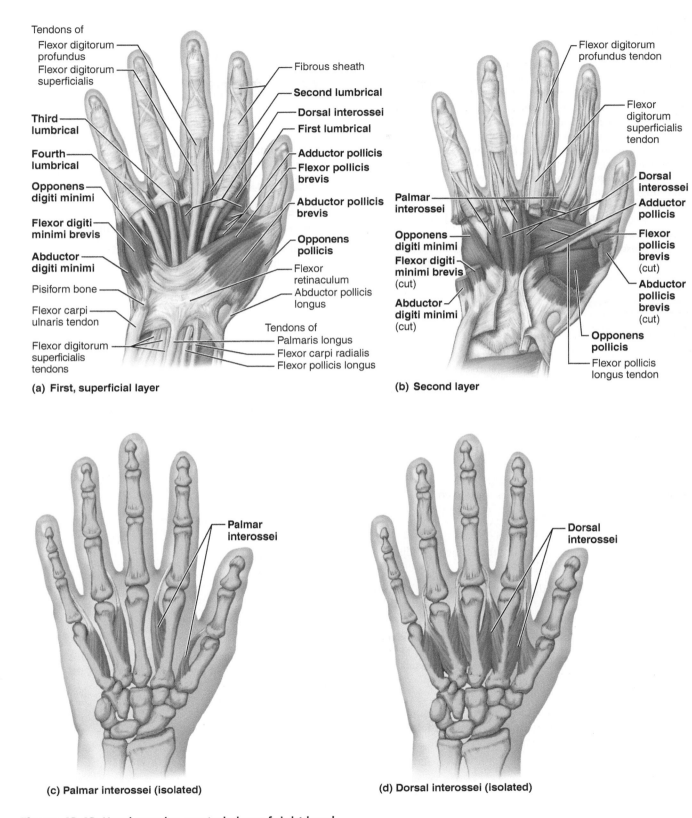

Tendons of
Flexor digitorum profundus
Flexor digitorum superficialis

Fibrous sheath

Second lumbrical

Dorsal interossei

First lumbrical

Third lumbrical

Fourth lumbrical

Opponens digiti minimi

Flexor digiti minimi brevis

Abductor digiti minimi

Pisiform bone

Flexor carpi ulnaris tendon

Flexor digitorum superficialis tendons

Adductor pollicis

Flexor pollicis brevis

Abductor pollicis brevis

Opponens pollicis

Flexor retinaculum

Abductor pollicis longus

Tendons of
Palmaris longus
Flexor carpi radialis
Flexor pollicis longus

(a) First, superficial layer

Flexor digitorum profundus tendon

Flexor digitorum superficialis tendon

Palmar interossei

Opponens digiti minimi

Flexor digiti minimi brevis (cut)

Abductor digiti minimi (cut)

Dorsal interossei

Adductor pollicis

Flexor pollicis brevis (cut)

Abductor pollicis brevis (cut)

Opponens pollicis

Flexor pollicis longus tendon

(b) Second layer

Palmar interossei

(c) Palmar interossei (isolated)

Dorsal interossei

(d) Dorsal interossei (isolated)

Figure 12.12 Hand muscles, ventral view of right hand.

Muscles of the Lower Limb

Muscles that act on the lower limb cause movement at the hip, knee, and ankle joints. Since the human pelvic girdle is made up of heavy fused bones that allow very little movement, no special group of muscles is necessary to stabilize it. This is unlike the shoulder girdle, where several muscles (mainly trunk muscles) are needed to stabilize the scapulae.

Muscles acting on the thigh (femur) cause various movements at the multiaxial hip joint. These include the iliopsoas, the adductor group, and others.

Muscles acting on the leg form the major musculature of the thigh. (Anatomically, the term *leg* refers only to that portion between the knee and the ankle.) The thigh muscles cross the knee to allow flexion and extension of the leg. They include the hamstrings and the quadriceps.

The muscles originating on the leg cross the ankle joint. These muscles, as well as the intrinsic muscles of the foot, act on the foot and toes.

ACTIVITY 4

Identifying Muscles of the Lower Limb

Read the descriptions of specific muscles acting on the thigh and leg, and identify them in the figures (Table 12.8, Table 12.9, Figure 12.13, and Figure 12.14), trying to visualize their action when they contract. Because some of the muscles acting on the leg also have attachments on the pelvic girdle, they can cause movement at the hip joint.

Do the same for muscles acting on the foot and toes (Table 12.10, Table 12.11, Figure 12.15, Figure 12.16, and Figure 12.17).

Identify the muscles acting on the thigh, leg, foot, and toes as instructed previously for other muscle groups.

Demonstrating Operations of the Lower Limb Muscles

Complete this activity by performing the following palpation demonstrations with your lab partner.

1. Go into a deep knee bend, and palpate your own *gluteus maximus* muscle as you extend your thigh to resume the upright posture.

2. Demonstrate the contraction of the anterior *quadriceps femoris* by trying to extend your leg against resistance. Do this while seated, and note how the quadriceps tendon reacts. The *biceps femoris* of the posterior thigh comes into play when you flex your leg against resistance.

3. Now stand on your toes. Have your partner palpate the lateral and medial heads of the *gastrocnemius* and follow it to its insertion in the calcaneal tendon.

4. Dorsiflex and invert your foot while palpating your *tibialis anterior* muscle (which parallels the sharp anterior crest of the tibia laterally). ▬

ACTIVITY 5

Review of Human Musculature

Review the muscles by labeling the figures in the Review Sheet (questions 11 and 12, pages 238–239). ▬

ACTIVITY 6

Making a Muscle Painting

1. Choose a male student to be "muscle painted."

2. Obtain brushes and water-based paints from the supply area while the "volunteer" removes his shirt and rolls up his pant legs (if necessary).

3. Using paints of different colors, identify the muscles listed below by painting his skin. If a muscle covers a large body area, you may opt to paint only its borders.

- biceps brachii
- deltoid
- erector spinae
- pectoralis major
- rectus femoris
- tibialis anterior
- triceps brachii
- vastus lateralis
- biceps femoris
- extensor carpi radialis longus
- latissimus dorsi
- rectus abdominis
- sternocleidomastoid
- trapezius
- triceps surae
- vastus medialis

4. Check your "human painting" with your instructor before cleaning your bench and leaving the laboratory. ▬

Text continues on page 217.

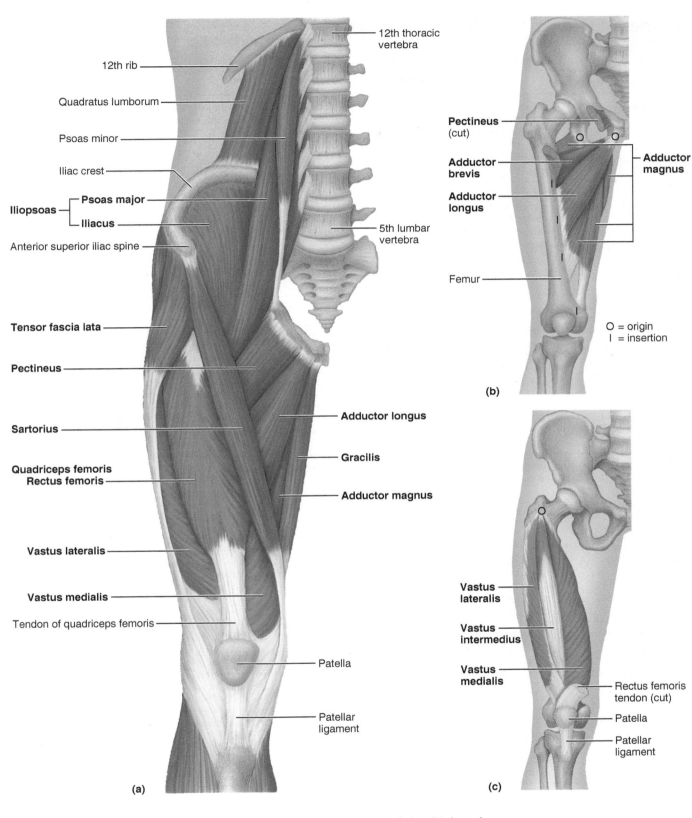

Figure 12.13 Anterior and medial muscles promoting movements of the thigh and leg. (a) Anterior view of the deep muscles of the pelvis and superficial muscles of the right thigh. **(b)** Adductor muscles of the medial compartment of the thigh. **(c)** The vastus muscles (isolated) of the quadriceps group.

Table 12.8   Muscles That Move the Human Thigh and Leg, Anterior and Medial Aspects (see Figure 12.13)

Muscle	Comments	Origin	Insertion	Action
Origin on the Pelvis				
Iliopsoas—iliacus and psoas major	Two closely related muscles; fibers pass under inguinal ligament to insert into femur via a common tendon; iliacus is more lateral	Iliacus—iliac fossa and crest, lateral sacrum; psoas major—transverse processes, bodies, and discs of T_{12} and lumbar vertebrae	On and just below lesser trochanter of femur	Prime mover of thigh flexion and flexing trunk on thigh; lateral flexion of vertebral column (psoas)
Sartorius	Straplike superficial muscle running obliquely across anterior surface of thigh to knee	Anterior superior iliac spine	By an aponeurosis into medial aspect of proximal tibia	Flexes, abducts, and laterally rotates thigh; flexes leg; known as "tailor's muscle" because it helps produce the crosslegged position in which tailors are often depicted
Medial Compartment				
Adductors—magnus, longus, and brevis	Large muscle mass forming medial aspect of thigh; arise from front of pelvis and insert at various levels on femur	Magnus—ischial and pubic rami and ischial tuberosity; longus—pubis near pubic symphysis; brevis—body and inferior pubic ramus	Magnus—linea aspera and adductor tubercle of femur; longus and brevis—linea aspera	Adduct and medially rotate and flex thigh; posterior part of magnus is also a synergist in thigh extension
Pectineus	Overlies adductor brevis on proximal thigh	Pectineal line of pubis (and superior pubic ramus)	Inferior from lesser trochanter to linea aspera of femur	Adducts, flexes, and medially rotates thigh
Gracilis	Straplike superficial muscle of medial thigh	Inferior ramus and body of pubis	Medial surface of tibia just inferior to medial condyle	Adducts thigh; flexes and medially rotates leg, especially during walking
Anterior Compartment				
*Quadriceps femoris**				
Rectus femoris	Superficial muscle of thigh; runs straight down thigh; only muscle of group to cross hip joint	Anterior inferior iliac spine and superior margin of acetabulum	Tibial tuberosity and patella	Extends leg and flexes thigh
Vastus lateralis	Forms lateral aspect of thigh; intramuscular injection site	Greater trochanter, intertrochanteric line, and linea aspera	Tibial tuberosity and patella	Extends leg and stabilizes knee
Vastus medialis	Forms inferomedial aspect of thigh	Linea aspera and intertrochanteric line	Tibial tuberosity and patella	Extends leg; stabilizes patella
Vastus intermedius	Obscured by rectus femoris; lies between vastus lateralis and vastus medialis on anterior thigh	Anterior and lateral surface of femur	Tibial tuberosity and patella	Extends leg
Tensor fascia lata	Enclosed between fascia layers of thigh	Anterior aspect of iliac crest and anterior superior iliac spine	Iliotibial tract (lateral portion of fascia lata)	Steadies the trunk on thigh; flexes, abducts, and medially rotates thigh

*The quadriceps form the flesh of the anterior thigh and have a common insertion in the tibial tuberosity via the patellar ligament. They are powerful leg extensors, enabling humans to kick a football, for example.

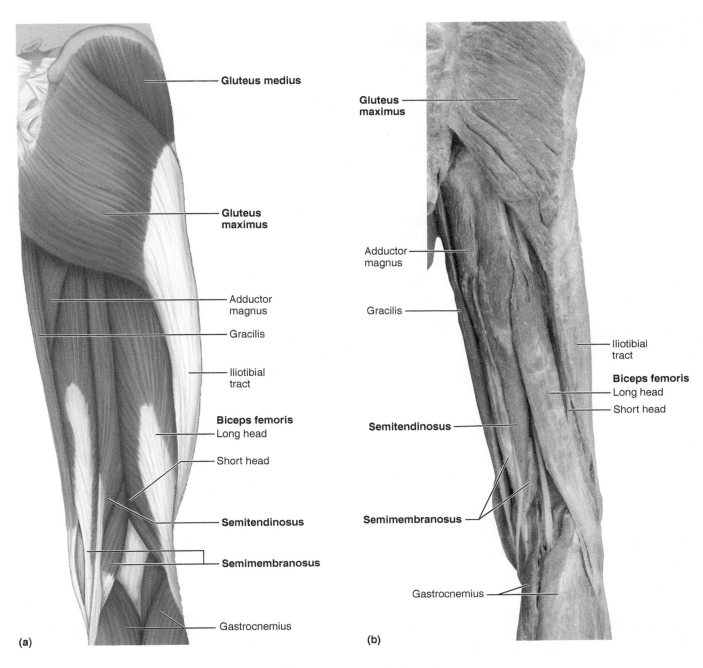

Figure 12.14 Muscles of the posterior aspect of the right hip and thigh.
(a) Superficial view showing the gluteus muscles of the buttock and hamstring muscles of the thigh. **(b)** Cadaver photo of muscles of the posterior thigh.

Table 12.9   Muscles That Move the Human Thigh and Leg, Posterior Aspect (see Figure 12.14)

Muscle	Comments	Origin	Insertion	Action
Origin on the Pelvis				
Gluteus maximus	Largest and most superficial of gluteal muscles (which form buttock mass); intramuscular injection site	Dorsal ilium, sacrum, and coccyx	Gluteal tuberosity of femur and iliotibial tract*	Major extensor of thigh; complex, powerful, and most effective when thigh is flexed, as in climbing stairs—but not as in walking; antagonist of iliopsoas; laterally rotates and abducts thigh
Gluteus medius	Partially covered by gluteus maximus; intramuscular injection site	Upper lateral surface of ilium	Greater trochanter of femur	Abducts and medially rotates thigh; steadies pelvis during walking
Gluteus minimus (not shown in figure)	Smallest and deepest gluteal muscle	External inferior surface of ilium	Greater trochanter of femur	Abducts and medially rotates thigh; steadies pelvis
Posterior Compartment				
Hamstrings†				
Biceps femoris	Most lateral muscle of group; arises from two heads	Ischial tuberosity (long head); linea aspera and distal femur (short head)	Tendon passes laterally to insert into head of fibula and lateral condyle of tibia	Extends thigh and flexes leg; laterally rotates leg
Semitendinosus	Medial to biceps femoris	Ischial tuberosity	Medial aspect of upper tibial shaft	Extends thigh; flexes leg; medially rotates leg
Semimembranosus	Deep to semitendinosus	Ischial tuberosity	Medial condyle of tibia; lateral condyle of femur	Extends thigh; flexes leg; medially rotates leg

*The iliotibial tract, a thickened lateral portion of the fascia lata, ensheathes all the muscles of the thigh. It extends as a tendinous band from the iliac crest to the knee.
†The hamstrings are the fleshy muscles of the posterior thigh. As a group, they are strong extensors of the thigh; they counteract the powerful quadriceps by stabilizing the knee joint when standing.

Table 12.10   Muscles That Move the Human Foot and Ankle (see Figures 12.15 and 12.16)

Muscle	Comments	Origin	Insertion	Action
Lateral Compartment (Figure 12.15a, b and Figure 12.16b)				
Fibularis (peroneus) longus	Superficial lateral muscle; overlies fibula	Head and upper portion of fibula	By long tendon under foot to metatarsal I and medial cuneiform	Plantar flexes and everts foot
Fibularis (peroneus) brevis	Smaller muscle; deep to fibularis longus	Distal portion of fibula shaft	By tendon running behind lateral malleolus to insert on proximal end of metatarsal V	Plantar flexes and everts foot

(Table continues on page 212.)

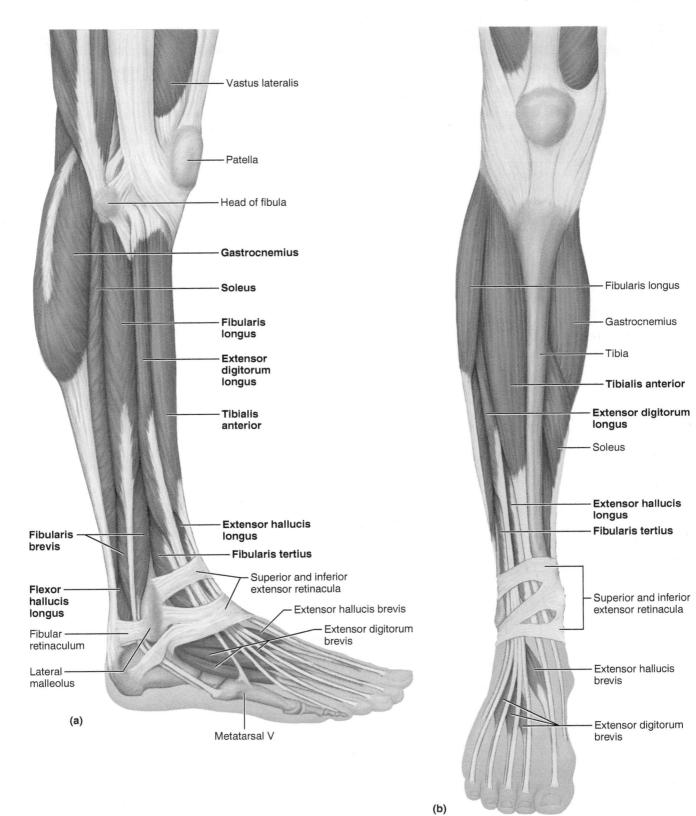

Figure 12.15 Muscles of the anterolateral aspect of the right leg. (a) Superficial view of lateral aspect of the leg, illustrating the positioning of the lateral compartment muscles (fibularis longus and brevis) relative to anterior and posterior leg muscles. **(b)** Superficial view of anterior leg muscles.

Table 12.10   Muscles That Move the Human Foot and Ankle (see Figures 12.15 and 12.16) *(continued)*

Muscle	Comments	Origin	Insertion	Action
Anterior Compartment (Figure 12.15a and b)				
Tibialis anterior	Superficial muscle of anterior leg; parallels sharp anterior margin of tibia	Lateral condyle and upper two-thirds of tibia; interosseous membrane	By tendon into inferior surface of first cuneiform and metatarsal I	Prime mover of dorsiflexion; inverts foot; supports longitudinal arch of foot
Extensor digitorum longus	Anterolateral surface of leg; lateral to tibialis anterior	Lateral condyle of tibia; proximal third-fourths of fibula; interosseous membrane	Tendon divides into four parts; inserts into middle and distal phalanges of toes II–V	Prime mover of toe extension; dorsiflexes foot
Fibularis (peroneus) tertius	Small muscle; often fused to distal part of extensor digitorum longus	Distal anterior surface of fibula and interosseous membrane	Tendon inserts on dorsum of metatarsal V	Dorsiflexes and everts foot
Extensor hallucis longus	Deep to extensor digitorum longus and tibialis	Anteromedial shaft of fibula and interosseous membrane	Tendon inserts on distal phalanx of great toe	Extends great toe; dorsiflexes foot
Posterior Compartment				
Superficial (Figures 12.15 and 12.16)				
Triceps surae	Muscle pair that shapes posterior calf		Via common tendon (calcaneal) into heel	Plantar flex foot
Gastrocnemius	Superficial muscle of pair; two prominent bellies	By two heads from medial and lateral condyles of femur	Calcaneus via calcaneal tendon	Plantar flexes foot when leg is extended; crosses knee joint; thus can flex leg (when foot is dorsiflexed)
Soleus	Deep to gastrocnemius	Proximal portion of tibia and fibula; interosseous membrane	Calcaneus via calcaneal tendon	Plantar flexes foot; is an important muscle for locomotion
Deep (Figure 12.16b)				
Popliteus	Thin muscle at posterior aspect of knee	Lateral condyle of femur and lateral meniscus	Proximal tibia	Flexes and rotates leg medially to "unlock" knee when leg flexion begins
Tibialis posterior	Thick muscle deep to soleus	Superior portion of tibia and fibula and interosseous membrane	Tendon passes obliquely behind medial malleolus and under arch of foot; inserts into several tarsals and metatarsals II–IV	Prime mover of foot inversion; plantar flexes foot; stabilizes longitudinal arch of foot
Flexor digitorum longus	Runs medial to and partially overlies tibialis posterior	Posterior surface of tibia	Distal phalanges of toes II–V	Flexes toes; plantar flexes and inverts foot
Flexor hallucis longus (see also Figure 12.15a)	Lies lateral to inferior aspect of tibialis posterior	Middle portion of fibula shaft; interosseous membrane	Tendon runs under foot to distal phalanx of great toe	Flexes great toe (*hallux* = great toe); plantar flexes and inverts foot; the "push-off muscle" during walking

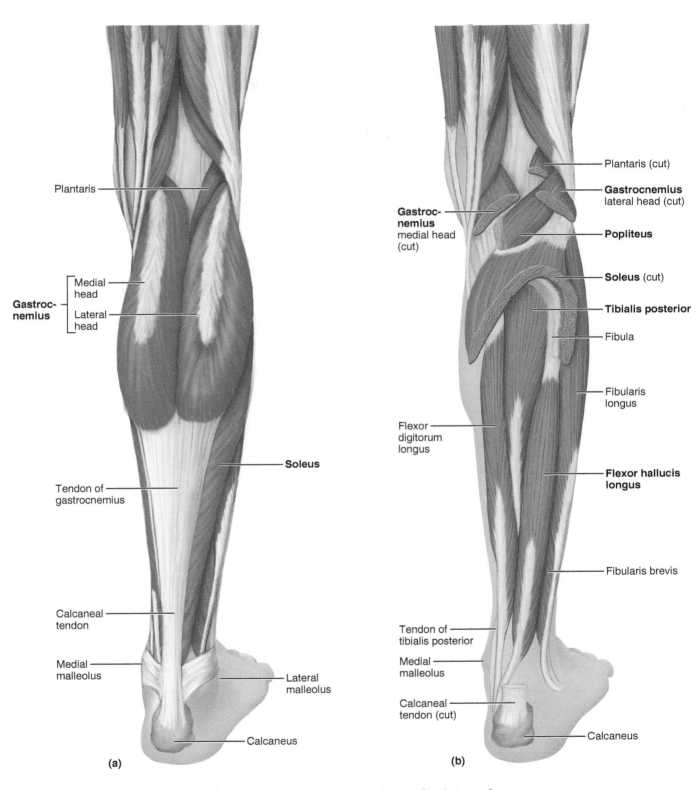

Figure 12.16 Muscles of the posterior aspect of the right leg. (a) Superficial view of the posterior leg. **(b)** The triceps surae has been removed to show the deep muscles of the posterior compartment.

Table 12.11   Intrinsic Muscles of the Foot: Toe Movement and Foot Support (see Figures 12.15 and 12.17)

Muscle	Comments	Origin	Insertion	Action
Muscles on Dorsum of Foot (Figure 12.15)				
Extensor digitorum brevis	Small, four-part muscle on dorsum of foot; deep to the tendons of extensor digitorum longus; corresponds to the extensor indicis and extensor pollicis muscles of forearm	Anterior part of calcaneus bone; extensor retinaculum	Base of proximal phalanx of big toe; extensor expansions on toes II–IV	Helps extend toes at metatarsophalangeal joints
Muscles on Sole of Foot (Figure 12.17)				
First layer				
Flexor digitorum brevis	Bandlike muscle in middle of sole; corresponds to flexor digitorum superficialis of forearm and inserts into digits in the same way	Calcaneal tuberosity	Middle phalanx of toes II–IV	Flexes toes
Abductor hallucis	Lies medial to flexor digitorum brevis; recall the similar thumb muscle, abductor pollicis brevis	Calcaneal tuberosity and flexor retinaculum	Proximal phalanx of great toe, via a sesamoid bone in tendon of flexor hallucis brevis (see below)	Abducts great toe
Abductor digiti minimi	Most lateral of the three superficial sole muscles; recall the similar abductor muscle in palm	Calcaneal tuberosity	Lateral side of base of little toe's proximal phalanx	Abducts and flexes little toe
Second layer				
Flexor accessorius (quadratus plantae)	Rectangular muscle just deep to flexor digitorum brevis in posterior half of sole; two heads	Medial and lateral sides of calcaneus	Tendon of flexor digitorum longus in midsole	Straightens out the oblique pull of flexor digitorum longus
Lumbricals	Four little "worms," like lumbricals in hand	From each tendon of flexor digitorum longus	Extensor expansion on proximal phalanx of toes II–V, medial side	By pulling on extensor expansion, flex toes at metatarsophalangeal joints and extend toes at interphalangeal joints
Third layer				
Flexor hallucis brevis	Covers metatarsal I; splits into two bellies; recall the flexor pollicis brevis of thumb	Mostly from cuboid bone	Via two tendons onto both sides of the base of the proximal phalanx of great toe	Flexes great toe's metatarsophalangeal joint
Adductor hallucis	Oblique and transverse heads; deep to lumbricals; recall adductor pollicis in thumb	From metatarsals II–IV, fibularis longus tendon, and a ligament across metatarsophalangeal joints	Base of proximal phalanx of great toe, lateral side	Helps maintain the transverse arch of foot; weak adductor of great toe

(Table continues on page 216.)

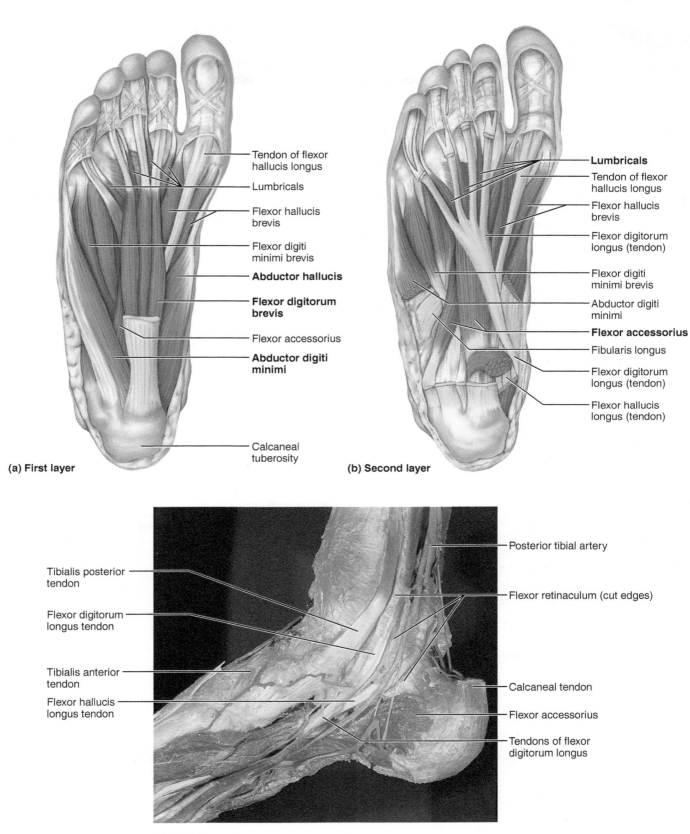

Tendon of flexor hallucis longus

Lumbricals

Flexor hallucis brevis

Flexor digiti minimi brevis

Abductor hallucis

Flexor digitorum brevis

Flexor accessorius

Abductor digiti minimi

Calcaneal tuberosity

(a) First layer

Lumbricals

Tendon of flexor hallucis longus

Flexor hallucis brevis

Flexor digitorum longus (tendon)

Flexor digiti minimi brevis

Abductor digiti minimi

Flexor accessorius

Fibularis longus

Flexor digitorum longus (tendon)

Flexor hallucis longus (tendon)

(b) Second layer

Tibialis posterior tendon

Flexor digitorum longus tendon

Tibialis anterior tendon

Flexor hallucis longus tendon

Posterior tibial artery

Flexor retinaculum (cut edges)

Calcaneal tendon

Flexor accessorius

Tendons of flexor digitorum longus

(c) Medial aspect

Figure 12.17 Muscles of the right foot, plantar and medial aspects.

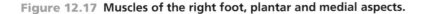

Table 12.11 Intrinsic Muscles of the Foot: Toe Movement and Foot Support (see Figures 12.15 and 12.17) *(continued)*

Muscle	Comments	Origin	Insertion	Action
Muscles on Sole of Foot *(continued)*				
Flexor digiti minimi brevis	Covers metatarsal V; recall same-named muscle in hand	Base of metatarsal V and tendon of fibularis longus	Base of proximal phalanx of toe V	Flexes little toe at metatarsophalangeal joint
Fourth layer				
Plantar and dorsal interossei	Three plantar and four dorsal interossei; similar to the palmar and dorsal interossei of hand in locations, attachments, and actions; however, the long axis of foot around which these muscles orient is the second digit, not the third	See palmar and dorsal interossei (Table 12.7)	See palmar and dorsal interossei (Table 12.7)	See palmar and dorsal interossei (Table 12.7)

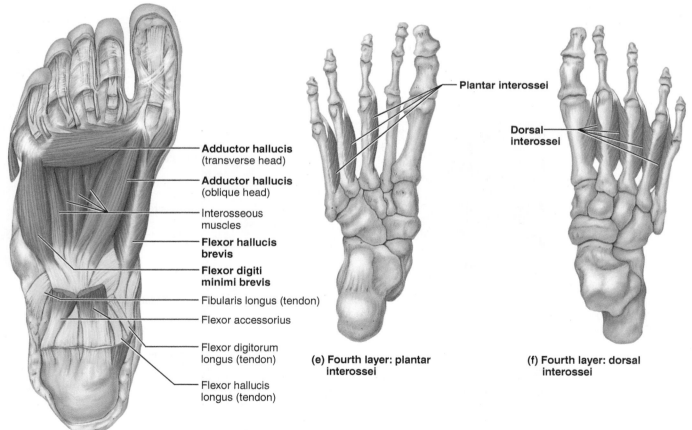

Adductor hallucis (transverse head)

Adductor hallucis (oblique head)

Interosseous muscles

Flexor hallucis brevis

Flexor digiti minimi brevis

Fibularis longus (tendon)

Flexor accessorius

Flexor digitorum longus (tendon)

Flexor hallucis longus (tendon)

(d) Third layer

Plantar interossei

Dorsal interossei

(e) Fourth layer: plantar interossei

(f) Fourth layer: dorsal interossei

Figure 12.17 *(continued)* **Muscles of the right foot, plantar and medial aspects.**

Muscle IDs

Work in groups of three or four to fill out the chart below. Do not look back at the tables; use the "brain power" of your group and the appropriate muscle models. One student in the group should act as the "muscle model." All members of the group should discuss where the origins and insertions in the table are located. The "model" will point to the appropriate attachment locations on his or her body and will perform the primary action as advised by the group. To help complete this task, recall that when a muscle contracts, the muscle's insertion moves toward the muscle's origin. Also, in the muscles of the limbs, the origin typically lies proximal to the insertion. Sometimes the origin and insertion are even part of the muscle's name!

Group Challenge: Muscle IDs

Origin	Insertion	Muscle	Primary action
Zygomatic arch and zygomatic bone	Angle and ramus of the mandible		
Anterior surface of ribs 3–5	Coracoid process of the scapula		
Inferior border of rib above	Superior border of rib below		
Distal portion of anterior humerus	Coronoid process of the ulna		
Anterior inferior iliac spine and superior margin of acetabulum	Tibial tuberosity and patella		
By two heads from medial and lateral condyles of femur	Calcaneus via calcaneal tendon		

DISSECTION AND IDENTIFICATION
Cat Muscles

The skeletal muscles of all mammals are named in a similar fashion. However, some muscles that are separate in other animals are fused in humans, and some muscles present in other animals are absent in humans. In this exercise, you will dissect the cat musculature to enhance your knowledge of the human muscular system. Because the aim is to become familiar with the muscles of the human body, you should pay particular attention to the similarities between cat and human muscles. However, pertinent differences will be pointed out as they are encountered.

Wear safety glasses and a lab coat or apron when dissecting to protect your eyes and your clothes. ■

Preparing the Cat for Dissection

The preserved laboratory animals purchased for dissection have been embalmed with a solution that prevents the tissues from deteriorating. The animals are generally delivered in plastic bags that contain a small amount of the embalming fluid. *Do not dispose of this fluid* when you remove the cat; the fluid prevents the cat from drying out. It is very important to keep the cat's tissues moist because you will probably use the same cat from now until the end of the course.

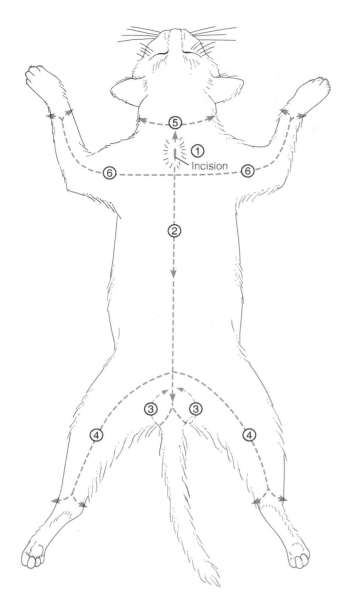

Figure 12.18 Incisions to be made in skinning a cat. Numbers indicate sequence.

1. Don disposable gloves and safety glasses. Then obtain a cat, dissection tray, dissection instruments, and a name tag. Mark the name tag with the names of the members of your group, and set it aside. The name tag will be attached to the plastic bag at the end of the dissection so that you may identify your animal in subsequent laboratories.

2. To begin removing the skin, place the cat ventral side down on the dissecting tray. Cutting away from yourself with a scalpel, make a short, shallow incision in the midline of the neck, just to penetrate the skin. From this point on, use

scissors. Continue the cut the length of the back to the sacro-lumbar region, stopping at the tail (Figure 12.18).

3. From the dorsal surface of the tail region, continue the incision around the tail, encircling the anus and genital organs. Do not remove the skin from this region.

4. Beginning again at the dorsal tail region, make an incision through the skin down each hind leg nearly to the ankle. Continue the cuts completely around the ankles.

5. Return to the neck. Cut the skin around the neck.

6. Cut down each foreleg to the wrist. Completely cut through the skin around the wrists.

7. Now free the skin from the loose connective tissue (superficial fascia) that binds it to the underlying structures. With one hand, grasp the skin on one side of the midline dorsal incision. Then, using your fingers or a blunt probe, break through the "cottony" connective tissue fibers to release the skin from the muscle beneath. Work toward the ventral surface and then toward the neck. As you pull the skin from the body, you should see small, white, cordlike structures extending from the skin to the muscles at fairly regular intervals. These are the cutaneous nerves that serve the skin. You will also see (particularly as you approach the ventral surface) that a thin layer of muscle fibers remains adhered to the skin. This is the **cutaneous maximus** muscle, which enables the cat to move its skin, rather like our facial muscles allow us to express emotion. Where the cutaneous maximus fibers cling to those of the deeper muscles, carefully cut them free. Along the ventral surface of the trunk notice the two lines of nipples associated with the mammary glands. These are more prominent in females.

8. You will notice as you start to free the skin in the neck that it is more difficult to remove. Take extra care and time in this area. The large flat **platysma** muscle in the ventral neck region (a skin muscle like the cutaneous maximus) will remain attached to the skin. You will not remove the skin from the top of the head because the cat's head muscles are not sufficiently similar to human head muscles to merit study.

9. Complete the skinning process by freeing the skin from the forelimbs, the lower torso, and the hindlimbs in the same manner. The skin is more difficult to remove as you approach the paws, so you may need to spend additional time on these areas. *Do not discard the skin.*

10. Inspect your skinned cat. Notice that it is difficult to see any cleavage lines between the muscles because of the overlying connective tissue, which is white or yellow. If time allows, carefully remove as much of the fat and fascia from the surface of the muscles as possible, using forceps or your fingers. The muscles, when exposed, look grainy or threadlike and are light brown. If you carry out this clearing process carefully and thoroughly, you will be ready to begin your identification of the superficial muscles.

11. If the muscle dissection exercises are to be done at a later laboratory session, follow the cleanup instructions noted in the following box. *Prepare your cat for storage in this way every time you use the cat.* ▄

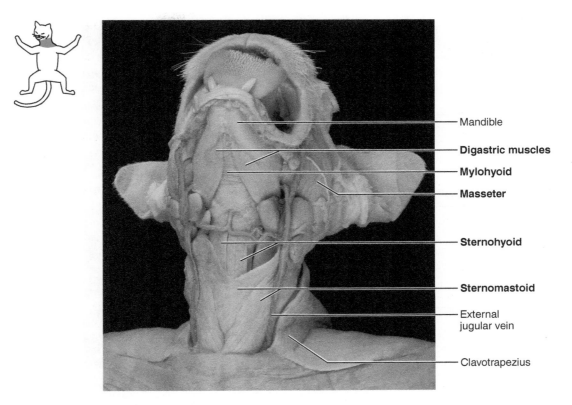

- Mandible
- **Digastric muscles**
- **Mylohyoid**
- **Masseter**
- **Sternohyoid**
- **Sternomastoid**
- External jugular vein
- Clavotrapezius

Figure 12.19 Superficial muscles of the anterior neck of the cat.

Preparing the Dissection Animal for Storage

Before leaving the lab, prepare your animal for storage as follows:

1. To prevent the internal organs from drying out, dampen a layer of folded paper towels with embalming fluid, and wrap them snugly around the animal's torso. (Do not use *water-soaked* paper towels because this will encourage the growth of mold.) Make sure the dissected areas are completely enveloped.

2. Return the animal's skin flaps to their normal position over the ventral cavity body organs.

3. Place the animal in a plastic storage bag. Add more embalming fluid if necessary, press out excess air, and securely close the bag with a rubber band or twine.

4. Make sure your name tag is securely attached, and place the animal in the designated storage container.

5. Clean all dissecting equipment with soapy water, and rinse and dry it for return to the storage area. Wash down the lab bench, and properly dispose of organic debris and your gloves before leaving the laboratory. Return safety glasses to the appropriate location.

ACTIVITY 8

Dissecting Trunk and Neck Muscles

To dissect muscles properly, you must be sure to carefully separate one muscle from another and transect the superficial muscles in order to study those lying deeper. In general, when you are directed to transect a muscle, you should first make sure the muscle is completely freed from all adhering connective tissue, and then cut through it about halfway between its origin and insertion points. *Use caution when working around points of muscle origin or insertion, and do not remove the fascia associated with such attachments.*

As a rule, all the fibers of one muscle are held together by a connective tissue sheath (epimysium) and run in the same general direction. Before you begin dissection, observe your skinned cat. If you look carefully, you can see changes in the direction of the muscle fibers, which will help you to locate the muscle borders. Pulling in slightly different directions on two adjacent muscles will usually allow you to expose the subtle white cleavage line between them. Once you have identified cleavage lines, *use a blunt probe* to break the connective tissue between muscles and to separate them. If the muscles separate as clean, distinct bundles, your procedure is probably correct. If they appear ragged or chewed up, you are probably tearing a muscle apart rather than separating it from adjacent muscles. Because of time considerations, in this exercise you will identify only the muscles that are easiest to identify and separate out.

Anterior Neck Muscles

1. Examine the anterior neck surface of the cat, and identify the following superficial neck muscles. The platysma belongs in this group but was probably removed during the skinning process. Refer to **Figure 12.19** as you work. The **sternomastoid** muscle and the more lateral and deeper **cleidomastoid** muscle (not visible in Figure 12.19) are joined in humans to form the sternocleidomastoid. The large external jugular vein, which drains the head, should be obvious crossing the anterior aspect of these muscles. The **mylohyoid** muscle parallels the bottom aspect of the chin, and the **digastric** muscles form a V over the mylohyoid muscle. Although it is not one of

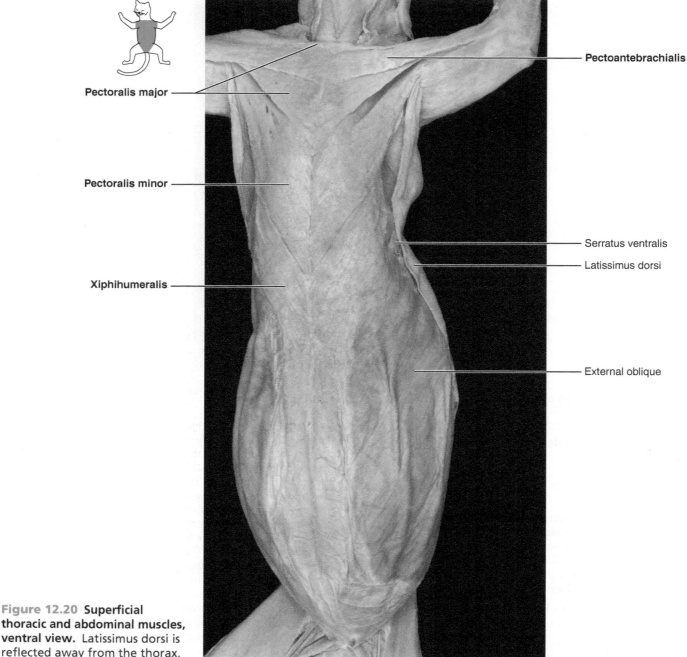

Figure 12.20 Superficial thoracic and abdominal muscles, ventral view. Latissimus dorsi is reflected away from the thorax.

the neck muscles, you can now identify the fleshy **masseter** muscle, which flanks the digastric muscle laterally. Finally, the **sternohyoid** is a narrow muscle between the mylohyoid (superiorly) and the inferior sternomastoid.

2. The deeper muscles of the anterior neck of the cat are small and straplike and hardly worth the effort of dissection. However, one of these deeper muscles can be seen with a minimum of extra effort. Transect the sternomastoid and sternohyoid muscles approximately at midbelly. Reflect the cut ends to reveal the **sternothyroid** muscle (not visible in Figure 12.19), which runs along the anterior surface of the throat just deep and lateral to the sternohyoid muscle. The

cleidomastoid muscle, which lies deep to the sternomastoid, is also more easily identified now.

Superficial Chest Muscles

In the cat, the chest or pectoral muscles adduct the arm, just as they do in humans. However, humans have only two pectoral muscles, and cats have four—the pectoralis major, pectoralis minor, xiphihumeralis, and pectoantebrachialis (Figure 12.20). However, because of their relatively great degree of fusion, the cat's pectoral muscles appear to be a single muscle. The pectoral muscles are rather difficult to dissect and identify because they do not separate from one another easily.

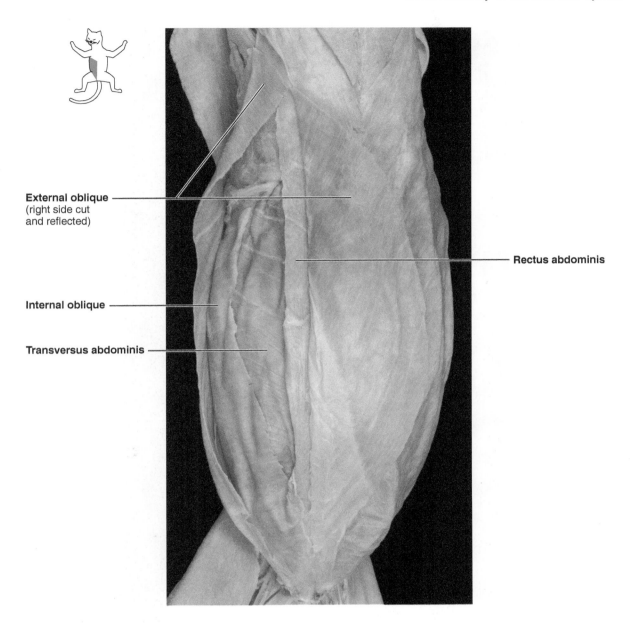

External oblique
(right side cut
and reflected)

Rectus abdominis

Internal oblique

Transversus abdominis

Figure 12.21 Muscles of the abdominal wall of the cat.

The **pectoralis major** is 5 to 8 cm (2 to 3 inches) wide and can be seen arising on the manubrium, just inferior to the sternomastoid muscle of the neck, and running to the humerus. Its fibers run at right angles to the longitudinal axis of the cat's body.

The **pectoralis minor** lies beneath the pectoralis major and extends posterior to it on the abdominal surface. It originates on the sternum and inserts on the humerus. Its fibers run obliquely to the long axis of the body, which helps to distinguish it from the pectoralis major. Contrary to what its name implies, the pectoralis minor is a larger and thicker muscle than the pectoralis major.

The **xiphihumeralis** can be distinguished from the posterior edge of the pectoralis minor only by virtue of the fact that its origin is lower—on the xiphoid process of the sternum. Its fibers run parallel to and are fused with those of the pectoralis minor.

The **pectoantebrachialis** is a thin, straplike muscle, about 1.3 cm (½ inch) wide, lying over the pectoralis major. It originates from the manubrium, passes laterally over the pectoralis major, and merges with the muscles of the forelimb approximately halfway down the humerus. It has no homologue in humans.

Identify, free, and trace out the origin and insertion of the cat's chest muscles. (Refer to Figure 12.20 as you work.)

Muscles of the Abdominal Wall

The superficial trunk muscles include those of the abdominal wall (see Figure 12.20 and **Figure 12.21**). Cat musculature in this area is quite similar in function to that of humans.

1. Complete the dissection of the more superficial anterior trunk muscles of the cat by identifying the origins and insertions of the muscles of the abdominal wall. Work carefully here. These muscles are very thin, and it is easy to miss their

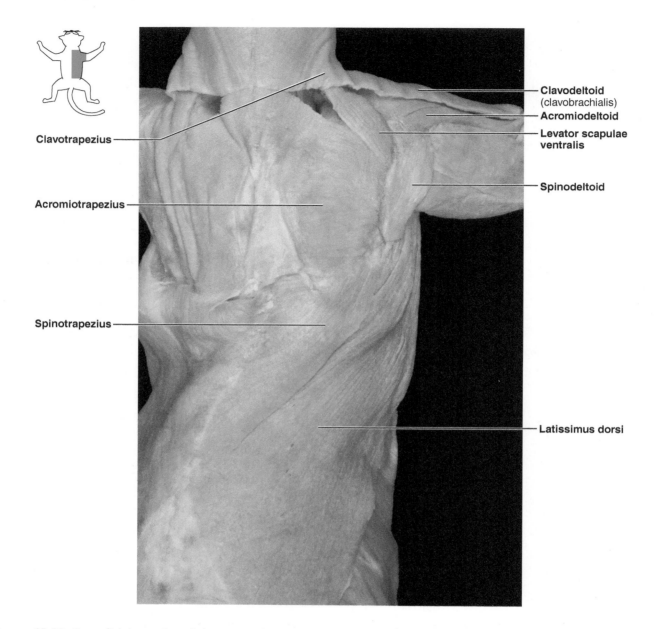

Figure 12.22 Superficial muscles of the anterodorsal aspect of the shoulder, trunk, and neck of the cat.

boundaries. Begin with the **rectus abdominis,** a long band of muscle approximately 2½ cm (1 inch) wide running immediately lateral to the midline of the body on the abdominal surface. Humans have four transverse *tendinous intersections* in the rectus abdominis (Figure 12.8a), but they are absent or difficult to identify in the cat. Identify the **linea alba,** which separates the rectus abdominis muscles. Note the relationship of the rectus abdominis to the other abdominal muscles and their fascia.

2. The **external oblique** is a sheet of muscle immediately beside the rectus abdominis (Figure 12.21). Carefully free and then transect the external oblique to reveal the anterior attachment of the rectus abdominis. Reflect the external oblique; observe the **internal oblique,** the deeper muscle. Notice which way the fibers run.

How does the fiber direction of the internal oblique compare to that of the external oblique?

_____

_____

3. Free and then transect the internal oblique muscle to reveal the fibers of the **transversus abdominis,** whose fibers run transversely across the abdomen.

Superficial Muscles of the Shoulder and Dorsal Trunk and Neck

Dissect the superficial muscles of the dorsal surface of the trunk. Refer to **Figure 12.22**.

1. Turn your cat on its ventral surface, and start your observations with the **trapezius group.** Humans have a single large trapezius muscle, but the cat has three separate muscles—the clavotrapezius, acromiotrapezius, and spinotrapezius—that together perform a similar function. The prefix (*clavo-, acromio-,* and *spino-*) in each case reveals the muscle's site of insertion. The **clavotrapezius,** the most superior muscle of the group, is homologous to the part of the human trapezius that inserts into the clavicle. Slip a probe under this muscle and follow it to its apparent origin.

Where does the clavotrapezius appear to originate?

_____

Is this similar to its origin in humans? _____

The fibers of the clavotrapezius are continuous inferiorly with those of the clavicular part of the cat's deltoid muscle (clavodeltoid), and the two muscles work together to flex the humerus. Release the clavotrapezius muscle from adjoining muscles. The **acromiotrapezius** is a large, nearly square muscle easily identified by its aponeurosis, which passes over the vertebral border of the scapula. It originates from the cervical and T_1 vertebrae and inserts into the scapular spine. The triangular **spinotrapezius** runs from the thoracic vertebrae to the scapular spine. This is the most posterior of the trapezius muscles in the cat. Now that you know where they are located, pull on the three trapezius muscles to mimic their action.

Do the trapezius muscles appear to have the same functions

in cats as in humans? _____

2. The **levator scapulae ventralis,** a flat, straplike muscle, can be located in the triangle created by the division of the fibers of the clavotrapezius and acromiotrapezius. Its anterior fibers run underneath the clavotrapezius from its origin at the occipital bone, and it inserts on the vertebral border of the scapula. In the cat, the levator scapulae ventralis helps to hold the upper edges of the scapulae together (drawing them toward the head).

What is the function of the levator scapulae in humans?

_____

3. The **deltoid group:** like the trapezius, the human deltoid muscle is represented by three separate muscles in the cat—the clavodeltoid, acromiodeltoid, and spinodeltoid. The **clavodeltoid** (also called the *clavobrachialis*), the most superficial muscle of the shoulder, is a continuation of the clavotrapezius below the clavicle, which is this muscle's point of origin (see Figure 12.22). Follow its path down the forelimb to the point where it merges along a white line with the pectoantebrachialis. Separate it from the pectoantebrachialis, and then transect it and pull it back.

Where does the clavodeltoid insert? _____

What do you think the function of this muscle is? _____

_____

The **acromiodeltoid** lies posterior to the clavodeltoid and runs over the top of the shoulder. This small triangular muscle originates on the acromion of the scapula. It inserts into the spinodeltoid (a muscle of similar size) posterior to it. The **spinodeltoid** is covered with fascia near the anterior end of the scapula. Its tendon extends under the acromiodeltoid muscle and inserts on the humerus. Notice that its fibers run obliquely to those of the acromiodeltoid. Like the human deltoid muscle, the acromiodeltoid and clavodeltoid muscles in the cat abduct and rotate the humerus.

4. The **latissimus dorsi** is a large flat muscle covering most of the lateral surface of the posterior trunk. Its anterior edge is covered by the spinotrapezius. As in humans, it inserts into the humerus. But before inserting, its fibers merge with the fibers of many other muscles, among them the xiphihumeralis of the pectoralis group. Latissimus dorsi extends and adducts the arm.

Deep Muscles of the Laterodorsal Trunk and Neck

1. In preparation, transect the latissimus dorsi, the muscles of the pectoralis group, and the spinotrapezius, and reflect them back. Be careful not to damage the large brachial nerve plexus, which lies in the axillary space beneath the pectoralis group.

2. The **serratus ventralis** represents two separate muscles in humans. The posterior portion, homologous to the *serratus anterior* of humans, arises deep to the pectoral muscles and covers the lateral surface of the rib cage. It is easily identified by its fingerlike muscular origins, which arise on the first 9 or 10 ribs. It inserts into the scapula. The anterior portion of the serratus ventralis, which arises from the cervical vertebrae, is homologous to the *levator scapulae* in humans. Both portions pull the scapula toward the sternum. Trace this muscle to its insertion. In general, in the cat, this muscle pulls the scapula posteriorly and downward.

3. Reflect the upper limb to reveal the **subscapularis,** which occupies most of the ventral surface of the scapula (Figure 12.23). Humans have a homologous muscle.

4. Locate the anterior, posterior, and middle **scalene** muscles on the lateral surface of the cat's neck and trunk. The most prominent and longest of these muscles is the middle scalene, which lies between the anterior and posterior members. The scalenes originate on the ribs and run cephalad over the serratus ventralis to insert in common on the cervical vertebrae. These muscles draw the ribs anteriorly and bend the neck downward; thus they are homologous to the human scalene muscles, which elevate the ribs and flex the neck. (Notice that the difference is only one of position. Humans walk erect, but cats are quadrupeds.)

5. Reflect the flaps of the transected latissimus dorsi, spinodeltoid, acromiodeltoid, and levator scapulae ventralis. The **splenius** is a large flat muscle occupying most of the side of the

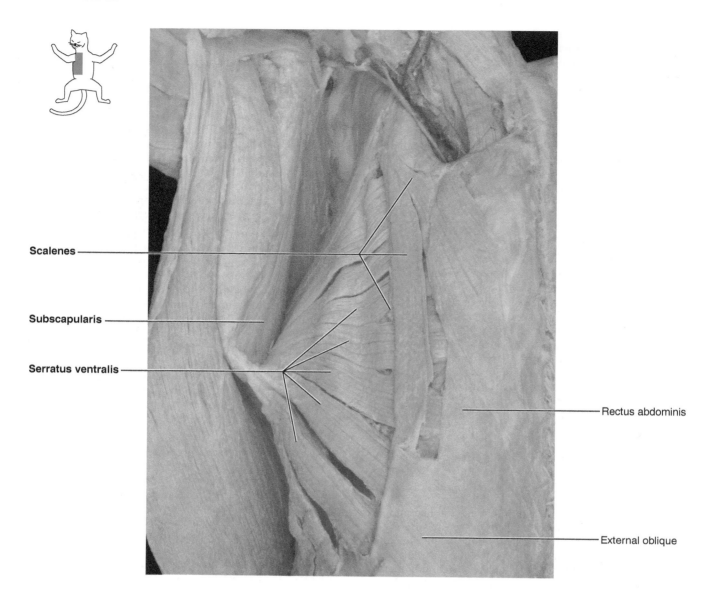

Scalenes

Subscapularis

Serratus ventralis

Rectus abdominis

External oblique

Figure 12.23 Deep muscles of the inferolateral thorax of the cat.

neck close to the vertebrae. As in humans, it originates on the ligamentum nuchae at the back of the neck and inserts into the occipital bone. It raises the head. (Refer to **Figure 12.24**.)

6. To view the rhomboid muscles, lay the cat on its side, and hold its forelegs together to spread the scapulae apart. The rhomboid muscles lie between the scapulae and beneath the acromiotrapezius. All the rhomboid muscles originate on the vertebrae and insert on the scapula. They hold the dorsal part of the scapula to the cat's back.

There are three rhomboids in the cat. The ribbonlike **rhomboid capitis,** the most anterolateral muscle of the group, has no counterpart in the human body. The **rhomboid minor,** located posterior to the rhomboid capitis, is much larger. Its fibers run transversely to those of the rhomboid capitis. The most posterior muscle of the group, the **rhomboid major,** is so closely fused to the rhomboid minor that many consider them to be one muscle—the **rhomboideus,** which is homologous to human *rhomboid muscles.*

7. The **supraspinatus** and **infraspinatus** muscles are similar to the same muscles in humans. The supraspinatus can be found under the acromiotrapezius, and the infraspinatus is deep to the spinotrapezius. Both originate on the lateral scapular surface and insert on the humerus. ◼

ACTIVITY 9

Dissecting Forelimb Muscles

Cat forelimb muscles fall into the same three categories as human upper limb muscles, but in this section the muscles of the entire forelimb are considered together. (Refer to **Figure 12.25** as you study these muscles.)

Muscles of the Lateral Surface

1. The **triceps brachii** of the cat is easily identified if the cat is placed on its side. It is a large, fleshy muscle covering the posterior aspect and much of the side of the humerus. As in

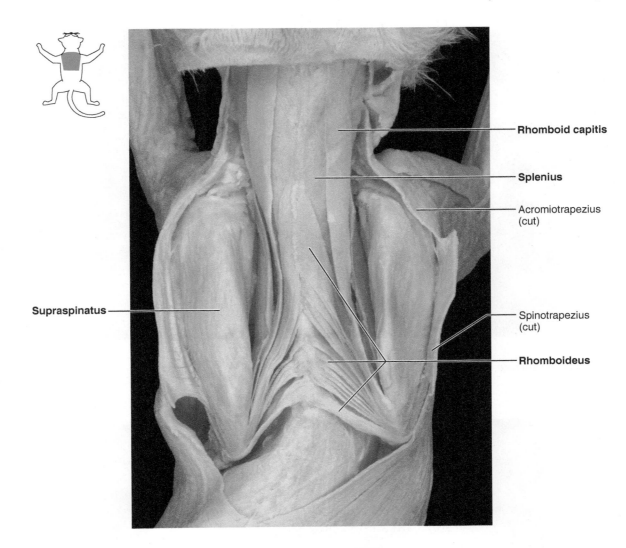

Figure 12.24 Deep muscles of the superior aspect of the dorsal thorax of the cat.

humans, this muscle arises from three heads, which originate from the humerus and scapula and insert jointly into the olecranon of the ulna. Remove the fascia from the superior region of the lateral arm surface to identify the lateral and long heads of the triceps. The long head is approximately twice as long as the lateral head and lies medial to it on the posterior arm surface. The medial head can be exposed by transecting the lateral head and pulling it aside. Now pull on the triceps muscle.

How does the function of the triceps muscle compare in cats and in humans?

Anterior and distal to the medial head of the triceps is the tiny **anconeus** muscle (not visible in Figure 12.25), sometimes called the fourth head of the triceps muscle. Notice its darker color and the way it wraps the tip of the elbow.

2. The **brachialis** can be located anterior to the lateral head of the triceps muscle. Identify its origin on the humerus, and

trace its course as it crosses the elbow and inserts on the ulna. It flexes the cat's foreleg.

Identifying the forearm muscles is difficult because of the tough fascia sheath that encases them, but give it a try.

3. Remove as much of the connective tissue as possible, and cut through the ligaments that secure the tendons at the wrist (transverse carpal ligaments) so that you will be able to follow the muscles to their insertions. Begin your identification with the muscles at the lateral surface of the forearm. The muscles of this region look very similar and are difficult to identify accurately unless you follow a definite order. Thus you will begin with the most anterior muscles and proceed to the posterior aspect. Remember to check carefully the tendons of insertion to verify your muscle identifications.

4. The ribbonlike muscle on the lateral surface of the humerus is the **brachioradialis.** Observe how it passes down the forearm to insert on the radial styloid process. (If you did not remove the fascia carefully, you may have also removed this muscle.)

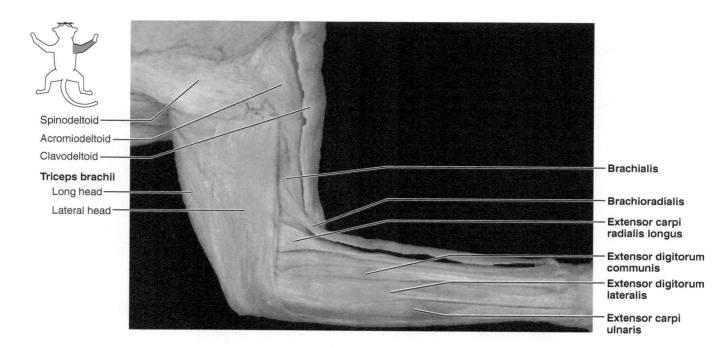

Spinodeltoid

Acromiodeltoid

Clavodeltoid

Triceps brachii

Long head

Lateral head

Brachialis

Brachioradialis

Extensor carpi radialis longus

Extensor digitorum communis

Extensor digitorum lateralis

Extensor carpi ulnaris

Figure 12.25 Lateral surface of the forelimb of the cat.

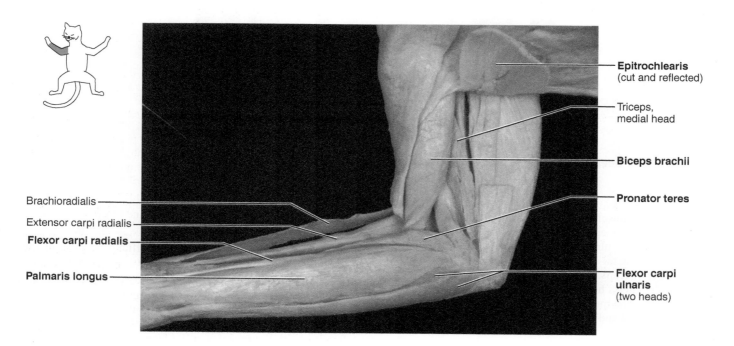

Epitrochlearis (cut and reflected)

Triceps, medial head

Biceps brachii

Pronator teres

Flexor carpi ulnaris (two heads)

Brachioradialis

Extensor carpi radialis

Flexor carpi radialis

Palmaris longus

Figure 12.26 Medial surface of the forelimb of the cat.

5.   The **extensor carpi radialis longus** has a broad origin and is larger than the brachioradialis. It extends down the anterior surface of the radius (see Figure 12.25). Transect this muscle to view the **extensor carpi radialis brevis** (not shown in Figure 12.25), which is partially covered by and sometimes fused with the extensor carpi radialis longus. Both muscles have origins, insertions, and actions similar to their human counterparts.

6.   You can see the entire **extensor digitorum communis** along the lateral surface of the forearm. Trace it to its four tendons, which insert on the second to fifth digits. This muscle extends these digits. The **extensor digitorum lateralis** (absent in humans) also extends the digits. This muscle lies immediately posterior to the extensor digitorum communis.

7.   Follow the **extensor carpi ulnaris** from the lateral epicondyle of the humerus to the ulnar side of the fifth metacarpal. Often this muscle has a shiny tendon, which helps you identify it.

Muscles of the Medial Surface

1.   The **biceps brachii** (refer to **Figure 12.26**) is a large spindle-shaped muscle medial to the brachialis on the anterior surface of the humerus. Pull back the cut ends of the pectoral muscles to get a good view of the biceps. This muscle is much more prominent in humans, but its origin, insertion, and action are very similar in cats and in humans. Follow the muscle to its origin.

Does the biceps have two heads in the cat? _____

2.   The broad, flat, exceedingly thin muscle on the posteromedial surface of the arm is the **epitrochlearis.** Its tendon originates from the fascia of the latissimus dorsi, and the muscle inserts into the olecranon of the ulna. This muscle extends the forelimb of the cat; it is not found in humans.

3.   The **coracobrachialis** of the cat is insignificant (approximately 1.3 cm, or ½ inch long) and can be seen as a very small muscle crossing the ventral aspect of the shoulder joint. It runs beneath the biceps brachii to insert on the humerus and has the same function as the human coracobrachialis.

4.   Turn the cat so that the ventral forearm muscles (mostly flexors and pronators) can be observed (refer to Figure 12.26). As in humans, most of these muscles arise from the medial epicondyle of the humerus. The **pronator teres** runs from the medial epicondyle of the humerus and declines in size as it approaches its insertion on the radius. Do not bother to trace it to its insertion.

5.   Like its human counterpart, the **flexor carpi radialis** runs from the medial epicondyle of the humerus to insert into the second and third metacarpals.

6.   The large flat muscle in the center of the medial surface is the **palmaris longus.** Its origin on the medial epicondyle of the humerus abuts that of the pronator teres and is shared with the flexor carpi radialis. The palmaris longus extends down the forelimb to terminate in four tendons on the digits. Comparatively speaking, this muscle is much larger in cats than in humans.

The **flexor carpi ulnaris** arises from a two-headed origin (medial epicondyle of the humerus and olecranon of the ulna). Its two bellies (fleshy parts) pass downward to the wrist, where they are united by a single tendon that inserts into the carpals of the wrist. As in humans, this muscle flexes the wrist. ■

ACTIVITY 10

Dissecting Hindlimb Muscles

Remove the fat and fascia from all thigh surfaces, but do not cut through or remove the **fascia lata** (or iliotibial band), which is a tough white aponeurosis covering the anterolateral surface of the thigh from the hip to the leg. If the cat is a male, the cordlike sperm duct will be embedded in the fat near the pubic symphysis. Carefully clear around, but not in, this region.

Posterolateral Hindlimb Muscles

1.   Turn the cat on its ventral surface, and identify the following superficial muscles of the hip and thigh (refer to **Figure 12.27**). Viewing the lateral aspect of the hindlimb, you will identify these muscles in sequence from the anterior to the posterior aspects of the hip and thigh. Most anterior is the **sartorius,** seen in this view as a thin band (Figure 12.27). Approximately 4 cm (1½ inches) wide, it extends around the lateral aspect of the thigh to the anterior surface, where the major portion of it lies (see Figure 12.29a). Free it from the adjacent muscles, and pass a blunt probe under it to trace its origin and insertion. Homologous to the sartorius muscle in humans, it adducts and rotates the thigh, but in addition, the cat sartorius acts as a knee extensor. Transect this muscle.

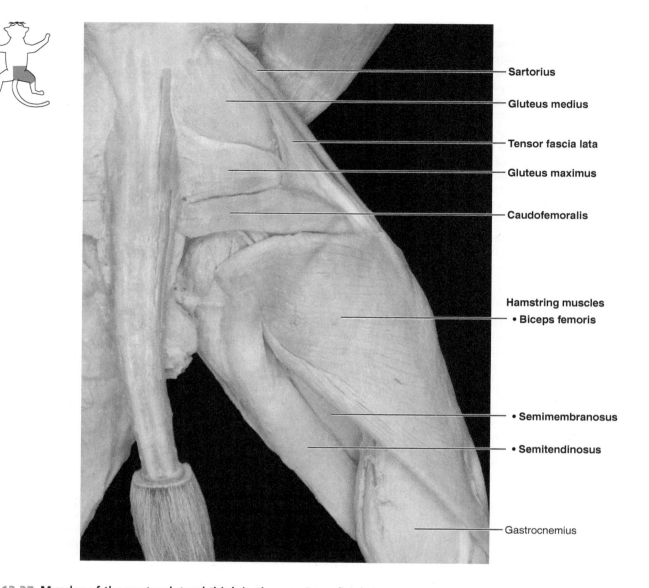

Sartorius

Gluteus medius

Tensor fascia lata

Gluteus maximus

Caudofemoralis

Hamstring muscles
• Biceps femoris

• Semimembranosus

• Semitendinosus

Gastrocnemius

Figure 12.27 Muscles of the posterolateral thigh in the cat. Superficial view.

2. The **tensor fascia lata** (Figure 12.27) is posterior to the sartorius. It is wide at its superior end, where it originates on the iliac crest, and narrows as it approaches its insertion into the fascia lata, which runs to the proximal tibial region. Transect its superior end and pull it back to expose the **gluteus medius** lying beneath it. This is the largest of the gluteus muscles in the cat. It originates on the ilium and inserts on the greater trochanter of the femur. The gluteus medius overlays and obscures the gluteus minimus, pyriformis, and gemellus muscles, which will not be identified here.

3. The **gluteus maximus** is a small triangular hip muscle posterior to the superior end of the tensor fasciae latae and paralleling it. In humans the gluteus maximus is a large fleshy muscle forming most of the buttock mass. In the cat it is only about 1.3 cm (½ inch) wide and 5 cm (2 inches) long, and it is

smaller than the gluteus medius. The gluteus maximus covers part of the gluteus medius as it extends from the sacral region and the end of the femur. It abducts the thigh.

4. Posterior to the gluteus maximus, identify the triangular **caudofemoralis,** which originates on the caudal vertebrae and inserts into the patella via an aponeurosis. There is no homologue to this muscle in humans; in cats it abducts the thigh and flexes the vertebral column.

5. The **hamstring muscles** of the hindlimb include the biceps femoris, the semitendinosus, and the semimembranosus muscles. The **biceps femoris** is a large, powerful muscle that covers most of the posterolateral surface of the thigh. It is 4–5 cm (1½ to 2 inches) wide throughout its length. Trace it from its origin on the ischial tuberosity to its insertion on

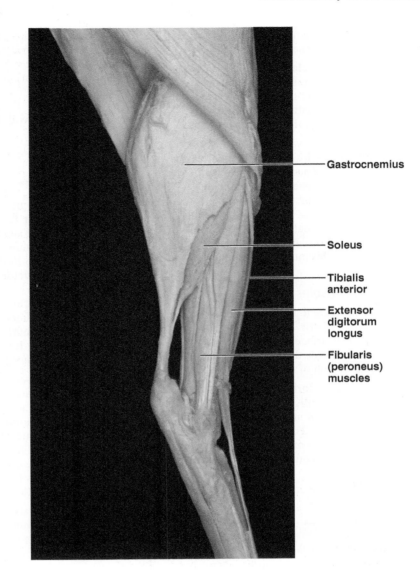

— Gastrocnemius

— Soleus

— **Tibialis anterior**

— **Extensor digitorum longus**

— **Fibularis (peroneus) muscles**

Figure 12.28 Superficial muscles of the posterolateral aspect of the shank (leg).

the tibia. Part of the **semitendinosus** can be seen beneath the posterior border of the biceps femoris. Transect and reflect the biceps muscle to reveal the whole length of the semitendinosus and the large sciatic nerve positioned under the biceps. Contrary to what its name implies ("half-tendon"), this muscle is muscular and fleshy except at its insertion. It is uniformly about 2 cm (¾ inch) wide as it runs down the thigh from the ischial tuberosity to the medial side of the tibia. It flexes the knee. The **semimembranosus,** a large muscle lying medial to the semitendinosus and largely obscured by it, is best seen in an anterior view of the thigh (Figure 12.29b). If desired, however, the semitendinosus can be transected to view it from the posterior aspect. The semimembranosus is larger and broader than the semitendinosus. Like the other hamstrings, it originates on the ischial tuberosity and inserts on the medial epicondyle of the femur and the medial tibial surface.

How does the semimembranosus compare with its human homologue?

_____

_____

6. Remove the heavy fascia covering the lateral surface of the shank (leg). Moving from the posterior to the anterior aspect, identify the following muscles on the posterolateral shank. Refer to **Figure 12.28**. First reflect the lower portion of the biceps femoris to see the origin of the **triceps surae,** the large composite muscle of the calf. Humans also have a triceps surae. The **gastrocnemius,** part of the triceps surae, is the largest muscle on the shank. As in humans, it has two heads and inserts via the calcaneal tendon into the calcaneus.

Run a probe beneath this muscle and then transect it to reveal the **soleus,** which is deep to the gastrocnemius.

7.   Another important group of muscles in the leg is the **fibularis (peroneus) muscles,** which collectively appear as a slender, evenly shaped superficial muscle lying anterior to the triceps surae (Figure 12.28). Originating on the fibula and inserting on the digits and metatarsals, the fibularis muscles flex the foot.

8.   The **extensor digitorum longus** lies anterior to the fibularis muscles. Its origin, insertion, and action in cats are similar to the homologous human muscle. The **tibialis anterior** is anterior to the extensor digitorum longus. The tibialis anterior is roughly triangular in cross section and heavier at its proximal end. Locate its origin on the proximal fibula and tibia and its insertion on the first metatarsal. You can see the sharp edge of the tibia at the anterior border of this muscle. As in humans, it is a foot flexor.

Anteromedial Hindlimb Muscles

1.   Turn the cat onto its dorsal surface to identify the muscles of the anteromedial hindlimb **(Figure 12.29)**. Note once again the straplike sartorius at the surface of the thigh, which you have already identified and transected. It originates on the ilium and inserts on the medial region of the tibia.

2.   Reflect the cut ends of the sartorius to identify the **quadriceps** muscles. The **vastus medialis** lies just beneath the sartorius. Resting close to the femur, it arises from the ilium and inserts into the patellar ligament. The small cylindrical muscle anterior and lateral to the vastus medialis is the **rectus femoris.** In cats this muscle originates entirely from the femur.

What is the origin of the rectus femoris in humans?

_____

_____

Free the rectus femoris from the most lateral muscle of this group, the large, fleshy **vastus lateralis,** which lies deep to the tensor fascia lata. The vastus lateralis arises from the lateral femoral surface and inserts, along with the other vasti muscles, into the patellar ligament. Transect this muscle to identify the deep **vastus intermedius,** the smallest of the vasti muscles. It lies medial to the vastus lateralis and merges superiorly with the vastus medialis. The vastus intermedius is not shown in the figure.

3.   The **gracilis** is a broad muscle that covers the posterior portion of the medial aspect of the thigh (see Figure 12.29a). It originates on the pubic symphysis and inserts on the medial proximal tibial surface. In cats, the gracilis adducts the leg and draws it posteriorly.

How does this compare with the human gracilis?

_____

_____

4.   Free and transect the gracilis to view the adductor muscles deep to it. The **adductor femoris** is a large muscle that lies beneath the gracilis and abuts the semimembranosus medially. Its origin is the pubic ramus and the ischium, and its fibers pass downward to insert on most of the length of the femoral shaft. The adductor femoris is homologous to the human *adductor magnus* and *adductor brevis*. Its function is to extend the thigh after it has been drawn forward and to adduct the thigh. A small muscle about 2.5 cm (1 inch) long—the **adductor longus**—touches the superior margin of the adductor femoris. It originates on the pubic bone and inserts on the proximal surface of the femur.

5.   Before continuing your dissection, locate the **femoral triangle,** an important area bordered by the proximal edge of the sartorius and the adductor muscles. It is usually possible to identify the femoral artery (injected with red latex) and the femoral vein (injected with blue latex), which span the triangle (Figure 12.29a). (You will identify these vessels again in your study of the circulatory system.) If your instructor wishes you to identify the pectineus and iliopsoas, remove these vessels and go on to steps 6 and 7.

6.   Examine the superolateral margin of the adductor longus to locate the small **pectineus.** It is sometimes covered by the gracilis (which you have cut and reflected). The pectineus, which originates on the pubis and inserts on the proximal end of the femur, is similar in all ways to its human homologue.

7.   Just lateral to the pectineus you can see a small portion of the **iliopsoas,** a long and cylindrical muscle. Its origin is on the transverse processes of T_1 through T_{12} and the lumbar vertebrae, and it passes posteriorly toward the body wall to insert on the medial aspect of the proximal femur. The iliopsoas flexes and laterally rotates the thigh. It corresponds to the human iliopsoas.

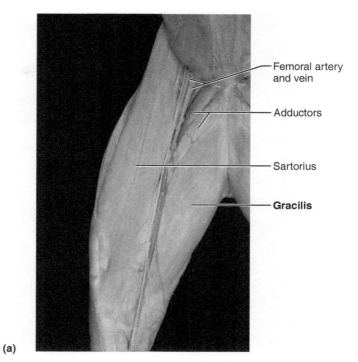

Femoral artery
and vein

Adductors

Sartorius

Gracilis

(a)

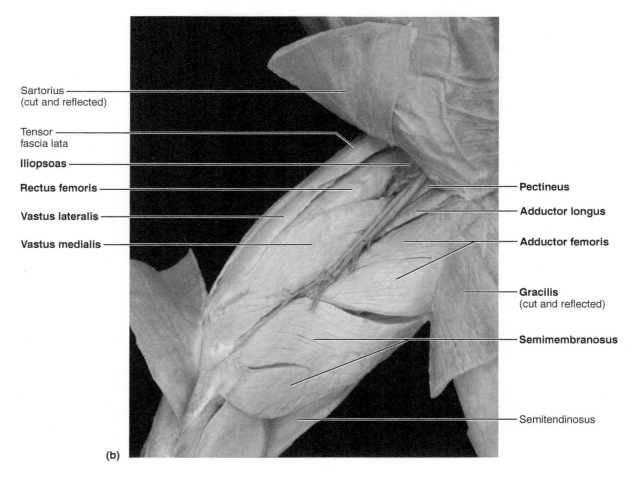

Sartorius
(cut and reflected)

Tensor
fascia lata

Iliopsoas

Rectus femoris

Vastus lateralis

Vastus medialis

Pectineus

Adductor longus

Adductor femoris

Gracilis
(cut and reflected)

Semimembranosus

Semitendinosus

(b)

Figure 12.29 Muscles of the anteromedial thigh. (a) The gracilis and sartorius
muscles are intact in this superficial view. **(b)** The gracilis and sartorius are transected
and reflected to show deeper muscles.

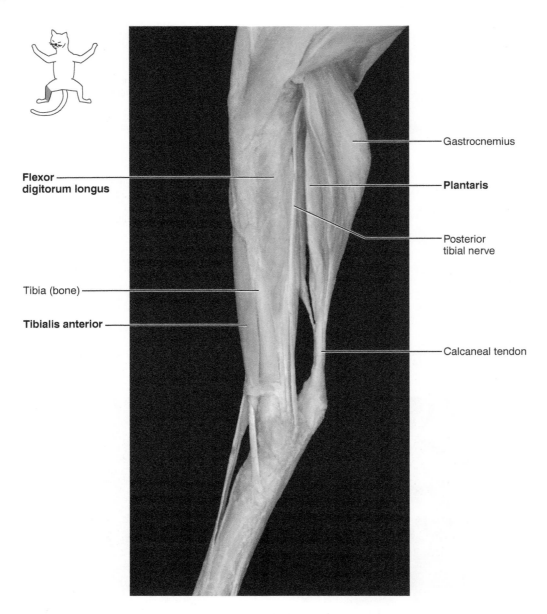

Figure 12.30 Superficial muscles of the anteromedial shank (leg) of the cat.

Flexor digitorum longus

Tibia (bone)

Tibialis anterior

Gastrocnemius

Plantaris

Posterior tibial nerve

Calcaneal tendon

8. Reidentify the gastrocnemius of the shank and then the **plantaris,** which is fused with the lateral head of the gastrocnemius **(Figure 12.30)**. It originates from the lateral aspect of the femur and patella, and its tendon passes around the calcaneus to insert on the second phalanx. Working with the triceps surae, it flexes the digits and extends the foot.

9. Anterior to the plantaris is the **flexor digitorum longus,** a long, tapering muscle with two heads. It originates on the lateral surfaces of the proximal fibula and tibia and inserts via four tendons into the terminal phalanges. As in humans, it flexes the toes.

10. The **tibialis posterior** is a long, flat muscle lateral and deep to the flexor digitorum longus (not shown in Figure 12.30). It originates on the medial surface of the head of the fibula and the ventral tibia. It merges with a flat, shiny tendon to insert into the tarsals.

11. The **flexor hallucis longus** (also not illustrated) is a long muscle that lies lateral to the tibialis posterior. It originates from the posterior tibia and passes downward to the ankle. It is a uniformly broad muscle in the cat. As in humans, it is a flexor of the great toe. ■

Gross Anatomy of the Muscular System

Classification of Skeletal Muscles

1. Several criteria were given relative to the naming of muscles. For each of the criteria below, list at least two muscles that are named for the given criterion.

1. Muscle location: _____

2. Muscle shape: _____

3. Muscle size: _____

4. Direction of muscle fibers: _____

5. Number of origins: _____

6. Location of attachments: _____

7. Muscle action: _____

2. Match the key terms to the descriptions below.

Key: a. prime mover (agonist)   b. antagonist   c. synergist   d. fixator   e. origin   f. insertion

_____ 1. term for the biceps brachii during forearm flexion

_____ 2. term that describes the relation of the brachioradialis to the biceps brachii during forearm flexion

_____ 3. term for the triceps brachii during forearm flexion

_____ 4. term for the more movable muscle attachment

_____ 5. term for the more fixed muscle attachment

_____ 6. terms for the rotator cuff muscles and deltoid when the forearm is flexed and the hand grabs a tabletop to lift the table

Muscles of the Head and Neck

3. Using choices from the key on the right, correctly identify muscles provided with leader lines.

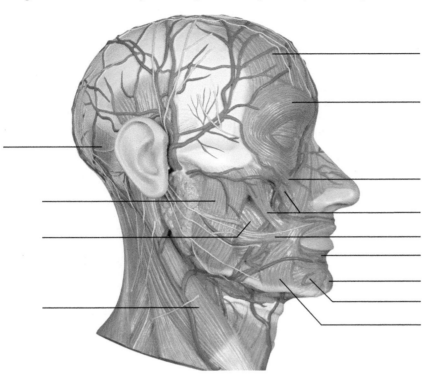

Key:

a. buccinator

b. depressor anguli oris

c. depressor labii inferioris

d. frontal belly of the epicranius

e. levator labii superioris

f. masseter

g. mentalis

h. occipital belly of epicranius

i. orbicularis oculi

j. orbicularis oris

k. risorius

l. sternocleidomastoid

m. zygomaticus minor and major

4. Using the key provided in question 3, identify the muscles described below. (Not all the terms from question 3 will be used.)

_____ 1. used in smiling

_____ 2. used to suck in your cheeks

_____ 3. used in blinking and squinting

_____ 4. used to pout (pulls the corners of the mouth downward)

_____ 5. raises your eyebrows for a questioning expression

_____ 6. used to turn and tilt the head toward the shoulder

_____ 7. your kissing muscle

_____ 8. prime mover of jaw closure

_____ 9. draw corners of the lip back (laterally)

Muscles of the Trunk

5. Correctly identify both intact and transected (cut) muscles depicted in the diagram, using the key at the right.

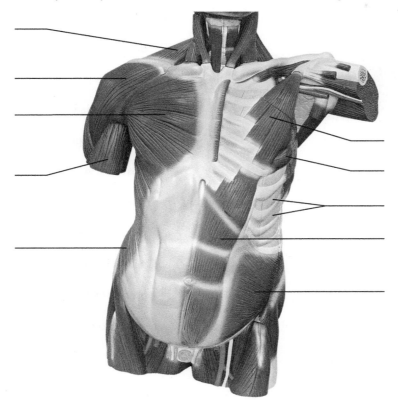

Key:

a.   biceps brachii (cut)

b.   deltoid

c.   external intercostals

d.   external oblique

e.   internal oblique

f.   pectoralis major

g.   pectoralis minor

h.   rectus abdominis

i.   serratus anterior

j.   trapezius

6. Using the key provided in question 5, identify the major muscles described below.

_____ 1. a major flexor of the vertebral column

_____ 2. prime mover for forearm flexion

_____ 3. prime mover for arm flexion

_____ 4. assume major responsibility for forming the abdominal wall (three pairs of muscles)

_____ 5. prime mover of arm abduction

_____ 6. with ribs fixed, pulls scapula forward and downward

_____ 7. moves the scapula forward and rotates scapula upward

_____ 8. small, inspiratory muscles between the ribs; elevate the rib cage

_____ 9. extends the head

Muscles of the Upper Limb

7. Using the terms from the key on the right, correctly identify all muscles provided with leader lines.

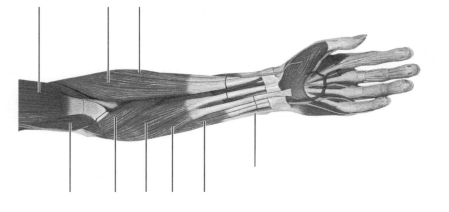

Key:

a. biceps brachii

b. brachialis

c. brachioradialis

d. extensor carpi radialis longus

e. flexor carpi radialis

f. flexor carpi ulnaris

g. flexor digitorum superficialis

h. palmaris longus

i. pronator teres

8. Use the key provided in question 7 to identify the muscles described below. (Some choices from the key will be used more than once.)

_____ 1. flexes and supinates the forearm

_____ 2. muscle located in the posterior compartment of the forearm

_____ 3. forearm flexors; no role in supination (two muscles)

_____ 4. muscle located medial to the palmaris longus

_____ 5. flexes and abducts the hand

_____ 6. flexes the hand and middle phalanges

_____ 7. pronates the forearm

_____ 8. flexes and adducts the hand

_____ 9. extends and abducts the hand

_____ 10. flat muscle that is a weak hand flexor, tenses skin of the palm

Muscles of the Lower Limb

9. Using the terms from the key on the right, correctly identify all muscles provided with leader lines.

   Key: a. adductor longus   e. rectus femoris   h. tensor facia lata
           b. extensor digitorum longus   f. sartorius   i. vastus lateralis
           c. fibularis longus   g. soleus   j. vastus medialis
           d. gastrocnemius

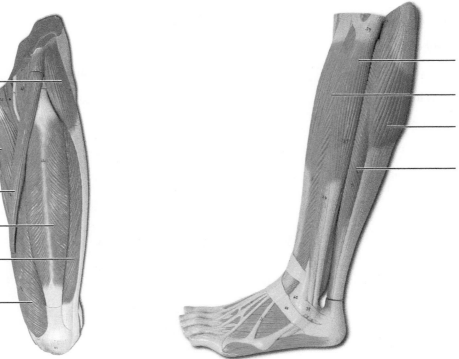

10. Use the key provided in question 9 to respond to the descriptions below.

    _____ 1. "tailor's muscle"

    _____ 2. lateral compartment muscle that plantar flexes and everts the foot

    _____ 3. abducts the thigh to take the "at ease" stance

    _____ 4. extend leg and stabilize knee (two muscles)

    _____ 5. posterior compartment muscles that plantar flex foot (two muscles)

    _____ 6. adducts the thigh, as when standing at attention

    _____ 7. extends the toes

    _____ 8. extends leg and flexes thigh

General Review: Muscle Recognition

11. Identify each lettered muscle in this diagram of the human anterior superficial musculature by matching its letter with one of the following muscle names:

_____ 1. adductor longus

_____ 2. biceps brachii

_____ 3. brachioradialis

_____ 4. deltoid

_____ 5. extensor digitorum longus

_____ 6. external oblique

_____ 7. fibularis longus

_____ 8. flexor carpi radialis

_____ 9. flexor carpi ulnaris

_____ 10. frontal belly of epicranius

_____ 11. gastrocnemius

_____ 12. gracilis

_____ 13. iliopsoas

_____ 14. intercostals

_____ 15. internal oblique

_____ 16. masseter

_____ 17. orbicularis oculi

_____ 18. orbicularis oris

_____ 19. palmaris longus

_____ 20. pectineus

_____ 21. pectoralis major

_____ 22. pectoralis minor

_____ 23. platysma

_____ 24. pronator teres

_____ 25. rectus abdominis

_____ 26. rectus femoris

_____ 27. sartorius

_____ 28. serratus anterior

_____ 29. soleus

_____ 30. sternocleidomastoid

_____ 31. sternohyoid

_____ 32. temporalis

_____ 33. tensor fascia lata

_____ 34. tibialis anterior

_____ 35. transversus abdominis

_____ 36. trapezius

_____ 37. triceps brachii

_____ 38. vastus lateralis

_____ 39. vastus medialis

_____ 40. zygomaticus

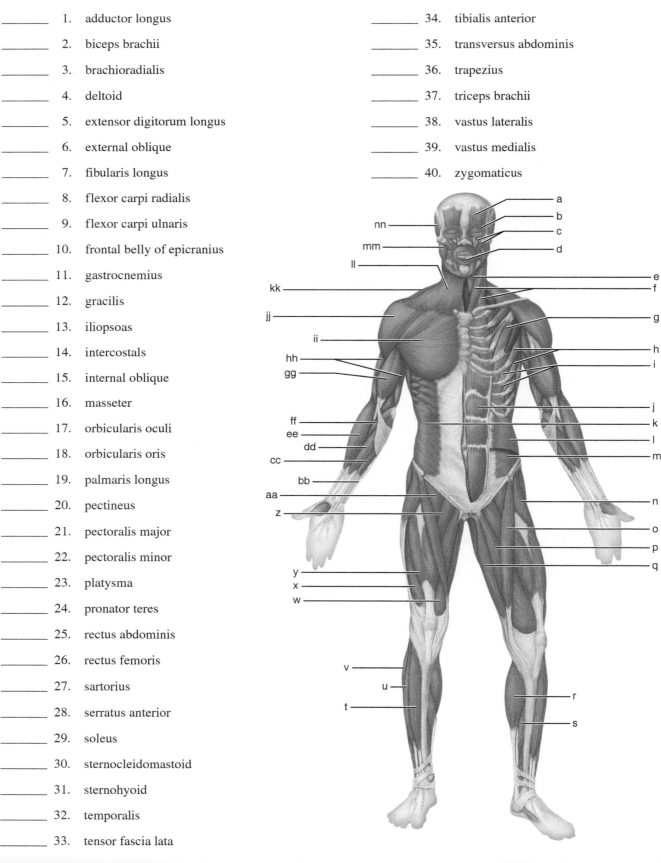

12. Identify each lettered muscle in this illustration of the human posterior superficial musculature by matching its letter with one of the following muscle names:

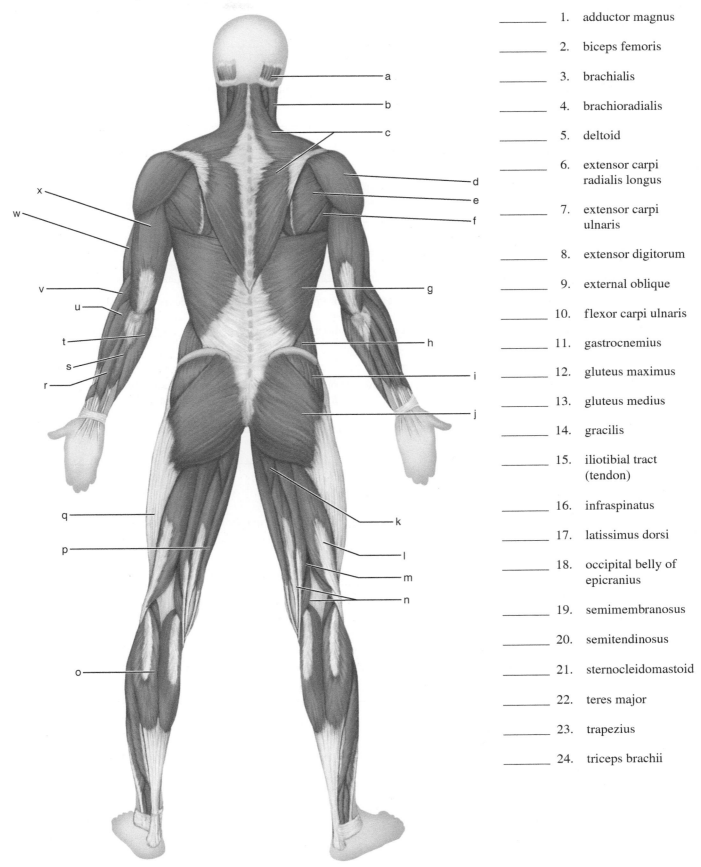

_____   1.   adductor magnus

_____   2.   biceps femoris

_____   3.   brachialis

_____   4.   brachioradialis

_____   5.   deltoid

_____   6.   extensor carpi radialis longus

_____   7.   extensor carpi ulnaris

_____   8.   extensor digitorum

_____   9.   external oblique

_____   10.   flexor carpi ulnaris

_____   11.   gastrocnemius

_____   12.   gluteus maximus

_____   13.   gluteus medius

_____   14.   gracilis

_____   15.   iliotibial tract (tendon)

_____   16.   infraspinatus

_____   17.   latissimus dorsi

_____   18.   occipital belly of epicranius

_____   19.   semimembranosus

_____   20.   semitendinosus

_____   21.   sternocleidomastoid

_____   22.   teres major

_____   23.   trapezius

_____   24.   triceps brachii

General Review: Muscle Descriptions

13. Identify the muscles described by completing the following statements. Use an appropriate reference as needed.

1. The _____, _____, _____, and _____ are commonly used for intramuscular injections (four muscles).

2. The insertion tendon of the _____ group contains a large sesamoid bone, the patella.

3. The triceps surae insert in common into the _____ tendon.

4. The bulk of the tissue of a muscle tends to lie _____ to the part of the body it causes to move.

5. The extrinsic muscles of the hand originate on the _____.

6. Most flexor muscles are located on the _____ aspect of the body;

   most extensors are located _____. An exception to this generalization

   is the extensor-flexor musculature of the _____.

14. ✚ Bruxism is a condition in which individuals clench and/or grind their teeth. It often occurs as they sleep, leading to jaw

pain and damaged teeth. Which muscles contract during this nocturnal event? _____

_____

15. ✚ Repetitive extension of the hand at the wrist and abduction of the hand can lead to lateral epicondylitis. Although sometimes called "tennis elbow," it more often affects individuals who don't play tennis. Based on the name *lateral epicondylitis* and

the action described above, which muscle would most likely have microscopic tears in the tendon? _____

_____

Dissection and Identification: Cat Muscles

Many human muscles are modified from those of the cat (or any quadruped). The following questions refer to these differences.

16. How does the human trapezius muscle differ from the cat's?

_____

17. How does the deltoid differ?

_____

18. How do the size and orientation of the human gluteus maximus muscle differ from that in the cat?

_____

_____

19. Explain how the tendinous intersections of the rectus abdominis of the cat differ from those found in humans.

_____

_____

Histology of Nervous Tissue

MATERIALS

- Model of a "typical" neuron (if available)
- Compound microscope
- Immersion oil
- Prepared slides of an ox spinal cord smear and teased myelinated nerve fibers
- Prepared slides of Purkinje cells (cerebellum), pyramidal cells (cerebrum), and a dorsal root ganglion
- Prepared slide of a nerve (x.s.)
- Prepared slides (l.s.) of lamellar corpuscles, tactile corpuscles, tendon organs, and muscle spindles

LEARNING OUTCOMES

☐ Discuss the functional differences between neurons and neuroglia.

☐ List six types of neuroglia, and indicate where each is found.

☐ Identify the important anatomical features of a neuron on an appropriate image.

☐ List the functions of dendrites, axons, and terminal boutons.

☐ Explain how a nerve impulse is transmitted from one neuron to another.

☐ State the function of myelin sheaths, and explain how Schwann cells myelinate nerve fibers (axons) in the peripheral nervous system.

☐ Classify neurons structurally and functionally.

☐ Differentiate a nerve from a tract, and a ganglion from a CNS nucleus.

☐ Identify endoneurium, perineurium, and epineurium microscopically or in an appropriate image, and cite their functions.

☐ Recognize and describe the various types of general sensory receptors as studied in the laboratory, and list the functions and locations of each.

PRE-LAB QUIZ

1. _____ are the functional units of nervous tissue.
2. Neuroglia of the peripheral nervous system include:
   a. ependymal cells and satellite cells
   b. oligodendrocytes and astrocytes
   c. satellite cells and Schwann cells
3. These branching neuron processes serve as receptive regions and transmit electrical signals toward the cell body. They are:
   a. axons   b. collaterals   c. dendrites   d. neuroglia
4. Most axons are covered with a fatty material called _____, which insulates the fibers and increases the speed of neurotransmission.
5. Circle the correct underlined term. Axons running through the central nervous system form <u>tracts</u> / <u>nerves</u> of white matter.
6. Neurons can be classified according to structure. _____ neurons have many processes that issue from the cell body.
   a. Bipolar   b. Multipolar   c. Unipolar
7. Within a nerve, each axon is surrounded by a covering called the:
   a. endoneurium   b. epineurium   c. perineurium
8. Sensory receptors can be classified according to their source of stimulus. _____ are found close to the body surface and react to stimuli in the external environment.
   a. Exteroceptors   b. Interoceptors   c. Proprioceptors   d. Visceroceptors
9. Tactile corpuscles respond to light touch. Where would you expect to find tactile corpuscles?
   a. deep within the dermal layer of hairy skin
   b. in the dermal papillae of hairless skin
   c. in the hypodermis of hairless skin
   d. in the uppermost portion of the epidermis
10. _____ are sensory receptors that detect stretch in skeletal muscles, tendons, and joints.

The nervous system is the master integrating and coordinating system, continuously monitoring and processing sensory information both from the external environment and from within the body. Every thought, action, and sensation is a reflection of its activity. Like a computer, it processes and integrates new "inputs" with information previously fed into it to produce a response.

Two primary divisions make up the nervous system: the central nervous system, or CNS, consisting of the brain and spinal cord; and the peripheral nervous system, or PNS, which includes all the nervous elements located outside the central nervous system. PNS structures include nerves, sensory receptors, and some clusters of neuron cell bodies.

Despite its complexity, nervous tissue is made up of just two principal cell types: neurons and neuroglia.

Neuroglia

The **neuroglia** ("nerve glue"), or **glial cells,** of the CNS include *astrocytes, oligodendrocytes, microglial cells,* and *ependymal cells* (**Figure 13.1**). The neuroglia found in the PNS include *Schwann cells,* also called neurolemmocytes, and *satellite cells.*

Neuroglia serve the needs of the delicate neurons by supporting and protecting them. In addition, they act as phagocytes (microglial cells), myelinate the cytoplasmic extensions of the neurons (oligodendrocytes and Schwann cells), play a role in capillary-neuron exchanges, and control the chemical environment around neurons (astrocytes). Although neuroglia resemble neurons in some ways (many have branching cellular extensions), they are not capable of generating and transmitting nerve impulses. In this exercise, we focus on the highly excitable neurons.

Neurons

Neurons, or nerve cells, are the basic functional units of nervous tissue. They are highly specialized to transmit messages from one part of the body to another in the form of nerve impulses. Although neurons differ structurally, they have many identifiable features in common (**Figure 13.2a** and **b**). All have a **cell body** from which slender processes extend. The cell body is both the *biosynthetic center* of the neuron and part of its *receptive region*. Neuron cell bodies make up the gray matter of the CNS and form clusters there that are called **nuclei.** In the PNS, clusters of neuron cell bodies are called **ganglia.**

The neuron cell body contains a large round nucleus surrounded by cytoplasm. Two prominent structures are found in the cytoplasm: (1) cytoskeletal elements called **neurofibrils**, which provide support for the cell and a means to transport substances throughout the neuron; and (2) darkly staining

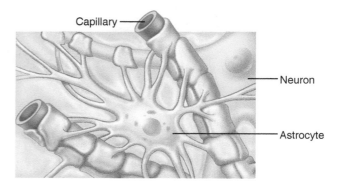

(a) **Astrocytes are the most abundant CNS neuroglia.**

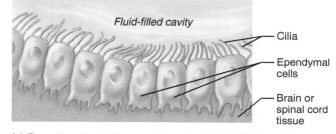

(b) **Microglial cells are defensive cells in the CNS.**

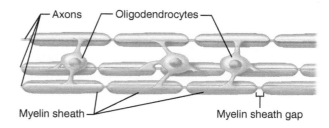

(c) **Ependymal cells line cerebrospinal fluid–filled cavities.**

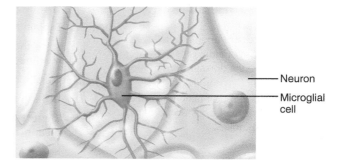

(d) **Oligodendrocytes have processes that form myelin sheaths around CNS nerve fibers.**

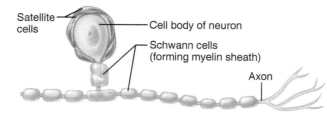

(e) **Satellite cells and Schwann cells (which form myelin) surround neurons in the PNS.**

Figure 13.1 Neuroglia. (a–d) The four types of neuroglia of the central nervous system. **(e)** Neuroglia of the peripheral nervous system.

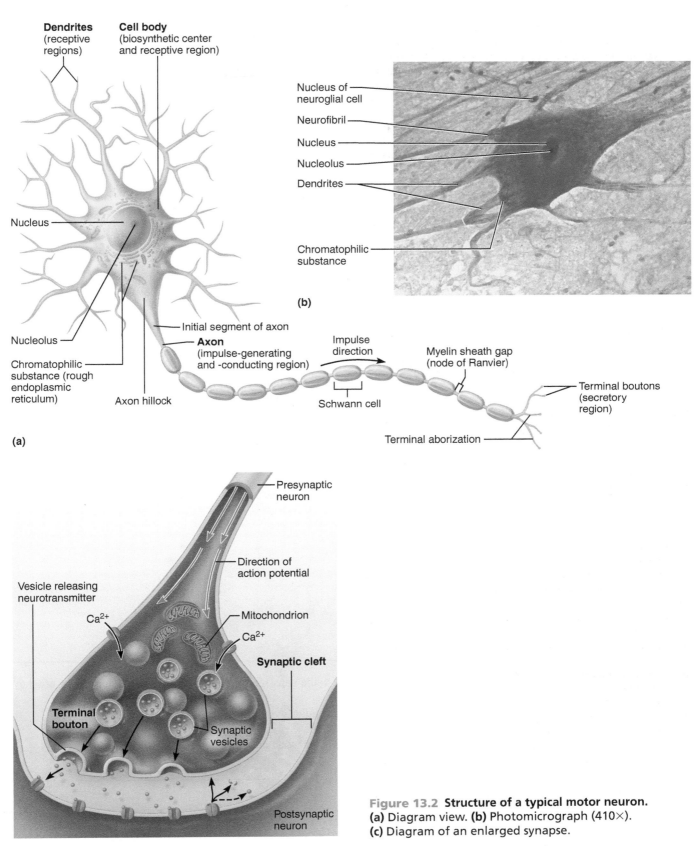

Dendrites (receptive regions)

Cell body (biosynthetic center and receptive region)

Nucleus

Nucleolus

Chromatophilic substance (rough endoplasmic reticulum)

Axon hillock

Initial segment of axon

Axon (impulse-generating and -conducting region)

Impulse direction

Schwann cell

Myelin sheath gap (node of Ranvier)

Terminal boutons (secretory region)

Terminal aborization

(a)

Nucleus of neuroglial cell

Neurofibril

Nucleus

Nucleolus

Dendrites

Chromatophilic substance

(b)

Presynaptic neuron

Direction of action potential

Vesicle releasing neurotransmitter

Ca^{2+}

Mitochondrion

Ca^{2+}

Synaptic cleft

Terminal bouton

Synaptic vesicles

Postsynaptic neuron

(c)

Figure 13.2 Structure of a typical motor neuron.
(a) Diagram view. **(b)** Photomicrograph (410×).
(c) Diagram of an enlarged synapse.

structures called **chromatophilic substance**—clusters of rough endoplasmic reticulum involved in protein synthesis.

Neurons have two types of processes. **Dendrites** are *receptive regions* that bear receptors for neurotransmitters released by the terminal boutons of other neurons. **Axons,** also called *nerve fibers,* form the *impulse-generating and impulse-conducting region* of the neuron. The white matter of the nervous system is made up of axons. Neurons may have many dendrites, but they have only a single axon. The axon may branch, forming one or more processes called **axon collaterals.**

In general, a neuron is excited by other neurons when their axons release neurotransmitters close to its dendrites or cell body. The electrical signal produced travels across the cell body, and if it is great enough, it elicits a regenerative electrical signal, a *nerve impulse* or *action potential,* that travels down the axon. The axon in motor neurons begins just distal to a slightly enlarged cell body structure called the **axon hillock** (Figure 13.2a). The axon ends in many small structures called **terminal boutons,** or **axon terminals,** which form **synapses,** or junctions, with neurons or effector cells. These terminals store the neurotransmitter chemical in tiny vesicles. Each terminal bouton of the presynaptic neuron is separated from the cell body or dendrites of the next (postsynaptic) neuron by a tiny gap called the **synaptic cleft** (Figure 13.2c). Thus, although they are close, there is no actual physical contact between neurons.

Most long nerve fibers are covered with a fatty material called *myelin,* and such fibers are referred to as **myelinated fibers.** Because of its chemical composition, myelin insulates the fibers and greatly increases the speed of neurotransmission by neuron fibers. Nerve fibers in the peripheral nervous system are typically heavily myelinated by special supporting cells called **Schwann cells,** which wrap themselves tightly around the axon in jelly-roll fashion (**Figure 13.3**). The wrapping is the **myelin sheath.** Since the myelin sheath is formed by many individual Schwann cells, it is a discontinuous sheath. The gaps or indentations in the sheath are called **myelin sheath gaps** or *nodes of Ranvier* (see Figure 13.2).

Within the CNS, myelination is accomplished by neuroglia called **oligodendrocytes** (see Figure 13.1d). These CNS sheaths do not exhibit the outer collar of perinuclear cytoplasm seen in fibers myelinated by Schwann cells. Because of its chemical composition, myelin insulates the fibers and greatly increases the transmission speed of nerve impulses.

ACTIVITY 1

Identifying Parts of a Neuron

1. Study the illustration of a typical motor neuron (Figure 13.2), noting the structural details described above, and then identify these structures on a neuron model.

2. Obtain a prepared slide of the ox spinal cord smear, which has large, easily identifiable neurons. Study one representative neuron under oil immersion, and identify the cell body; the nucleus; the large, prominent "owl's eye" nucleolus; and the granular chromatophilic substance. If possible, distinguish the axon from the many dendrites.

Sketch the neuron in the space provided below, and label the important anatomical details you have observed. (Compare your sketch to Figure 13.2c.)

3. Obtain a prepared slide of teased myelinated nerve fibers. Identify the following (use **Figure 13.4** as a guide): myelin sheath gaps, axon, Schwann cell nuclei, and myelin sheath.

Sketch a portion of a myelinated nerve fiber in the space provided below, illustrating a myelin sheath gap. Label the axon, myelin sheath, a myelin sheath gap, and Schwann cell nucleus.

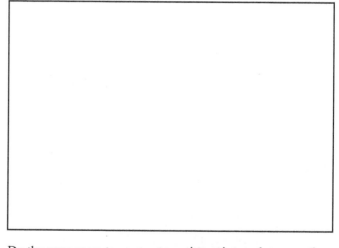

Do the gaps seem to occur at consistent intervals, or are they

irregularly distributed? _____

Explain the functional significance of this finding: _____

_____

_____ ▬

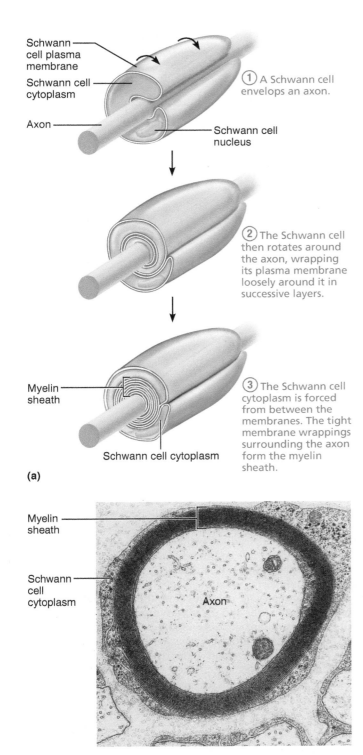

Neuron Classification

Neurons may be classified on the basis of structure or of function.

Classification by Structure

Structurally, neurons may be differentiated according to the number of processes attached to the cell body **(Figure 13.5a)**. In **unipolar neurons,** one very short process, which divides into *peripheral* and *central processes,* extends from the cell body. Functionally, only the most distal portions of the peripheral process act as receptive endings; the rest acts as an axon along with the central process. Unipolar neurons are move accurately called **pseudounipolar neurons** because they are derived from bipolar neurons. Nearly all neurons that conduct impulses toward the CNS are unipolar.

Bipolar neurons have two processes attached to the cell body. This neuron type is quite rare, typically found only as part of the receptor apparatus of the eye, ear, and olfactory mucosa.

Many processes issue from the cell body of **multipolar neurons,** all classified as dendrites except for a single axon. Most neurons in the brain and spinal cord (CNS neurons) and those whose axons carry impulses away from the CNS fall into this last category.

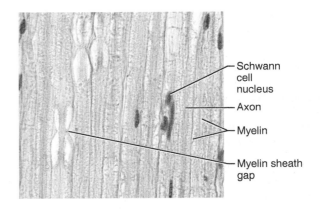

Figure 13.4 Photomicrograph of a small portion of a peripheral nerve in longitudinal section (265×).

Figure 13.3 Myelination of a nerve fiber (axon) by individual Schwann cells. (a) Nerve fiber myelination. **(b)** Electron micrograph of cross section through a myelinated axon (11,000×).

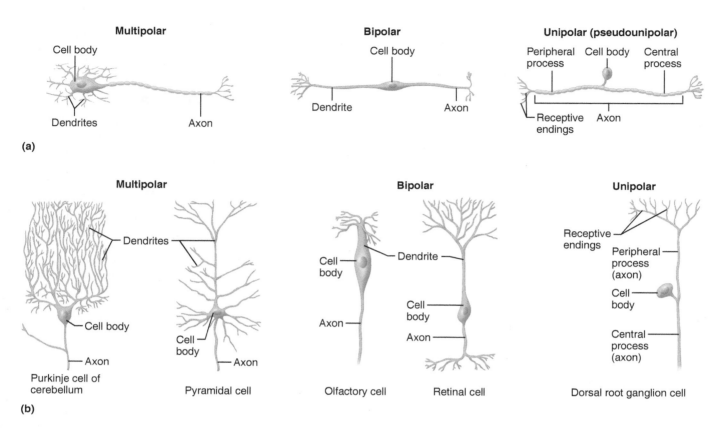

Figure 13.5 Classification of neurons according to structure. (a) Classification of neurons based on structure (number of processes extending from the cell body). **(b)** Structural variations within the classes.

ACTIVITY 2

Studying the Microscopic Structure of Selected Neurons

Obtain prepared slides of pyramidal cells of the cerebral cortex, Purkinje cells of the cerebellar cortex, and a dorsal root ganglion. As you observe them under the microscope, try to pick out the anatomical details, and compare the cells to Figure 13.5b and **Figure 13.6**. Notice that the neurons of the cerebral and cerebellar tissues (both brain tissues) are extensively branched; in contrast, the neurons of the dorsal root ganglion are more rounded. The many small nuclei visible surrounding the neurons are those of bordering neuroglia.

Which of these neuron types would be classified as multipolar neurons?

_____

As unipolar? _____ ▬

Classification by Function

In general, neurons carrying impulses from sensory receptors in the internal organs (viscera), the skin, skeletal muscles, joints, or special sensory organs are termed **sensory,** or **afferent, neurons (Figure 13.7)**. The receptive endings of sensory neurons are often equipped with specialized receptors that are stimulated by specific changes in their immediate environment. The cell bodies of sensory neurons are always found in a ganglia outside the CNS, and these neurons are typically unipolar.

Neurons carrying impulses from the CNS to the viscera and/or body muscles and glands are termed **motor,** or **efferent, neurons.** Motor neurons are most often multipolar, and their cell bodies are almost always located in the CNS.

The third functional category of neurons is the **interneurons,** which are situated between and contribute to pathways that connect sensory and motor neurons. Their cell bodies are always located within the CNS, and they are multipolar neurons structurally.

Structure of a Nerve

In the CNS, bundles of axons are called **tracts.** In the PNS, bundles of axons are called **nerves.** Wrapped in connective tissue coverings, nerves extend to and/or from the CNS and visceral organs or structures of the body periphery, such as skeletal muscles, glands, and skin.

Like neurons, nerves are classified according to the direction in which they transmit impulses. **Sensory (afferent) nerves** conduct impulses only toward the CNS. A few of the cranial nerves are pure sensory nerves. **Motor (efferent) nerves** carry impulses only away from the CNS. The ventral roots of the spinal cord are motor nerves. Nerves carrying both sensory (afferent) and motor (efferent) fibers are called **mixed nerves;** most nerves of the body, including all spinal nerves, are mixed nerves.

Within a nerve, each axon is surrounded by a delicate connective tissue sheath called an **endoneurium,** which insulates it from the other neuron processes adjacent to it. The endoneurium is often mistaken for the myelin sheath; it is instead an additional sheath that surrounds the myelin sheath. Groups of axons are bound by a coarser connective tissue, called the **perineurium,** to form bundles of fibers called **fascicles.** Finally, all the fascicles are bound together by a white, fibrous connective tissue sheath called the **epineurium,** forming the cordlike nerve **(Figure 13.8)**. In addition to the connective tissue wrappings, blood vessels and lymphatic vessels serving the fibers also travel within a nerve.

ACTIVITY 3

Examining the Microscopic Structure of a Nerve

Use the compound microscope to examine a prepared cross section of a peripheral nerve. Identify axons, myelin sheaths, fascicles, and endoneurium, perineurium, and epineurium sheaths. If desired, sketch the nerve in the space below. ▬

Structure of General Sensory Receptors

You cannot become aware of changes in the environment unless your sensory neurons and their receptors are operating properly. Sensory receptors are either modified dendritic endings or specialized cells associated with the dendrites that are sensitive to certain environmental stimuli. They respond to stimuli by initiating a nerve impulse. Sensory receptors may be classified by the type of stimulus they detect (for example, touch, pain, or temperature), their structure (free nerve endings or complex encapsulated structures), or their body location. **Exteroceptors** respond to stimuli in the external environment, and are typically found close to the body surface. **Interoceptors** respond to stimuli arising within the body (including the visceral organs). **Proprioceptors,** a subclass of interoceptors, are found in skeletal muscles, joints, tendons, and ligaments and report on the amount of stretch in these structures.

The receptors of the special sense organs are complex and deserve considerable study. (The special senses are covered separately in Exercises 17 through 20.) Only the anatomically

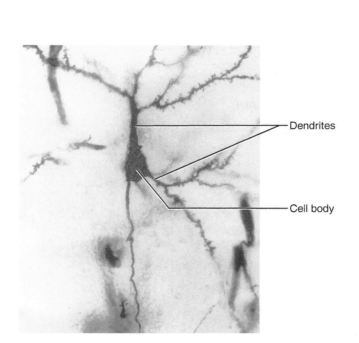

(a)

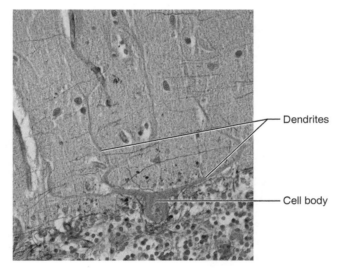

(b)

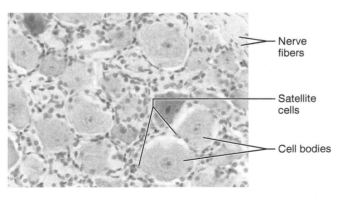

(c)

Figure 13.6 Photomicrographs of neurons. (a) Pyramidal neuron from cerebral cortex (195×). **(b)** Purkinje cell from the cerebellar cortex (200×). **(c)** Dorsal root ganglion cells (235×).

Dendrites

Cell body

Dendrites

Cell body

Nerve fibers

Satellite cells

Cell bodies

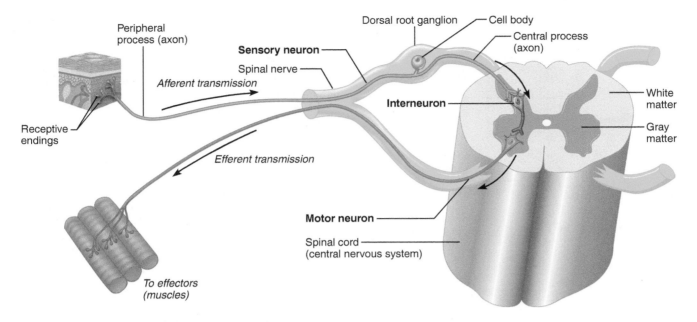

Figure 13.7 Classification of neurons on the basis of function. Sensory (afferent) neurons conduct impulses from the body's sensory receptors to the central nervous system; most are unipolar neurons with their cell bodies in ganglia in the peripheral nervous system (PNS). Motor (efferent) neurons transmit impulses from the CNS to effectors such as muscles and glands. Interneurons complete the communication line between sensory and motor neurons. They are typically multipolar, and their cell bodies reside in the CNS.

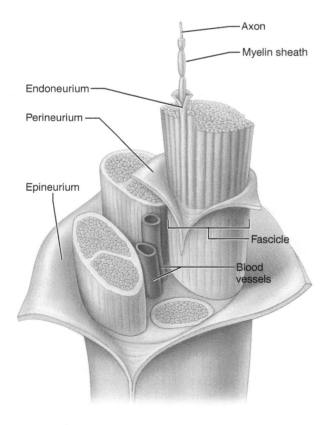

Figure 13.8 Three-dimensional view of a portion of a nerve showing connective tissue wrappings.

simpler **general sensory receptors** (**Figure 13.9** and **Figure 13.10**)—cutaneous receptors and proprioceptors—will be studied in this section. Cutaneous receptors reside in the skin (Figure 13.9). Proprioceptors are located in skeletal muscles, tendons, and joint capsules.

Anatomically, general sensory receptors are nerve endings that are either nonencapsulated or encapsulated. General sensory receptors are summarized in **Table 13.1**.

ACTIVITY 4

Studying the Structure of Selected Sensory Receptors

1. Obtain histologic slides of lamellar and tactile corpuscles. Locate, under low power, a tactile corpuscle in the dermal layer of the skin. As mentioned above, these are usually found in the dermal papillae. Then switch to the oil immersion lens for a detailed study. Notice that the free nerve fibers within the corpuscle are aligned parallel to the skin surface. Compare your observations to Figure 13.9.

2. Next, observe a lamellar corpuscle located much deeper in the dermis. Try to identify the slender free dendrite ending in the center of the receptor and the many layers of connective tissue surrounding it (which looks rather like an onion cut lengthwise). Also, notice how much larger the lamellar corpuscles are than the tactile corpuscles. Compare your observations to Figure 13.9.

3.   Obtain slides of muscle spindles and tendon organs, the two major types of proprioceptors (see Figure 13.10). In the

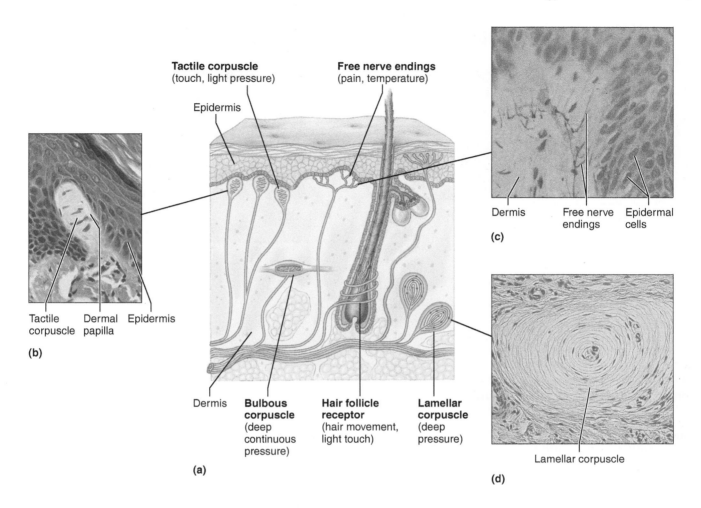

Figure 13.9 Examples of cutaneous receptors. Drawing (a) and photomicrographs (b–d). **(a)** Free nerve endings, hair follicle receptor, tactile corpuscle, lamellar corpuscle, and bulbous corpuscle. Epithelial tactile complexes are not illustrated. **(b)** Tactile corpuscle in a dermal papilla (400×). **(c)** Free nerve endings at dermal-epidermal junction (330×). **(d)** Cross section of a lamellar corpuscle in the dermis (220×).

Table 13.1   Receptors of the General Senses (Figures 13.9 and 13.10)

Structural class	Body location(s)	Stimulus type
Nonencapsulated		
Free nerve endings	Most body tissues; especially the epithelia and connective tissues	Primarily pain, heat, and cold
Epithelial tactile complexes (Merkel discs)	Stratum basale of the epidermis	Light pressure
Hair follicle receptors	Wrapped around hair follicles	Bending of hairs
Encapsulated		
Tactile (Meissner's) corpuscles	Dermal papillae of hairless skin	Light pressure, discriminative touch
Bulbous corpuscles	Deep in the dermis, subcutaneous tissue, and joint capsules	Deep pressure and stretch
Lamellar corpuscles	Dermis, subcutaneous tissue, periosteum, tendons, and joint capsules	Deep pressure, stretch, and vibration
Muscle spindles	Skeletal muscles	Muscle stretch (proprioception)
Tendon organs	Tendons	Tendon stretch, tension (proprioception)

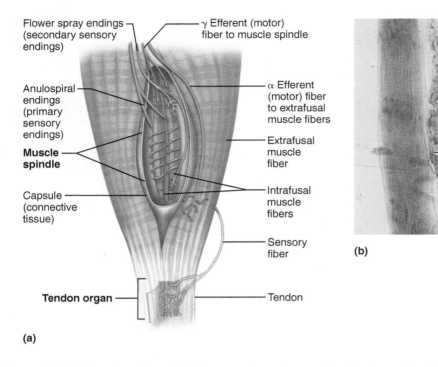

Flower spray endings (secondary sensory endings)

γ Efferent (motor) fiber to muscle spindle

Anulospiral endings (primary sensory endings)

Muscle spindle

Capsule (connective tissue)

α Efferent (motor) fiber to extrafusal muscle fibers

Extrafusal muscle fiber

Intrafusal muscle fibers

Sensory fiber

Tendon organ

Tendon

(a)

Extrafusal muscle fibers

Capsule

Intrafusal muscle fibers

(b)

Figure 13.10 Proprioceptors. (a) Diagram view of a muscle spindle and tendon organ. **(b)** Photomicrograph of a muscle spindle (80×).

slide of **muscle spindles,** note that minute extensions of the dendrites of the sensory neurons coil around specialized slender skeletal muscle cells called **intrafusal muscle fibers.** The **tendon organs** are composed of dendrites that ramify through the tendon tissue close to the muscle tendon attachment. Stretching the muscles or tendons excites these receptors, which then transmit impulses that ultimately reach the cerebellum for interpretation. Compare your observations to Figure 13.10. ▄▄

Histology of Nervous Tissue

1. The basic functional unit of the nervous system is the neuron. What is the major function of this cell type?

_____

2. Match each statement with the correct type of neuroglia by filling in the blank.

_____ 1. forms the myelin sheath in the CNS

_____ 2. lines CSF-filled cavities

_____ 3. surrounds the cell body of a neuron found in the PNS

_____ 4. act as a phagocyte in the CNS

_____ 5. forms the myelin sheath in the PNS

_____ 6. controls the chemical environment around neurons in the CNS

3. For each description, choose the appropriate term from the key. (Not all terms will be used.)

Key: a. afferent neuron   e. interneuron   i. nucleus
      b. central nervous system   f. neuroglia   j. peripheral nervous system
      c. efferent neuron   g. neurotransmitters   k. synaptic cleft
      d. ganglion   h. nerve   l. tract

_____ 1. the brain and spinal cord collectively

_____ 2. specialized supporting cells in the nervous system

_____ 3. junction or point of close contact between neurons

_____ 4. a bundle of axons inside the PNS

_____ 5. neuron serving as part of the conduction pathway between sensory and motor neurons

_____ 6. ganglia and spinal and cranial nerves

_____ 7. collection of neuron cell bodies found within the CNS

_____ 8. neuron that conducts impulses away from the CNS to muscles and glands

_____ 9. neuron that conducts impulses toward the CNS from the body periphery

_____ 10. chemicals released by neurons that stimulate or inhibit other neurons or effectors

_____ 11. collection of neuron cell bodies found in the PNS

_____ 12. bundle of axons inside the CNS

Neuron Anatomy

4. Match each anatomical term in column B with its description or function in column A.

Column A

_____ 1. region of the cell body from which the axon originates

_____ 2. secretes neurotransmitters

_____ 3. receptive regions of a neuron (2 terms)

_____ 4. insulates the nerve fibers

_____ 5. site of the nucleus and is the most important metabolic area

_____ 6. involved in the transport of substances within the neuron

_____ 7. essentially rough endoplasmic reticulum, important metabolically

_____ 8. impulse generator and transmitter

Column B

a. axon

b. axon hillock

c. cell body

d. chromatophilic substance

e. dendrite

f. myelin sheath

g. neurofibril

h. terminal bouton

5. Label the following structures on the diagram of a multipolar neuron shown below: cell body, nucleus, nucleolus, chromatophilic substance, dendrites, initial segment of axon, myelin sheath, myelin sheath gaps, and terminal boutons.

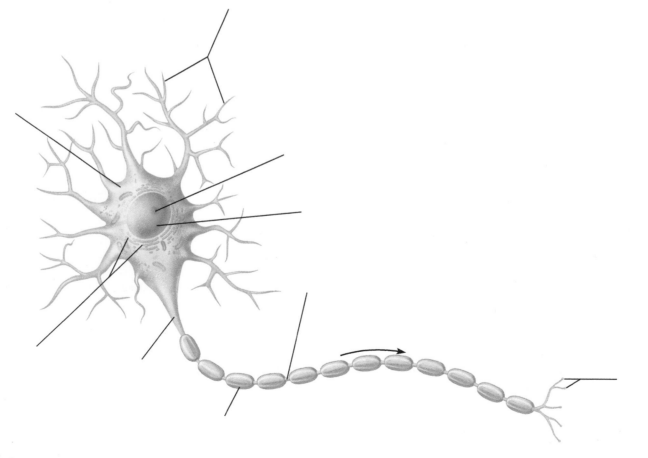

6. What substance is found in synaptic vesicles of the terminal bouton? _____

7. What anatomical characteristic determines whether a particular neuron is classified as unipolar, bipolar, or multipolar?

_____

8. Correctly identify the sensory (afferent) neuron, interneuron, and motor (efferent) neuron in the figure below.

Which of these neuron types is/are unipolar? _____

Which is/are most likely multipolar? _____

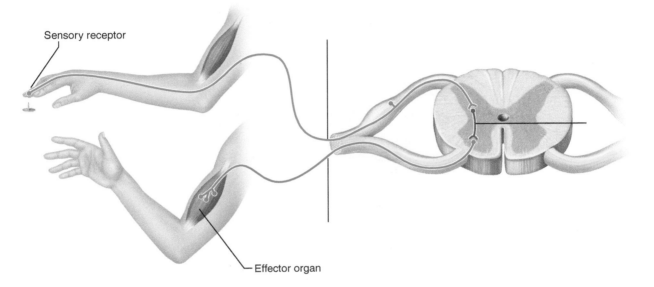

Sensory receptor

Effector organ

9. Describe how the Schwann cells form the myelin sheath encasing the nerve fibers.

_____

_____

_____

Structure of a Nerve

10. What is a nerve? _____

_____

11. State the location of each of the following connective tissue coverings.

endoneurium: _____

perineurium: _____

epineurium: _____

12. What is the function of the connective tissue wrappings found in a nerve? _____

_____

13. Define _mixed nerve._ _____

_____

14. Identify all indicated parts of the nerve section.

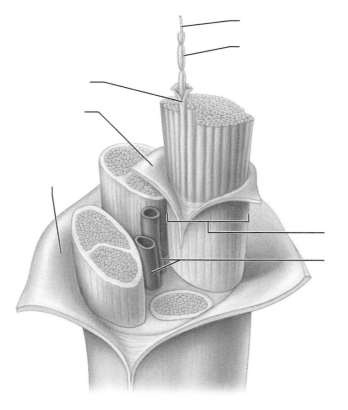

Structure of General Sensory Receptors

15. A number of activities and sensations are listed in the chart below. For each, check whether the receptors would be exteroceptors or interoceptors; and then name the specific receptor types.

Activity or sensation	Exteroceptor	Interoceptor	Specific receptor type
Backing into a sun-heated iron railing			
Someone steps on your foot			
Reading a street sign			
Leaning on your elbows			
Doing sit-ups			

16. ✚ Amyotrophic lateral sclerosis is a neurodegenerative disease in which motor neurons are progressively destroyed. Excess levels of the neurotransmitter glutamate have been implicated in this process. Which type of neuroglia would play a role in controlling glutamate levels in the chemical environment of neurons?

_____

17. ✚ Peripheral neuropathy has a variety of causes. Worldwide, the most common cause is leprosy, also known as Hansen's

disease. Would you expect peripheral neuropathy to cause damage to tracts or to nerves? _____

Why? _____

Gross Anatomy of the Brain and Cranial Nerves

MATERIALS

- Human brain model (dissectible)
- Preserved human brain (if available)
- Three-dimensional model of ventricles
- Frontally sectioned or cross-sectioned human brain slice (if available)
- Materials as needed for cranial nerve testing (see Table 14.2): aromatic oils (e.g., vanilla and cloves); eye chart; ophthalmoscope; penlight; safety pin; blunt probe (hot and cold); cotton; solutions of sugar, salt, vinegar, and quinine; ammonia; tuning fork, and tongue depressor
- Preserved sheep brain (meninges and cranial nerves intact)
- Dissecting instruments and tray
- Disposable gloves

LEARNING OUTCOMES

- ☐ List the elements of the central and peripheral divisions of the nervous system.
- ☐ Discuss the difference between the sensory and motor portions of the nervous system, and name the two divisions of the motor portion.
- ☐ Recognize the terms that describe the development of the human brain, and discuss the relationships between the terms.
- ☐ As directed by your instructor, identify the bold terms associated with the cerebral hemispheres, diencephalon, brain stem, and cerebellum on a dissected human brain, brain model, or appropriate image, and state their functions.
- ☐ State the differences among gyri, fissures, and sulci.
- ☐ Describe the composition of gray matter and white matter in the nervous system.
- ☐ Name and describe the three meninges that cover the brain, state their functions, and locate the falx cerebri, falx cerebelli, and tentorium cerebelli.
- ☐ Discuss the formation, circulation, and drainage of cerebrospinal fluid.
- ☐ Identify the cranial nerves by number and name on a model or image, stating the origin and function of each.
- ☐ Identify at least four key anatomical differences between the human brain and sheep brain.

PRE-LAB QUIZ

1. Circle the correct underlined term. The <u>central nervous system</u> / <u>peripheral nervous system</u> consists of the brain and spinal cord.
2. Circle the correct underlined term. The most superior portion of the brain includes the <u>cerebral hemispheres</u> / <u>brain stem</u>.
3. Circle True or False. Deep grooves within the cerebral hemispheres are known as gyri.
4. On the ventral surface of the brain, you can observe the optic nerves and chiasma, the pituitary gland, and the mammillary bodies. These externally visible structures form the floor of the:
   a. brain stem   c. frontal lobe
   b. diencephalon   d. occipital lobe
5. Circle the correct underlined term. The inferior region of the brain stem, the <u>medulla oblongata</u> / <u>cerebellum</u> houses many vital autonomic centers involved in the control of heart rate, respiratory rhythm, and blood pressure.
6. Directly under the occipital lobe of the cerebrum is a large cauliflower-like structure known as the:
   a. brain stem   b. cerebellum   c. diencephalon

Text continues on next page.

7. Circle the correct underlined term. The outer cortex of the brain contains the cell bodies of cerebral neurons and is known as white matter / gray matter.
8. The brain and spinal cord are covered and protected by three connective tissue layers called:
   a. lobes   c. sulci
   b. meninges   d. ventricles

9. Circle True or False. Cerebrospinal fluid is produced by the frontal lobe of the cerebrum and is unlike any other body fluid.
10. How many pairs of cranial nerves are there? _____

When viewed alongside all Earth's animals, humans are indeed unique, and the key to our uniqueness is found in the brain. Each of us is a reflection of our brain's experience. If all past sensory input could mysteriously and suddenly be "erased," we would be unable to walk, talk, or communicate in any manner. Spontaneous movement would occur, as in a fetus, but no voluntary integrated function of any type would be possible. Clearly we would cease to be the same individuals.

For convenience, the nervous system, is considered in terms of two principal divisions: the central nervous system and the peripheral nervous system. The **central nervous system (CNS)** consists of the brain and spinal cord, which primarily interpret incoming sensory information and issue instructions based on that information and on past experience. The **peripheral nervous system (PNS)** consists of the cranial and spinal nerves, ganglia, and sensory receptors.

The PNS has two major subdivisions: the **sensory portion,** which consists of nerve fibers that conduct impulses from sensory receptors toward the CNS, and the **motor portion,** which contains nerve fibers that conduct impulses away from the CNS. The motor arm, in turn, consists of the **somatic division** (sometimes called the *voluntary system*), which controls the skeletal muscles, and the other subdivision, the **autonomic nervous system (ANS),** which controls smooth and cardiac muscles and glands.

This exercise focuses on the brain (CNS) and cranial nerves (PNS) because of their close anatomical relationship.

The Human Brain

During embryonic development of all vertebrates, the CNS first makes its appearance as a simple tubelike structure, the **neural tube,** that extends down the dorsal median plane. By the fourth week, the human brain begins to form as an expansion of the anterior or rostral end of the neural tube (the end toward the head). Shortly thereafter, constrictions appear, dividing the developing brain into three major regions—**forebrain, midbrain,** and **hindbrain** (Figure 14.1). The remainder of the neural tube becomes the spinal cord.

The central canal of the neural tube, which remains continuous throughout the brain and cord, enlarges in four regions of the brain, forming chambers called **ventricles** (see Figure 14.8a and b, page 265).

(see Figure 14.8a and b, page 265)

ACTIVITY 1

Identifying External Brain Structures

Identify external brain structures using the figures cited. Also use a model of the human brain and other learning aids as they are mentioned.

Generally, the brain is studied in terms of four major regions: the cerebral hemispheres, diencephalon, brain stem, and cerebellum.

Cerebral Hemispheres

The **cerebral hemispheres** are the most superior portion of the brain (Figure 14.2). Their entire surface is thrown into elevated ridges of tissue called **gyri** that are separated by shallow grooves called **sulci** or deeper grooves called **fissures.** Many of the fissures and gyri are important anatomical landmarks.

The cerebral hemispheres are divided by a single deep fissure, the **longitudinal fissure.** The **transverse cerebral fissure** separates the cerebral hemispheres from the cerebellum below. The **central sulcus** divides the **frontal lobe** from the **parietal lobe,** and the **lateral sulcus** separates the **temporal lobe** from the parietal lobe. The **parieto-occipital sulcus** on the medial surface of each hemisphere divides the **occipital lobe** from the parietal lobe. Notice that these cerebral hemisphere lobes are named for the cranial bones that lie over them. A fifth lobe of each cerebral hemisphere, the **insula,** is buried deep within the lateral sulcus and is covered by portions of the temporal, parietal, and frontal lobes.

Some important functional areas of the cerebral hemispheres have also been located (Figure 14.2d).

The cell bodies of cerebral neurons involved in these functions are found only in the outermost gray matter of the cerebrum, the area called the **cerebral cortex.** Most of the balance of cerebral tissue—the deeper **cerebral white matter**—is composed of myelinated fibers bundled into tracts carrying impulses to or from the cortex. Some of the functional areas are summarized in Table 14.1.

Using a model of the human brain (and a preserved human brain, if available), identify the areas and structures of the cerebral hemispheres described above and in the table.

Then continue using the model and preserved brain along with the figures as you read about other structures.

Diencephalon

The **diencephalon** is embryologically part of the forebrain, along with the cerebral hemispheres. See Figure 14.1.

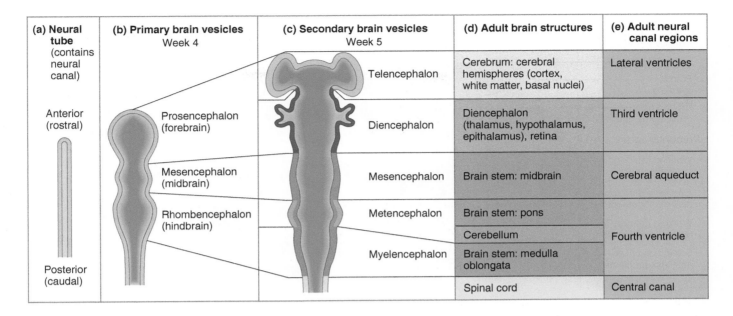

(a) Neural tube (contains neural canal)	(b) Primary brain vesicles Week 4	(c) Secondary brain vesicles Week 5	(d) Adult brain structures	(e) Adult neural canal regions
Anterior (rostral)	Prosencephalon (forebrain)	Telencephalon	Cerebrum: cerebral hemispheres (cortex, white matter, basal nuclei)	Lateral ventricles
		Diencephalon	Diencephalon (thalamus, hypothalamus, epithalamus), retina	Third ventricle
	Mesencephalon (midbrain)	Mesencephalon	Brain stem: midbrain	Cerebral aqueduct
	Rhombencephalon (hindbrain)	Metencephalon	Brain stem: pons	Fourth ventricle
			Cerebellum	
Posterior (caudal)		Myelencephalon	Brain stem: medulla oblongata	
			Spinal cord	Central canal

Figure 14.1 Embryonic development of the human brain. (a) The neural tube becomes subdivided into **(b)** the primary brain vesicles, which subsequently form **(c)** the secondary brain vesicles, which differentiate into **(d)** the adult brain structures. **(e)** The adult structures derived from the neural canal.

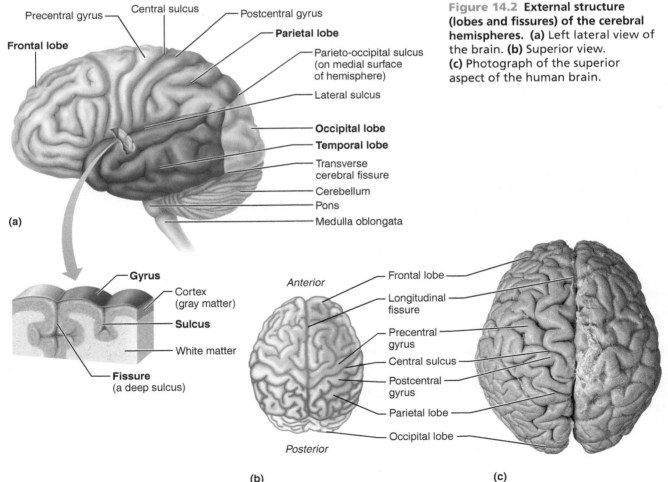

Figure 14.2 External structure (lobes and fissures) of the cerebral hemispheres. (a) Left lateral view of the brain. **(b)** Superior view. **(c)** Photograph of the superior aspect of the human brain.

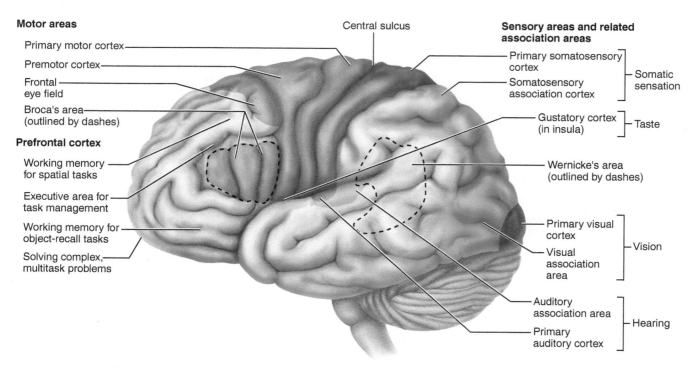

(d)

Figure 14.2 (*continued*) **External structure (lobes and fissures) of the cerebral hemispheres. (d)** Functional areas of the left cerebral cortex. The olfactory cortex, which is deep within the temporal lobe on the medial hemispheric surface, is not identified here.

Turn the brain model so the ventral surface of the brain can be viewed. Starting superiorly (and using **Figure 14.3** as a guide), identify the externally visible structures that mark the position of the floor of the diencephalon. These are the **olfactory bulbs** (synapse point of cranial nerve I) and **tracts, optic nerves** (cranial nerve II), **optic chiasma** (where the medial fibers of the optic nerves cross over), **optic tracts, pituitary gland,** and **mammillary bodies.**

Brain Stem

Continue to identify the **brain stem** structures—the **cerebral peduncles** (fiber tracts in the **midbrain** connecting the pons

Table 14.1	Important Functional Areas of the Cerebral Cortex	
Functional sensory areas	**Location**	**Functions**
Primary somatosensory cortex	Postcentral gyrus of the parietal lobe	Receives information from the body's sensory receptors in the skin and from proprioceptors in the skeletal muscles, joints, and tendons
Primary visual cortex	Occipital lobe	Receives visual information that originates in the retina of the eye
Primary auditory cortex	Temporal lobe in the gyrus bordering the lateral sulcus	Receives sound information from the receptors for hearing in the internal ear
Olfactory cortex	Medial surface of the temporal lobe, in a region called the uncus	Receives information from olfactory (smell) receptors in the superior nasal cavity
Functional motor areas	**Location**	**Functions**
Primary motor cortex	Precentral gyrus of the frontal lobe	Conscious control of voluntary movement of skeletal muscles
Broca's area	Anterior to the inferior region of the premotor area in the frontal lobe in only one hemisphere	Controls the muscles involved in speech production and also plays a role in the planning of nonspeech motor functions

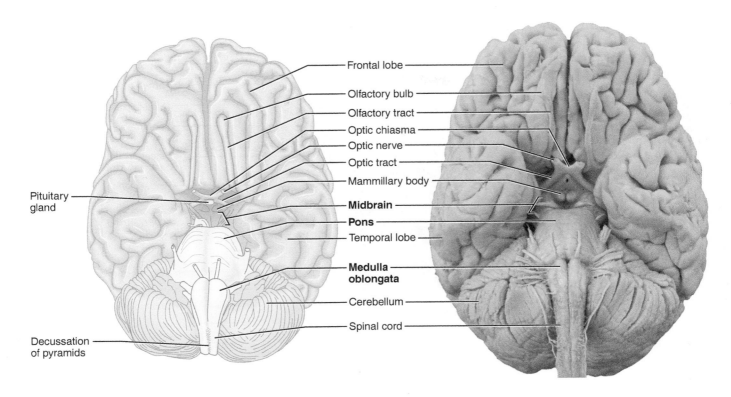

Frontal lobe
Olfactory bulb
Olfactory tract
Optic chiasma
Optic nerve
Optic tract
Mammillary body
Midbrain
Pons
Temporal lobe
Medulla oblongata
Cerebellum
Spinal cord

Pituitary gland

Decussation of pyramids

Figure 14.3 Ventral (inferior) aspect of the human brain, showing the three regions of the brain stem. Only a small portion of the midbrain can be seen; the rest is surrounded by other brain regions.

below with the cerebrum above), the pons, and the medulla oblongata. *Pons* means "bridge," and the **pons** consists primarily of motor and sensory fiber tracts connecting the brain with lower CNS centers. The lowest brain stem region, the **medulla oblongata,** often shortened to just **medulla,** is also composed primarily of fiber tracts. You can see the **decussation of pyramids,** a crossover point for the major motor tracts (pyramidal tracts) descending from the motor areas of the cerebrum to the spinal cord, on the medulla's surface. The medulla also houses many vital autonomic centers involved in the control of heart rate, respiratory rhythm, and blood pressure as well as involuntary centers involved in vomiting, and swallowing.

Cerebellum

1. Turn the brain model so you can see the dorsal aspect. Identify the large cauliflower-shaped **cerebellum,** which projects dorsally from under the occipital lobe of the cerebrum. Notice that, like the cerebrum, the cerebellum has two major hemispheres and a convoluted surface (see Figure 14.6). It also has an outer cortex made up of gray matter with an inner region of white matter.

2. Remove the cerebellum to view the **corpora quadrigemina (Figure 14.4),** located on the posterior aspect of the midbrain, a brain stem structure. The two superior prominences are the **superior colliculi** (visual reflex centers); the

two smaller inferior prominences are the **inferior colliculi** (auditory reflex centers). ■

ACTIVITY 2

Identifying Internal Brain Structures

The deeper structures of the brain have also been well mapped. As the internal brain areas are described, identify them on the figures cited. Also, use the brain model as indicated to help you in this study.

Cerebral Hemispheres

1. Take the brain model apart so you can see a medial view of the internal brain structures (see Figure 14.4). Observe the model closely to see the extent of the outer cortex (gray matter), which contains the cell bodies of cerebral neurons. The pyramidal cells of the cerebral motor cortex (Exercise 13 and Figure 13.5) are representative of the neurons seen in the precentral gyrus.

2. Observe the deeper area of white matter, which is made up of fiber tracts. The fiber tracts found in the cerebral hemisphere white matter are called *association tracts* if they connect two portions of the same hemisphere, *projection tracts* if they run between the cerebral cortex and lower brain structures or spinal cord, and *commissures* if they run from one hemisphere to another. Observe the large **corpus callosum,**

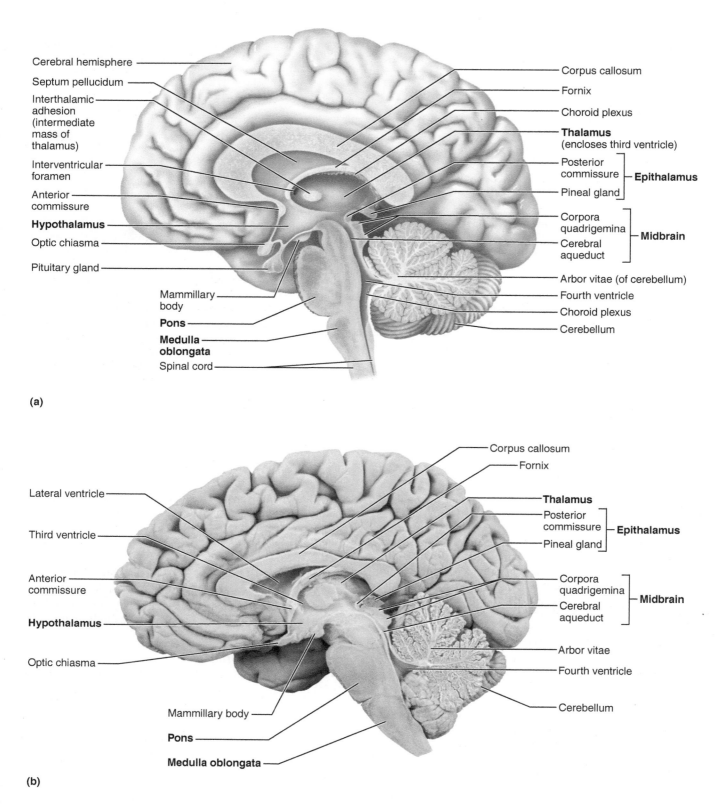

(a)

(b)

Figure 14.4 **Diencephalon and brain stem structures as seen in a median section of the brain.** **(a)** Diagram. **(b)** Photograph.

the major commissure connecting the cerebral hemispheres. The corpus callosum arches above the structures of the diencephalon and roofs over the lateral ventricles. Notice also the **fornix,** a bandlike fiber tract concerned with olfaction as well as limbic system functions, and the membranous **septum pellucidum,** which separates the lateral ventricles of the cerebral hemispheres.

3.   In addition to the gray matter of the cerebral cortex, there are several clusters of neuron cell bodies called **nuclei** buried deep within the white matter of the cerebral hemispheres. One important group of cerebral nuclei, called the **basal nuclei,** or **basal ganglia,*** flank the lateral and third ventricles. You can see these nuclei if you have an appropriate dissectible model or a coronally or cross-sectioned human brain slice. Otherwise, the figure **(Figure 14.5)** will suffice.

The basal nuclei are involved in regulating voluntary motor activities. The most important of them are the arching, comma-shaped **caudate nucleus,** the **putamen,** and **globus pallidus.**

The **corona radiata,** a spray of projection fibers coursing down from the precentral (motor) gyrus, combines with sensory fibers traveling to the primary somatosensory cortex to form a broad band of fibrous material called the **internal capsule.** The internal capsule passes between the thalamus and the basal nuclei and through the caudate and putamen, giving them a striped appearance. Hence they are called the **striatum,** or "striped" body (Figure 14.5a).

4.   Examine the relationship of the lateral ventricles and corpus callosum to the thalamus and third ventricle—from the cross-sectional viewpoint (see Figure 14.5b).

Diencephalon

1.   The major internal structures of the diencephalon are the thalamus, hypothalamus, and epithalamus (see Figure 14.4). The **thalamus** consists of two large lobes of gray matter that laterally enclose the narrow third ventricle of the brain. A slender stalk of thalamic tissue, the **interthalamic adhesion,** or **intermediate mass,** connects the two thalamic lobes and bridges the ventricle. The thalamus is a major integrating and relay station for sensory impulses passing upward to the cortical sensory areas for localization and interpretation. Locate also the **interventricular foramen,** a tiny opening connecting the third ventricle with the lateral ventricle on the same side.

2.   The **hypothalamus** makes up the floor and the inferolateral walls of the third ventricle. It is an important autonomic center involved in regulating body temperature, water balance, and fat and carbohydrate metabolism as well as many other activities and drives (sex, hunger, thirst). Locate again the pituitary gland, which hangs from the anterior floor of the hypothalamus by a slender stalk, the **infundibulum.** The pituitary rests in the hypophyseal fossa of the sella turcica of the sphenoid bone.

*We follow the guidelines of *Terminologia Anatomica* and use the term *basal nuclei* in this lab manual. However, the use of the term *basal ganglia* is widespread in clinical settings.

Anterior to the pituitary, identify the optic chiasma portion of the optic pathway to the brain. The **mammillary bodies,** relay stations for olfaction, bulge exteriorly from the floor of the hypothalamus just posterior to the pituitary gland.

3.   The **epithalamus** forms the roof of the third ventricle and is the most dorsal portion of the diencephalon. Important structures in the epithalamus are the **pineal gland** and the **choroid plexus** of the third ventricle.

WHY THIS MATTERS | Unresponsive Wakefulness Syndrome (UWS)

We appreciate the thalamus more when we look at what happens when it is damaged. This is the problem that afflicts patients with unresponsive wakefulness syndrome (UWS). UWS is described as a disorder of consciousness that lasts more than 4 weeks in which the patient has lost awareness of self and environment. Unlike coma patients, who cannot be awakened, UWS patients can be aroused—they display wakefulness without awareness. This syndrome is characterized by clinical loss of cerebral cortex function with unaffected brain stem function. In many cases, there is no structural damage to the cerebral cortex, but its function is lost because the relay of signals to the cortex—the thalamus—is not working. ▪

Brain Stem

1.   Now trace the short midbrain from the mammillary bodies to the rounded pons below. (Continue to refer to Figure 14.4.) The **cerebral aqueduct** is a slender canal traveling through the midbrain; it connects the third ventricle to the fourth ventricle. The cerebral peduncles and the rounded corpora quadrigemina make up the midbrain tissue anterior and posterior (respectively) to the cerebral aqueduct.

2.   Locate the hindbrain structures. Trace the rounded pons to the medulla oblongata below, and identify the fourth ventricle posterior to these structures. Attempt to identify the single median aperture and the two lateral apertures, three openings found in the walls of the fourth ventricle. These apertures serve as passageways for cerebrospinal fluid to circulate into the subarachnoid space from the fourth ventricle.

Cerebellum

Examine the cerebellum. Notice that it is composed of two lateral hemispheres, each with three lobes (*anterior, posterior,* and a deep *flocculonodular*) connected by a midline lobe called the **vermis (Figure 14.6).** As in the cerebral hemispheres, the cerebellum has an outer cortical area of gray matter and an inner area of white matter. The treelike branching of the cerebellar white matter is referred to as the **arbor vitae,** or "tree of life." The cerebellum controls the unconscious coordination of skeletal muscle activity along with balance and equilibrium. ▬

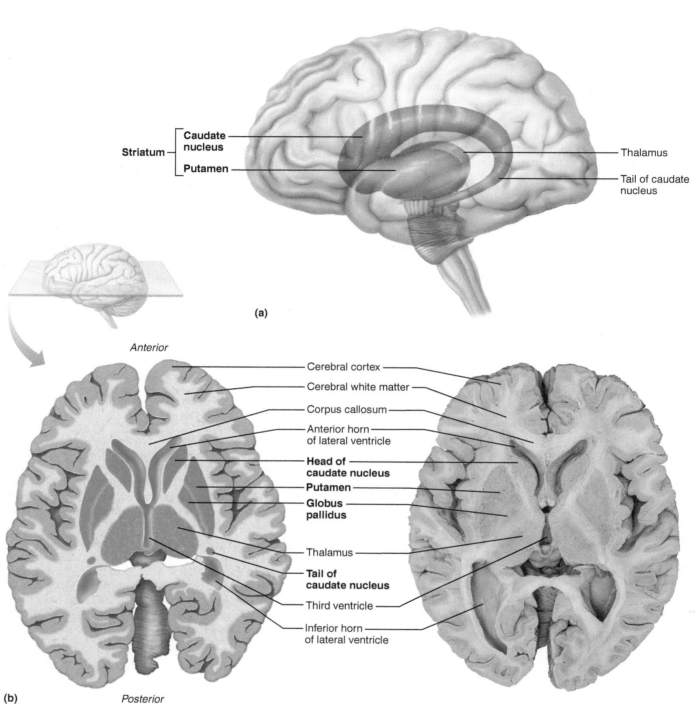

Striatum — Caudate nucleus
Putamen

Thalamus

Tail of caudate nucleus

(a)

Anterior

Cerebral cortex
Cerebral white matter
Corpus callosum
Anterior horn of lateral ventricle
Head of caudate nucleus
Putamen
Globus pallidus
Thalamus
Tail of caudate nucleus
Third ventricle
Inferior horn of lateral ventricle

(b)   Posterior

Figure 14.5 Basal nuclei. (a) Three-dimensional view of the basal nuclei showing their positions within the cerebrum. Globus pallidus lies medial and deep to the putamen. **(b)** A transverse section of the cerebrum and diencephalon showing the relationship of the basal nuclei to the thalamus and the lateral and third ventricles.

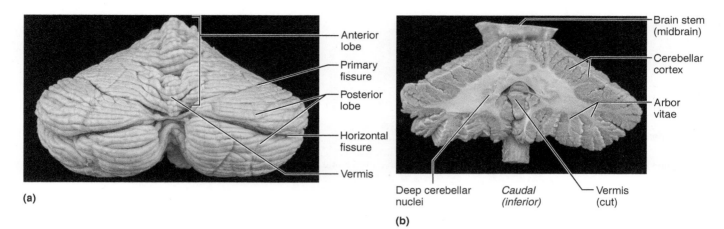

(a)

(b)

Figure 14.6 Cerebellum. (a) Posterior (dorsal) view. **(b)** The cerebellum, sectioned to reveal its cortex and medullary regions. (Note that the cerebellum is sectioned frontally and the brain stem is sectioned transversely in this posterior view.)

Meninges of the Brain

The brain and spinal cord are covered and protected by three connective tissue membranes called **meninges (Figure 14.7)**. The outermost meninx is the leathery **dura mater,** a double-layered membrane. One of its layers (the *periosteal layer*) is attached to the inner surface of the skull, forming the periosteum. The other layer (the *meningeal layer*) forms the outermost brain covering and is continuous with the dura mater of the spinal cord.

The dural layers are fused together except in three places where the inner membrane extends inward to form a septum that secures the brain to structures inside the cranial cavity. One such extension, the **falx cerebri,** dips into the longitudinal fissure between the cerebral hemispheres to attach to the crista galli of the ethmoid bone of the skull (Figure 14.7a). The cavity created at this point is the large **superior sagittal sinus,** which collects blood draining from the brain tissue. The **falx cerebelli,** separating the two cerebellar hemispheres, and the **tentorium cerebelli,** separating the cerebrum from the cerebellum below, are two other important inward folds of the inner dural membrane.

The middle meninx, the weblike **arachnoid mater,** underlies the dura mater and is partially separated from it by the **subdural space.** Threadlike projections bridge the **subarachnoid space** to attach the arachnoid to the innermost meninx, the **pia mater.** The delicate pia mater is highly vascular and clings tenaciously to the surface of the brain, following its gyri.

The subarachnoid space is filled with cerebrospinal fluid. Specialized projections of the arachnoid tissue called **arachnoid granulations,** or *arachnoid villi,* protrude through the dura mater. These granulations allow the cerebrospinal fluid to drain back into the venous circulation via the superior sagittal sinus and other dural venous sinuses.

Meningitis, inflammation of the meninges, is a serious threat to the brain because of the intimate association between the brain and meninges. Should infection spread to the neural tissue of the brain itself, life-threatening **encephalitis** may occur. Meningitis is often diagnosed by taking a sample of cerebrospinal fluid (via a lumbar puncture) from the subarachnoid space. ✚

Cerebrospinal Fluid

The cerebrospinal fluid (CSF), much like plasma in composition, is continually formed by the **choroid plexuses,** small capillary knots hanging from the roof of the ventricles of the brain. The cerebrospinal fluid in and around the brain forms a watery cushion that protects the delicate brain tissue against blows to the head.

Within the brain, the cerebrospinal fluid circulates from the two lateral ventricles (in the cerebral hemispheres) into the third ventricle via the **interventricular foramina,** and then through the cerebral aqueduct of the midbrain into the fourth ventricle **(Figure 14.8).** CSF enters the subarachnoid space through the paired **lateral apertures** in the side walls of the fourth ventricle and the **median aperture** in its roof. There it bathes the outer surfaces of the brain and spinal cord. The fluid returns to the blood in the dural venous sinuses via the arachnoid granulations.

Ordinarily, cerebrospinal fluid forms and drains at a constant rate. However, under certain conditions— for example, obstructed drainage or circulation resulting from tumors or anatomical deviations—cerebrospinal fluid accumulates and exerts increasing pressure on the brain which, uncorrected, causes neurological damage in adults. In infants, **hydrocephalus** (literally, "water on the brain") is indicated by a gradually enlarging head. The infant's skull is still flexible and contains fontanelles, so it can expand to accommodate the increasing size of the brain. ✚

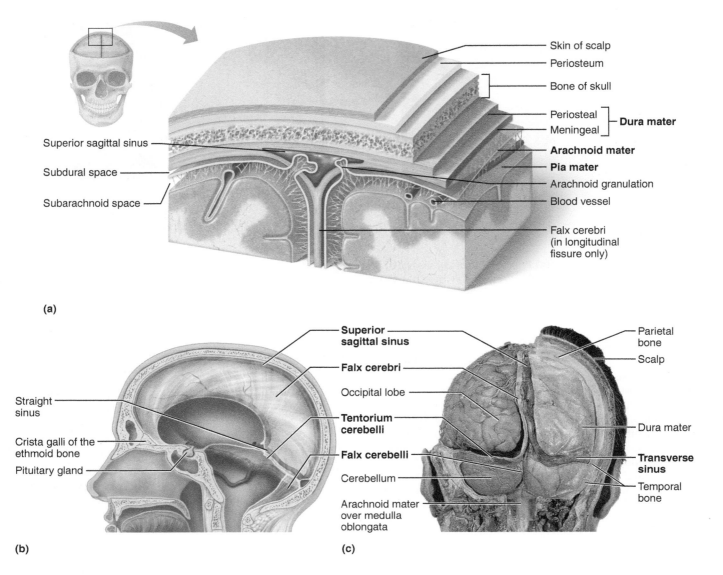

(a)

(b)

(c)

Figure 14.7 Meninges of the brain. (a) Three-dimensional frontal section showing the relationship of the dura mater, arachnoid mater, and pia mater. The meningeal dura forms the falx cerebri fold, which extends into the longitudinal fissure and attaches the brain to the ethmoid bone of the skull. The superior sagittal sinus is enclosed by the dural membranes superiorly. Arachnoid granulations, which return cerebrospinal fluid to the dural sinus, are also shown. **(b)** Midsagittal view showing the position of the dural folds, the falx cerebri, tentorium cerebelli, and falx cerebelli. **(c)** Posterior view of the brain in place, surrounded by the dura mater. Sinuses between periosteal and meningeal dura contain venous blood.

Cranial Nerves

The **cranial nerves (Figure 14.9)** are part of the peripheral nervous system and not part of the brain proper, but they are most appropriately identified while studying brain anatomy. The 12 pairs of cranial nerves primarily serve the head and neck. Only one pair, the vagus nerves, extends into the thoracic and abdominal cavities. All but the first two pairs (olfactory and optic nerves) arise from the brain stem and pass through foramina in the base of the skull to reach their destination.

The cranial nerves are numbered consecutively, and in most cases their names reflect the major structures they control. The cranial nerves are described by name, number (Roman numeral), origin, course, and function in **Table 14.2**. You should memorize this information. A mnemonic device that might be helpful for remembering the cranial nerves in order is "**O**n **O**ccasion, **O**ur **T**rusty **T**ruck **A**cts **F**unny—**V**ery **G**ood **V**ehicle **A**ny**H**ow." The first letter of each word and the "a" and "h" of the final word "anyhow" will remind you of the first letter of each cranial nerve name.

Text continues on page 269.

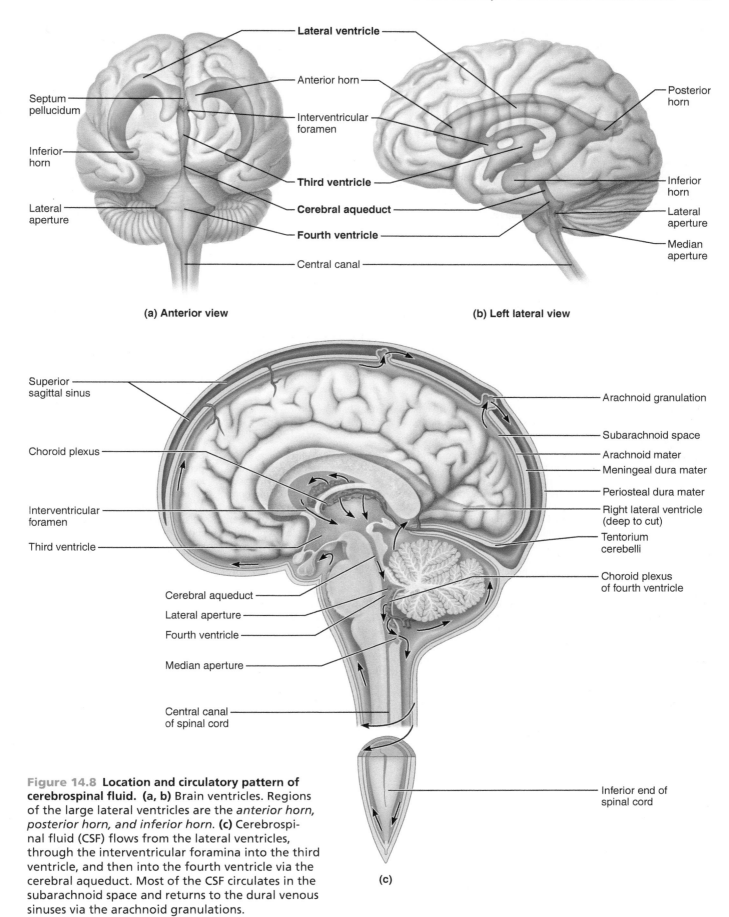

(a) Anterior view

(b) Left lateral view

(c)

Figure 14.8 Location and circulatory pattern of cerebrospinal fluid. (a, b) Brain ventricles. Regions of the large lateral ventricles are the *anterior horn, posterior horn, and inferior horn.* **(c)** Cerebrospinal fluid (CSF) flows from the lateral ventricles, through the interventricular foramina into the third ventricle, and then into the fourth ventricle via the cerebral aqueduct. Most of the CSF circulates in the subarachnoid space and returns to the dural venous sinuses via the arachnoid granulations.

Table 14.2 The Cranial Nerves (see Figure 14.9)

Number and name	Origin and course	Function*	Testing
I. Olfactory	Fibers arise from olfactory epithelium and run through cribriform foramina of ethmoid bone to synapse in olfactory bulbs.	Purely sensory—carries afferent impulses for sense of smell.	Person is asked to sniff aromatic substances, such as oil of cloves and vanilla, and to identify each.
II. Optic	Fibers arise from retina of eye and pass through optic canal of the sphenoid bone. Fibers partially cross over at the optic chiasma and continue on to the thalamus as the optic tracts. Final fibers of this pathway travel from the thalamus to the primary visual cortex as the optic radiation.	Purely sensory—carries afferent impulses associated with vision.	Vision and visual field are determined with eye chart and by testing the point at which the person first sees an object (finger) moving into the visual field. Fundus of eye viewed with ophthalmoscope to detect papilledema (swelling of optic disc, the point at which optic nerve leaves the eye) and to observe blood vessels.
III. Oculomotor	Fibers emerge from ventral midbrain and course ventrally to enter the orbit. They exit from skull via superior orbital fissure.	Primarily motor—somatic motor fibers to inferior oblique and superior, inferior, and medial rectus muscles, which direct eyeball, and to levator palpebrae muscles of the superior eyelid; parasympathetic fibers to iris and smooth muscle controlling lens shape and pupil size.	Pupils are examined for size, shape, and equality. Pupillary reflex is tested with penlight (pupils should constrict when illuminated). Convergence for near vision is tested, as is subject's ability to follow objects with the eyes.
IV. Trochlear	Fibers emerge from midbrain and exit from skull via superior orbital fissure.	Primarily motor—provides somatic motor fibers to superior oblique muscle that moves the eyeball.	Tested with cranial nerve III.
V. Trigeminal	Fibers run from face to pons and form three divisions, which exit separately from skull: mandibular division fibers pass through foramen ovale in sphenoid bone, maxillary division fibers pass via foramen rotundum in sphenoid bone, and ophthalmic division fibers pass through superior orbital fissure of sphenoid bone.	Mixed—major sensory nerve of face; conducts sensory impulses from skin of face and anterior scalp, from mucosae of mouth and nose, and from surface of eyes; mandibular division also contains motor fibers that innervate muscles of mastication and muscles of floor of mouth.	Sensations of pain, touch, and temperature are tested with safety pin and hot and cold probes. Corneal reflex tested with wisp of cotton. Motor branch assessed by asking person to clench the teeth, open mouth against resistance, and move jaw side to side.
VI. Abducens	Fibers leave inferior region of pons and exit from skull via superior orbital fissure.	Primarily motor—carries somatic motor fibers to lateral rectus muscle that abducts the eyeball.	Tested with cranial nerve III.

*Does not include sensory impulses from proprioceptors.

Table 14.2   *(continued)*

Number and name	Origin and course	Function*	Testing
VII. Facial	Fibers leave pons and travel through temporal bone via internal acoustic meatus, exiting via stylomastoid foramen to reach the face.	Mixed—supplies somatic motor fibers to muscles of facial expression and the posterior belly of the digastric muscle; supplies parasympathetic motor fibers to lacrimal and salivary glands; carries sensory fibers from taste receptors of anterior portion of tongue.	Anterior two-thirds of tongue is tested for ability to taste sweet (sugar), salty, sour (vinegar), and bitter (quinine) substances. Symmetry of face is checked. Subject is asked to close eyes, smile, whistle, and so on. Tearing is assessed with ammonia fumes.
VIII. Vestibulocochlear	Fibers run from inner ear equilibrium and hearing apparatus, housed in temporal bone, through internal acoustic meatus to enter pons.	Mostly sensory—vestibular branch transmits impulses associated with sense of equilibrium from vestibular apparatus and semicircular canals; cochlear branch transmits impulses associated with hearing from cochlea. Small motor component adjusts the sensitivity of sensory receptors.	Hearing is checked by air and bone conduction using tuning fork.
IX. Glossopharyngeal	Fibers emerge from medulla and leave skull via jugular foramen to run to throat.	Mixed—somatic motor fibers serve pharyngeal muscles, and parasympathetic motor fibers serve salivary glands; sensory fibers carry impulses from pharynx, tonsils, posterior tongue (taste buds), and pressure receptors of carotid artery.	A tongue depressor is used to check the position of the uvula. Gag and swallowing reflexes are checked. Subject is asked to speak and cough. Posterior third of tongue may be tested for taste.
X. Vagus	Fibers emerge from medulla and pass through jugular foramen and descend through neck region into thorax and abdomen.	Mixed—fibers carry somatic motor impulses to pharynx and larynx and sensory fibers from same structures; very large portion is composed of parasympathetic motor fibers, which supply heart and smooth muscles of abdominal visceral organs; transmits sensory impulses from viscera.	As for cranial nerve IX (IX and X are tested together, because they both innervate muscles of throat and mouth).
XI. Accessory	Fibers arise from the superior aspect of spinal cord and travel through jugular foramen to reach muscles of neck and back.	Mixed (but primarily motor in function)—provides somatic motor fibers to sternocleido-mastoid and trapezius muscles.	Sternocleidomastoid and trapezius muscles are checked for strength by asking person to rotate head and shoulders against resistance.
XII. Hypoglossal	Fibers arise from medulla and exit from skull via hypoglossal canal to travel to tongue.	Mixed (but primarily motor in function)—carries somatic motor fibers to muscles of tongue.	Person is asked to protrude and retract tongue. Any deviations in position are noted.

*Does not include sensory impulses from proprioceptors.

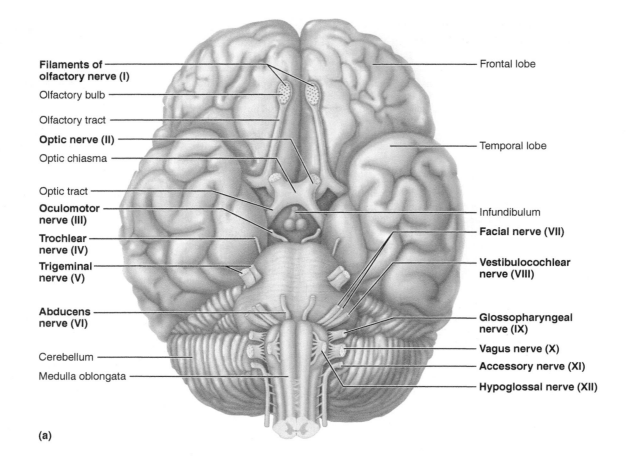

Filaments of olfactory nerve (I)
Olfactory bulb
Olfactory tract
Optic nerve (II)
Optic chiasma
Optic tract
Oculomotor nerve (III)
Trochlear nerve (IV)
Trigeminal nerve (V)
Abducens nerve (VI)
Cerebellum
Medulla oblongata

Frontal lobe
Temporal lobe
Infundibulum
Facial nerve (VII)
Vestibulocochlear nerve (VIII)
Glossopharyngeal nerve (IX)
Vagus nerve (X)
Accessory nerve (XI)
Hypoglossal nerve (XII)

(a)

Cranial nerves	Sensory function		Motor function	
	Somatic sensory (SS)	Visceral sensory (VS)	Somatic motor (SM)	Visceral motor: parasympathetic (VM)
I Olfactory		Smell		
II Optic	Vision			
III Oculomotor			SM	VM
IV Trochlear			SM	
V Trigeminal	General		SM	
VI Abducens			SM	

Cranial nerves	Sensory function		Motor function	
	Somatic sensory (SS)	Visceral sensory (VS)	Somatic motor (SM)	Visceral motor: parasympathetic (VM)
VII Facial	General	General; taste	SM	VM
VIII Vestibulocochlear	Hearing; equilibrium		Some	
IX Glossopharyngeal	General	General; taste	SM	VM
X Vagus	General	General; taste	SM	VM
XI Accessory			SM	
XII Hypoglossal			SM	

(b)

Figure 14.9 **Ventral aspect of the human brain, showing the cranial nerves.** (See also Figure 14.3.)

Most cranial nerves are mixed nerves (containing both motor and sensory fibers). But close scrutiny of the table (Table 14.2) will reveal that three pairs of cranial nerves (optic, olfactory, and vestibulocochlear) are primarily or exclusively sensory in function.

ACTIVITY 3

Identifying and Testing the Cranial Nerves

1. Observe the anterior surface of the brain model to identify the cranial nerves (Figure 14.9). Notice that the first (olfactory) cranial nerves are not visible on the model because they consist only of short axons that run from the nasal mucosa through the cribriform foramina of the ethmoid bone. (However, the synapse points of the first cranial nerves, the *olfactory bulbs,* are visible on the model.)

2. Testing cranial nerves is an important part of any neurological examination. (See the last column of Table 14.2 for techniques you can use for such tests.) The results may help you understand cranial nerve function, especially as it pertains to some aspects of brain function.

3. Several cranial nerve ganglia are named in the accompanying chart. *Using your textbook or another appropriate reference,* fill in the **Activity 3 chart** by naming the cranial nerve the ganglion is associated with and stating the ganglion's location. ▄▄

Activity 3: Cranial Nerve Ganglia		
Cranial nerve ganglion	Cranial nerve	Site of ganglion
Trigeminal		
Geniculate		
Inferior		
Superior		
Spiral		
Vestibular		

DISSECTION
The Sheep Brain

The sheep brain is enough like the human brain to warrant comparison. Obtain a sheep brain, disposable gloves, dissecting tray, and instruments, and bring them to your laboratory bench.

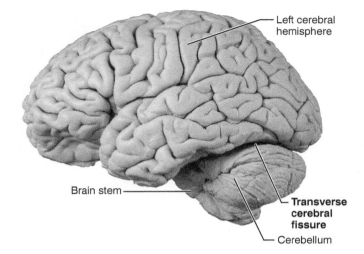

Figure 14.10 Photo of lateral aspect of the human brain.

1. If the dura mater is present, remove it as described here. Don disposable gloves. Place the intact sheep brain ventral surface down on the dissecting pan, and observe the dura mater. Feel its consistency, and note its toughness. Cut through the dura mater along the line of the longitudinal fissure (which separates the cerebral hemispheres) to enter the superior sagittal sinus. Gently force the cerebral hemispheres apart laterally to expose the corpus callosum deep to the longitudinal fissure.

2. Carefully remove the dura mater, and examine the superior surface of the brain. Notice that like the human brain, its surface is thrown into convolutions (fissures and gyri). Locate the arachnoid mater, which appears on the brain surface as a delicate "cottony" material spanning the fissures. In contrast, the innermost meninx, the pia mater, closely follows the cerebral contours.

3. Before beginning the dissection, turn your sheep brain so that you are viewing its left lateral aspect. Compare the various areas of the sheep brain (cerebrum, brain stem, cerebellum) to the photo of the human brain (Figure 14.10). Relatively speaking, which of these structures is obviously much larger in the human brain?

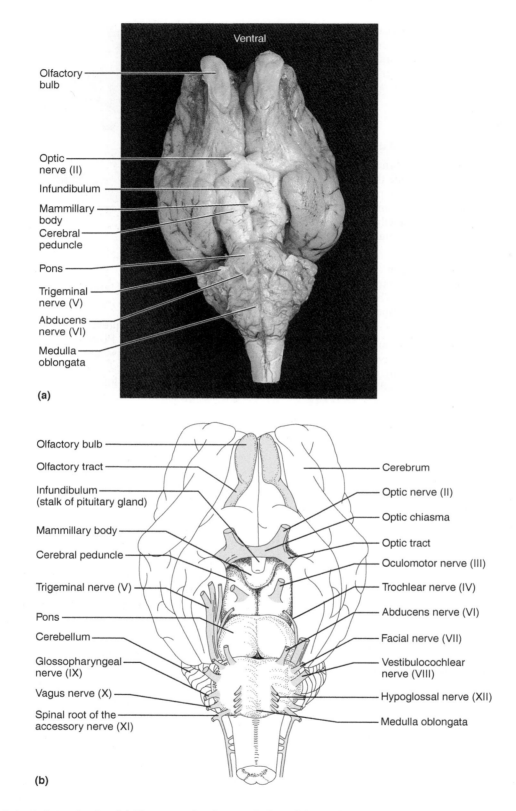

Figure 14.11 Intact sheep brain. (a) Photograph of ventral view. **(b)** Diagram of ventral view.

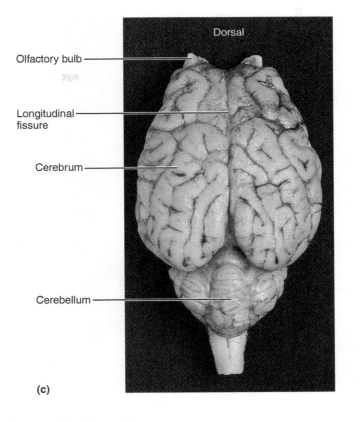

Olfactory bulb

Longitudinal fissure

Cerebrum

Cerebellum

Dorsal

(c)

Figure 14.11 (*continued*) **(c)** Photograph of dorsal view.

Ventral Structures

Turn the brain so that its ventral surface is uppermost. (**Figure 14.11a** and **b** show the important features of the ventral surface of the brain.)

1. Look for the clublike olfactory bulbs anteriorly, on the inferior surface of the frontal lobes of the cerebral hemispheres. Axons of olfactory neurons run from the nasal mucosa through the cribriform foramina of the ethmoid bone to synapse with the olfactory bulbs.

How does the size of these olfactory bulbs compare with those of humans?

_____

Is the sense of smell more important for protection and foraging in sheep or in humans?

_____

2. The optic nerve (II) carries sensory impulses from the retina of the eye. Thus this cranial nerve is involved in the sense of vision. Identify the optic nerves, optic chiasma, and optic tracts.

3. Posterior to the optic chiasma, two structures protrude from the ventral aspect of the hypothalamus—the infundibulum (stalk of the pituitary gland) immediately posterior to

the optic chiasma and the mammillary body. Notice that the sheep's mammillary body is a single rounded eminence. In humans, it is a double structure.

4. Identify the cerebral peduncles on the ventral aspect of the midbrain, just posterior to the mammillary body of the hypothalamus. The cerebral peduncles are fiber tracts connecting the cerebrum and medulla oblongata. Identify the large oculomotor nerves (III), which arise from the ventral midbrain surface, and the tiny trochlear nerves (IV), which can be seen at the junction of the midbrain and pons. Both of these cranial nerves provide motor fibers to extrinsic muscles of the eyeball.

5. Move posteriorly from the midbrain to identify first the pons and then the medulla oblongata, structures composed primarily of ascending and descending fiber tracts.

6. Return to the junction of the pons and midbrain, and proceed posteriorly to identify the following cranial nerves, all arising from the pons. Check them off as you locate them. (Figure 14.11b):

☐ Trigeminal nerves (V), which are involved in chewing and sensations of the head and face.

☐ Abducens nerves (VI), which abduct the eye (and thus work in conjunction with cranial nerves III and IV).

☐ Facial nerves (VII), large nerves involved in taste sensation, gland function (salivary and lacrimal glands), and facial expression.

7. Continue posteriorly to identify:

☐ Vestibulocochlear nerves (VIII), mostly sensory nerves that are involved with hearing and equilibrium.

☐ Glossopharyngeal nerves (IX), which contain motor fibers innervating throat structures and sensory fibers transmitting taste stimuli (in conjunction with cranial nerve VII).

☐ Vagus nerves (X), often called "wanderers," which serve many organs of the head, thorax, and abdominal cavity.

☐ Accessory nerves (XI), which serve muscles of the neck, larynx, and shoulder; notice that the accessory nerves arise from both the medulla and the spinal cord.

☐ Hypoglossal nerves (XII), which stimulate tongue and neck muscles.

It is likely that some of the cranial nerves will have been broken off during brain removal. If so, observe sheep brains of other students to identify those missing from your specimen, using your check marks as a guide.

Dorsal Structures

1. Refer to Figure 14.11c as a guide in identifying the following structures. Re-identify the now exposed cerebral hemispheres. How does the depth of the fissures in the sheep's cerebral hemispheres compare to that in the human brain?

_____

2. Examine the cerebellum. Notice that in contrast to the human cerebellum, it is not divided longitudinally and that its fissures are oriented differently. What dural fold (falx cerebri or falx cerebelli) is missing that is present in humans?

_____

3. Locate the three pairs of cerebellar peduncles, fiber tracts that connect the cerebellum to other brain structures, by lifting the cerebellum dorsally away from the brain stem. The most posterior pair, the inferior cerebellar peduncles, connect the cerebellum to the medulla. The middle cerebellar peduncles attach the cerebellum to the pons, and the superior cerebellar peduncles run from the cerebellum to the midbrain.

4. To expose the dorsal surface of the midbrain, gently separate the cerebrum and cerebellum (Figure 14.12). Identify the corpora quadrigemina, which appear as four rounded prominences on the dorsal midbrain surface.

What is the function of the corpora quadrigemina?

_____

_____

Also locate the pineal gland, which appears as a small oval protrusion in the midline just anterior to the corpora quadrigemina.

Internal Structures

1. The internal structure of the brain can be examined only after further dissection. Place the brain ventral side down on

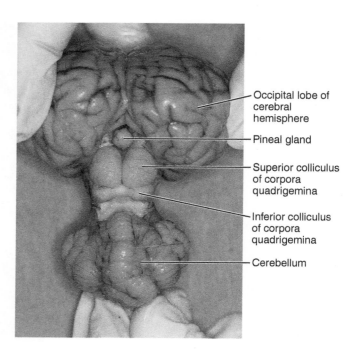

Occipital lobe of cerebral hemisphere

Pineal gland

Superior colliculus of corpora quadrigemina

Inferior colliculus of corpora quadrigemina

Cerebellum

Figure 14.12 Means of exposing the dorsal midbrain structures of the sheep brain.

the dissecting tray, and make a cut completely through it in a superior-to-inferior direction. Cut through the longitudinal fissure, corpus callosum, and midline of the cerebellum. Refer to **Figure 14.13** as you work.

2. The thin nervous tissue membrane immediately ventral to the corpus callosum that separates the lateral ventricles is the septum pellucidum. Pierce this membrane, and probe the lateral ventricle cavity. The fiber tract ventral to the septum pellucidum and anterior to the third ventricle is the fornix.

How does the relative size of the fornix in this brain compare with the human fornix?

_____

_____

Why do you suppose this is so? (Hint: What is the function of this band of fibers?)

_____

_____

3. Identify the thalamus, which forms the walls of the third ventricle and is located posterior and ventral to the fornix. The interthalamic adhesion spanning the ventricular cavity appears as an oval protrusion of the thalamic wall.

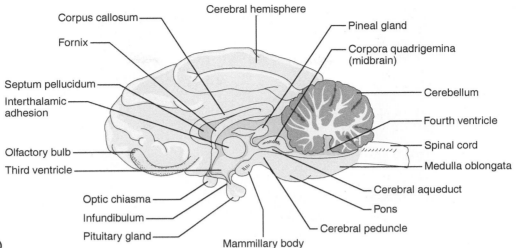

(a)

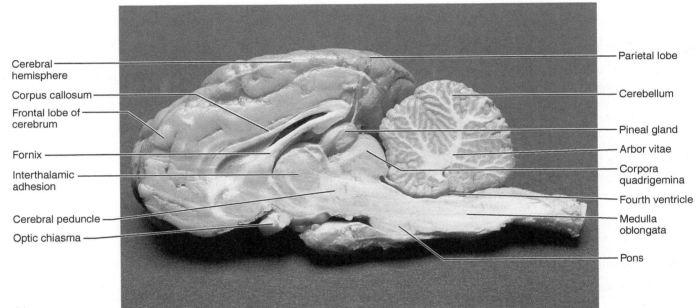

(b)

Figure 14.13 Median section of the sheep brain showing internal structures.
(a) Diagram. **(b)** Photograph.

4. The hypothalamus forms the floor of the third ventricle. Identify the optic chiasma, infundibulum, and mammillary body on its exterior surface. The pineal gland is just beneath the junction of the corpus callosum and fornix.

5. Locate the midbrain by identifying the corpora quadrigemina that form its dorsal roof. Follow the cerebral aqueduct through the midbrain tissue to the fourth ventricle. Identify the cerebral peduncles, which form its anterior walls.

6. Identify the pons and medulla oblongata, which lie anterior to the fourth ventricle. The medulla continues into the spinal cord without any obvious anatomical change, but the point at which the fourth ventricle narrows to a small canal is generally accepted as the beginning of the spinal cord.

7. Identify the cerebellum posterior to the fourth ventricle. Notice its internal treelike arrangement of white matter, the arbor vitae.

8. If time allows, obtain another sheep brain, and section it along the frontal plane so that the cut passes through the infundibulum. Compare your specimen to **Figure 14.14**, and attempt to identify all the structures shown in the figure.

9. Ask your instructor whether you should save a small portion of spinal cord from your brain specimen for the spinal cord studies (Exercise 15). Otherwise, dispose of all the organic debris in the appropriate laboratory containers, and clean the laboratory bench, the dissection instruments, and the tray before leaving the laboratory. ■

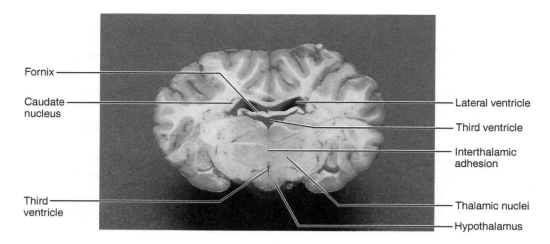

Figure 14.14 Frontal section of a sheep brain. Major structures include the thalamus, hypothalamus, and lateral and third ventricles.

Odd (Cranial) Nerve Out

The following boxes each contain four cranial nerves. One of the listed nerves does not share a characteristic with the other three. Working in groups of three, discuss the characteristics of the four cranial nerves in each set. On a separate piece of paper, one student will record the characteristics for each nerve for the group. For each set of four nerves, discuss the possible candidates for the "odd nerve" and which characteristic it lacks, based upon your notes. Once you have come to a consensus within your group, circle the cranial nerve that doesn't belong with the others, and explain why it is singled out. What characteristic is missing? Sometimes there may be multiple reasons why the cranial nerve doesn't belong with the others.

1. Which is the "odd" nerve?	Why is it the odd one out?
Optic nerve (II) Oculomotor nerve (III) Olfactory nerve (I) Vestibulocochlear nerve (VIII)	
2. Which is the "odd" nerve?	**Why is it the odd one out?**
Oculomotor nerve (III) Trochlear nerve (IV) Abducens nerve (VI) Hypoglossal nerve (XII)	
3. Which is the "odd" nerve?	**Why is it the odd one out?**
Facial nerve (VII) Hypoglossal nerve (XII) Trigeminal nerve (V) Glossopharyngeal nerve (IX)	

Gross Anatomy of the Brain and Cranial Nerves

The Human Brain

1. Using the terms from the key, identify the structures of the brain.

   Key:

   a. brain stem

   b. central sulcus

   c. cerebellum

   d. frontal lobe

   e. lateral sulcus

   f. occipital lobe

   g. parietal lobe

   h. parieto-occipital sulcus

   i. postcentral gyrus

   j. precentral gyrus

   k. temporal lobe

   l. transverse cerebral fissure

2. In which of the cerebral lobes are the following functional areas found?

   primary auditory cortex: _____   olfactory cortex: _____

   primary motor cortex: _____   primary visual cortex: _____

   primary somatosensory cortex: _____   Broca's area: _____

3. Which of the following structures are *not* part of the brain stem? (Circle the appropriate response or responses.)

   cerebral hemispheres   pons   midbrain   cerebellum   medulla oblongata   diencephalon

4. Complete the following statements by writing the proper word or phrase on the corresponding blanks on the right.

A(n) __1__ is an elevated ridge of cerebral tissue. The convolutions seen in the cerebrum are important because they increase the __2__. Gray matter is composed of __3__. White matter is composed of __4__. A fiber tract that provides for communication between different parts of the same cerebral hemisphere is called a(n) __5__ tract, whereas one that carries impulses from the cerebrum to lower CNS areas is called a(n) __6__ tract. The caudate nucleus and putamen are collectively called the __7__.

1. _____

2. _____

3. _____

4. _____

5. _____

6. _____

7. _____

5. Using the terms from the key, identify the structures on the following midsagittal view of the human brain.

Key: a.   anterior commissure   h.   fornix   o.   optic chiasma

b.   cerebellum   i.   fourth ventricle   p.   pineal gland

c.   cerebral aqueduct   j.   hypothalamus   q.   pituitary gland

d.   cerebral hemisphere   k.   interthalamic adhesion   r.   pons

e.   choroid plexus   l.   mammillary body   s.   septum pellucidum

f.   corpora quadrigemina   m.   medulla oblongata   t.   thalamus

g.   corpus callosum   n.   midbrain

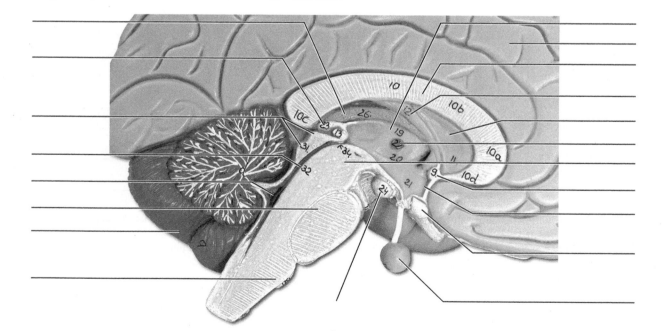

6. Using the letters in front of the terms from question 5, match the appropriate structures with the descriptions given below.

_____ 1.   site of regulation of body temperature and water balance; most important autonomic center

_____ 2.   site where medial fibers of the optic nerve cross

_____ 3.   located in the midbrain; contains reflex centers for vision and hearing

_____ 4. responsible for regulation of posture and coordination of complex muscular movements

_____ 5. important synapse site for afferent fibers traveling to the primary somatosensory cortex

_____ 6. contains autonomic centers regulating blood pressure, heart rate, and respiratory rhythm, as well as coughing, sneezing, and swallowing centers

_____ 7. large fiber tract connecting the cerebral hemispheres

_____ 8. relay station for olfactory pathways

_____ 9. canal that connects the third and fourth ventricles

_____ 10. portion of the brain stem where the cerebral peduncles are located

7. Embryologically, the brain arises from the rostral end of a tubelike structure that quickly becomes divided into three major regions. Groups of structures that develop from the embryonic brain are listed below. Designate the embryonic origin of each group as the hindbrain, midbrain, or forebrain.

_____ 1. the diencephalon, including the thalamus, optic chiasma, and hypothalamus

_____ 2. the medulla oblongata, pons, and cerebellum

_____ 3. the cerebral hemispheres

8. What is the function of the basal nuclei? _____

_____

WHY THIS **MATTERS**   9. Explain how patients with unresponsive wakefulness syndrome (UWS) can have no damage to their cerebral cortex and yet lack awareness of their environment. _____

_____

10. Patients with unresponsive wakefulness syndrome (UWS) will often reflexively respond to visual and auditory stimuli. Where in the brain are the centers for these reflexes located? _____

Explain how this phenomenon relates to the unaffected parts of their brain involved in sensory input.

_____

Meninges of the Brain

11. Identify the meningeal (or associated) structures described below:

_____ 1. outermost meninx covering the brain; composed of tough fibrous connective tissue

_____ 2. innermost meninx covering the brain; delicate and highly vascular

_____ 3. structures instrumental in returning cerebrospinal fluid to the venous blood in the dural venous sinuses

_____ 4. structure that produces the cerebrospinal fluid

_____ 5. middle meninx; like a cobweb in structure

_____ 6. its outer layer forms the periosteum of the skull

_____ 7. a dural fold that attaches the cerebrum to the crista galli of the skull

_____ 8. a dural fold separating the cerebrum from the cerebellum

Cerebrospinal Fluid

12. Label the structures involved with circulation of cerebrospinal fluid on the accompanying diagram.

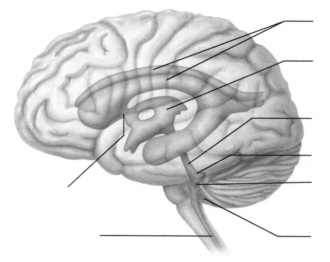

Add arrows to the figure above to indicate the flow of cerebrospinal fluid from its formation in the lateral ventricles to the site of its exit from the fourth ventricle. Then fill in the blanks in the following paragraph.

Cerebrospinal fluid flows from the fourth ventricle into the central canal of the spinal cord and the __1__ space surrounding the brain and spinal cord. From this space it drains through the __2__ into the __3__ .

1. _____

2. _____

3. _____

13. Provide the name and number of the cranial nerves involved in each of the following activities, sensations, or disorders.

_____ 1. rotating the head

_____ 2. smelling a flower

_____ 3. raising the eyelids; constricting the pupils of the eye

_____ 4. slowing the heart; increasing the motility of the digestive tract

_____ 5. involved in Bell's palsy (facial paralysis)

_____ 6. chewing food

_____ 7. listening to music; seasickness

_____ 8. secreting saliva; tasting well-seasoned food

_____ 9. involved in "rolling" the eyes (three nerves—provide numbers only)

_____ 10. feeling a toothache

_____ 11. reading the newspaper

_____ 12. primarily sensory in function (three nerves—provide numbers only)

Cranial Nerves

14. Using the terms below, correctly identify all structures indicated by leader lines on the diagram.

a. abducens nerve (VI)

b. accessory nerve (XI)

c. facial nerve (VII)

d. glossopharyngeal nerve (IX)

e. hypoglossal nerve (XII)

f. longitudinal fissure

g. mammillary body

h. medulla oblongata

i. oculomotor nerve (III)

j. olfactory bulb

k. olfactory tract

l. optic chiasma

m. optic nerve (II)

n. optic tract

o. pons

p. trigeminal nerve (V)

q. trochlear nerve (IV)

r. vagus nerve (X)

s. vestibulocochlear nerve (VIII)

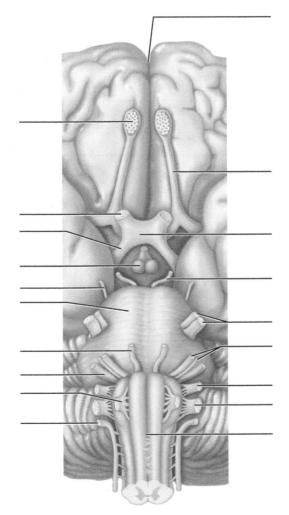

Dissection of the Sheep Brain

15. Describe the firmness and texture of the sheep brain tissue as observed when you cut into it. _____

_____

Given that formalin hardens all tissue, what conclusions might you draw about the firmness and texture of living brain tissue?

_____

16. When comparing human and sheep brains, you observed some profound differences between them. Record your observations in the chart below.

Structure	Human	Sheep
Olfactory bulb		
Pons-medulla relationship		
Location of cranial nerve III		
Mammillary body		
Corpus callosum		
Interthalamic adhesion		
Relative size of superior and inferior colliculi		
Pineal gland		

17. ✚ A brain hemorrhage within the region of the right internal capsule results in paralysis of the left side of the body. Explain why the left side (rather than the right side) is affected. _____

_____

18. ✚ Explain why trauma to the brain stem is often much more dangerous than trauma to the frontal lobes.

_____

_____

_____

The Spinal Cord and Spinal Nerves

MATERIALS

- Spinal cord model (cross section)
- Three-dimensional models or laboratory charts of the spinal cord and spinal nerves
- Red and blue pencils
- Preserved cow spinal cord sections with meninges and nerve roots intact (or spinal cord segment saved from the brain dissection in Exercise 14)
- Dissecting instruments and tray
- Disposable gloves
- Safety glasses
- Dissecting microscope
- Prepared slide of spinal cord (x.s.)
- Compound microscope
- Animal specimen from previous dissections
- Embalming fluid
- Organic debris container

LEARNING OUTCOMES

☐ List two major functions of the spinal cord.

☐ Define *conus medullaris, cauda equina,* and *filum terminale.*

☐ Name the meninges of the spinal cord, and state their functions.

☐ Indicate two major areas where the spinal cord is enlarged, and explain the reasons for the enlargement.

☐ Identify important anatomical areas on a model or image of a cross section of the spinal cord. Where applicable, name the neuron type found in these areas.

☐ Locate on a diagram the fiber tracts in the spinal cord, and state their functions.

☐ Note the number of pairs of spinal nerves that arise from the spinal cord, describe their division into groups, and identify the number of pairs in each group.

☐ Describe the origin and fiber composition of the spinal nerves, differentiating among roots, the spinal nerve proper, and rami. Discuss the result of transecting these structures.

☐ Discuss the distribution of the dorsal and ventral rami of the spinal nerves.

☐ Identify the four major nerve plexuses on a model or image, name the major nerves of each plexus, and describe the destination and function of each.

☐ Identify on a dissected animal the musculocutaneous, radial, median, and ulnar nerves of the upper limb and the femoral, saphenous, sciatic, common fibular, and tibial nerves of the lower limb.

PRE-LAB QUIZ

1. The spinal cord extends from the foramen magnum of the skull to the first or second lumbar vertebra, where it terminates in the:
   a. conus medullaris
   b. denticulate ligament
   c. filum terminale
   d. gray matter
2. How many pairs of spinal nerves do humans have?
   a. 10   c. 31
   b. 12   d. 47
3. Circle the correct underlined term. In cross section, the <u>gray</u> / <u>white</u> matter of the spinal cord looks like a butterfly or the letter H.
4. Circle True or False. The cell bodies of sensory neurons are found in an enlarged area of the dorsal root called the gray commissure.
5. Circle the correct underlined term. Fiber tracts conducting impulses to the brain are called ascending or <u>sensory</u> / <u>motor</u> tracts.

Text continues on next page.

6. Circle True or False. Because the spinal nerves arise from fusion of the ventral and dorsal roots of the spinal cord and contain motor and sensory fibers, all spinal nerves are considered mixed nerves.

7. The ventral rami of all spinal nerves except for T_2 through T_{12} form complex networks of nerves known as:
   a. fissures   c. plexuses
   b. ganglia   d. sulci

8. Severe injuries to the _____ plexus cause weakness or paralysis of the entire upper limb.
   a. brachial   c. lumbar
   b. cervical   d. sacral

9. Circle True or False. The femoral nerve is the largest nerve from the sacral plexus.

10. Circle the correct underlined term. The sciatic nerve divides into the tibial and <u>posterior femoral cutaneous</u> / <u>common fibular</u> nerves.

The cylindrical **spinal cord,** a continuation of the brain stem, is a communication center. It plays a major role in spinal reflex activity and provides neural pathways to and from the brain.

Anatomy of the Spinal Cord

Enclosed within the vertebral canal of the spinal column, the spinal cord extends from the foramen magnum of the skull to the first or second lumbar vertebra, where it terminates in the cone-shaped **conus medullaris.** Like the brain, the spinal cord is cushioned and protected by meninges **(Figure 15.1)**. The dura mater and arachnoid mater extend beyond the conus medullaris, approximately to the level of S_2. The **filum terminale,** a fibrous extension of the pia mater, extends even farther into the coccygeal canal to attach to the posterior coccyx **(Figure 15.2)**. **Denticulate ligaments,** saw-toothed shelves of pia mater, secure the spinal cord to the dura mater.

The cerebrospinal fluid–filled meninges extend beyond the end of the spinal cord, providing an excellent site for removing cerebrospinal fluid without endangering the delicate spinal cord. Analysis of the fluid can provide information about suspected bacterial or viral infections of the meninges.

This procedure, called a *lumbar puncture,* is usually performed below L_3.

In humans, 31 pairs of spinal nerves arise from the spinal cord and pass through intervertebral foramina to serve the body area at their approximate level of emergence. The cord is about the width of a thumb for most of its length, but there are obvious enlargements in the cervical and lumbar areas where the nerves serving the upper and lower limbs arise.

Because the spinal cord does not extend to the end of the vertebral column, the lumbar and sacral spinal nerve roots must travel through the vertebral canal for some distance before exiting at the appropriate intervertebral foramina. This collection of spinal nerves passing through the inferior end of the vertebral canal is called the **cauda equina** (Figure 15.2a and d) because of its similarity to a horse's tail. **Figure 15.3** illustrates the spinal cord in cross section. Notice that the gray matter looks like a butterfly.

ACTIVITY 1

Identifying Structures of the Spinal Cord

Obtain a three-dimensional model or laboratory chart of a cross section of a spinal cord, and identify its structures as they are described in **Table 15.1** and shown in Figure 15.3. ■

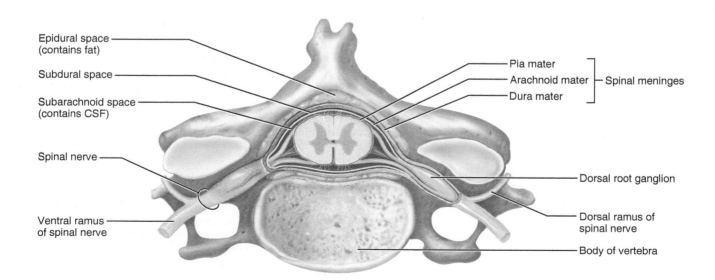

Figure 15.1 Cross section through the spinal cord illustrating its relationship to the surrounding vertebra, cervical region.

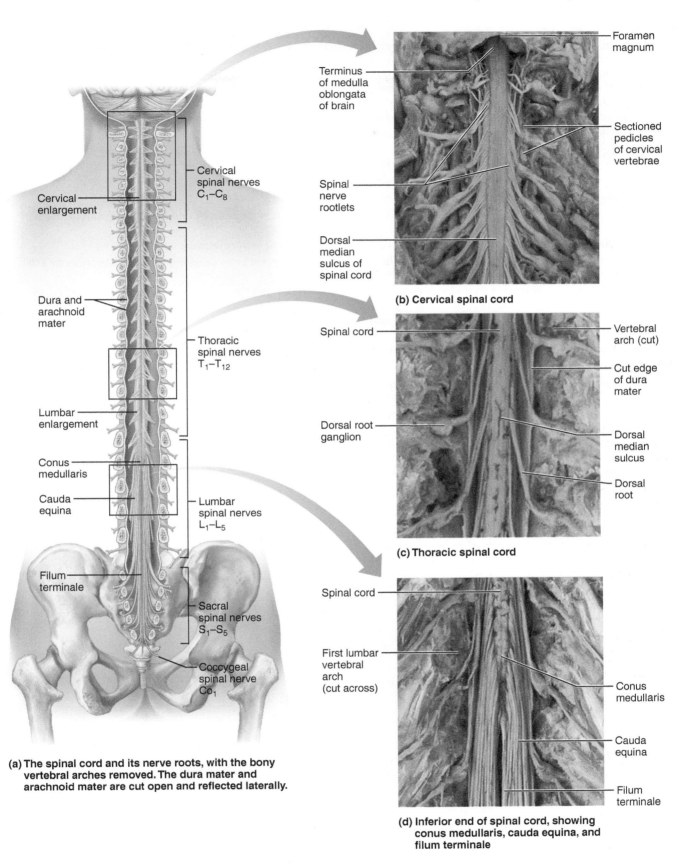

Cervical enlargement

Cervical spinal nerves C₁–C₈

Dura and arachnoid mater

Thoracic spinal nerves T₁–T₁₂

Lumbar enlargement

Conus medullaris

Cauda equina

Lumbar spinal nerves L₁–L₅

Filum terminale

Sacral spinal nerves S₁–S₅

Coccygeal spinal nerve Co₁

(a) The spinal cord and its nerve roots, with the bony vertebral arches removed. The dura mater and arachnoid mater are cut open and reflected laterally.

Terminus of medulla oblongata of brain

Spinal nerve rootlets

Dorsal median sulcus of spinal cord

Foramen magnum

Sectioned pedicles of cervical vertebrae

(b) Cervical spinal cord

Spinal cord

Dorsal root ganglion

Vertebral arch (cut)

Cut edge of dura mater

Dorsal median sulcus

Dorsal root

(c) Thoracic spinal cord

Spinal cord

First lumbar vertebral arch (cut across)

Conus medullaris

Cauda equina

Filum terminale

(d) Inferior end of spinal cord, showing conus medullaris, cauda equina, and filum terminale

Figure 15.2 Gross structure of the spinal cord, dorsal view.

Table 15.1 Anatomy of the Spinal Cord in Cross Section (Figure 15.3)

Structure	Description
Gray matter	Located in the center of the spinal cord and shaped like a butterfly or the letter H. Areas of the gray matter are individually named and described below.
Dorsal (posterior) horns	Posterior projections of the gray matter that contain primarily interneurons and the axons of sensory neurons.
Ventral (anterior) horns	Anterior projections of the gray matter that contain the cell bodies of somatic motor neurons and some interneurons.
Lateral horns	Small lateral projections that are present only in the thoracic and lumbar regions of the gray matter. When present, they contain the cell bodies of motor neurons of the autonomic nervous system.
Gray commissure	The cross bar of the H that surrounds the central canal.
Central canal	A narrow central cavity that is continuous with the ventricles of the brain.
Dorsal root	A nerve root that fans out into dorsal rootlets to connect to the posterior spinal cord. It contains the axons of sensory neurons.
Dorsal root ganglion	A bulge on the dorsal root that contains the cell bodies of sensory neurons.
Ventral root	A nerve root that is formed by the ventral rootlets connected to the anterior spinal cord. It contains the axons of motor neurons.
Spinal nerve	Formed by the fusion of dorsal and ventral roots. They are mixed nerves because they contain both sensory and motor fibers.
White matter	Forms the outer region of the spinal cord and is composed of myelinated and nonmyelinated axons organized into tracts.
Ventral median fissure	The anterior, more open of the two grooves that partially divide the spinal cord into left and right halves.
Dorsal median sulcus	The posterior, shallower of the two grooves that partially divide the spinal cord into left and right halves.
White columns	Each side of the spinal cord has three funiculi (columns): dorsal (posterior) funiculus, lateral funiculus, and ventral (anterior) funiculus, which are further divided into tracts.
Ascending (sensory) tracts	Carry sensory information from the sensory neurons to the brain.
Descending (motor) tracts	Carry motor instructions from the brain to the body's muscles and glands.

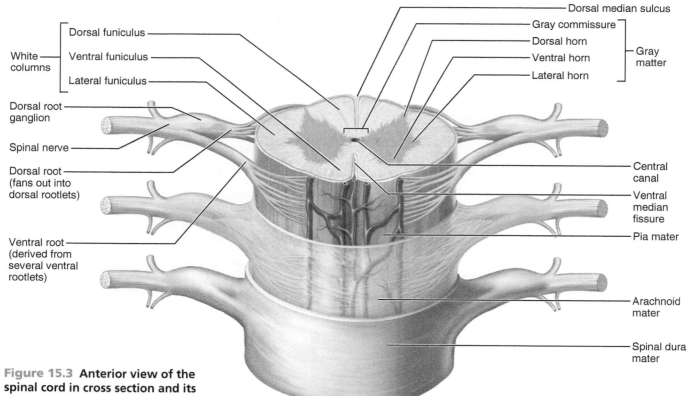

Figure 15.3 Anterior view of the spinal cord in cross section and its meninges, thoracic region.

WHY THIS MATTERS | Shingles

Did you have chickenpox as a child or perhaps were immunized with the vaccine? If you had chickenpox, you have a one in three chance of experiencing shingles in your lifetime. Receiving the vaccine appears to prevent both chickenpox and shingles. The virus that causes both diseases is the varicella-zoster virus. After a person recovers from chickenpox, the virus lies dormant (latent) in the cell body of sensory neurons in the dorsal root ganglion. The virus can lie dormant in multiple levels of the spinal cord and in cranial nerves, as well. Later in life, the virus can be reactivated and travel within a sensory neuron to the surface of the skin, usually in response to stress or a weakened immune response. The later activation of the virus, called *shingles,* results in a rash and extreme pain. ■

Even though the spinal cord is protected by meninges and cerebrospinal fluid in the vertebral canal, it is highly vulnerable to traumatic injuries, such as might occur in an automobile accident.

When the cord is damaged, both motor and sensory functions are lost in body areas normally served by that region and lower regions of the spinal cord. Injury to certain spinal cord areas may even result in a permanent flaccid paralysis of both legs, called **paraplegia,** or of all four limbs, called **quadriplegia.** ✛

ACTIVITY 2

Identifying Spinal Cord Tracts

With the help of your textbook, label the spinal cord diagram (Figure 15.4) with the tract names that follow. Each tract is represented on both sides of the cord, but for clarity, label the motor tracts on the right side of the diagram and the sensory tracts on the left side of the diagram. *Color ascending (sensory) tracts blue and descending (motor) tracts red.* Then fill in the functional importance of each tract beside its name in the list that follows. As you work, try to be aware of how the naming of the tracts is related to their anatomical distribution.

Dorsal columns

    Fasciculus gracilis _____

    Fasciculus cuneatus _____

Dorsal spinocerebellar _____

Ventral spinocerebellar _____

Lateral spinothalamic _____

Ventral spinothalamic _____

Lateral corticospinal _____

Ventral corticospinal _____

Rubrospinal _____

Tectospinal _____

Vestibulospinal _____

Medial reticulospinal _____

Lateral reticulospinal _____ ■

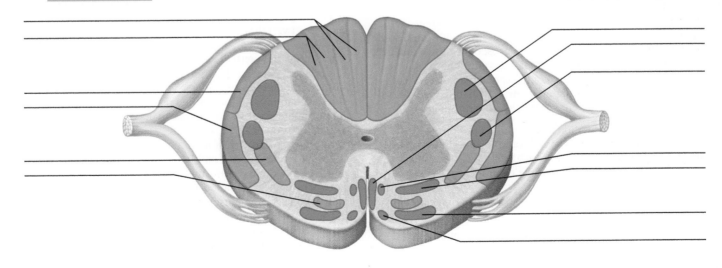

Ascending tracts

Descending tracts

Figure 15.4 Cross section of the spinal cord showing the relative positioning of its major tracts.

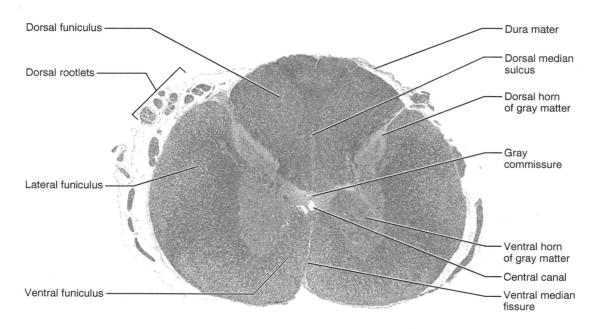

Dorsal funiculus

Dorsal rootlets

Lateral funiculus

Ventral funiculus

Dura mater

Dorsal median sulcus

Dorsal horn of gray matter

Gray commissure

Ventral horn of gray matter

Central canal

Ventral median fissure

Figure 15.5 **Light micrograph of cross section of the spinal cord (8×), cervical region.**

DISSECTION
Spinal Cord

1. Obtain a dissecting tray and instruments, disposable gloves, and a segment of preserved spinal cord (from a cow or saved from the brain specimen used in Exercise 14). Identify the dura mater and the weblike arachnoid mater.

What name is given to the third meninx, and where is it found?

_____

_____

Peel back the dura mater, and observe the fibers making up the dorsal and ventral roots. If possible, identify a dorsal root ganglion.

2. Cut a thin cross section of the cord, and identify the ventral and dorsal horns of the gray matter with the naked eye or with the aid of a dissecting microscope. Notice that the dorsal horns are more tapered than the ventral horns and that they extend closer to the edge of the spinal cord. Also identify the central canal, white matter, ventral median fissure, dorsal median sulcus, and dorsal, ventral, and lateral funiculi.

3. Obtain a prepared slide of the spinal cord (cross section) and a compound microscope. Examine the slide carefully under low power (refer to **Figure 15.5** to identify spinal cord features). Observe the shape of the central canal.

Is it basically circular or oval? _____

Name the neuroglial cell type that lines this canal. _____

_____ ■

Spinal Nerves and Nerve Plexuses

The 31 pairs of human spinal nerves arise from the fusions of the ventral and dorsal roots of the spinal cord and are therefore *mixed nerves* containing both sensory and motor fibers (see Figure 15.2). There are 8 pairs of cervical nerves (C_1-C_8), 12 pairs of thoracic nerves (T_1-T_{12}), 5 pairs of lumbar nerves (L_1-L_5), 5 pairs of sacral nerves (S_1-S_5), and 1 pair of coccygeal nerves (Co_1) **(Figure 15.6a)**. The first pair of spinal nerves leaves the vertebral canal between the base of the skull and the atlas, but all the rest exit via the intervertebral foramina. The first through seventh pairs of cervical nerves emerge *above* the vertebra for which they are named; C_8 emerges between C_7 and T_1. (Notice that there are 7 cervical vertebrae, but 8 pairs of cervical nerves.) The remaining spinal nerve pairs emerge from the spinal cord area *below* the same-numbered vertebra.

Almost immediately after emerging, each nerve divides into **dorsal** and **ventral rami.** Thus, each spinal nerve is only about 1 or 2 cm long. The rami, like the spinal nerves, contain both motor and sensory fibers. The smaller dorsal rami serve the skin and musculature of the posterior body trunk at their approximate level of emergence. The ventral rami of spinal nerves T_2 through T_{12} pass anteriorly as the **intercostal nerves** to supply the muscles of intercostal spaces, and the skin and muscles of the anterior and lateral trunk (see Figure 15.6b). The ventral rami of all other spinal nerves form complex networks of nerves called **nerve plexuses.** These plexuses primarily serve the muscles and skin of the limbs. The fibers of the ventral rami unite in the plexuses (with a few rami supplying fibers to more than one plexus). From the plexuses the fibers diverge again to form peripheral nerves, each of which contains fibers from more than one spinal nerve. The four major nerve plexuses and their chief peripheral nerves are

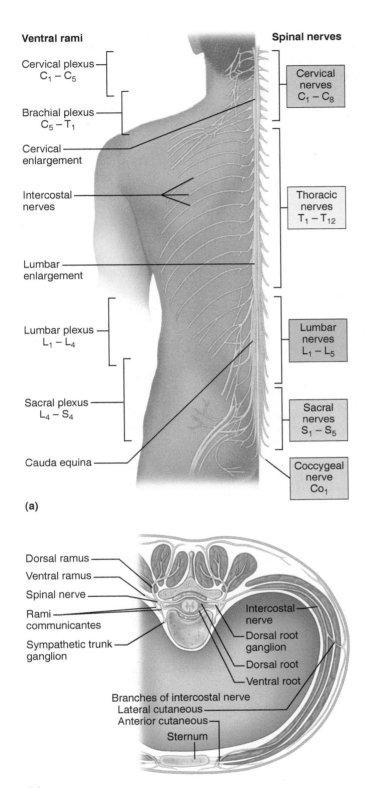

Ventral rami

- Cervical plexus $C_1 - C_5$
- Brachial plexus $C_5 - T_1$
- Cervical enlargement
- Intercostal nerves
- Lumbar enlargement
- Lumbar plexus $L_1 - L_4$
- Sacral plexus $L_4 - S_4$
- Cauda equina

Spinal nerves

- Cervical nerves $C_1 - C_8$
- Thoracic nerves $T_1 - T_{12}$
- Lumbar nerves $L_1 - L_5$
- Sacral nerves $S_1 - S_5$
- Coccygeal nerve Co_1

(a)

- Dorsal ramus
- Ventral ramus
- Spinal nerve
- Rami communicantes
- Sympathetic trunk ganglion
- Intercostal nerve
- Dorsal root ganglion
- Dorsal root
- Ventral root
- Branches of intercostal nerve
- Lateral cutaneous
- Anterior cutaneous
- Sternum

(b)

Figure 15.6 Human spinal nerves. (a) Spinal nerves are shown at right; ventral rami and the major nerve plexuses are shown at left. **(b)** The distribution of dorsal and ventral rami. In the thoracic region, each ventral ramus continues as an intercostal nerve. Dorsal rami innervate the intrinsic muscles and skin of the back.

described in the tables (**Tables 15.2, 15.3, 15.4,** and **15.5**) and illustrated in the figures (**Figures 15.7, 15.8, 15.9,** and **15.10**). Their names and sites of origin should be committed to memory. The tiny S_5 and Co_1 spinal nerves contribute to a small plexus that serves part of the pelvic floor.

Cervical Plexus and the Neck

The **cervical plexus** (Figure 15.7 and Table 15.2) arises from the ventral rami of C_1 through C_5 to supply muscles of the shoulder and neck. The major motor branch of this plexus is the **phrenic nerve,** which arises from C_3 through C_4 (plus some fibers from C_5) and passes into the thoracic cavity in front of the first rib to innervate the diaphragm. The primary danger of a broken neck is that the phrenic nerve may be severed, leading to paralysis of the diaphragm and cessation of breathing. A jingle to help you remember the rami (roots) forming the phrenic nerves is "C_3, C_4, C_5 keep the diaphragm alive."

Brachial Plexus and the Upper Limb

The **brachial plexus** is large and complex, arising from the ventral rami of C_5 through C_8 and T_1 (Table 15.3). The plexus, after being rearranged consecutively into *trunks, divisions,*

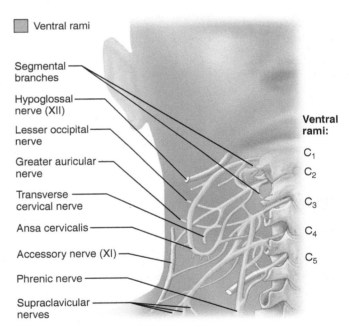

☐ Ventral rami

- Segmental branches
- Hypoglossal nerve (XII)
- Lesser occipital nerve
- Greater auricular nerve
- Transverse cervical nerve
- Ansa cervicalis
- Accessory nerve (XI)
- Phrenic nerve
- Supraclavicular nerves

Ventral rami:
- C_1
- C_2
- C_3
- C_4
- C_5

Figure 15.7 The cervical plexus. Note cranial nerves XI and XII do not belong to the cervical plexus. (See Table 15.2.)

Text continues on page 290.

Table 15.2   Branches of the Cervical Plexus (see Figure 15.7)

Nerves	Ventral rami	Structures served
Cutaneous Branches (Superficial)		
Lesser occipital	C_2 (C_3)	Skin on posterolateral aspect of head and neck
Greater auricular	C_2, C_3	Skin of ear, skin over parotid gland
Transverse cervical	C_2, C_3	Skin on anterior and lateral aspect of neck
Supraclavicular (medial, intermediate, and lateral)	C_3, C_4	Skin of shoulder and clavicular region
Motor Branches (Deep)		
Ansa cervicalis (superior and inferior roots)	C_1–C_3	Infrahyoid muscles of neck (omohyoid, sternohyoid, and sternothyroid)
Segmental and other muscular branches	C_1–C_5	Deep muscles of neck (geniohyoid and thyrohyoid) and portions of scalenes, levator scapulae, trapezius, and sternocleidomastoid muscles
Phrenic	C_3–C_5	Diaphragm (sole motor nerve supply)

Table 15.3   Branches of the Brachial Plexus (see Figure 15.8)

Nerves	Ventral rami	Structures served
Axillary	Posterior cord (C_5, C_6)	Muscular branches: deltoid and teres minor muscles Cutaneous branches: some skin of shoulder region
Musculocutaneous	Lateral cord (C_5–C_7)	Muscular branches: flexor muscles in anterior arm (biceps brachii, brachialis, coracobrachialis) Cutaneous branches: skin on anterolateral forearm (extremely variable)
Median	By two branches, one from medial cord (C_8, T_1) and one from the lateral cord (C_5–C_7)	Muscular branches to flexor group of anterior forearm (palmaris longus, flexor carpi radialis, flexor digitorum superficialis, flexor pollicis longus, lateral half of flexor digitorum profundus, and pronator muscles); intrinsic muscles of lateral palm and digital branches to the fingers Cutaneous branches: skin of lateral two-thirds of hand on ventral side and dorsum of fingers II and III
Ulnar	Medial cord (C_8, T_1)	Muscular branches: flexor muscles in anterior forearm (flexor carpi ulnaris and medial half of flexor digitorum profundus); most intrinsic muscles of the hand Cutaneous branches: skin of medial third of hand, both anterior and posterior aspects
Radial	Posterior cord (C_5–C_8, T_1)	Muscular branches: posterior muscles of arm and forearm (triceps brachii, anconeus, supinator, brachioradialis, extensors carpi radialis longus and brevis, extensor carpi ulnaris, and several muscles that extend the fingers) Cutaneous branches: skin of posterolateral surface of entire limb (except dorsum of fingers II and III)
Dorsal scapular	Branches of C_5 rami	Rhomboid muscles and levator scapulae
Long thoracic	Branches of C_5–C_7 rami	Serratus anterior muscle
Subscapular	Posterior cord; branches of C_5 and C_6 rami	Teres major and subscapularis muscles
Suprascapular	Upper trunk (C_5, C_6)	Shoulder joint; supraspinatus and infraspinatus muscles
Pectoral (lateral and medial)	Branches of lateral and medial cords (C_5–T_1)	Pectoralis major and minor muscles

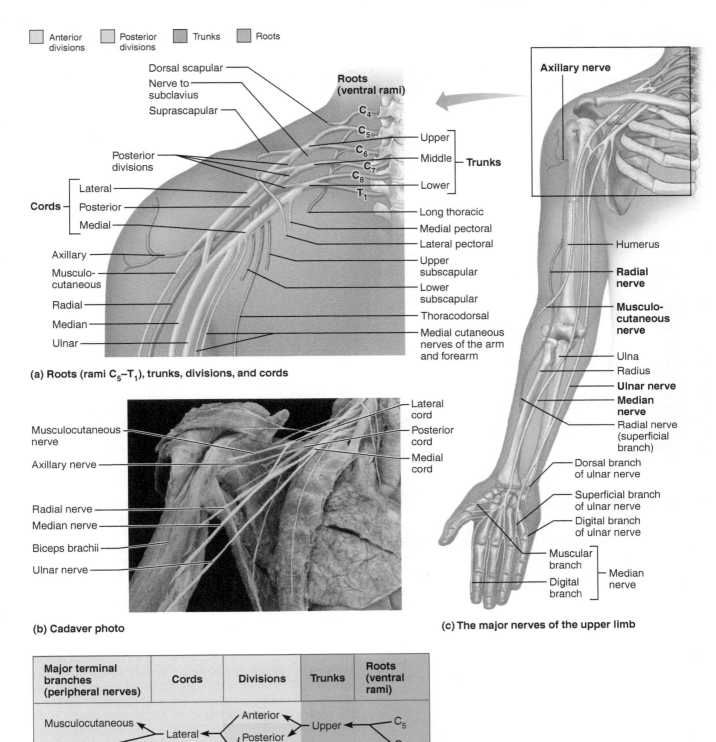

Anterior divisions | **Posterior divisions** | **Trunks** | **Roots**

(a) Roots (rami C₅–T₁), trunks, divisions, and cords

Dorsal scapular
Nerve to subclavius
Suprascapular
Posterior divisions
Cords — Lateral, Posterior, Medial
Axillary
Musculo-cutaneous
Radial
Median
Ulnar

Roots (ventral rami)
C₄
C₅
C₆
C₇
C₈
T₁

Upper
Middle — **Trunks**
Lower

Long thoracic
Medial pectoral
Lateral pectoral
Upper subscapular
Lower subscapular
Thoracodorsal
Medial cutaneous nerves of the arm and forearm

(b) Cadaver photo

Musculocutaneous nerve
Axillary nerve
Radial nerve
Median nerve
Biceps brachii
Ulnar nerve

Lateral cord
Posterior cord
Medial cord

(c) The major nerves of the upper limb

Axillary nerve
Humerus
Radial nerve
Musculo-cutaneous nerve
Ulna
Radius
Ulnar nerve
Median nerve
Radial nerve (superficial branch)
Dorsal branch of ulnar nerve
Superficial branch of ulnar nerve
Digital branch of ulnar nerve
Muscular branch
Digital branch — Median nerve

(d) Flowchart summarizing relationships within the brachial plexus

Major terminal branches (peripheral nerves)	Cords	Divisions	Trunks	Roots (ventral rami)
Musculocutaneous	Lateral	Anterior	Upper	C₅
Median		Posterior		C₆
Ulnar	Medial	Anterior	Middle	C₇
Radial		Posterior		C₈
Axillary	Posterior	Anterior	Lower	T₁
		Posterior		

Figure 15.8 The brachial plexus. (a and c) Distribution of the major peripheral nerves of the upper limb. **(b)** Photograph of the brachial plexus from a cadaver, anterior view. **(d)** Flowchart of consecutive branches within the brachial plexus. (See Table 15.3.)

Table 15.4   Branches of the Lumbar Plexus (see Figure 15.9)

Nerves	Ventral rami	Structures served
Femoral	L_2–L_4	Skin of anterior and medial thigh via *anterior femoral cutaneous* branch; skin of medial leg and foot, hip and knee joints via *saphenous* branch; motor to anterior muscles (quadriceps and sartorius) of thigh and to pectineus, iliacus
Obturator	L_2–L_4	Motor to adductor magnus (part), longus, and brevis muscles, gracilis muscle of medial thigh, obturator externus; sensory for skin of medial thigh and for hip and knee joints
Lateral femoral cutaneous	L_2, L_3	Skin of lateral thigh; some sensory branches to peritoneum
Iliohypogastric	L_1	Skin of lower abdomen and hip; muscles of anterolateral abdominal wall (internal obliques and transversus abdominis)
Ilioinguinal	L_1	Skin of external genitalia and proximal medial aspect of the thigh; inferior abdominal muscles
Genitofemoral	L_1, L_2	Skin of scrotum in males, of labia majora in females, and of anterior thigh inferior to middle portion of inguinal region; cremaster muscle in males

and *cords,* finally becomes subdivided into five major *peripheral nerves* (Figure 15.8).

The **axillary nerve,** which serves the muscles and skin of the shoulder, has the most limited distribution. The large **radial nerve** passes down the posterolateral surface of the arm and forearm, supplying all the extensor muscles of the arm, forearm, and hand and the skin along its course. The radial nerve is often injured in the axillary region by the pressure of a crutch or by hanging one's arm over the back of a chair. The **median nerve** passes down the anteromedial surface of the arm to supply most of the flexor muscles in the forearm and several muscles in the hand (plus the skin of the lateral surface of the palm of the hand).

☐ Hyperextend at the wrist to identify the long, obvious tendon of your palmaris longus muscle, which crosses the exact midline of the anterior wrist. Your median nerve lies immediately deep to that tendon, and the radial nerve lies just *lateral* to it.

The **musculocutaneous nerve** supplies the arm muscles that flex the forearm and the skin of the lateral surface of the forearm. The **ulnar nerve** travels down the posteromedial surface of the arm. It courses around the medial epicondyle of the humerus to supply the flexor carpi ulnaris, the ulnar head of the flexor digitorum profundus of the forearm, and all intrinsic muscles of the hand not served by the median nerve. It supplies the skin of the medial third of the hand, both the anterior and posterior surfaces. Trauma to the ulnar nerve, which often occurs when the elbow is hit, is commonly referred to as "hitting the funny bone" because of the odd, painful, tingling sensation that it causes.

Severe injuries to the brachial plexus cause weakness or paralysis of the entire upper limb. Such injuries may occur when the upper limb is pulled hard and the plexus is stretched (as when a football tackler yanks the arm of the halfback), and by blows to the shoulder that force the humerus inferiorly (as when a cyclist is pitched headfirst off his motorcycle and grinds his shoulder into the pavement). ✚

Lumbosacral Plexus and the Lower Limb

The **lumbosacral plexus,** which serves the pelvic region of the trunk and the lower limbs, is actually a complex of two plexuses, the lumbar plexus and the sacral plexus (Figures 15.9 and 15.10). These plexuses interweave considerably, and many fibers of the lumbar plexus contribute to the sacral plexus.

The Lumbar Plexus

The **lumbar plexus** arises from ventral rami of L_1 through L_4 (and sometimes T_{12}). Its nerves serve the lower abdominopelvic region and the anterior thigh (Table 15.4 and Figure 15.9). The largest nerve of this plexus is the **femoral nerve,** which passes beneath the inguinal ligament to innervate the anterior thigh muscles. The cutaneous branches of the femoral nerve (median and anterior femoral cutaneous and the saphenous nerves) supply the skin of the anteromedial surface of the entire lower limb.

The Sacral Plexus

Arising from L_4 through S_4, the nerves of the **sacral plexus** supply the buttock, the posterior surface of the thigh, and virtually all sensory and motor fibers of the leg and foot (Table 15.5 and Figure 15.10). The major peripheral nerve of this plexus is the **sciatic nerve,** the largest nerve in the body. The sciatic nerve leaves the pelvis through the greater sciatic notch and travels down the posterior thigh, serving its flexor muscles and skin. In the popliteal region, the sciatic nerve divides into the **common fibular nerve** and the **tibial nerve,** which together supply the balance of the leg muscles and skin, both directly and via several branches.

Injury to the proximal part of the sciatic nerve, as might follow a fall or disc herniation, results in a number of lower limb impairments. **Sciatica** characterized by stabbing pain radiating over the course of the sciatic nerve, is common. When the sciatic nerve is completely severed, the leg is nearly useless. The leg cannot be flexed, and the foot drops into plantar flexion (dangles), a condition called **footdrop.** ✚

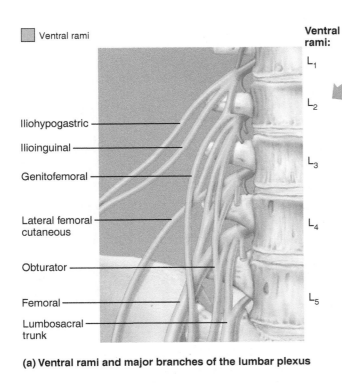

Ventral rami

Ventral rami:

L₁

L₂

L₃

L₄

L₅

Iliohypogastric

Ilioinguinal

Genitofemoral

Lateral femoral cutaneous

Obturator

Femoral

Lumbosacral trunk

(a) Ventral rami and major branches of the lumbar plexus

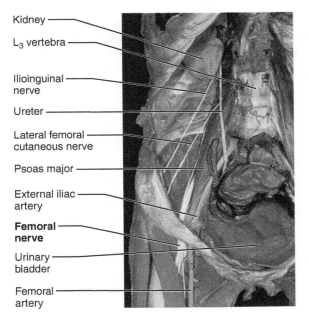

Kidney

L₃ vertebra

Ilioinguinal nerve

Ureter

Lateral femoral cutaneous nerve

Psoas major

External iliac artery

Femoral nerve

Urinary bladder

Femoral artery

(b) Nerves of the lumbar plexus, anterior view

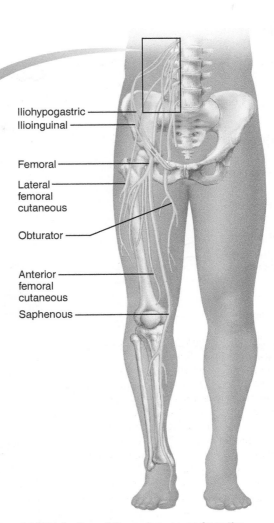

Iliohypogastric
Ilioinguinal

Femoral

Lateral femoral cutaneous

Obturator

Anterior femoral cutaneous

Saphenous

(c) Distribution of the major nerves from the lumbar plexus to the lower limb

Figure 15.9 The lumbar plexus (anterior view). Illustrations **(a)** and **(c).**
(b) Photograph of the lumbar plexus from a cadaver. (See Table 15.4)

Ventral rami

Ventral rami:

L₄

L₅

S₁

S₂

S₃

S₄

S₅

Co₁

Superior gluteal

Lumbosacral trunk

Inferior gluteal

Common fibular

Tibial

Posterior femoral cutaneous

Pudendal

Sciatic

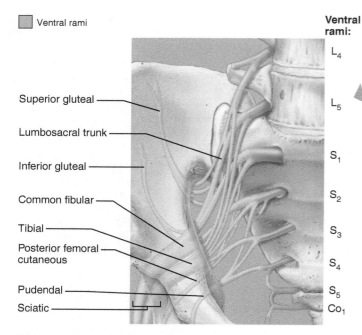

(a) Ventral rami and major branches of the sacral plexus, anterior view

Gluteus maximus

Piriformis

Inferior gluteal nerve

Common fibular nerve

Tibial nerve

Pudendal nerve

Posterior femoral cutaneous nerve

Sciatic nerve

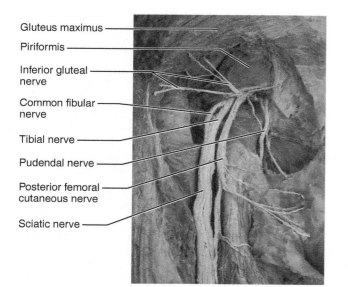

(b) Dissection of the gluteal region, posterior view

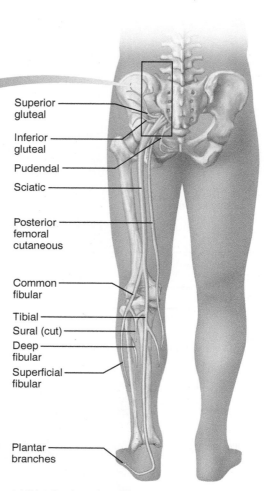

Superior gluteal

Inferior gluteal

Pudendal

Sciatic

Posterior femoral cutaneous

Common fibular

Tibial

Sural (cut)

Deep fibular

Superficial fibular

Plantar branches

(c) Distribution of the major nerves from the sacral plexus to the lower limb

Figure 15.10 The sacral plexus (posterior view). Illustrations **(a)** and **(c)**. **(b)** Photograph of the sacral plexus from a cadaver. (See Table 15.5.)

Table 15.5   Branches of the Sacral Plexus (see Figure 15.10)

Nerves	Ventral rami	Structures served
Sciatic nerve	L_4, L_5, S_1–S_3	Composed of two nerves (tibial and common fibular) in a common sheath; they diverge just proximal to the knee
• Tibial (including sural branch and medial and lateral plantar branches)	L_4–S_3	Cutaneous branches to skin of posterior surface of leg and sole of foot Motor branches to muscles of back of thigh, leg, and foot (hamstrings [except short head of biceps femoris], posterior part of adductor magnus, triceps surae, tibialis posterior, popliteus, flexor digitorum longus, flexor hallucis longus, and intrinsic muscles of foot)
• Common fibular (superficial and deep branches)	L_4–S_2	Cutaneous branches to skin of anterior and lateral surface of leg and dorsum of foot Motor branches to short head of biceps femoris of thigh, fibularis muscles of lateral leg, tibialis anterior, and extensor muscles of toes (extensor hallucis longus, extensors digitorum longus and brevis)
Superior gluteal	L_4, L_5, S_1	Motor branches to gluteus medius and minimus and tensor fascia lata
Inferior gluteal	L_5–S_2	Motor branches to gluteus maximus
Posterior femoral cutaneous	S_1–S_3	Skin of buttock, posterior thigh, and popliteal region; length variable; may also innervate part of skin of calf and heel
Pudendal	S_2–S_4	Supplies most of skin and muscles of perineum (region encompassing external genitalia and anus and including clitoris, labia, and vaginal mucosa in females, and scrotum and penis in males); external anal sphincter

ACTIVITY 3

Identifying the Major Nerve Plexuses and Peripheral Nerves

Identify each of the four major nerve plexuses and its major nerves (Figures 15.7 to 15.10) on a large laboratory chart or model. Trace the courses of the nerves, and relate those observations to the information provided (Tables 15.2 to 15.5). ▦

DISSECTION AND IDENTIFICATION
Cat Spinal Nerves

The cat has 38 or 39 pairs of spinal nerves (as compared to 31 in humans). Of these, 8 are cervical, 13 thoracic, 7 lumbar, 3 sacral, and 7 or 8 caudal.

A complete dissection of the cat's spinal nerves would be extraordinarily time-consuming and exacting and is not warranted in a basic anatomy course. However, it is desirable for you to have some dissection work to complement your study of the anatomical charts. Thus, at this point you will carry out a partial dissection of the brachial plexus and lumbosacral plexus and identify some of the major nerves. ▪

ACTIVITY 4

Dissecting Nerves of the Brachial Plexus

1. Don disposable gloves and safety glasses. Place your cat specimen on the dissecting tray, dorsal side down. Reflect the cut ends of the left pectoralis muscles to expose the large brachial plexus in the axillary region **(Figure 15.11)**.

Carefully clean the exposed nerves as far back toward their points of origin as possible.

2. The **musculocutaneous nerve** is the most superior nerve of this group. It splits into two subdivisions that run under the margins of the coracobrachialis and biceps brachii muscles. Trace its fibers into the ventral muscles of the arm it serves.

3. Locate the large **radial nerve** inferior to the musculocutaneous nerve. The radial nerve serves the dorsal muscles of the arm and forearm. Follow it into the three heads of the triceps brachii muscle.

4. In the cat, the **median nerve** is closely associated with the brachial artery and vein (Figure 15.11). It courses through the arm to supply the ventral muscles of the forearm (with the exception of the flexor carpi ulnaris and the ulnar head of the flexor digitorum profundus). It also innervates some of the intrinsic hand muscles, as in humans.

5. The **ulnar nerve** is the most posterior of the large brachial plexus nerves. Follow it as it travels down the forelimb, passing over the medial epicondyle of the humerus, to supply the flexor carpi ulnaris, the ulnar head of the flexor digitorum profundus, and the hand muscles. ▦

ACTIVITY 5

Dissecting Nerves of the Lumbosacral Plexus

1. To locate the **femoral nerve** arising from the lumbar plexus, first identify the right *femoral triangle,* which is

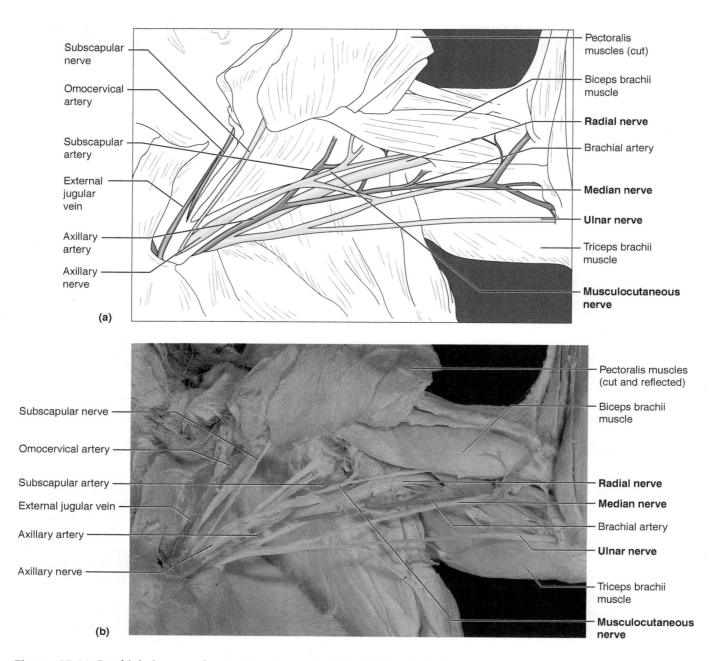

Figure 15.11 Brachial plexus and major blood vessels of the left forelimb of the cat, ventral aspect. (a) Diagram. **(b)** Photograph.

bordered by the sartorius and adductor muscles of the anterior thigh **(Figure 15.12)**. The large femoral nerve travels through this region after emerging from the psoas major muscle in close association with the femoral artery and vein. Follow the nerve into the muscles and skin of the anterior thigh, which it supplies. Notice also its cutaneous branch in the cat, the **saphenous nerve,** which continues down the anterior medial surface of the thigh with the great saphenous artery and vein to supply the skin of the anterior shank and foot.

2. Turn the cat ventral side down so you can view the posterior aspect of the lower limb **(Figure 15.13)**. Reflect the ends of the transected biceps femoris muscle to view the large

cordlike sciatic nerve. The **sciatic nerve** arises from the sacral plexus and serves the dorsal thigh muscles and all the muscles of the leg and foot. Follow the nerve as it travels down the posterior thigh lateral to the semimembranosus muscle. Note that just superior to the gastrocnemius muscle of the calf, it divides into its two major branches, which serve the leg.

3. Identify the **tibial nerve** medially and the **common fibular (peroneal) nerve,** which curves over the lateral surface of the gastrocnemius.

4. Before you leave the laboratory, follow the boxed instructions to prepare your cat for storage and to clean the area (page 219). ■

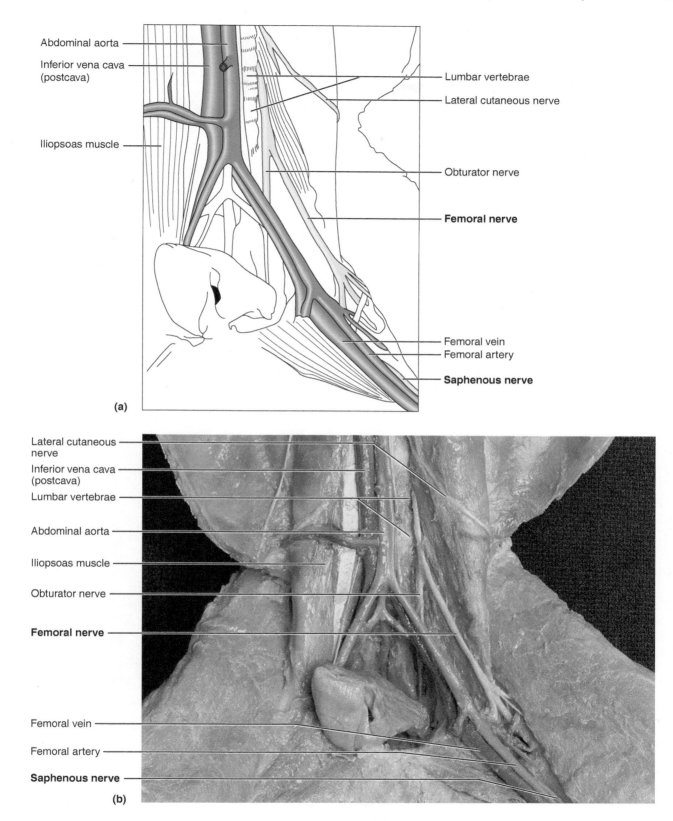

Abdominal aorta

Inferior vena cava (postcava)

Iliopsoas muscle

Lumbar vertebrae

Lateral cutaneous nerve

Obturator nerve

Femoral nerve

Femoral vein
Femoral artery
Saphenous nerve

(a)

Lateral cutaneous nerve

Inferior vena cava (postcava)

Lumbar vertebrae

Abdominal aorta

Iliopsoas muscle

Obturator nerve

Femoral nerve

Femoral vein

Femoral artery

Saphenous nerve

(b)

Figure 15.12 Lumbar plexus of the cat, ventral aspect. (a) Diagram. **(b)** Photograph.

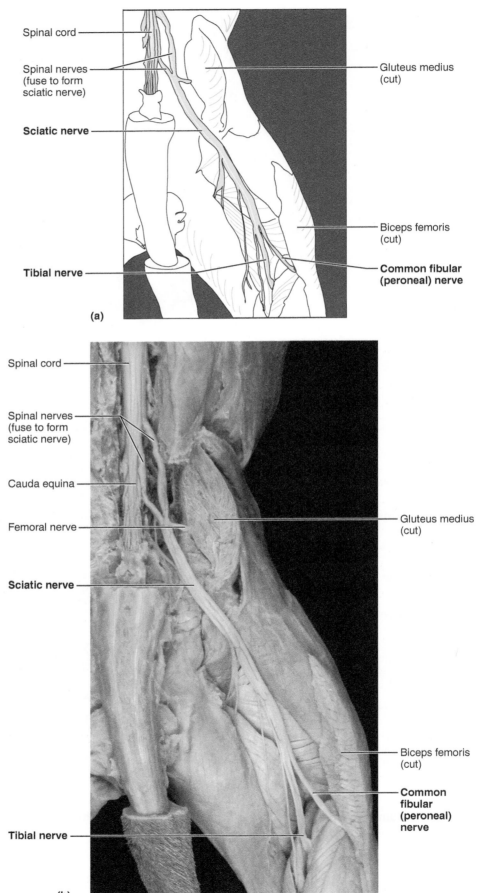

Figure 15.13 Sacral plexus of the cat, dorsal aspect.
(a) Diagram.
(b) Photograph.

Name _____

Lab Time/Date _____

The Spinal Cord and Spinal Nerves

Anatomy of the Spinal Cord

1. Match each description with the proper anatomical term from the key.

Key: a. cauda equina   b. conus medullaris   c. filum terminale   d. foramen magnum

_____ 1. most superior boundary of the spinal cord

_____ 2. meningeal extension beyond the spinal cord terminus

_____ 3. spinal cord terminus

_____ 4. collection of spinal nerves traveling in the vertebral canal below the terminus of the spinal cord

2. Match the key letters on the diagram with the following terms.

_____ 1. central canal

_____ 2. dorsal horn

_____ 3. dorsal median sulcus

_____ 4. dorsal root ganglion

_____ 5. dorsal root of spinal nerve

_____ 6. gray commissure

_____ 7. lateral horn

_____ 8. spinal nerve

_____ 9. ventral horn

_____ 10. ventral median fissure

_____ 11. ventral root of spinal nerve

k. ─────

j. ─────

i. ─────

h. ─────

g. ─────

a. ─────

b. ─────

c. ─────

d. ─────

e. ─────

f. ─────

3. Choose the proper answer from the key to respond to the descriptions relating to spinal cord anatomy. (Some terms are used more than once.)

Key: a. sensory   b. motor   c. both sensory and motor   d. interneurons

_____ 1. primary neuron type found in dorsal horn

_____ 2. primary neuron type found in ventral horn

_____ 3. neuron type in dorsal root ganglion

_____ 4. fiber type in ventral root

_____ 5. fiber type in dorsal root

_____ 6. fiber type in spinal nerve

4. Where in the vertebral column is a lumbar puncture generally done? _____.

Why is this the site of choice? _____

_____

5. The spinal cord is enlarged in two regions, the _____ and the _____ regions.

What is the significance of these enlargements? _____

_____

6. How does the position of the gray and white matter differ in the spinal cord and the cerebral hemispheres?

_____

_____

WHY THIS MATTERS

7. Where in the body does the varicella-zoster virus lie dormant? _____

_____

8. Do you think it is possible to get shingles more than once? _____ Explain your answer.

_____

_____

9. From the key on the right, choose the name of the tract that might be damaged when the following conditions are observed. (More than one choice may apply. Some terms are used more than once.)

_____ 1. uncoordinated movement

_____ 2. lack of voluntary movement

_____ 3. tremors, jerky movements

_____ 4. diminished pain perception

_____ 5. diminished sense of touch

Key: a. dorsal columns (fasciculus cuneatus and fasciculus gracilis)
b. lateral corticospinal tract
c. ventral corticospinal tract
d. tectospinal tract
e. rubrospinal tract
f. vestibulospinal tract
g. lateral spinothalamic tract
h. ventral spinothalamic tract

Dissection of the Spinal Cord

10. Compare and contrast the meninges of the spinal cord and the brain. _____

_____

_____

_____

11. How can you distinguish the ventral from the dorsal horns? _____

_____

Spinal Nerves and Nerve Plexuses

12. In the human, there are 31 pairs of spinal nerves, named according to the region of the vertebral column from which they issue. The spinal nerves are named below. Indicate how they are numbered.

cervical nerves _____   sacral nerves _____

lumbar nerves _____   thoracic nerves _____

13. The ventral rami of spinal nerves C_1 through T_1 and T_{12} through S_4 take part in forming _____,

which serve the _____ of the body. The ventral rami of T_2 through T_{12} run

between the ribs to serve the _____. The dorsal rami of the spinal nerves

serve _____.

14. What would happen if the following structures were damaged or transected? (Use the key choices for responses.)

Key: a.   loss of motor function   b.   loss of sensory function   c.   loss of both motor and sensory function

_____ 1.   dorsal root of a spinal nerve

_____ 2.   ventral root of a spinal nerve

_____ 3.   ventral ramus of a spinal nerve

15. Define *nerve plexus*. _____

_____

16. Name the major nerves that serve the following body areas.

_____ 1.   biceps brachii

_____ 2.   diaphragm

_____ 3.   posterior thigh

_____ 4.   fibularis muscles

_____ 5.   flexor carpi radialis

_____ 6.   deltoid

_____   7.   gracilis

_____   8.   anterior thigh

_____   9.   muscles of the perineum

17. ➕ Wrist drop results in an inability to extend the hand at the wrist. Which nerve would most likely be affected in this injury,

and why? _____

_____

Dissection and Identification: Cat Spinal Nerves

18. From anterior to posterior, put the nerves issuing from the brachial plexus in their proper order (i.e., the median, musculo-cutaneous, radial, and ulnar nerves).

_____

19. Which of the nerves named above serves most of the cat's forearm extensor muscles? _____

Which serves the forearm flexors? _____

20. Just superior to the gastrocnemius muscle, the sciatic nerve divides into its two main branches, the _____

and _____ nerves.

21. What name is given to the cutaneous nerve of the cat's thigh? _____

The Autonomic Nervous System

MATERIALS

- Laboratory chart or three-dimensional model of the sympathetic trunk (chain)

LEARNING OUTCOMES

☐ Identify the site of origin and the function of the sympathetic and parasympathetic divisions of the autonomic nervous system.

☐ State how the autonomic nervous system differs from the somatic nervous system.

☐ Identify the neurotransmitters associated with the sympathetic and parasympathetic fibers.

PRE-LAB QUIZ

1. The _____ nervous system is the subdivision of the peripheral nervous system that regulates body activities that are generally *not* under conscious control.
   a. autonomic
   b. cephalic
   c. somatic
   d. vascular
2. Circle the correct underlined term. The parasympathetic division of the autonomic nervous system is also known as the craniosacral / thoracolumbar division.
3. Circle True or False. Cholinergic fibers release epinephrine.
4. The _____ division of the autonomic nervous system is responsible for the "fight-or-flight" response because it adapts the body for extreme conditions such as exercise.
5. Circle True or False. Preganglionic fibers of the autonomic nervous system release acetylcholine.

The **autonomic nervous system (ANS)** is the subdivision of the peripheral nervous system (PNS) that regulates body activities that are generally not under conscious control. For this reason, the ANS is also called the *involuntary nervous system.*

The motor pathways of the **somatic** (voluntary) **nervous system** innervate the skeletal muscles, and the motor pathways of the autonomic nervous system innervate smooth muscle, cardiac muscle, and glands. In the somatic division, the cell bodies of the motor neurons reside in the brain stem or ventral horns of the spinal cord, and their axons extend directly to the skeletal muscles they serve. However, the autonomic nervous system consists of chains of two motor neurons. The first motor neuron of each pair, called the *preganglionic neuron,* resides in the brain stem or the spinal cord. Its axon leaves the central nervous system to synapse with the second motor neuron *(postganglionic neuron),* whose cell body is located in an autonomic ganglion outside the CNS. The axon of the postganglionic neuron then extends to the organ it serves.

The ANS has two major functional subdivisions, the sympathetic and parasympathetic divisions. Both serve most of the same organs, but generally cause opposing or antagonistic effects.

Table 16.1 compares the parasympathetic and sympathetic divisions. Refer also to **Figure 16.1** and **Figure 16.2.**

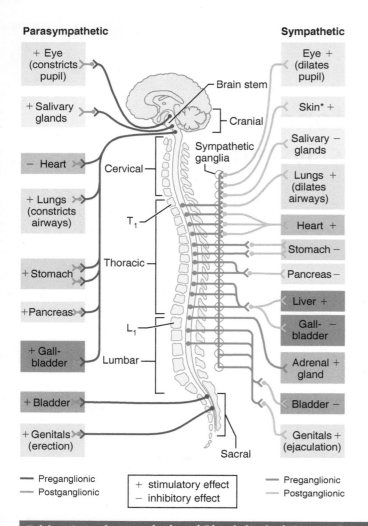

Parasympathetic

- \+ Eye (constricts pupil)
- \+ Salivary glands
- − Heart
- \+ Lungs (constricts airways)
- \+ Stomach
- \+ Pancreas
- \+ Gall-bladder
- \+ Bladder
- \+ Genitals (erection)

Brain stem — Cranial — Cervical — T₁ — Thoracic — L₁ — Lumbar — Sacral

Sympathetic ganglia

Sympathetic

- Eye + (dilates pupil)
- Skin* +
- Salivary − glands
- Lungs + (dilates airways)
- Heart +
- Stomach −
- Pancreas −
- Liver +
- Gall-bladder −
- Adrenal + gland
- Bladder −
- Genitals + (ejaculation)

— Preganglionic
— Postganglionic

+ stimulatory effect
− inhibitory effect

— Preganglionic
— Postganglionic

Figure 16.1 Overview of the subdivisions of the autonomic nervous system. Although sympathetic innervation to the skin (*) is shown here mapped to the cervical area, all nerves to the periphery carry postganglionic sympathetic fibers.

As we grow older, our sympathetic nervous system gradually becomes less and less efficient, particularly in causing vasoconstriction of blood vessels. When elderly people stand up quickly after sitting or lying down, they often become light-headed or faint. This is because the sympathetic nervous system is not able to react quickly enough to counteract the pull of gravity by activating the vasoconstrictor fibers. So, blood pools in the feet. This condition, **orthostatic hypotension,** is a type of low blood pressure resulting from changes in body position as described. Orthostatic hypotension can be prevented to some degree if the person changes position *slowly*. This gives the sympathetic nervous system a little more time to react and adjust. ✚

ACTIVITY 1

Locating the Sympathetic Trunk (Chain)

Locate the sympathetic trunk on the spinal nerve chart or three-dimensional model. Notice that the trunk resembles a string of beads. There is a trunk on either side of the vertebral column. ▬

Table 16.1	Anatomical and Physiological Comparison of the Parasympathetic and Sympathetic Divisions (Figure 16.1)	
Characteristic	**Parasympathetic (craniosacral)**	**Sympathetic (thoracolumbar)**
Origin in the CNS (contains the preganglionic cell body)	Brain stem nuclei of cranial nerves III, VII, IX, and X; spinal cord segments S_2 through S_4	Lateral horns of the gray matter of the spinal cord T_1 through L_2
Location of ganglia (contains the postganglionic cell body)	Located close to the target organ (**terminal ganglia**) or within the wall of the target organ (**intramural ganglia**)	Located close to the CNS: alongside the vertebral column (**sympathetic trunk ganglia**) or anterior to the vertebral column (**collateral ganglia**)
Axon lengths	Long preganglionic axons	Short preganglionic axons
	Short postganglionic axons	Long postganglionic axons
White and gray rami communicantes (see Figure 16.2)	None	Each white ramus communicans contains myelinated preganglionic axons. Each gray ramus communicans contains nonmyelinated postganglionic axons.
Degree of branching of preganglionic axons	Minimal	Extensive
Functional role	Performs maintenance functions; conserves and stores energy; rest-and-digest response	Prepares the body for emergency situations and vigorous physical activity; fight-or-flight response
Neurotransmitters	Preganglionic axons release acetylcholine	Preganglionic axons release acetylcholine
	Postganglionic axons release acetylcholine	Most postganglionic axons release norepinephrine (adrenergic fibers); postganglionic axons serving sweat glands and the blood vessels of skeletal muscle release acetylcholine; neurotransmitter activity is supplemented by the release of epinephrine and norepinephrine by the adrenal medulla
Effects of the division	More specific and local	More general and widespread

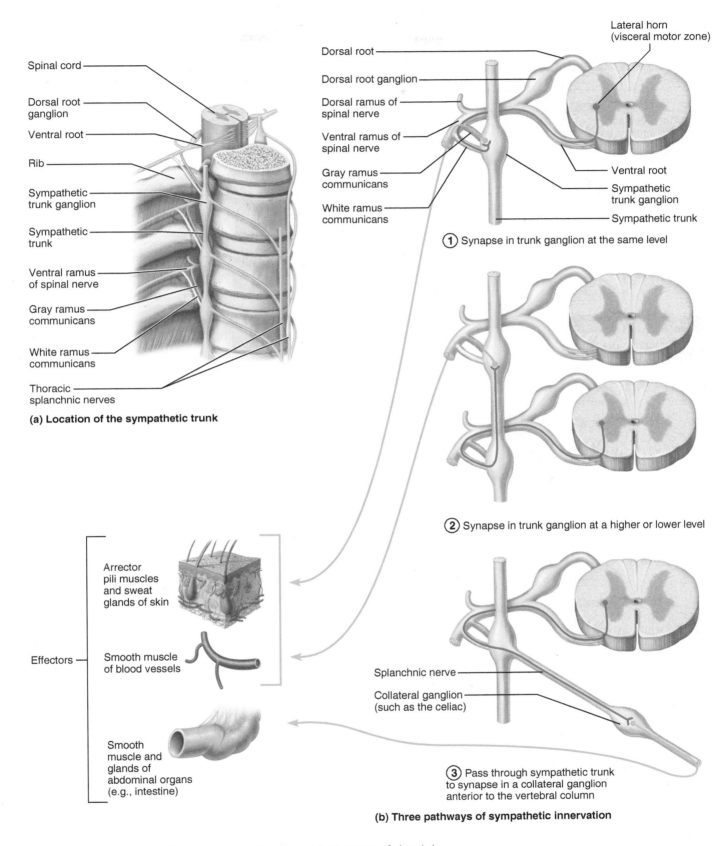

Spinal cord

Dorsal root ganglion

Ventral root

Rib

Sympathetic trunk ganglion

Sympathetic trunk

Ventral ramus of spinal nerve

Gray ramus communicans

White ramus communicans

Thoracic splanchnic nerves

(a) Location of the sympathetic trunk

Dorsal root

Dorsal root ganglion

Dorsal ramus of spinal nerve

Ventral ramus of spinal nerve

Gray ramus communicans

White ramus communicans

Lateral horn (visceral motor zone)

Ventral root

Sympathetic trunk ganglion

Sympathetic trunk

(1) Synapse in trunk ganglion at the same level

(2) Synapse in trunk ganglion at a higher or lower level

Splanchnic nerve

Collateral ganglion (such as the celiac)

(3) Pass through sympathetic trunk to synapse in a collateral ganglion anterior to the vertebral column

Arrector pili muscles and sweat glands of skin

Smooth muscle of blood vessels

Effectors

Smooth muscle and glands of abdominal organs (e.g., intestine)

(b) Three pathways of sympathetic innervation

Figure 16.2 Sympathetic trunks and pathways. (a) Diagram of the right sympathetic trunks in the posterior thorax. **(b)** Synapses between preganglionic and postganglionic neurons can occur at three different places.

Activity 2: Parasympathetic and Sympathetic Effects		
Organ	Parasympathetic effect	Sympathetic effect
Heart		
Bronchioles of lungs		
Digestive tract		
Urinary bladder		
Iris of the eye		
Blood vessels (most)		
Penis/clitoris		
Sweat glands		
Adrenal medulla		
Pancreas		

ACTIVITY 2

Comparing Sympathetic and Parasympathetic Effects

Several body organs are listed in the **Activity 2 chart**. Using your textbook as a reference, list the effect of the sympathetic and parasympathetic divisions on each. ■

Name ______________________________

Lab Time/Date ______________________________

EXERCISE 16

REVIEW SHEET

The Autonomic Nervous System

Parasympathetic and Sympathetic Divisions

1. List the names of the two motor neurons of the autonomic nervous system.

____

2. List the names and numbers of the four cranial nerves from which the parasympathetic division of the ANS arises.

____

3. List two types of sympathetic ganglia that contain postganglionic cell bodies.

____

4. List two types of parasympathetic ganglia that contain postganglionic cell bodies.

____

5. Which part of the rami communicantes contains nonmyelinated fibers? ______________________________

____

6. The following chart states a number of conditions. Use a check mark to show which division of the autonomic nervous system is involved in each.

Sympathetic division	Characteristic	Parasympathetic division
	Postganglionic axons secrete norepinephrine; adrenergic fibers	
	Postganglionic axons secrete acetylcholine; cholinergic fibers	
	Long preganglionic axon; short postganglionic axon	
	Short preganglionic axon; long postganglionic axon	
	Arises from cranial and sacral nerves	
	Arises from spinal nerves T_1 through L_3	
	Normally in control	
	"Fight-or-flight" system	
	Has more specific effects	
	Has rami communicantes	
	Has extensive branching of preganglionic axons	

7. ✚ Ogilvie syndrome is a condition that mimics a bowel obstruction. The patient experiences abdominal bloating and constipation in the absence of a mechanical blockage. Ogilvie syndrome is usually preceded by surgery and results in extreme loss of motor activity in the bowel. Which division of the autonomic nervous system is underactive in this case, especially with respect to the gastrointestinal system?

_____

8. ✚ Neostigmine is a drug that is classified as an acetylcholinesterase inhibitor. Explain how neostigmine could reverse the effects of Ogilvie syndrome.

_____

_____

Special Senses: Anatomy of the Visual System

- Chart of eye anatomy
- Dissectible eye model
- Prepared slide of longitudinal section of an eye showing retinal layers
- Compound microscope
- Preserved cow or sheep eye
- Dissecting instruments and tray
- Disposable gloves

L E A R N I N G   O U T C O M E S

☐ Identify the anatomy of the eye and its accessory anatomical structures on a model or appropriate image, and list the function(s) of each; identify the structural components that are present in a preserved sheep or cow eye (if available).

☐ Define *conjunctivitis, cataract,* and *glaucoma.*

☐ Describe the cellular makeup of the retina.

☐ Explain the difference between rods and cones with respect to visual perception and retinal distribution.

☐ Trace the visual pathway to the primary visual cortex, and indicate the effects of damage to various parts of this pathway.

P R E - L A B   Q U I Z

1. Name the mucous membrane that lines the internal surface of the eyelids and continues over the anterior surface of the eyeball. _____

2. How many extrinsic eye muscles are attached to the exterior surface of each eyeball?
   a. three   c. five
   b. four   d. six

3. The wall of the eye has three layers. The outermost fibrous layer is made up of the opaque white sclera and the transparent:
   a. choroids   c. cornea
   b. ciliary gland d. lacrima

4. Circle the correct underlined term. The <u>aqueous humor</u> / <u>vitreous humor</u> is a clear, watery fluid that helps to maintain the intraocular pressure of the eye and provides nutrients for the avascular lens and cornea.

5. Circle True or False. At the optic chiasma, the fibers from the medial side of each eye cross over to the opposite side.

Anatomy of the Eye

Accessory Structures

The adult human eye is a sphere measuring about 2.5 cm (1 inch) in diameter. Only about one-sixth of the eye's anterior surface is observable (Figure 17.1); the remainder is enclosed and protected by a cushion of fat and the walls of the bony orbit.

The accessory structures of the eye include the eyebrows, eyelids, conjunctivae, lacrimal apparatus, and extrinsic eye muscles (Table 17.1 and Figure 17.2).

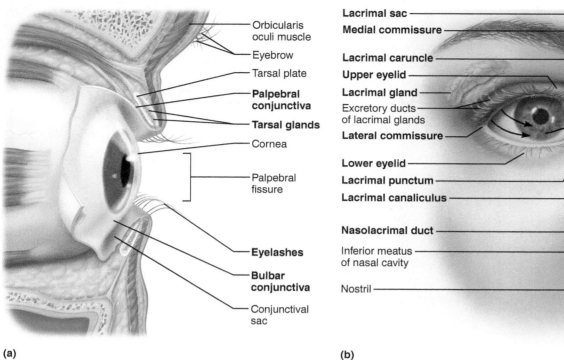

(a)

(b)

Figure 17.1 The eye and accessory structures. (a) Lateral view. **(b)** Anterior view with lacrimal apparatus. Arrows indicate the direction of the flow of lacrimal fluid.

Table 17.1	Accessory Structures of the Eye (Figure 17.1 and 17.2)	
Structure	**Description**	**Function**
Eyebrows	Short hairs located on the supraorbital margins	Shade and prevent sweat from entering the eyes.
Eyelids (palpebrae)	Skin-covered upper and lower lids, with eyelashes projecting from their free margin	Protect the eyes and spread lacrimal fluid (tears) with blinking.
Tarsal glands	Modified sebaceous glands embedded in the tarsal plate of the eyelid	Secrete an oily secretion that lubricates the surface of the eye.
Ciliary glands	Typical sebaceous and modified sweat glands that lie between the eyelash follicles	Secrete an oily secretion that lubricates the surface of the eye and the eyelashes. An infection of a ciliary gland is called a **sty**.
Conjunctivae	A clear mucous membrane that lines the eyelids (palpebral conjunctivae) and lines the anterior white of the eye (bulbar conjunctiva)	Secrete mucus to lubricate the eye. Inflammation of the conjunctiva results in conjunctivitis, (commonly called "pinkeye").
Medial and lateral commissures	Junctions where the eyelids meet medially and laterally	Form the corners of the eyes. The medial commissure contains the lacrimal caruncle.
Lacrimal caruncle	Fleshy reddish elevation that contains sebaceous and sweat glands	Secretes a whitish oily secretion for lubrication of the eye (can dry and form "eye sand").
Lacrimal apparatus	Includes the lacrimal gland and a series of ducts that drain the lacrimal fluid into the nasal cavity	Protects the eye by keeping it moist. Blinking spreads the lacrimal fluid.
Lacrimal gland	Located in the superior and lateral aspect of the orbit of the eye	Secretes lacrimal fluid, which contains mucus, antibodies, and lysozyme.
Lacrimal puncta	Two tiny openings on the medial margin of each eyelid	Allow lacrimal fluid to drain into the superior and inferiorly located lacrimal canaliculi.
Lacrimal canaliculi	Two tiny canals that are located in the eyelids	Allow lacrimal fluid to drain into the lacrimal sac.
Lacrimal sac	A single pouch located in the medial orbital wall	Allows lacrimal fluid to drain into the nasolacrimal duct.
Nasolacrimal duct	A single tube that empties into the nasal cavity	Allows lacrimal fluid to flow into the nasal cavity.
Extrinsic eye muscles	Six muscles for each eye; four recti and two oblique muscles (see Figure 17.2)	Control the movement of each eyeball and hold the eyes in the orbits.

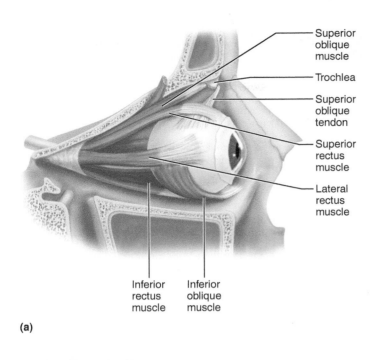

Superior oblique muscle

Trochlea

Superior oblique tendon

Superior rectus muscle

Lateral rectus muscle

Inferior rectus muscle

Inferior oblique muscle

(a)

Muscle	Action
Lateral rectus	Moves eye laterally
Medial rectus	Moves eye medially
Superior rectus	Elevates eye and turns it medially
Inferior rectus	Depresses eye and turns it medially
Inferior oblique	Elevates eye and turns it laterally
Superior oblique	Depresses eye and turns it laterally

(b)

Figure 17.2 Extrinsic muscles of the eye. (a) Lateral view of the right eye. **(b)** Summary of actions of the extrinsic eye muscles.

ACTIVITY 1

Identifying Accessory Eye Structures

Using a chart of eye anatomy or Figure 17.1, observe the eyes of another student, and identify as many of the accessory structures as possible. Ask the student to look to the left. What extrinsic eye muscles are responsible for this action?

Right eye: _____

Left eye: _____ ■

Internal Anatomy of the Eye

Anatomically, the wall of the eye is constructed of three layers: the **fibrous layer,** the **vascular layer,** and the **inner layer** (Table 17.2 and Figure 17.3).

Distribution of Photoreceptors

The photoreceptor cells are distributed over the entire neural retina, except where the optic nerve leaves the eyeball. This site, called the **optic disc,** or *blind spot,* is located in a weak spot in the **fundus** (posterior wall). Lateral to each blind spot, and directly posterior to the lens, is an area called the **macula lutea,** "yellow spot," an area of high cone density. In its center is the **fovea centralis,** a small pit, which contains only cones and is the area of greatest visual acuity. Focusing for detailed color vision occurs in the fovea centralis.

Internal Chambers and Fluids

The lens divides the eye into two segments: the **anterior segment** anterior to the lens, which contains a clear watery fluid called the **aqueous humor,** and the **posterior segment** behind the lens, filled with a gel-like substance, the **vitreous humor.** The anterior segment is further divided into **anterior** and **posterior chambers,** located before and after the iris, respectively. The aqueous humor is continually formed by the capillaries of the **ciliary processes** of the ciliary body. It helps to maintain the intraocular pressure of the eye and provides nutrients for the avascular lens and cornea. The aqueous humor is drained into the **scleral venous sinus.** The vitreous humor provides the major internal reinforcement of the posterior part of the eyeball, and helps to keep the retina pressed firmly against the wall of the eyeball. It is formed *only* before birth and is not renewed. In addition to the cornea, the light-bending media of the eye include the lens, the aqueous humor, and the vitreous humor.

Anything that interferes with drainage of the aqueous fluid increases intraocular pressure. When intraocular pressure reaches dangerously high levels, the retina and optic nerve are compressed, resulting in pain and possible blindness, a condition called **glaucoma.** ✚

ACTIVITY 2

Identifying Internal Structures of the Eye

Obtain a dissectible eye model and identify its internal structures described previously. As you work, also refer to Figure 17.3. ■

Microscopic Anatomy of the Retina

Cells of the retina include the pigment cells of the outer pigmented layer and neurons of the neural layer (Figure 17.4). The inner neural layer is composed of three major populations. These are, from outer to inner aspect, the **photoreceptors,** the **bipolar cells,** and the **ganglion cells.**

The **rods** are the specialized receptors for dim light. Visual interpretation of their activity is in gray tones. The **cones** are color receptors that permit high levels of visual acuity, but they function only under conditions of high light intensity; thus, for example, no color vision is possible in moonlight. The fovea contains only cones; the macula contains mostly cones; and from the edge of the macula to the

Table 17.2   Layers of the Eye (Figure 17.3)

Structure	Description	Function
Fibrous Layer (External Layer)		
Sclera	Opaque white connective tissue that forms the "white of the eye."	Helps to maintain the shape of the eyeball and provides an attachment point for the extrinsic eye muscles.
Cornea	Structurally continuous with the sclera; modified to form a transparent layer that bulges anteriorly; contains no blood vessels.	Forms a clear window that is the major light bending (refracting) medium of the eye.
Vascular Layer (Middle Layer)		
Choroid	A blood vessel–rich, dark membrane.	The blood vessels nourish the other layers of the eye, and the melanin helps to absorb excess light.
Cilary body	Modification of the choroid that encircles the lens.	Contains the ciliary muscle and the ciliary process.
Ciliary muscle	Smooth muscle found within the ciliary body.	Alters the shape of the lens with contraction and relaxation.
Ciliary process	Radiating folds of the ciliary muscle.	Capillaries of the ciliary process form the aqueous humor by filtering plasma.
Ciliary zonule (Suspensory ligament)	A halo of fine fibers that extends from the lens and attaches to the ciliary process.	Attaches the lens to the ciliary process.
Iris	The anterior portion of the vascular layer that is pigmented. It contains two layers of smooth muscle (sphincter pupillae and dilator pupillae).	Controls the amount of light entering the eye by changing the size of the pupil diameter. The sphincter pupillae contract to constrict the pupil. The dilator pupillae contract to dilate the pupil.
Pupil	The round central opening of the iris.	Allows light to enter the eye.
Inner Layer (Retina)		
Pigmented layer of the retina	The outer layer that is composed of only a single layer of pigment cells (melanocytes).	Absorbs light and prevents it from scattering in the eye. Pigment cells act as phagocytes for cleaning up cell debris and also store vitamin A needed for photoreceptor renewal.
Neural layer of the retina	The thicker inner layer composed of three main types of neurons: photoreceptors (rods and cones), bipolar cells, and ganglion cells.	Photoreceptors respond to light and convert the light energy into action potentials that travel to the primary visual cortex of the brain.

retina periphery, cone density declines gradually. By contrast, rods are most numerous in the periphery, and their density decreases as the macula is approached.

Light must pass through the ganglion cell layer and the bipolar cells to reach and excite the rods and cones. As a result of a light stimulus, the photoreceptors undergo changes in their membrane potential that influence the bipolar neurons. These in turn stimulate the ganglion cells, whose axons leave the retina in the tight bundle of fibers known as the **optic nerve** (Figure 17.3). In addition to these three major types of neurons, the retina also contains other types of neurons, the *horizontal cells* and *amacrine cells* (both are interneurons), which play a role in visual processing.

ACTIVITY 3

Studying the Microscopic Anatomy of the Retina

Use a compound microscope to examine a histologic slide of a longitudinal section of the eye. Identify the retinal layers by comparing your view to Figure 17.4b. ■

DISSECTION
The Cow (Sheep) Eye

1. Obtain a preserved cow or sheep eye, dissecting instruments, and a dissecting tray. Don disposable gloves.

2. Examine the external surface of the eye, noting the thick cushion of adipose tissue. Identify the optic nerve as it leaves the eyeball, the remnants of the extrinsic eye muscles, the conjunctiva, the sclera, and the cornea. Refer to **Figure 17.5** as you work.

3. Trim away most of the fat and connective tissue, but leave the optic nerve intact. Holding the eye with the cornea facing downward, carefully make an incision with a sharp scalpel into the sclera about 6 mm (¼ inch) above the cornea. Using scissors, complete the incision around the circumference of the eyeball paralleling the corneal edge.

4. Carefully lift the anterior part of the eyeball away from the posterior portion. The vitreous humor should remain with the posterior part of the eyeball.

5. Examine the anterior part of the eye, and identify the following structures:

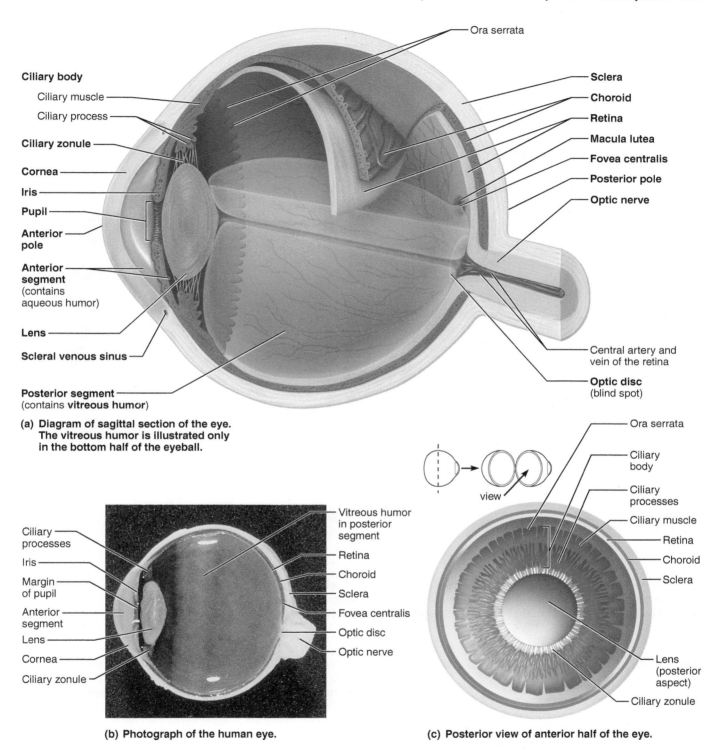

(a) Diagram of sagittal section of the eye. The vitreous humor is illustrated only in the bottom half of the eyeball.

(b) Photograph of the human eye.

(c) Posterior view of anterior half of the eye.

Figure 17.3 Internal anatomy of the eye.

Ciliary body: Black pigmented body that appears to be a halo encircling the lens.

Lens: Biconvex structure that is opaque in preserved specimens.

Carefully remove the lens and identify the adjacent structures:

Iris: Anterior continuation of the ciliary body penetrated by the pupil.

Cornea: More convex anteriormost portion of the sclera; normally transparent, but cloudy in preserved specimens.

6. Examine the posterior portion of the eyeball. Carefully remove the vitreous humor, and identify the following structures:

Retina: Appears as a delicate tan crumpled membrane that separates easily from the choroid.

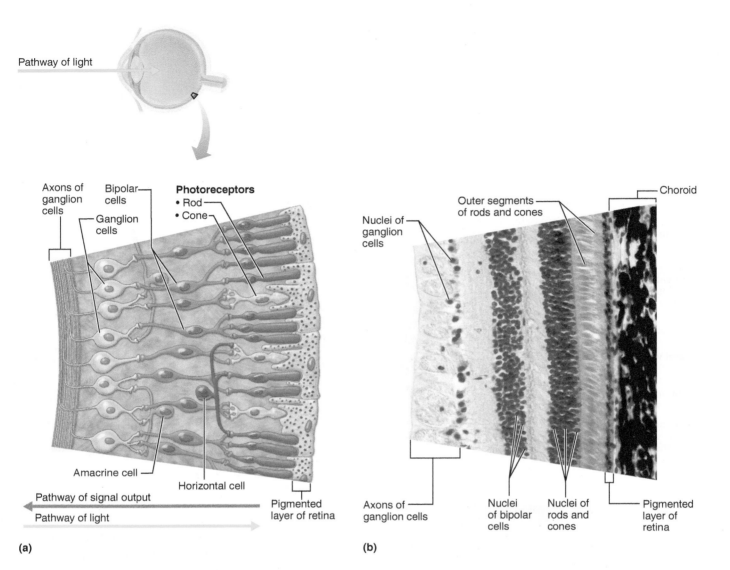

Figure 17.4 Microscopic anatomy of the retina. (a) Diagram of cells of the retina. Note the pathway of light through the retina. Neural signals (output of the retina) flow in the opposite direction. **(b)** Photomicrograph of the retina (185 ×).

Note its point of attachment. What is this point called?

_____

Pigmented choroid coat: Appears iridescent in the cow or sheep eye owing to a modification called the **tapetum lucidum.** This specialized surface reflects the light within the eye and is found in the eyes of animals that live under conditions of dim light. It is not found in humans. ■

Visual Pathways to the Brain

The axons of the ganglion cells of the retina converge at the posterior aspect of the eyeball and exit from the eye as the optic nerve. At the **optic chiasma,** the fibers from the medial side of each eye cross over to the opposite side

(Figure 17.6). The fiber tracts thus formed are called the **optic tracts.** Each optic tract contains fibers from the lateral side of the eye on the same side and from the medial side of the opposite eye.

The optic tract fibers synapse with neurons in the **lateral geniculate nucleus** of the thalamus, whose axons form the **optic radiation,** terminating in the **primary visual cortex** in the occipital lobe of the brain. Here they synapse with cortical neurons, and visual interpretation begins.

ACTIVITY 4

Predicting the Effects of Visual Pathway Lesions

After examining Figure 17.6a, determine what effects lesions in the following areas would have on vision:

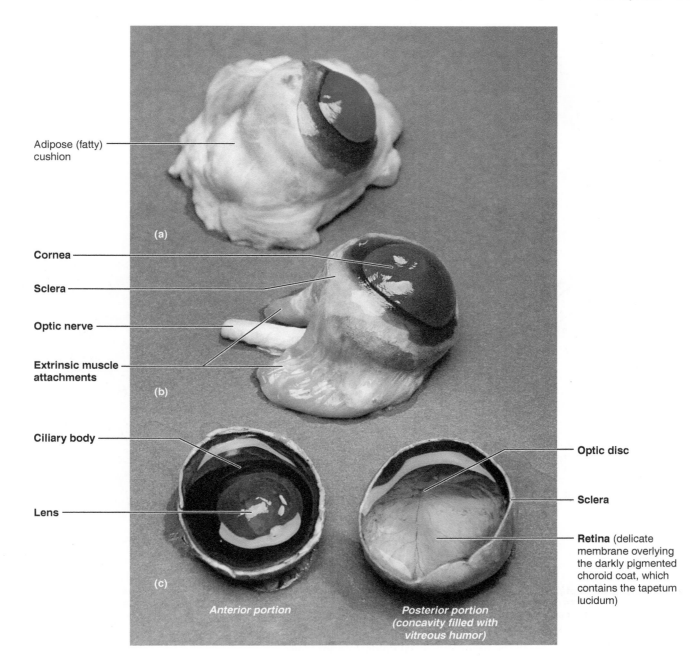

Figure 17.5 Anatomy of the cow eye. (a) Cow eye (entire) removed from the bony orbit (notice the large amount of fat cushioning the eyeball). **(b)** Cow eye (entire) with fat removed to show the extrinsic muscle attachments and optic nerve. **(c)** Cow eye cut along the frontal plane to reveal internal structures.

In the right optic nerve: _____

_____

Through the optic chiasma: _____

_____

In the left optic tract: _____

_____

In the right cerebral cortex (primary visual cortex): _____

_____

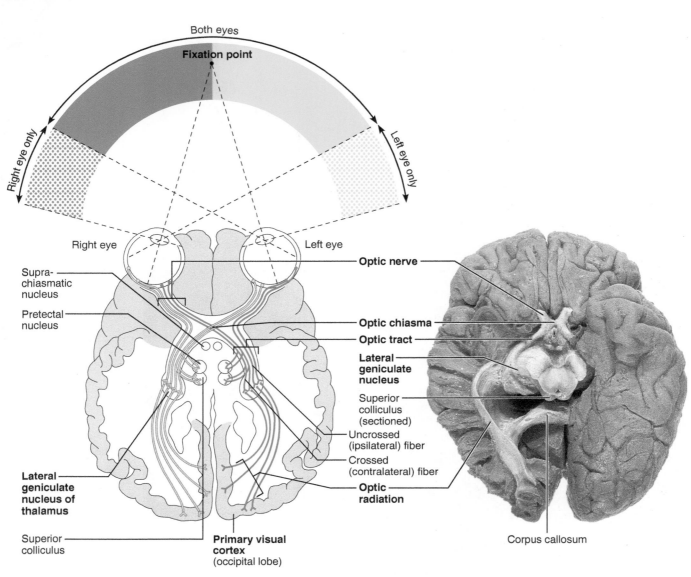

(a) **The visual fields of the two eyes overlap considerably.** Note that fibers from the lateral portion of each retinal field do not cross at the optic chiasma.

(b) **Photograph of human brain, with the right side dissected to reveal internal structures**

Figure 17.6 **Visual pathway to the brain and visual fields, inferior view.**

Name ______________________________

Lab Time/Date ______________________

EXERCISE 17

REVIEW SHEET

Special Senses: Anatomy of the Visual System

Anatomy of the Eye

1. Name five accessory eye structures that contribute to the formation of lacrimal fluid (tears) and/or help lubricate the eyeball, and then describe the major secretory product of each.

Accessory structures	Product

2. The eyeball is wrapped in adipose tissue within the bony orbit. What is the function of the adipose tissue?

____

3. Why does one often have to blow one's nose after crying? ______________________________________

____

4. Identify the extrinsic eye muscle predominantly responsible for the actions described below.

______________________ 1. turns the eye laterally

______________________ 2. turns the eye medially

______________________ 3. turns the eye up and laterally

______________________ 4. turns the eye down and medially

______________________ 5. turns the eye up and medially

______________________ 6. turns the eye down and laterally

5. What is a sty? ____

____

What is conjunctivitis? ____

____

6. Correctly identify each lettered structure in the diagram by writing the letter next to its name in the numbered list. Use an appropriate reference if necessary.

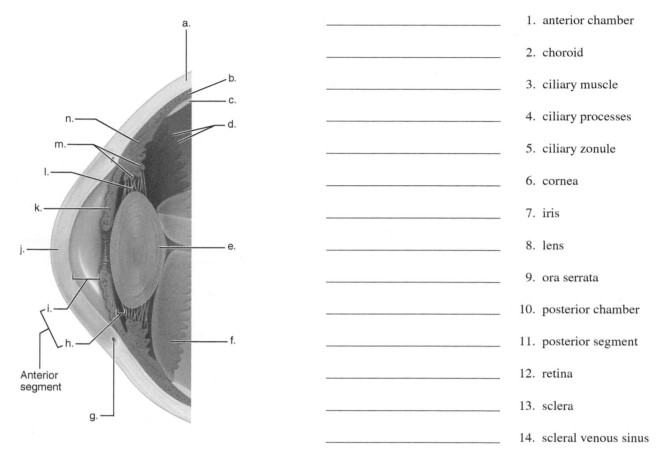

_____ 1. anterior chamber

_____ 2. choroid

_____ 3. ciliary muscle

_____ 4. ciliary processes

_____ 5. ciliary zonule

_____ 6. cornea

_____ 7. iris

_____ 8. lens

_____ 9. ora serrata

_____ 10. posterior chamber

_____ 11. posterior segment

_____ 12. retina

_____ 13. sclera

_____ 14. scleral venous sinus

7. The iris is composed primarily of two smooth muscle layers.

Which muscle layer dilates the pupil? _____

8. In which of the following circumstances would you expect the pupil to be dilated?

a. in bright light

b. in dim light

9. Which of the following controls the intrinsic eye muscles?

autonomic nervous system   somatic nervous system

10. Match the key terms with the descriptions that follow. (Some choices will be used more than once.)

Key: a. aqueous humor   e. cornea   j. retina
     b. choroid   f. fovea centralis   k. sclera
     c. ciliary body   g. iris   l. scleral venous sinus
     d. ciliary processes of the   h. lens   m. vitreous humor
       ciliary body   i. optic disc

_____ 1. fluid filling the anterior segment of the eye

_____ 2. the "white" of the eye

_____ 3. part of the retina that lacks photoreceptors

_____ 4. modification of the choroid that contains the ciliary muscle

_____ 5. drains the aqueous humor from the eye

_____ 6. layer containing the rods and cones

_____ 7. substance occupying the posterior segment of the eyeball

_____ 8. forms the bulk of the heavily pigmented vascular layer

_____, _____ 9. composed of smooth muscle structures (2)

_____ 10. area of critical focusing and detailed color vision

_____ 11. form (by filtration) the aqueous humor

_____, _____, _____, _____ 12. light-bending media of the eye (4)

_____ 13. anterior continuation of the sclera—your "window on the world"

_____ 14. composed of tough, white, opaque, fibrous connective tissue

Microscopic Anatomy of the Retina

11. The two major layers of the retina are the pigmented and neural layers. In the neural layer, the neuron populations are arranged as follows from the pigmented layer to the vitreous humor. (Circle the proper response.)

bipolar cells, ganglion cells, photoreceptors   photoreceptors, ganglion cells, bipolar cells

ganglion cells, bipolar cells, photoreceptors   photoreceptors, bipolar cells, ganglion cells

12. The axons of the _____ cells form the optic nerve, which exits from the eyeball.

13. Complete the following statements by writing either *rods* or *cones* on each blank.

The dim light receptors are the _____. Only _____ are found in the fovea centralis, whereas mostly

_____ are found in the periphery of the retina. _____ are the photoreceptors that operate best in

bright light and allow for color vision.

Dissection of the Cow (Sheep) Eye

14. What modification of the choroid that is *not* present in humans is found in the cow eye? _____

_____ What is its function? _____

15. What does the retina look like? _____

_____

At what point is the retina attached to the posterior aspect of the eyeball? _____

Visual Pathways to the Brain

16. The visual pathway to the occipital lobe of the brain consists most simply of a chain of five neurons. Beginning with the photoreceptor cell of the retina, name them, and note their location in the pathway.

1. _____   4. _____

2. _____   5. _____

3. _____

17. How is the right optic *tract* anatomically different from the right optic *nerve*? _____

_____

_____

18. ✚ Visual field tests are done to reveal destruction along the visual pathway from the retina to the optic region of the brain. Note where the lesion is likely to be in the following cases.

Normal vision in left eye visual field; absence of vision in right eye visual field: _____

Normal vision in both eyes for right half of the visual field; absence of vision in both eyes for left half of the visual field:

_____

19. ✚ A cornea transplant involves the grafting of a donor cornea into a recipient's anterior eye. The sutures to hold the graft in place must stay in place for a long period of time because the cornea is slow to heal. Explain why the healing process is

so slow and also why graft rejection is unlikely with a cornea transplant. _____

_____

_____

Special Senses: Visual Tests and Experiments

MATERIALS

- Metric ruler; meter stick
- Common straight pins
- Snellen eye chart, floor marked with chalk or masking tape to indicate 20-ft. distance from posted Snellen chart
- Ishihara's color plates
- Two pencils
- Test tubes, each large enough to accommodate a pencil
- Laboratory lamp or penlight
- Ophthalmoscope (if available)

LEARNING OUTCOMES

- ☐ Discuss the mechanism by which images form on the retina.
- ☐ Define the following terms: *accommodation, astigmatism, emmetropic, hyperopia, myopia, refraction,* and *presbyopia;* describe several simple visual tests to which the terms apply.
- ☐ Discuss the benefits of binocular vision.
- ☐ Define *convergence*, and discuss the importance of the pupillary and convergence reflexes.
- ☐ State the importance of an ophthalmoscopic examination.

PRE-LAB QUIZ

1. Circle the correct underlined term. Photoreceptors are distributed over the entire neural retina, except where the optic nerve leaves the eyeball. This site is called the macula lutea / optic disc.
2. Circle True or False. Persons with difficulty seeing objects at a distance are said to have myopia.
3. A condition that results in the loss of elasticity of the lens and difficulty focusing on a close object is called:
   a. myopia   c. hyperopia
   b. presbyopia   d. astigmatism
4. Photoreceptors of the eye include rods and cones. Which is responsible for interpreting color; which can function only under conditions of high light intensity? _____
5. Circle the correct underlined term. Extrinsic / Intrinsic eye muscles are controlled by the autonomic nervous system.

I n this exercise, you will perform several visual tests and experiments focusing on the physiology of vision. The first test involves demonstrating the blind spot (optic disc), the site where the optic nerve exits the eyeball.

The Optic Disc

ACTIVITY 1

Demonstrating the Blind Spot

1. Hold the figure for the blind spot test (**Figure 18.1**) about 46 cm (18 inches) from your eyes. Close your left eye, and focus your right eye on the **X**, which should be positioned so that it is directly in line with your right eye. Move the figure slowly toward your face, keeping your right eye focused on the **X**. When the dot focuses on the blind spot, which lacks photoreceptors, it will disappear.

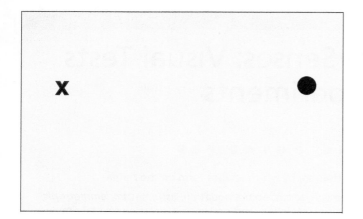

Figure 18.1 Bind spot test figure.

2. Have your laboratory partner obtain a metric ruler and record in metric units the distance at which this occurs. The dot will reappear as the figure is moved closer. Distance at which the dot disappears:

Right eye _____

Repeat the test for the left eye, this time closing the right eye and focusing the left eye on the dot. Record the distance at which the **X** disappears:

Left eye _____ ▪

Refraction, Visual Acuity, and Astigmatism

When light rays pass from one medium to another, their speed changes, and the rays are bent, or **refracted.** Thus, the light rays in the visual field are refracted as they encounter the cornea, lens, and aqueous and vitreous humors of the eye.

The refractive index (bending power) of the cornea and humors are constant. But the lens's refractive index can be varied by changing the lens's shape. The greater the lens convexity, or bulge, the more the light will be bent. Conversely, the less convex the lens (the flatter it is), the less it bends light.

In general, light from a distant source (over 6 m, or 20 feet) approaches the eye as parallel rays, and no change in lens convexity is necessary for it to focus properly on the retina. However, light from a close source tends to diverge, and the convexity of the lens must increase to make close vision possible. To achieve this, the ciliary muscle contracts, decreasing the tension on the ciliary zonule attached to the lens and allowing the elastic lens to bulge. Thus, a lens capable of bringing a *close* object into sharp focus is more convex than a lens focusing on a more distant object. The ability of the eye to focus differentially for objects of close vision (less than 6 m, or 20 feet) is called **accommodation.** It should be noted that the image formed on the retina as a result of the refractory activity of the lens **(Figure 18.2)** is a **real image** (reversed from left to right, inverted, and smaller than the object).

The normal eye, or **emmetropic eye,** is able to accommodate properly. However, visual problems may result from (1) lenses that are too strong or too "lazy" (overconverging and underconverging, respectively), (2) from structural problems such as an eyeball that is too long or too short to provide for proper focusing by the lens, or (3) a cornea or lens with improper curvatures. Problems of refraction are summarized in **Figure 18.3.**

Irregularities in the curvatures of the lens and/or the cornea lead to a blurred vision problem called **astigmatism.** Cylindrically ground lenses are prescribed to correct the condition.

Accommodation

The elasticity of the lens decreases dramatically with age, resulting in difficulty in focusing for near or close vision, especially when reading. This condition is called **presbyopia**—literally, "old vision." Lens elasticity can be tested by measuring the **near point of vision.** The near point of vision is about 10 cm from the eye in young adults. It is closer in children and farther in elderly people.

ACTIVITY 2

Determining Near Point of Vision

To determine your near point of vision, hold a common straight pin (or other object) at arm's length in front of one eye. Slowly move the pin toward that eye until the pin image becomes distorted. Have your lab partner use a metric ruler to measure the distance in centimeters from your eye to the pin at this point, and record the distance. Repeat the procedure for the other eye.

Near point for right eye: _____

Near point for left eye: _____ ▪

Visual Acuity

Visual acuity, or sharpness of vision, is generally tested with a Snellen eye chart, which consists of letters of various sizes printed on a white card. This test is based on the fact that letters of a certain size can be seen clearly by eyes with normal vision at a specific distance. The distance at which the normal, or emmetropic, eye can read a line of letters is printed at the end of that line.

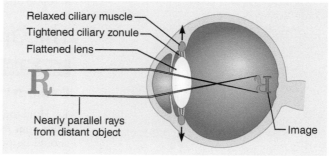

Figure 18.2 Refraction and real images. The refraction of light in the eye produces a real image (reversed, inverted, and reduced) on the retina.

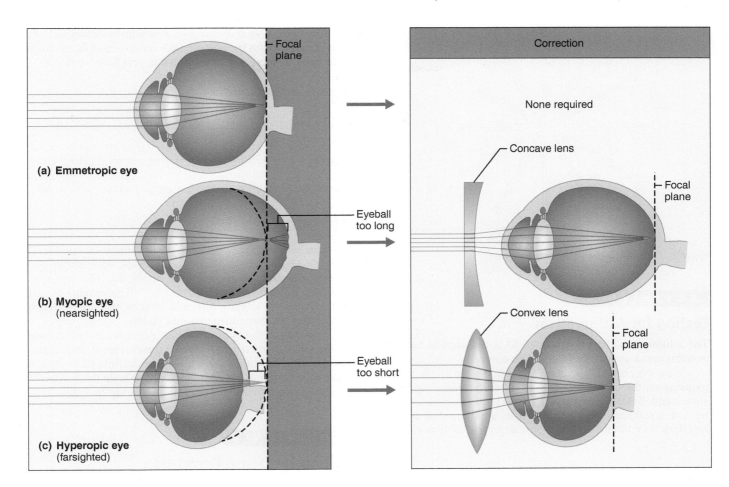

Figure 18.3 **Problems of refraction.** **(a)** In the emmetropic (normal) eye, light from both near and far objects is focused properly on the retina. **(b)** In a myopic (nearsighted) eye, light from distant objects is brought to a focal point before reaching the retina. It then diverges. Applying a concave lens focuses objects properly on the retina. **(c)** In the hyperopic (farsighted) eye, light from a near object is brought to a focal point behind the retina. Applying a convex lens focuses objects properly on the retina. (Refractory effect of cornea is ignored here.)

ACTIVITY 3

Testing Visual Acuity

1. Have your partner stand 6 m (20 feet) from the posted Snellen eye chart and cover one eye with a card or hand. As your partner reads each consecutive line aloud, check for accuracy. If this individual wears glasses, give the test twice—first with glasses off and then with glasses on. *Do not remove contact lenses, but note that they were in place during the test.*

2. Record the number of the line with the smallest-sized letters read. If it is 20/20, the person's vision for that eye is normal. If it is 20/40, or any ratio with a value less than one, he or she has less than the normal visual acuity. (Such an individual is myopic.) If the visual acuity is 20/15, vision is better than normal, because this person can stand at 6 m (20 feet) from the chart and read letters that are only

discernible by the normal eye at 4.5 m (15 feet). Give your partner the number of the line corresponding to the smallest letters read, to record in step 4.

3. Repeat the process for the other eye.

4. Have your partner test and record your visual acuity. If you wear glasses, the test results *without* glasses should be recorded first.

Visual acuity, right eye without glasses: _____

Visual acuity, right eye with glasses:_____

Visual acuity, left eye without glasses: _____

Visual acuity, left eye with glasses:_____ ▪

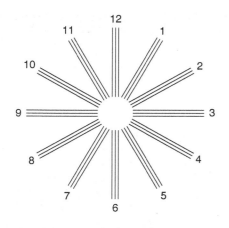

Figure 18.4 Astigmatism testing chart.

ACTIVITY 4

Testing for Astigmatism

The astigmatism chart (Figure 18.4) is designed to test for unequal curvatures of the lens and/or cornea.

View the chart first with one eye and then with the other, focusing on the center of the chart. If all the radiating lines appear equally dark and distinct, there is no distortion of your refracting surfaces. If some of the lines are blurred or appear less dark than others, at least some degree of astigmatism is present.

Is astigmatism present in your left eye?_____

Is it present in your right eye?_____ ■

Color Blindness

Ishihara's color plates are designed to test for deficiencies in the color photoreceptor cells, the cones. There are three cone types, each containing a different light-absorbing pigment. One type primarily absorbs the red wavelengths of the visible light spectrum, another the blue wavelengths, and a third the green wavelengths. Nerve impulses reaching the brain from these different photoreceptor types are then interpreted (seen) as red, blue, and green, respectively. Overlapping input from more than one cone type leads the brain to interpret the intermediate colors of the visible light spectrum.

ACTIVITY 5

Testing for Color Blindness

1. Find the interpretation table that accompanies the Ishihara color plates, and prepare a sheet to record data for the test. Note which plates are patterns rather than numbers.

2. View the color plates in bright light or sunlight while holding them about 0.8 m (30 inches) away and at right angles to your line of vision. Report to your laboratory partner what you see in each plate. Take no more than 3 seconds for each decision.

3. Your partner should record your responses and then check their accuracy with the correct answers provided in the color plate book. Is there any indication that you have some

degree of color blindness?_____

If so, what type? _____

_____

Repeat the procedure to test your partner's color vision. ■

Binocular Vision

Humans, cats, predatory birds, and most primates are endowed with *binocular vision*. Their visual fields, each about 170 degrees, overlap to a considerable extent, and each eye sees a slightly different view (Figure 18.5). The primary visual cortex fuses the slightly different images, providing **depth perception** (or **three-dimensional vision**).

In contrast, the eyes of rabbits, pigeons, and others are on the sides of their head. Such animals see in two different directions and thus have a panoramic field of view and panoramic vision. A mnemonic device to keep these straight is "Eyes in the front—likes to hunt. Eyes on the side—likes to hide."

ACTIVITY 6

Testing for Depth Perception

1. To demonstrate that each eye sees a slightly different view, perform the following simple experiment.

Close your left eye. Hold a pencil at arm's length directly in front of your right eye. Position another pencil directly beneath it, and then move the lower pencil about half the distance toward you. As you move the lower pencil, make sure it remains in the *same plane* as the stationary pencil, so that the two pencils continually form a straight line. Then,

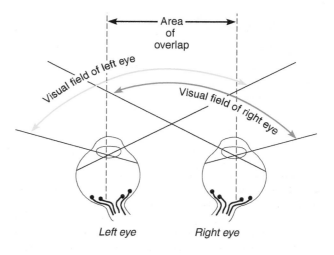

Figure 18.5 Overlapping of the visual fields.

without moving the pencils, close your right eye, and open your left eye. Notice that with only the right eye open, the moving pencil stays in the same plane as the fixed pencil, but that when viewed with the left eye, the moving pencil is displaced laterally away from the plane of the fixed pencil.

2. To demonstrate the importance of two-eyed binocular vision for depth perception, perform this second simple experiment.

Have your laboratory partner hold a test tube vertical about arm's length in front of you. With both eyes open, quickly insert a pencil into the test tube. Remove the pencil, bring it back close to your body, close one eye, and quickly and without hesitation insert the pencil into the test tube. *(Do not feel for the test tube with the pencil!)* Repeat with the other eye closed.

Was it as easy to dunk the pencil with one eye closed as with both eyes open?

_____ ▬

Eye Reflexes

Both intrinsic (internal) and extrinsic (external) muscles are necessary for proper eye functioning. The *intrinsic muscles,* controlled by the autonomic nervous system, are those of the ciliary body (which alters the lens curvature in focusing) and the sphincter pupillae and dilator pupillae muscles of the iris (which control pupil size and thus regulate the amount of light entering the eye). The *extrinsic muscles* are the rectus and oblique muscles, which are attached to the eyeball exterior (see Figure 17.2). These muscles control eye movement and make it possible to keep moving objects focused on the fovea centralis. They are also responsible for **convergence,** or medial eye movements, which is essential for near vision. When convergence occurs, both eyes are directed toward the near object viewed. The extrinsic eye muscles are controlled by the somatic nervous system.

ACTIVITY 7

Demonstrating Reflex Activity of Intrinsic and Extrinsic Eye Muscles

Involuntary activity of both the intrinsic and extrinsic muscle types is brought about by reflex actions that can be observed in the following experiments.

Pupillary Light Reflex

Sudden illumination of the retina by a bright light causes the pupil to constrict reflexively in direct proportion to the light intensity. This protective response prevents damage to the delicate photoreceptor cells.

Obtain a laboratory lamp or penlight. Have your laboratory partner sit with eyes closed and hands over the eyes. Turn on the light and position it so that it shines on the subject's right hand. After 1 minute, ask your partner to uncover and open the right eye. Quickly observe the pupil of that eye. What happens to the pupil?

_____

Shut off the light, and ask your partner to uncover and open the opposite eye. What are your observations of the pupil?

_____

_____

Accommodation Pupillary Reflex

Have your partner gaze for approximately 1 minute at a distant object in the lab—*not* toward the windows or another light source. Observe your partner's pupils. Then hold some printed material 15 to 25 cm (6 to 10 inches) from his or her face, and direct him or her to focus on it.

How does pupil size change as your partner focuses on the printed material?

_____

Explain the value of this reflex. _____

_____

_____

_____

Convergence Reflex

Repeat the previous experiment, this time using a pen or pencil as the close object to be focused on. Note the position of your partner's eyeballs while he or she gazes at the distant object, and then at the close object. Do they change position as the object of focus is changed?

_____ In what way? _____

_____

_____

_____

Ophthalmoscopic Examination of the Eye (Optional)

The ophthalmoscope is an instrument used to examine the *fundus,* or eyeball interior, to determine visually the condition of the retina, optic disc, and internal blood vessels. Such an examination can detect certain pathological conditions such as diabetes mellitus, arteriosclerosis, and degenerative changes of the optic nerve and retina. The ophthalmoscope consists of a set of lenses mounted on a rotating disc (the **lens selection disc**), a light source regulated by a **rheostat control,** and a mirror that reflects the light so that the eye interior can be illuminated.

The lens selection disc is positioned in a small slit in the mirror, and the examiner views the eye interior through this slit, appropriately called the **viewing window.** The focal length of each lens is indicated in diopters preceded by a plus (+) sign if the lens is convex and by a negative (−) sign if the lens is concave. When the zero (0) is seen in the **diopter window,** on the examiner's side of the instrument, there is no lens positioned in the slit. Changing the lens will change the depth of focus for viewing the eye interior.

The light is turned on by depressing the red **rheostat lock button** and then rotating the rheostat control in the clockwise direction. The **aperture selection dial** on the front of the instrument allows the nature of the light beam to be altered. The **filter switch,** also on the front, allows the choice of a green, unfiltered, or polarized light beam. Generally, green light allows for clearest viewing of the blood vessels in the eye interior and is most comfortable for the subject.

Once you have examined the ophthalmoscope and have become familiar with it, you are ready to conduct an eye examination.

ACTIVITY 8

Conducting an Ophthalmoscopic Examination

1. Conduct the examination in a dimly lighted or darkened room with the subject comfortably seated and gazing straight ahead. To examine the right eye, sit face-to-face with the subject, hold the instrument in your right hand, and use your right eye to view the interior of the eye. You may want to steady yourself by resting your left hand on the subject's shoulder. To view the left eye, use your left eye, hold the instrument in your left hand, and steady yourself with your right hand.

2. Begin the examination with the 0 (no lens) in position. Grasp the instrument so that you can rotate the lens disc with the index finger. Holding the ophthalmoscope about 15 cm (6 inches) from the subject's eye, direct the light into the pupil at a slight angle—through the pupil edge rather than directly through its center. You will see a red circular area that is the illuminated eye interior.

3. Move in as close as possible to the subject's cornea (to within 5 cm, or 2 inches) as you continue to observe the area. Steady your instrument-holding hand on the subject's cheek if necessary. If both your eye and that of the subject are normal,

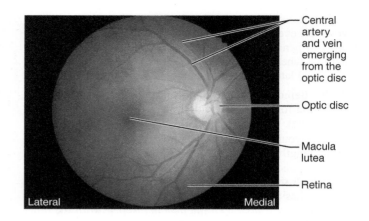

Figure 18.6 Posterior wall (fundus) of right retina.

you will be able to view the fundus clearly without further adjustment of the ophthalmoscope. If you cannot focus on the posterior wall, slowly rotate the lens disc counterclockwise until you can clearly see the posterior wall. When the ophthalmoscope is correctly set, the posterior wall of the right eye should appear as shown in **Figure 18.6.** (**Note:** If a positive [convex] lens is required and your eyes are normal, the subject has hyperopia. If a negative [concave] lens is necessary to view the posterior wall and your eyes are normal, the subject is myopic.)

When the examination is proceeding correctly, the subject can often see images of retinal vessels in his or her own eye that appear rather like cracked glass. If you are unable to achieve a sharp focus or to see the optic disc, move medially or laterally and begin again.

4. Examine the optic disc for color, elevation, and sharpness of outline, and observe the blood vessels radiating from near its center. Locate the macula lutea, lateral to the optic disc. It is a darker area in which blood vessels are absent, and the fovea appears to be a slightly lighter area in its center. The macula lutea is most easily seen when the subject looks directly into the light of the ophthalmoscope.

! Do not examine the macula lutea for longer than 1 second at a time.

5. When you have finished examining your partner's retina, shut off the ophthalmoscope. Change places with your partner (become the subject), and repeat steps 1–4. ∎

Special Senses: Visual Tests and Experiments

Blind Spot

1. Explain why vision is lost when light hits the blind spot._____

_____

Refraction, Visual Acuity, and Astigmatism

2. Match the terms in column B with the descriptions in column A.

Column A

Column B

_____ 1. light bending

a. accommodation

_____ 2. ability to focus for close (less than 20 feet) vision

b. astigmatism

_____ 3. normal vision

c. emmetropia

_____ 4. inability to focus well on close objects (farsightedness)

d. hyperopia

_____ 5. nearsightedness

e. myopia

_____ 6. blurred vision due to unequal curvatures of the lens or cornea

f. refraction

3. Complete the following statements:

In farsightedness, the light is focused __1__ the retina. The lens required to treat myopia is a __2__ lens. The near point of vision increases with age because the __3__ of the lens decreases as we get older. A convex lens, like that of the eye, produces an image that is upside down and reversed from left to right. Such an image is called a __4__ image.

1. _____

2. _____

3. _____

4. _____

4. Use terms from the key to complete the statements concerning near and distance vision. (Some terms will be used more than once.)

Key:

    a. contracted   b. decreased   c. increased   d. loose   e. relaxed   f. taut

During distance vision, the ciliary muscle is _____, the ciliary zonule is _____, the convexity of the lens

is _____, and light refraction is _____. During close vision, the ciliary muscle is _____, the ciliary

zonule is _____, lens convexity is _____, and light refraction is _____.

5. Using your Snellen eye test results, answer the following questions.

Is your visual acuity normal, less than normal, or better than normal? _____

Explain your answer. _____

_____

6. Define *astigmatism*. _____

_____

How can it be corrected? _____

7. Define *presbyopia*. _____

_____

What causes it? _____

Color Blindness

8. To which wavelengths of light do the three cone types of the retina respond maximally?

_____, _____, and _____

9. How can you explain the fact that we see a great range of colors even though only three cone types exist?

_____

Binocular Vision

10. What is the advantage of binocular vision? _____

What factor(s) are responsible for binocular vision? _____

_____

11. ⊞ Macular degeneration is an eye disease in which the macula lutea deteriorates. Explain why this would have a more

profound effect on vision than deterioration of other parts of the retina. _____

_____

12. ⊞ The "near triad" refers to the fact that three reflexes are required for near-point vision: accommodation of the lenses, constriction of the pupils, and convergence of the eyeballs. Strabismus is a misalignment of the eyes that can lead to double vision.

Which reflex is affected in this condition? _____

_____

Are intrinsic or extrinsic eye muscles affected? _____

Special Senses: Hearing and Equilibrium

MATERIALS

- Three-dimensional dissectible ear model and/or chart of ear anatomy
- Otoscope (if available)
- Disposable otoscope tips (if available) and autoclave bag
- Alcohol swabs
- Compound microscope
- Prepared slides of the cochlea of the ear
- Absorbent cotton
- Pocket watch or clock that ticks
- Metric ruler
- Tuning forks (range of frequencies)
- Rubber mallet
- Demonstration: Microscope focused on a slide of a crista ampullaris receptor of a semicircular canal
- Blackboard and chalk, or whiteboard and markers

LEARNING OUTCOMES

- ☐ Identify the anatomical structures of the external, middle, and internal ear on a model or appropriate diagram, and explain their functions.
- ☐ Describe the anatomy of the organ of hearing (spiral organ in the cochlea), and explain its function in sound reception.
- ☐ Discuss how one is able to localize the source of sounds.
- ☐ Define sensorineural deafness and conduction deafness, and relate these conditions to the Weber and Rinne tests.
- ☐ Describe the anatomy of the organs of equilibrium in the internal ear, and explain their relative function in maintaining equilibrium.
- ☐ State the locations and functions of endolymph and perilymph.
- ☐ Discuss the effects of acceleration on the semicircular canals.
- ☐ Define nystagmus, and relate this event to the balance test.
- ☐ State the purpose of the Romberg test.
- ☐ Explain the role of vision in maintaining equilibrium.

PRE-LAB QUIZ

1. Circle the correct underlined term. The ear is divided into <u>three</u> / <u>four</u> major areas.
2. The external ear is composed primarily of the _____ and the external acoustic meatus.
   a. auricle   c. eardrum   b. cochlea   d. stapes
3. Circle the correct underlined term. Sound waves that enter the external acoustic meatus eventually encounter the <u>tympanic membrane</u> / <u>oval window</u>, which then vibrates at the same frequency as the sound waves hitting it.
4. Three small bones found within the middle ear are the malleus, incus, and _____.
5. The snail-like _____, found in the internal ear, contains sensory receptors for hearing.
6. Circle the correct underlined term. Today you will use an <u>ophthalmoscope</u> / <u>otoscope</u> to examine the ear.
7. The _____ test is used for comparing bone and air-conduction hearing.
   a. balance   b. Rinne   c. Weber
8. The equilibrium apparatus of the ear, the vestibule, is found in the:
   a. external ear   b. internal ear   c. middle ear
9. Circle the correct underlined terms. The <u>crista ampullaris</u> / <u>macula</u> located in the <u>semicircular duct</u> / <u>vestibule</u> is essential for detecting static equilibrium.
10. Nystagmus is:
    a. he ability to hear only high-frequency tones
    b. the ability to hear only low-frequency tones
    c. involuntary trailing of eyes in one direction, then rapid movement in the other
    d. the sensation of dizziness

Anatomy of the Ear

Gross Anatomy

The ear is a complex structure containing sensory receptors for hearing and equilibrium. The ear is divided into three major areas: the *external ear,* the *middle ear,* and the *internal ear* (Figure 19.1). The external and middle ear structures serve the needs of the sense of hearing *only,* whereas internal ear structures function both in equilibrium and hearing.

ACTIVITY 1

Identifying Structures of the Ear

Obtain a dissectible ear model or chart of ear anatomy, and identify the structures summarized in Table 19.1.

Because the mucosal membranes of the middle ear cavity and nasopharynx are continuous through the pharyngotympanic tube, **otitis media,** or inflammation of the middle ear, is a fairly common condition, especially among children prone to sore throats. In cases where large amounts of fluid or pus accumulate in the middle ear cavity, an emergency myringotomy (lancing of the eardrum) may be necessary to relieve the pressure. Frequently, tiny ventilating tubes are put in during the procedure. ✛

The **internal ear** consists of a system of mazelike chambers and is therefore also referred to as the **labyrinth** (Figure 19.2). The internal ear (labyrinth) has two main divisions: the bony labyrinth and the membranous labyrinth. The **bony labyrinth** is a cavity in the temporal bone that includes the semicircular canals, the vestibule, and cochlea. The **membranous labyrinth** is a collection of ducts and sacs including the semicircular ducts, the utricle, the saccule, and the cochlear duct. The bony labyrinth is filled with an aqueous fluid called **perilymph.** The membranous labyrinth is filled with a more viscous fluid called **endolymph.** The ducts and sacs of the membranous labyrinth a re suspended in the perilymph. The snail-shaped **cochlea** contains the sensory receptors for hearing. The **vestibule** and **semicircular canals** are involved with equilibrium. ▬

The membranous **cochlear duct** (Figure 19.3) is a soft wormlike tube about 3.8 cm long. It winds through the full two and three-quarter turns of the cochlea and separates the perilymph-containing cochlear cavity into upper and lower chambers, the **scala vestibuli** and **scala tympani.** The scala vestibuli begins at the oval window, which "seats" the foot plate of the stapes located laterally in the tympanic cavity. The scala tympani is bounded by a membranous area called the **round window.** The round window can bulge into the tympanic cavity and acts as a pressure relief valve for the exit of pressure waves. The cochlear duct is the middle **scala media.** It is filled with endolymph and supports the **spiral organ,** which contains the receptors for hearing—the sensory hair cells and nerve endings of the **cochlear nerve,** a division of the vestibulocochlear nerve (VIII).

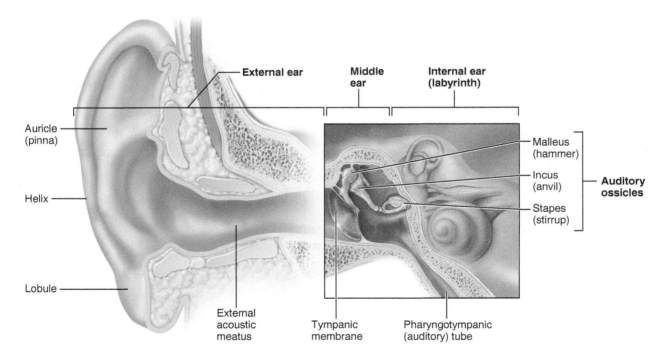

Figure 19.1 **Anatomy of the ear.**

Table 19.1   Structures of the External, Middle, and Internal Ear (Figure 19.1)

External Ear

Structure	Description	Function
Auricle (pinna)	Elastic cartilage covered with skin	Collects and directs sound waves into the external acoustic meatus
Lobule ("earlobe")	Portion of the auricle that is inferior to the external acoustic meatus	Completes the formation of the auricle
External acoustic meatus	Short, narrow canal carved into the temporal bone; lined with ceruminous glands	Transmits sound waves from the auricle to the tympanic membrane
Tympanic membrane (eardrum)	Thin membrane that separates the external ear from the middle ear	Vibrates at exactly the same frequency as the sound wave(s) hitting it and transmits vibrations to the auditory ossicles

Middle Ear

A small air-filled chamber—the tympanic cavity — Contains the auditory ossicles (malleus, incus, and stapes)

Structure	Description	Function
Malleus (hammer)	Tiny bone shaped like a hammer; its "handle" is attached to the eardrum	Transmits and amplifies vibrations from the tympanic membrane to the incus
Incus (anvil)	Tiny bone shaped like an anvil that articulates with the malleus and the stapes	Transmits and amplifies vibrations from the malleus to the stapes
Stapes (stirrup)	Tiny bone shaped like a stirrup; its "base" fits into the oval window	Transmits and amplifies vibrations from the incus to the oval window
Oval window	Oval-shaped membrane located deep to the stapes	Transmits vibrations from the stapes to the perilymph of the scala vestibuli
Pharyngotympanic (auditory) tube	A tube that connects the middle ear to the superior portion of the pharynx (throat)	Equalizes the pressure in the middle ear cavity with the external air pressure so that the tympanic membrane can vibrate properly

Internal Ear

Bony labyrinth	Membranous labyrinth (within the bony labyrinth)	Structure that contains the receptors	Function of the receptors
Cochlea	Cochlea duct	Spiral organ	Hearing
Vestibule	Utricle and saccule	Maculae	Equilibrium: static equilibrium and linear acceleration of the head
Semicircular canals	Semicircular ducts	Ampullae	Equilibrium: rotational acceleration of the head

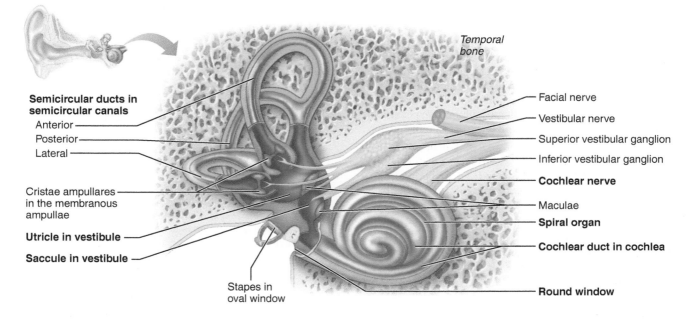

Figure 19.2 Internal ear. Right membranous labyrinth (blue) shown within the bony labyrinth (tan). The locations for sensory organs for hearing and equilibrium are shown in purple.

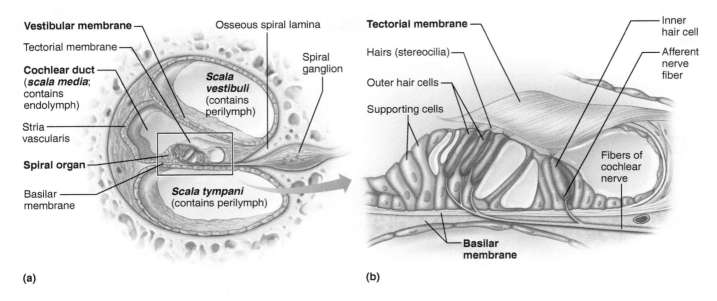

(a)

(b)

Figure 19.3 Anatomy of the cochlea. (a) Magnified cross-sectional view of one turn of the cochlea, showing the relationship of the three scalae. The scalae vestibuli and tympani contain perilymph; the cochlear duct (scala media) contains endolymph. **(b)** Detailed structure of the spiral organ.

Examining the Ear with an Otoscope (Optional)

1. Obtain an otoscope and two alcohol swabs. Inspect your partner's ear canal and then select the speculum with the largest diameter that will fit comfortably into his or her ear to permit full visibility. Clean the speculum thoroughly with an alcohol swab, and then attach the speculum to the battery-containing otoscope handle. Before beginning, check that the otoscope light beam is strong. If not, obtain another otoscope or new batteries. Some otoscopes come with disposable tips. Be sure to use a new tip for each ear examined. Dispose of these tips in an autoclave bag after use.

2. When you are ready to begin the examination, hold the lighted otoscope securely between your thumb and forefinger (like a pencil), and rest the little finger of the otoscope-holding hand against your partner's head. This maneuver forms a brace that allows the speculum to move as your partner moves and prevents the speculum from penetrating too deeply into the ear canal during unexpected movements.

3. Grasp the ear auricle firmly and pull it up, back, and slightly laterally. If your partner experiences pain or discomfort when the auricle is manipulated, an inflammation or infection of the external ear may be present. If this occurs, do not attempt to examine the ear canal.

4. Carefully insert the speculum of the otoscope into the external acoustic meatus in a downward and forward direction only far enough to permit examination of the tympanic membrane or eardrum. Note its shape, color, and vascular network. The healthy tympanic membrane is pearly white. During the examination, notice whether there is any discharge or redness in the canal, and identify earwax.

5. After the examination, thoroughly clean the speculum with the second alcohol swab before returning the otoscope to the supply area. ■■■

Microscopic Anatomy of the Spiral Organ and the Mechanism of Hearing

In the spiral organ, the auditory receptors are hair cells that rest on the **basilar membrane**, which forms the floor of the cochlear duct (Figure 19.3). Their "hairs" are stereocilia that project into a gelatinous membrane, the **tectorial membrane,** that overlies them. The roof of the cochlear duct is called the **vestibular membrane.** The endolymph-filled chamber of the cochlear duct is the **scala media.**

Examining the Microscopic Structure of the Cochlea

Obtain a compound microscope and a prepared microscope slide of the cochlea, and identify the areas shown in **Figure 19.4.** ■■■

The mechanism of hearing begins as sound waves pass through the external acoustic meatus and through the middle ear into the internal ear, where the vibration eventually reaches the spiral organ, which contains the receptors for hearing.

Vibration of the stapes at the oval window creates pressure waves in the perilymph of the scala vestibule, which are transferred to the endolymph in the cochlear duct (scala

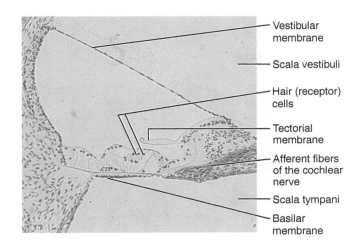

Figure 19.4 Histologic image of the spiral organ (100×).

media). As the waves travel through the cochlear duct, they displace the basilar membrane and bend the hairs (sterocilia) of the hair cells. Hair cells at any given location on the basilar membrane are stimulated by sounds of a specific frequency (pitch). High-frequency waves (high-pitched sounds) displace the basilar membrane near the base, where the fibers are short and stiff. Low-frequency waves (low-pitched sounds) displace the basilar membrane near the apex, where the fibers are long and flexible. Once the hair cells are stimulated, they depolarize and begin the chain of nervous impulses that travel along the cochlear nerve and eventually reach the auditory centers of the temporal lobe cortex. This series of events results in the phenomenon we call hearing (**Figure 19.5**).

Sensorineural deafness results from damage to neural structures anywhere from the cochlear hair cells through neurons of the auditory cortex. **Presbycusis** is a type of sensorineural deafness that occurs commonly in people by the time they are in their 60s. It results from a gradual deterioration and atrophy of the spiral organ, leading to a loss in the ability to hear high tones and speech sounds.

Although presbycusis is considered to be a disability of old age, early onset is becoming much more common because of headphone abuse. Prolonged or excessive noise tears the cilia from hair cells, and the damage is progressive and cumulative. +

A C T I V I T Y 4

Conducting Laboratory Tests of Hearing

Perform the following hearing tests in a quiet area.

Weber Test to Determine Conduction and Sensorineural Deafness

Strike a tuning fork on the heel of your hand or with a rubber mallet, and place the handle of the tuning fork medially on your partner's head (**Figure 19.6a**). Is the tone equally loud in both ears, or is it louder in one ear?

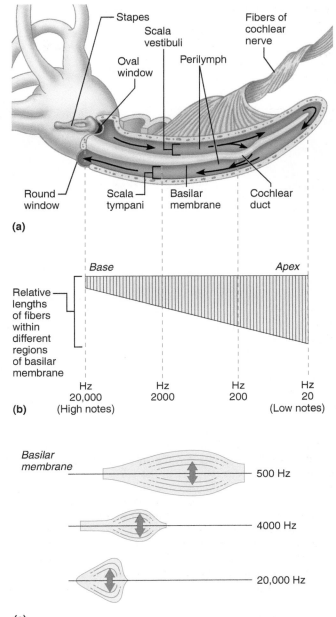

Figure 19.5 Resonance of the basilar membrane. The cochlea is depicted as if it has been uncoiled. **(a)** Perilymph movement in the cochlea following the stapes pushing on the oval window. The compressional wave thus created causes the round window to bulge into the middle ear. Pressure waves set up vibrations in the basilar membrane. **(b)** Fibers span the basilar membrane. The stiffness of the fibers "tunes" specific regions to vibrate at specific frequencies. **(c)** Different frequencies of pressure waves in the spiral organ stimulate particular hair cells located in the basilar membrane.

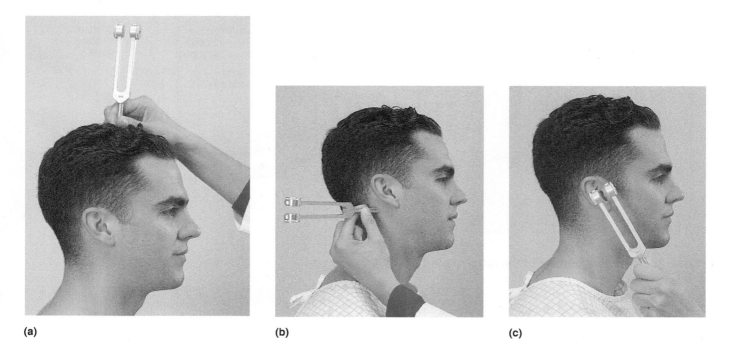

(a)   **(b)**   **(c)**

Figure 19.6 The Weber and Rinne tuning fork tests. (a) The Weber test to evaluate whether the sound remains centralized (normal) or lateralizes to one side or the other (indicative of some degree of conduction or sensorineural deafness). **(b, c)** The Rinne test to compare bone conduction and air conduction.

If it is equally loud in both ears, the subject has equal hearing or equal loss of hearing in both ears. If sensorineural deafness is present in one ear, the subject will hear the tone in the unaffected ear, but not in the ear with sensorineural deafness. *Conduction deafness* occurs when something prevents sound waves from reaching the fluids of the internal ear. Compacted earwax, a perforated eardrum, inflammation of the middle ear (otitis media), and damage to the ossicles are all causes of conduction deafness. If conduction deafness is present, the subject will hear the sound more strongly in the ear in which there is a hearing loss, because the bone of the skull is conducting the sound. Conduction deafness can be simulated by plugging one ear with cotton.

Rinne Test for Comparing Bone- and Air-Conduction Hearing

1. Strike the tuning fork, and place its handle on your partner's mastoid process (Figure 19.6b).

2. When your partner indicates that the sound is no longer audible, hold the still-vibrating prongs close to his or her external acoustic meatus (Figure 19.6c). If your partner hears the fork again (by air conduction) when it is moved to that position, hearing is not impaired, and you record the test result as positive (+). Record in step 5 below.

3. Repeat the test, but this time test air-conduction hearing first.

4. After the tone is no longer heard by air conduction, hold the handle of the tuning fork on the bony mastoid process. If the subject hears the tone again by bone conduction after hearing by air conduction is lost, there is some conduction deafness, and you record the result as negative (−).

5. Repeat the sequence for the opposite ear.

Right ear: _____ Left ear: _____

Does the subject hear better by bone or by air conduction?

_____

Acuity Test

Have your lab partner pack one ear with cotton and sit quietly with eyes closed. Obtain a ticking clock or pocket watch, and hold it very close to his or her *unpacked* ear. Then slowly move it away from the ear until your partner signals that the ticking is no longer audible. Record the distance in centimeters at which ticking is inaudible, and then remove the cotton from the packed ear.

Right ear: _____ Left ear: _____

Is the threshold of audibility sharp or indefinite?

_____

Sound Localization

Ask your partner to close both eyes. Hold the pocket watch at an audible distance (about 15 cm) from the ear, and move it to various locations (front, back, sides, and above the head). Have your partner locate the position by pointing in each instance. Can the sound be localized equally well at all

positions? _____

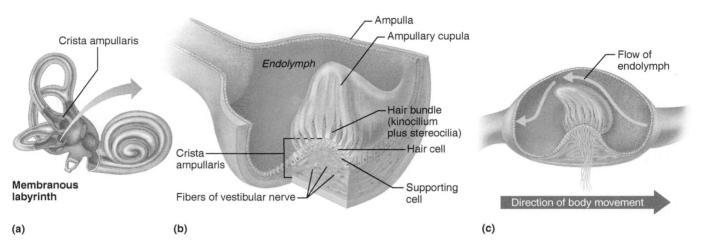

(a)   **(b)**   **(c)**

Figure 19.7 Structure and function of the crista ampullaris. (a) Arranged in the three spatial planes, the semicircular ducts in the semicircular canals each have a swelling called an ampulla at their base. **(b)** Each ampulla contains a crista ampullaris, a receptor that is essentially a cluster of hair cells with hairs projecting into a gelatinous cap called the ampullary cupula. **(c)** Movement of the ampullary cupula during rotational acceleration of the head.

If not, at what position(s) was the sound *less* easily located?

_____

_____

The ability to localize the source of a sound depends on two factors—the difference in the loudness of the sound reaching each ear and the time of arrival of the sound at each ear. How does this information help to explain your findings?

_____

_____

Frequency Range of Hearing

Obtain three tuning forks: one with a low frequency (75 to 100 hertz [Hz; 1 Hz = 1 cycle per second, or cps]), one with a frequency of approximately 1000 Hz, and one with a frequency of 4000 to 5000 Hz. Strike the lowest-frequency fork, and hold it close to your partner's ear. Repeat with the other two forks.

Which fork was heard most clearly and comfortably?

_____ Hz

Which was heard least well? _____ Hz ▬

Microscopic Anatomy of the Equilibrium Apparatus and Mechanisms of Equilibrium

The equilibrium receptors of the internal ear are collectively called the **vestibular apparatus** and are found in the vestibule

and semicircular canals of the bony labyrinth. The vestibule contains the saclike **utricle** and **saccule,** and the semicircular chambers contain membranous **semicircular ducts** (see Figure 19.2). Like the cochlear duct, these membranes are filled with endolymph and contain receptor cells that are activated by the bending of the hairs of their hair cells.

The semicircular canals monitor rotational acceleration of the head. This process is called **dynamic equilibrium.** The canals are 1.2 cm in circumference and are oriented in three planes—horizontal, frontal, and sagittal. At the base of each semicircular duct is an enlarged region, the **ampulla.** Within each ampulla is a receptor region called a **crista ampullaris,** which consists of a tuft of hair cells covered with a gelatinous cap, or **ampullary cupula (Figure 19.7).**

The cristae respond to changes in the velocity of rotational head movements. For example, consider what happens when you twirl around. During acceleration, when you begin to twirl around, inertia causes the endolymph in the canal to lag behind the head movement, pushing the ampullary cupula—like a swinging door—in the opposite direction. The head movement depolarizes the hair cells and results in enhanced impulse transmission in the vestibular division of the eighth cranial nerve to the brain (Figure 19.7c). If the body continues to rotate at a constant rate, the endolymph eventually comes to rest and moves at the same speed as the body. The ampullary cupula returns to its upright position, hair cells are no longer stimulated, and you lose the sensation of spinning. When rotational movement stops suddenly, the endolymph keeps on going in the direction of head movement. This pushes the ampullary cupula in the *same* direction as the previous head movement and hyperpolarizes the hair cells; as a result, fewer impulses are transmitted to the brain. This tells the brain that you have stopped moving and accounts for the reversed motion sensation you feel when you stop twirling suddenly.

Ampullary cupula

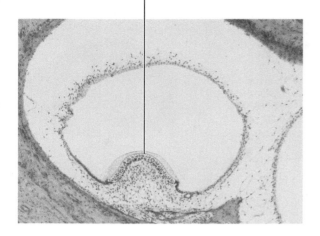

Figure 19.8 Micrograph of a crista ampullaris (**42×**).

ACTIVITY 5

Examining the Microscopic Structure of the Crista Ampullaris

Go to the demonstration area and examine the slide of a crista ampullaris. Identify the areas depicted in **Figure 19.8** and Figure 19.7b and c. ■

Maculae in the membranous utricle and saccule contain another set of **hair cells,** receptors that in this case monitor head position and acceleration in a straight line. This monitoring process is called **static equilibrium.** The maculae respond to gravitational pull, thus providing information on which way is up or down, and to linear or straightforward changes in speed. The hair cells in each macula have **sterocilia** plus one **kinocilium** that are embedded in the **otolith membrane,** a gelatinous material containing small grains of calcium carbonate called **otoliths.** When the head moves, the otoliths move in response to variations in gravitational pull. As they deflect different hair cells, they trigger hyperpolarization or depolarization of the hair cells and modify the rate of impulse transmission along the vestibular nerve (**Figure 19.9**).

Although the receptors of the semicircular canals and the vestibule are responsible for dynamic and static equilibrium respectively, they rarely act independently. Complex interaction of many of the receptors is the rule. Processing is also complex and involves the brain stem and cerebellum as well as input from proprioceptors and the eyes.

ACTIVITY 6

Conducting Laboratory Tests on Equilibrium

The function of the semicircular canals and vestibule is not routinely tested in the laboratory, but the following simple tests illustrate normal equilibrium apparatus function as well as some of the complex processing interactions.

Balance Tests

1. Have your partner walk a straight line, placing one foot directly in front of the other.

Is he or she able to walk without significant wobbling from

side to side? _____

Did he or she experience any dizziness? _____

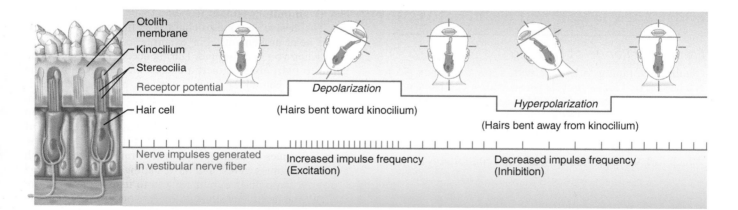

Otolith membrane
Kinocilium
Stereocilia
Receptor potential
Hair cell

Depolarization
(Hairs bent toward kinocilium)

Hyperpolarization
(Hairs bent away from kinocilium)

Nerve impulses generated in vestibular nerve fiber

Increased impulse frequency (Excitation)

Decreased impulse frequency (Inhibition)

Figure 19.9 The effect of gravitational pull on a macula receptor in the utricle. When movement of the otolith membrane bends the hair cells in the direction of the kinocilium, the hair cells depolarize, exciting the nerve fibers, which generate action potentials more rapidly. When the hairs are bent in the direction away from the kinocilium, the hair cells become hyperpolarized, inhibiting the nerve fibers and decreasing the action potential rate (i.e., below the resting rate of discharge).

The ability to walk with balance and without dizziness, unless subject to rotational forces, indicates normal function of the equilibrium apparatus.

Nystagmus is the involuntary rolling of the eyes in any direction or the trailing of the eyes slowly in one direction, followed by their rapid movement in the opposite direction. It is normal after rotation, abnormal otherwise. The direction of nystagmus is that of its quick phase on acceleration. +

Was nystagmus present? _____

2.   Place three coins of different sizes on the floor. Ask your lab partner to pick up the coins, and carefully observe his or her muscle activity and coordination.

Did your lab partner have any difficulty locating and picking

up the coins? _____

Describe your observations and your lab partner's observations during the test.

_____

_____

_____

_____

What kinds of complex interactions involving balance and coordination must occur for a person to move fluidly during this test?

_____

_____

3.   If a person has a depressed nervous system, mental concentration may result in a loss of balance. Ask your lab partner to stand up and count backward from 10 as rapidly as possible.

Did your lab partner lose balance? _____

Romberg Test

The Romberg test determines the integrity of the dorsal white column of the spinal cord, which transmits impulses to the brain from the proprioceptors involved with posture.

1.   Have your partner stand with his or her back to the blackboard or whiteboard.

2.   Draw one line parallel to each side of your partner's body. He or she should stand erect, with eyes open and staring straight ahead for 2 minutes while you observe any movements. Did you see any gross swaying movements?

_____

3.   Repeat the test. This time the subject's eyes should be closed. Note and record the degree of side-to-side movement.

_____

4.   Repeat the test with the subject's eyes first open and then closed. This time, however, the subject should be positioned with the left shoulder toward, but not touching, the board so that you may observe and record the degree of front-to-back swaying.

_____

_____

_____

Do you think the equilibrium apparatus of the internal ear was operating equally well in all these tests?

_____

_____

The proprioceptors? _____

Why was the observed degree of swaying greater when the eyes were closed?

_____

_____

What conclusions can you draw regarding the factors necessary for maintaining body equilibrium and balance?

_____

_____

_____

Role of Vision in Maintaining Equilibrium

To further demonstrate the role of vision in maintaining equilibrium, perform the following experiment. (Ask your lab partner to record observations and act as a "spotter.") Stand erect, with your eyes open. Raise your left foot approximately 30 cm off the floor, and hold it there for 1 minute.

1.   Record the observations: _____

_____

_____

2.   Rest for 1 or 2 minutes, and then repeat the experiment with the same foot raised but with your eyes closed. Record the observations:

_____

_____

Name _____

Lab Time/Date _____

Special Senses: Hearing and Equilibrium

Anatomy of the Ear

1. Identify all indicated structures and ear regions in the following photograph.

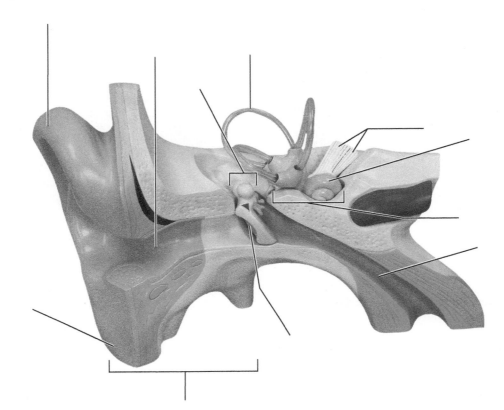

2. Identify the structures of the middle and internal ear indicated in the following photograph.

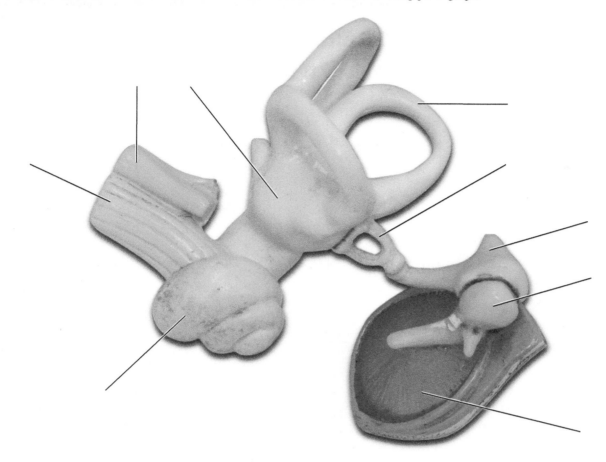

3. Select the terms from column B that match the column A descriptions.

Column A

Column B

_____, _____, _____   1. collectively called the auditory ossicles

_____   2. sacs found within the vestibule; sites of the maculae

_____   3. vibrates at the same frequency as the sound waves hitting it; transmits the vibrations to the ossicles

_____   4. fluid contained within the membranous labyrinth

_____   5. fluid contained within the bony labyrinth

_____   6. grains of calcium carbonate in the maculae

_____   7. location of the spiral organ

_____   8. involved in equalizing the pressure in the middle ear with the external air pressure

_____   9. gelatinous cap overlying hair cells of the crista ampullaris

a. ampulla

b. ampullary cupula

c. basilar membrane

d. cochlear duct

e. endolymph

f. incus (anvil)

g. malleus (hammer)

h. otoliths

i. oval window

j. perilymph

k. pharyngotympanic tube

l. saccule and utricle

m. stapes

n. tympanic membrane

4. Sound waves hitting the tympanic membrane initiate its vibratory motion. Trace the pathway through which vibrations and fluid currents are transmitted to finally stimulate the hair cells in the spiral organ. (Name the appropriate ear structures in their correct sequence.)

Tympanic membrane → _____

_____

_____

5. Explain how the basilar membrane allows us to differentiate sounds of different pitch. _____

_____

_____

_____

6. Explain the role of the endolymph of the semicircular ducts in activating the receptors during angular motion.

_____

_____

_____

7. Explain the role of the otoliths in perception of static equilibrium (head position). _____

_____

_____

_____

Laboratory Tests

8. Was the auditory acuity measurement made in Activity 4 (on page 332) the same or different for both ears?

_____ What factors might account for a difference in the acuity of the two ears?

_____

_____

9. During the sound localization experiment in Activity 4 (on page 332), in which position(s) was the sound least easily located?

_____

How can this phenomenon be explained? _____

_____

10. In the frequency experiment in Activity 4 (on page 332), which tuning fork was the most difficult to hear? _____ Hz

What conclusion can you draw? _____

_____

11. When the tuning fork handle was pressed to the forehead during the Weber test, where did the sound seem to originate?

_____

Where did it seem to originate when one ear was plugged with cotton? _____

12. Indicate whether the following conditions relate to conduction deafness (C), or sensorineural deafness (S), or both (C and S).

_____ 1. can result from the fusion of the ossicles

_____ 2. can result from a lesion on the cochlear nerve

_____ 3. sound heard in one ear but not in the other during bone and air conduction

_____ 4. can result from otitis media

_____ 5. can result from impacted cerumen or a perforated eardrum

_____ 6. can result from a blood clot in the primary auditory cortex

13. The Rinne test evaluates an individual's ability to hear sounds conducted by air or bone. Which is more indicative of normal

hearing? _____

14. What is the usual reason for conducting the Romberg test? _____

_____

Was the degree of sway greater with the eyes open or closed? Why? _____

_____

15. Normal balance, or equilibrium, depends on input from a number of sensory receptors. Name them.

_____

_____

16. ✚ Acute labyrinthitis is sudden onset of inflammation of the structures that form the membranous labyrinth. List the

structures that could be inflamed with this condition. _____

_____

17. ✚ Acute labyrinthitis causes temporary impairment of the receptors located in the membranous labyrinth. Describe the

symptoms that might accompany this condition. _____

_____

Special Senses: Olfaction and Taste

MATERIALS

- Prepared slides: the tongue showing taste buds; nasal olfactory epithelium (l.s.)
- Compound microscope
- Small mirror
- Paper towels
- Packets of granulated sugar
- Disposable autoclave bag
- Cotton-tipped swabs
- Prepared vials of oil of cloves, oil of peppermint, and oil of wintergreen, or corresponding flavorings found in the condiment section of a supermarket
- Nose clips
- Paper cups
- Flask of distilled or tap water
- Absorbent cotton
- Paper plates
- Foil-lined egg carton containing equal-sized cubes of foods, such as cheese, apple, raw potato, dried prunes, banana, raw carrot, and hard-cooked egg white
- Toothpicks
- Disposable gloves
- Chipped ice

LEARNING OUTCOMES

☐ State the location and cellular composition of the olfactory epithelium.
☐ Describe the structure of olfactory sensory neurons, and state their function.
☐ Discuss the locations and cellular composition of taste buds.
☐ Describe the structure of gustatory epithelial cells, and state their function.
☐ Identify the cranial nerves that carry the sensations of olfaction and taste.
☐ Name five basic qualities of taste sensation.
☐ Explain the interdependence between the senses of smell and taste.
☐ Name two factors other than olfaction that influence taste appreciation of foods.
☐ Define *olfactory adaptation*.

PRE-LAB QUIZ

1. Circle True or False. Receptors for olfaction and taste are classified as chemoreceptors because they respond to dissolved chemicals.
2. The organ of smell is the _____, located in the roof of the nasal cavity.
   a. nares
   b. nostrils
   c. olfactory epithelium
   d. olfactory nerve
3. Circle the correct underlined term. Olfactory receptors are <u>bipolar</u> / <u>unipolar</u> sensory neurons whose olfactory cilia extend outward from the epithelium.
4. Most taste buds are located in _____, peglike projections of the tongue mucosa.
   a. cilia
   b. conchae
   c. papillae
   d. supporting cells
5. Circle the correct underlined term. Vallate papillae are arranged in a V formation on the <u>anterior</u> / <u>posterior</u> surface of the tongue.
6. Circle the correct underlined term. Most taste buds are made of <u>two</u> / <u>three</u> types of modified epithelial cells.
7. There are five basic taste sensations. Name one. _____
8. Circle True or False. Taste buds typically respond optimally to one of the five basic taste sensations.
9. Circle True or False. Texture, temperature, and smell have little or no effect on the sensation of taste.
10. You will use absorbent cotton and oil of wintergreen, peppermint, or cloves to test for olfactory:
    a. accommodation
    b. adaptation
    c. identification
    d. recognition

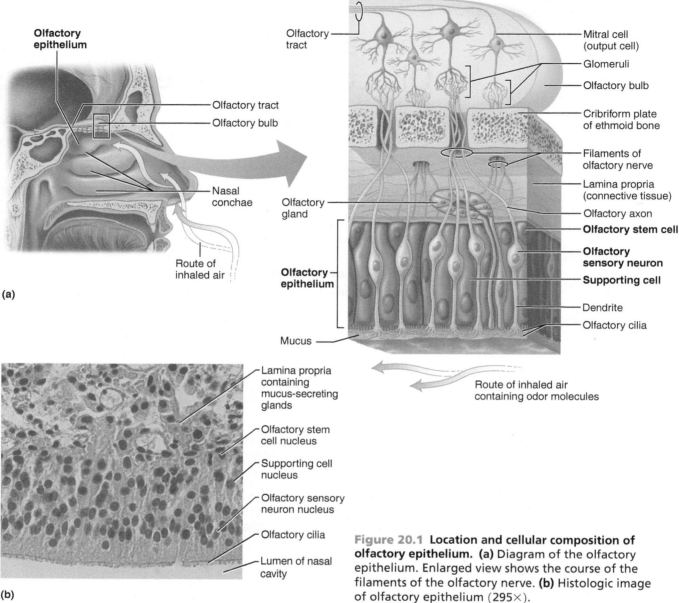

Olfactory epithelium

Olfactory tract
Olfactory bulb

Nasal conchae

Route of inhaled air

(a)

Olfactory tract

Mitral cell (output cell)

Glomeruli

Olfactory bulb

Cribriform plate of ethmoid bone

Filaments of olfactory nerve

Lamina propria (connective tissue)

Olfactory axon

Olfactory stem cell

Olfactory sensory neuron

Supporting cell

Dendrite

Olfactory cilia

Olfactory gland

Olfactory epithelium

Mucus

Route of inhaled air containing odor molecules

Lamina propria containing mucus-secreting glands

Olfactory stem cell nucleus

Supporting cell nucleus

Olfactory sensory neuron nucleus

Olfactory cilia

Lumen of nasal cavity

(b)

Figure 20.1 Location and cellular composition of olfactory epithelium. (a) Diagram of the olfactory epithelium. Enlarged view shows the course of the filaments of the olfactory nerve. **(b)** Histologic image of olfactory epithelium (295×).

The receptors for olfaction and taste are classified as **chemoreceptors** because they respond to chemicals in solution. Although five relatively specific types of taste receptors have been identified, the olfactory receptors are considered sensitive to a much wider range of chemical sensations. The sense of smell is the least understood of the special senses.

Olfactory Epithelium and Olfaction

A pseudostratified epithelium called the **olfactory epithelium** is the organ of smell. It occupies an area lining the roof of the nasal cavity **(Figure 20.1a)**. Because most of the air

entering the nasal cavity enters the respiratory passages below, the superiorly located nasal epithelium is in a rather poor position for performing its function. This is why sniffing, which brings more air into contact with the receptors, increases your ability to detect odors.

Three unique cell types are found within the olfactory epithelium:

- **Olfactory sensory neurons:** Specialized receptor cells that are bipolar neurons with nonmotile olfactory cilia.

- **Supporting cells:** Columnar cells that surround and support the olfactory sensory neurons. They form the bulk of the epithelium.

- **Olfactory stem cells:** Located near the basal surface of the epithelium, they divide to form new olfactory sensory neurons.

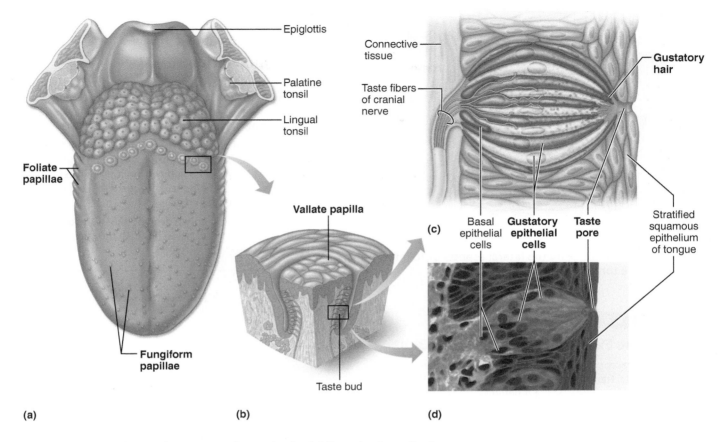

Figure 20.2 Location and structure of taste buds. (a) Taste buds on the tongue are associated with papillae, projections of the tongue mucosa. **(b)** An enlarged section of a vallate papilla shows the position of the taste buds in its lateral walls. **(c)** An enlarged view of a taste bud. **(d)** Photomicrograph of a taste bud (510×). (See also Figure 20.3.)

The axons of the olfactory sensory neurons form small fascicles called the *filaments of the olfactory nerve* (cranial nerve I), which penetrate the cribriform foramina and synapse in the olfactory bulbs.

ACTIVITY 1

Microscopic Examination of the Olfactory Epithelium

Obtain a longitudinal section of olfactory epithelium. Examine it closely using a compound microscope, comparing it to Figure 20.1b. ■■

Taste Buds and Taste

The **taste buds,** containing specific receptors for the sense of taste, are widely but not uniformly distributed in the oral cavity. Most are located in **papillae,** peglike projections of the mucosa, on the dorsal surface of the tongue. A few are found on the soft palate, epiglottis, pharynx, and inner surface of the cheeks.

The taste buds are located primarily on the sides of the large, **vallate papillae** (arranged in a V formation on the posterior surface of the tongue); in the side walls of the **foliate papillae;** and on the tops of the more numerous **fungiform papillae.** The latter look rather like minute mushrooms and are widely distributed on the tongue **(Figure 20.2)**.

Each taste bud consists largely of an arrangement of two types of modified epithelial cells:

• **Gustatory epithelial cells:** The receptors for taste; they have long microvilli called **gustatory hairs** that project through the epithelial surface through a **taste pore**.

• **Basal epithelial cells:** Precursor cells that divide to replace the gustatory epithelial cells.

When the gustatory hairs contact food molecules dissolved in saliva, the gustatory epithelial cells depolarize. The sensory (afferent) neurons that innervate the taste buds are located in three cranial nerves: the *facial nerve (VII)* serves the anterior two-thirds of the tongue; the *glossopharyngeal nerve (IX)* serves the posterior third of the tongue; and the *vagus nerve (X)* carries a few fibers from the pharyngeal region.

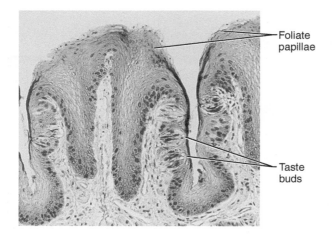

Figure 20.3 Taste buds on the lateral aspects of foliate papillae of the tongue (180×).

Microscopic Examination of the Taste Buds

Obtain a microscope and a prepared slide of a tongue cross section. Locate the taste buds on the tongue papillae (use Figure 20.2b as a guide). Make a detailed study of one taste bud. Identify the taste pore and gustatory hairs if observed. Compare your observations to **Figure 20.3**. ■■

When taste is tested with pure chemical compounds, most taste sensations can be grouped into one of five basic qualities—sweet, sour, bitter, salty, or umami (oo-mom-ē, "delicious"). Although all taste buds are believed to respond in some degree to all five classes of chemical stimuli, each type responds optimally to only one.

Laboratory Experiments

⚠️ *Notify instructor of any food or scent allergies or restrictions before beginning experiments.*

Stimulating Taste Buds

1. Obtain several paper towels, a sugar packet, and a disposable autoclave bag, and bring them to your bench.
2. With a paper towel, dry the dorsal surface of your tongue.

⚠️ Immediately dispose of the paper towel in the autoclave bag.

3. Tear off a corner of the sugar packet, and shake a few sugar crystals on your dried tongue. Do *not* close your mouth.

How long does it take to taste the sugar? _____ sec

Why couldn't you taste the sugar immediately?

_____ ■

Examining the Effect of Olfactory Stimulation

There is no question that what is commonly referred to as taste depends heavily on stimulation of the olfactory receptors, particularly in the case of strongly odoriferous substances. The following experiment should illustrate this fact.

1. Obtain vials of oil of wintergreen, peppermint, and cloves, a paper cup, a flask of water, paper towels, and some fresh cotton-tipped swabs. Ask the subject to sit so that he or she cannot see which vial is being used. Then ask the subject to dry the tongue and shut the nostrils.
2. Use a cotton swab to apply a drop of one of the oils to the subject's tongue. Can he or she distinguish the flavor?

_____

⚠️ Put the used swab in the autoclave bag. *Do not redip the swab into the oil.*

3. Have the subject open the nostrils, and record the change in sensation he or she reports.

_____

4. Have the subject rinse the mouth well and dry the tongue.
5. Prepare two swabs, each with one of the two remaining oils.
6. Hold one swab under the subject's open nostrils while touching the second swab to the tongue.

Record the reported sensations. _____

_____

_____

_____

7. ⚠️ Dispose of the used swabs and paper towels in the autoclave bag before continuing.

Which sense, taste or smell, appears to be more important in proper identification of a strongly flavored volatile substance?

_____ ■

Demonstrating Olfactory Adaptation

When olfactory receptors are subjected to the same odor for a prolonged period, they eventually stop responding to that stimulus, a phenomenon called **olfactory adaptation.** This activity allows you to demonstrate olfactory adaptation.

Obtain some absorbent cotton and two of the following oils: oil of wintergreen, peppermint, or cloves. Place several

Activity 6: Identification by Texture and Smell				
Food tested	Texture only	Chewing with nostrils pinched	Chewing with nostrils open	Identification not made

drops of one oil on the absorbent cotton. Press one nostril shut. Hold the cotton under the open nostril, and exhale through the mouth. Record the time required for the odor to disappear (for olfactory adaptation to occur).

_____ sec

Repeat the procedure with the other nostril.

_____ sec

Immediately test another oil with the nostril that has just experienced olfactory adaptation. What are the results?

_____

What conclusions can you draw? _____

_____ ■

A C T I V I T Y 6

Examining the Combined Effects of Smell, Texture, and Temperature on Taste
Effects of Smell and Texture

1. Ask the subject to sit with eyes closed and to pinch the nostrils shut.

2. Using a paper plate, obtain samples of the food items provided by your laboratory instructor. Do not let the subject see the foods being tested. Wear plastic gloves, and use toothpicks to handle food.

3. For each test, place a cube of food in the subject's mouth, and ask him or her to identify the food by using the following sequence of activities:

• First, manipulate the food with the tongue.

• Second, chew the food.

• Third, if the subject cannot make a positive identification with the first two techniques and the taste sense, ask the

subject to release the pinched nostrils and to continue chewing with the nostrils open. This may help the subject make a positive identification.

In the **Activity 6 chart,** record the type of food, and then put a check mark in the appropriate column for the result.

Was the sense of smell equally important in all cases?

_____

When did it seem to be important, and why?

_____

_____

Discard gloves in the autoclave bag.

Effect of Temperature

Olfaction and food texture are not the only influences on our taste sensations. The temperature of foods also helps determine whether we appreciate or even taste the foods we eat. To illustrate this, have your partner hold some chipped ice on the tongue for approximately a minute and then close his or her eyes. Immediately place in the person's mouth any of the foods previously identified, and ask for an identification.

What are the results?

_____

_____

_____

_____ ■

Special Senses:
Olfaction and Taste

Olfactory Epithelium and Olfaction

1. Match the terms in column B with the appropriate description in column A.

Column A

_____ 1. the organ of smell

_____ 2. cell type that forms most of the olfactory epithelium

_____ 3. cell type that differentiates to replace olfactory sensory neurons

_____ 4. bipolar neurons with nonmotile cilia

_____ 5. axons that pass through the cribriform foramina

Column B

a. filaments of the olfactory nerve
b. olfactory epithelium
c. olfactory sensory neurons
d. olfactory stem cells
e. supporting cells

2. How and why does sniffing increase your ability to detect an odor? _____

_____

Taste Buds and Taste

3. Name five sites where receptors for taste are found, and circle the predominant site.

_____, _____, _____,

_____, and _____

4. Describe the cellular makeup and arrangement of a taste bud. _____

_____

5. Taste and smell receptors are both classified as _____, because they both

respond to _____.

6. Why is it impossible to taste substances if your tongue is dry? _____

_____

7. Name the five basic taste sensations.

1. _____   4. _____

2. _____   5. _____

3. _____

Laboratory Experiments

8. Name three factors that influence our enjoyment of foods. Substantiate each choice with an example from the lab experience.

1. _____ Substantiation _____

_____

2. _____ Substantiation _____

_____

3. _____ Substantiation _____

_____

Expand on your choices by explaining why a cold, greasy hamburger is unappetizing to most people. _____

_____

9. ➕ One symptom of the common cold is loss of appetite. Explain why this occurs.

_____

_____

10. ➕ Invasive dental procedures can permanently or temporarily alter gustation. If taste sensations at the tip of the tongue are

absent, which cranial nerve is most likely to be affected? _____

_____

Functional Anatomy of the Endocrine Glands

MATERIALS

- Human torso model
- Anatomical chart of the human endocrine system
- Compound microscope
- Prepared slides of the anterior pituitary and pancreas (with differential staining), posterior pituitary, thyroid gland, parathyroid glands, and adrenal gland
- Dissection animal, tray, and instruments
- Bone cutters
- Embalming fluid
- Disposable gloves
- Safety glasses
- Organic debris container

LEARNING OUTCOMES

- ☐ Identify the major endocrine glands and tissues of the body using an appropriate image.
- ☐ List the major hormones and discuss the target and function of each.
- ☐ Explain how hormones contribute to body homeostasis.
- ☐ Discuss some mechanisms that stimulate release of hormones.
- ☐ Describe the structural and functional relationship between the hypothalamus and the pituitary gland.
- ☐ Correctly identify the histology of the thyroid, parathyroid, pancreas, anterior and posterior pituitary, adrenal cortex, and adrenal medulla by microscopic inspection or in an image.
- ☐ Identify the hormone-secreting cells in the tissues listed above.
- ☐ For several of the hormones studied, describe the pathology of hypersecretion and of hyposecretion.
- ☐ Identify and name the major endocrine organs on a dissected cat.

PRE-LAB QUIZ

1. Define *hormone*. _____
2. Circle the correct underlined term. An <u>endocrine</u> / <u>exocrine</u> gland is a ductless gland that empties its hormone into the extracellular fluid.
3. The pituitary gland, also known as the _____, is located in the sella turcica of the sphenoid bone.
   a. hypophysis   b. hypothalamus   c. thalamus
4. Circle True or False. The anterior pituitary gland is sometimes referred to as the master endocrine gland because it controls the activity of many other endocrine glands.
5. The _____ gland, composed of two lobes, is located in the throat, just inferior to the larynx.
   a. pancreas   c. thymus
   b. posterior pituitary   d. thyroid
6. The pancreas produces two hormones that are responsible for regulating blood sugar levels. Name the hormone that increases blood glucose levels.

   _____
7. Circle True or False. The gonads are considered to be both endocrine and exocrine glands.
8. This gland is rather large in an infant, but is inconspicuous by old age. It produces hormones that direct T cell maturation. It is the _____ gland.
   a. pineal   b. testes   c. thymus   d. thyroid
9. Circle the correct underlined term. <u>Pancreatic islets</u> / <u>Acinar cells</u> form the endocrine portion of the pancreas.
10. The outer cortex of the adrenal gland is divided into three areas or regions. Which zona produces aldosterone?
    a. fasciculata   b. glomerulosa   c. reticularis

The **endocrine system** is the second major control system of the body. Acting with the nervous system, it helps coordinate and integrate the activity of the body. The nervous system uses electrochemical impulses to bring about rapid control, whereas the more slowly acting endocrine system uses chemical messengers, or **hormones.**

The term *hormone* comes from a Greek word meaning "to arouse." The body's hormones, which are steroids or amino acid–based molecules, arouse the body's tissues and cells by stimulating changes in their metabolic activity. These changes lead to growth and development and to the physiological homeostasis of many body systems. Although hormones travel through the blood, a given hormone affects only a specific organ or organs. Cells within an organ that respond to a particular hormone are referred to as the **target cells** (also **target**) of that hormone. The ability of the target to respond depends on the ability of the hormone to bind with specific cellular receptors.

ACTIVITY 1

Identifying the Endocrine Organs

Locate the endocrine organs in **Figure 21.1.** Also locate these organs on the anatomical charts or torso model. As you locate the organs, read through Tables 21.1–21.4. ■

Endocrine Glands

Pituitary Gland (Hypophysis)

The **pituitary gland,** or **hypophysis,** is located in the sella turcica of the sphenoid bone. It consists of two lobes, the **anterior lobe,** or **adenohypophysis,** and the **posterior lobe,** or **neurohypophysis** and **infundibulum**—the stalk that attaches the pituitary gland to the hypothalamus **(Figure 21.2).**

The anterior lobe of the pituitary produces and secretes a number of hormones, four of which are **tropic hormones.** The target organ of a tropic hormone (tropin) is another endocrine gland.

Because the anterior pituitary controls the activity of many other endocrine glands, it has been called the *master endocrine gland.* However, it is now recognized that *releasing* or *inhibiting* hormones from neurons of the ventral hypothalamus control anterior pituitary cells; thus, the hypothalamus supersedes the anterior pituitary as the major controller of endocrine glands.

The ventral hypothalamic hormones control production and secretion of anterior pituitary hormones. The hypothalamic hormones reach the cells of the anterior pituitary through the **hypophyseal portal system** (Figure 21.2), a complex vascular arrangement of two capillary beds that are connected by the hypophyseal portal veins.

The posterior lobe is not an endocrine gland, because it does not synthesize the hormones it releases. Instead, it acts as a storage area for two *neurohormones* transported to it via the axons of neurons in the paraventricular and supraoptic nuclei of the hypothalamus. **Table 21.1** summarizes the hormones released by the pituitary gland.

Thyroid Gland

The *thyroid gland* is composed of two lobes joined by a central mass, or isthmus. It is located in the anterior neck, just inferior to the larynx.

Parathyroid Glands

The *parathyroid glands* are embedded in the posterior surface of the thyroid gland. Typically, there are two small oval glands on each lobe, but there may be more, and some may be located in other regions of the neck. **Table 21.2** summarizes the hormones secreted by the thyroid and parathyroid glands.

Adrenal Glands

The two *adrenal,* or *suprarenal, glands* are located atop the kidneys. Anatomically, the **adrenal medulla** develops from neural crest tissue, and it is directly controlled by the sympathetic nervous system. The medullary cells respond to this stimulation by releasing a hormone mix of **epinephrine** (80%) and **norepinephrine** (20%), which act in conjunction with the sympathetic nervous system to elicit the fight-or-flight response to stressors. **Table 21.3** summarizes the hormones secreted by the adrenal glands.

Pancreas

The *pancreas,* located posterior to the stomach and close to the small intestine, functions as both an endocrine and exocrine gland. It produces digestive enzymes as well as insulin and glucagon, important hormones concerned with regulating blood sugar levels. **Table 21.4** summarizes two of the hormones produced by the pancreas.

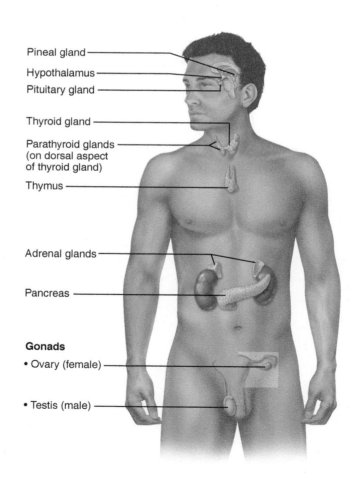

Pineal gland
Hypothalamus
Pituitary gland
Thyroid gland
Parathyroid glands
(on dorsal aspect
of thyroid gland)
Thymus
Adrenal glands
Pancreas
Gonads
• Ovary (female)
• Testis (male)

Figure 21.1 Human endocrine organs.

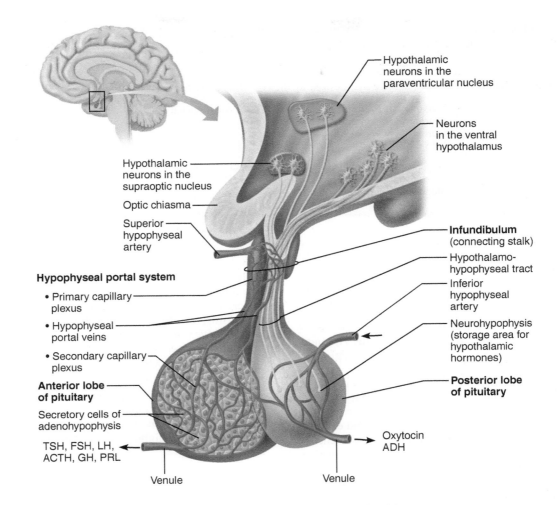

Figure 21.2 Hypothalamus and pituitary gland. Neural and vascular relationships between the hypothalamus and the anterior and posterior lobes of the pituitary are depicted.

Table 21.1A	Pituitary Gland Hormones (Figure 21.2)		
Hormone	**Stimulus for release**	**Target**	**Effects**
Anterior Pituitary Gland: Tropic Hormones			
Thyroid-stimulating hormone (TSH)	Thyrotropin-releasing hormone (TRH)*	Thyroid gland	Stimulates the secretion of thyroid hormones (T_3 and T_4)
Follicle-stimulating hormone (FSH)	Gonadotropin-releasing hormone (GnRH)*	Ovaries and testes (gonads)	**Females**—stimulates ovarian follicle maturation and estrogen production **Males**—stimulates sperm production
Luteinizing hormone (LH)	Gonadotropin–releasing hormone (GnRH)*	Ovaries and testes (gonads)	**Females**—triggers ovulation and stimulates ovarian production of estrogen and progesterone **Males**—stimulates testosterone production
Adrenocorticotropic hormone (ACTH)	Corticotropin-releasing hormone (CRH)*	Adrenal cortex	Stimulates the release of glucocorticoids and androgens (mineralocorticoids to a lesser extent)
Anterior Pituitary Gland: Other Hormones (Not Tropic)			
Growth hormone (GH)	Growth hormone–releasing hormone (GHRH)*	Liver, muscle, bone, and cartilage, mostly	Stimulates body growth and protein synthesis, mobilizes fat and conserves glucose
Prolactin (PRL)	A decrease in the amount of prolactin-inhibiting hormone (PIH)*	Mammary glands in the breasts	Stimulates milk production (lactation)

* Indicates hormones produced by the hypothalamus.

Table 21.1B Pituitary Gland Hormones (Figure 21.2)

Hormone	Stimulus for release	Target	Effects
Posterior Pituitary Gland (Hormones That Are Synthesized by the Hypothalamus and Stored in the Posterior Pituitary)			
Oxytocin*	Nerve impulses from hypothalamic neurons in response to cervical/uterine stretch or suckling of an infant	Uterus and mammary glands	Stimulates powerful uterine contractions during birth and stimulates milk ejection (let-down) in lactating mothers
Antidiuretic hormone (ADH)*	Nerve impulses from hypothalamic neurons in response to increased blood solute concentration or decreased blood volume	Kidneys	Stimulates the kidneys to reabsorb more water, reducing urine output and conserving body water

* Indicates hormones produced by the hypothalamus.

The Gonads

The *female gonads,* or *ovaries,* are paired, almond-sized organs located in the pelvic cavity. In addition to producing the female sex cells (ova), the ovaries produce two steroid hormone groups, the **estrogens** and **progesterone.** The endocrine and exocrine functions of the ovaries do not begin until the onset of puberty.

The paired oval *testes* of the male are suspended in a pouchlike sac, the scrotum, outside the pelvic cavity. In addition to the male sex cells (sperm), the testes produce the male sex hormone, **testosterone.** Both the endocrine and exocrine functions of the testes begin at puberty under the influence of the anterior pituitary gonadotropins. Table 21.4 summarizes the hormones produced by the gonads.

Two glands not mentioned earlier as major endocrine glands should also be briefly considered here, the thymus and the pineal gland.

Thymus

The *thymus* is a bilobed gland situated in the superior thorax, posterior to the sternum and overlying the heart. Conspicuous in the infant, it begins to atrophy at puberty, and by old age it is relatively inconspicuous. The thymus produces several different families of hormones, including **thymulin, thymosins,** and **thymopoietins.** These hormones are thought to be involved in the development of T lymphocytes and the immune response.

Pineal Gland

The *pineal gland* is a small, cone-shaped gland located in the roof of the third ventricle of the brain. Its major endocrine product is **melatonin.** Melatonin exhibits a diurnal (daily) cycle. It peaks at night, causing drowsiness, and is lowest around noon.

DISSECTION AND IDENTIFICATION
Selected Endocrine Organs of the Cat

If you have not previously opened the ventral body cavity, follow the directions provided in Activity 2. Otherwise, begin with Activity 3, "Identifying Organs." ■

Table 21.2 Thyroid and Parathyroid Gland Hormones

Hormone(s)	Stimulus for release	Target	Effects
Thyroid Gland			
Thyroxine (T_4) and Triiodothyronine (T_3), collectively referred to as thyroid hormone (TH)	Thyroid-stimulating hormone (TSH)	Most cells of the body	Increases basal metabolic rate (BMR); regulates tissue growth and development.
Calcitonin	High levels of calcium in the blood	Bones	No known physiological role in humans. When the hormone is supplemented at doses higher than normally found in humans, it does have some pharmaceutical applications.
Parathyroid Gland (Located on the Posterior Aspect of the Thyroid Gland)			
Parathyroid hormone (PTH)	Low levels of calcium in the blood	Bones and kidneys	Increases blood calcium by stimulating osteoclasts and by stimulating the kidneys to reabsorb more calcium. PTH also stimulates the kidneys to convert vitamin D to calcitriol, which is required for the absorption of calcium in the intestines.

Table 21.3   Adrenal Gland Hormones

Cortical area	Hormone(s)	Stimulus for release	Target	Effects
Adrenal Cortex				
Zona glomerulosa	Mineralcorticoids: mostly aldosterone	Angiotensin II release and increased potassium in the blood (ACTH only in times of severe stress)	Kidneys	Increases the reabsorption of sodium and water by the kidney tubules. Increases the secretion of potassium in the urine.
Zona fasciculata	Glucocorticoids: mostly cortisol	ACTH	Most body cells	Promotes the breakdown of fat and protein, promotes stress resistance, and inhibits the immune response.
Zona reticularis	Gonadocorticoids: androgens (most are converted to testosterone and some to estrogen)	ACTH	Bone, muscle, integument, and other tissues	In females, androgens contribute to body growth, contribute to the development of pubic and axillary hair, and enhance sex drive. They have insignificant effects in males.

Cells	Hormone(s)	Stimulus for release	Target	Effects
Adrenal Medulla				
Chromaffin cells	Catecholamines: epinephrine and norepinephrine	Nerve impulses from preganglionic sympathetic fibers	Most body cells	Mimics sympathetic nervous system activation, "fight-or-flight" response.

ACTIVITY 2

Opening the Ventral Body Cavity

1. Don gloves and safety glasses, and then obtain your dissection animal. Place the animal on the dissecting tray, ventral side up. Using scissors, make a longitudinal median incision through the ventral body wall. Begin your cut just superior to the midline of the pubis, and continue it anteriorly to the rib cage. Check the incision guide provided (Figure 21.3) as you work.

2. Angle the scissors slightly (1.3 cm or ½ in.) to the right or left of the sternum, and continue the cut through the rib cartilages, just lateral to the body midline, to the base of the throat.

3. Make two lateral cuts on both sides of the ventral body surface, anterior and posterior to the diaphragm, which

Table 21.4   Pancreas and Gonad Hormones

Hormone	Stimulus for release	Target(s)	Effects
Pancreas			
Insulin	Increased blood glucose levels, parasympathetic nervous system stimulation	Most cells of the body	Accelerates the transport of glucose into body cells; promotes glycogen, fat, and protein synthesis
Glucagon	Decreased blood glucose levels, sympathetic nervous system stimulation	Primarily the liver and adipose	Accelerates the breakdown of glycogen to glucose, stimulates the conversion of lactic acid into glucose, releases glucose into the blood from the liver
Ovaries (Female Gonads)			
Estrogens	Luteinizing hormone (LH) and follicle-stimulating hormone (FSH)	Most cells of the body	Promote the maturation of the female reproductive organs and the development of secondary sex characteristics
Estrogens and progesterone (together)	LH and FSH	Uterus and mammary glands	Regulate the menstrual cycle and promote breast development
Testes (Male Gonads)			
Testosterone	LH and FSH	Most cells of the body	Promotes the maturation of the male reproductive organs, the development of secondary sex characteristics, sperm production, and sex drive

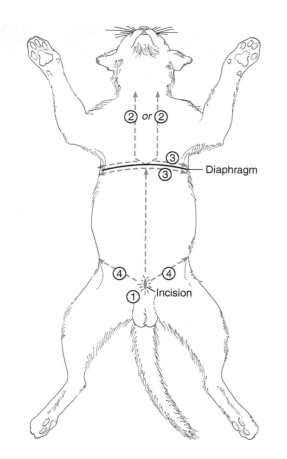

Figure 21.3 Endocrine organs of the cat. Incisions to be made in opening the ventral body cavity of a cat are shown. Numbers indicate sequence.

separates the thoracic and abdominal parts of the ventral body cavity. *Leave the diaphragm intact.* Spread the thoracic walls laterally to expose the thoracic organs.

4.   Make an angled lateral cut on each side of the median incision line just superior to the pubis, and spread the flaps to expose the abdominal cavity organs. ▆

ACTIVITY 3

Identifying Organs

A helpful adjunct to identifying selected endocrine organs of the cat is a general overview of ventral body cavity organs **(Figure 21.4)**. The objective here is simply to identify the most important organs and those that will help you to locate the desired endocrine organs (marked with a * in the following lists). Use **Figure 21.5** as a guide.

Neck and Thoracic Cavity Organs

Trachea: The windpipe; runs down the midline of the throat and then divides just anterior to the lungs to form the bronchi, which plunge into the lungs on either side.

*Thyroid: Its dark lobes straddle the trachea (Figure 21.5). Thyroid hormones are the main hormones regulating the body's metabolic rate. In general, the metabolic rate of a species of animals is inversely proportional to its size.

*Thymus: Glandular structure superior to and partly covering the heart (Figure 21.4). If you have a young cat, the thymus will be quite large. In old cats, most of this organ has been replaced by fat.

Heart: In the mediastinum enclosed by the pericardium.

Lungs: Paired organs flanking the heart.

Abdominal Cavity Organs

Liver: Large multilobed organ lying under the umbrella of the diaphragm.

Lift the drapelike, fat-infiltrated greater omentum covering the abdominal organs to expose the following organs:

Stomach: Dorsally located sac to the left of the liver.

Spleen: Flattened brown organ curving around the lateral aspect of the stomach.

Small intestine: Tubelike organ continuing posteriorly from the stomach.

*Pancreas: Diffuse gland lying deep to and between the small intestine and stomach (Figure 21.5). Lift the first section of the small intestine with your forceps; you should see the pancreas situated in the delicate mesentery posterior to the stomach.

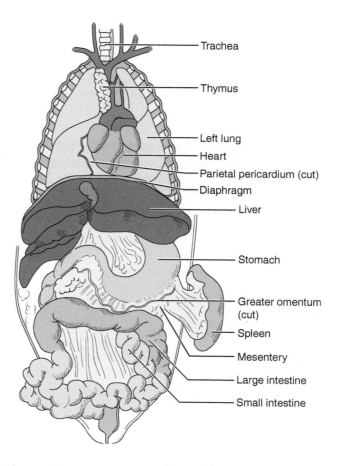

Figure 21.4 Ventral body cavity organs of the cat. Superficial view with greater omentum removed. (Also see Figure 27.20, page 478.)

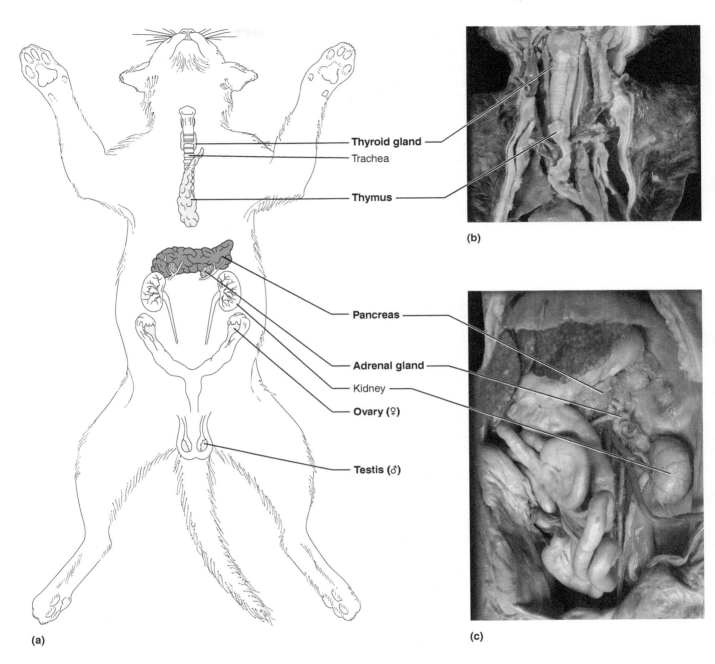

Thyroid gland

Trachea

Thymus

(b)

Pancreas

Adrenal gland

Kidney

Ovary (♀)

Testis (♂)

(a)

(c)

Figure 21.5 Endocrine organs in the cat. (a) Drawing. **(b, c)** Photographs.

Large intestine: Takes a U-shaped course around the small intestine to terminate in the rectum.

Push the intestines to one side with a probe to reveal the deeper organs in the abdominal cavity.

Kidneys: Bean-shaped organs located toward the dorsal body wall surface and posterior to the peritoneum (see Figure 21.5).

*Adrenal glands: Seen above and medial to each kidney (Figure 21.5).

*Gonads (ovaries or testes): The location of the gonads is illustrated (Figure 21.5a), but you will not identify them until the reproductive system organs are considered (Exercise 29).

Before you leave the laboratory, follow the boxed instructions to prepare your cat for storage and clean the area (page 219). ▄

Microscopic Anatomy of Selected Endocrine Glands

ACTIVITY 4

Examining the Microscopic Structure of Endocrine Glands

Obtain a microscope and one of each slide on the materials list (page 349). Compare your observations with **Figure 21.6a–f**.

Thyroid Gland

1. Scan the thyroid under low power, noting the **follicles,** generally spherical sacs containing a pink-stained material (*colloid*). Stored T_3 and T_4 are attached to the protein colloidal material stored in the follicles as **thyroglobulin** and are released gradually to the blood. Compare the tissue viewed to Figure 21.6a.

2. Observe the tissue under high power. Notice that the walls of the follicles are formed by simple cuboidal or squamous epithelial cells. The **parafollicular,** or **C, cells** you see between the follicles produce calcitonin.

When the thyroid gland is actively secreting, the follicles appear small. When the thyroid is hypoactive or inactive, the follicles are large and plump, and the follicular epithelium appears to be squamous.

Parathyroid Glands

Observe the parathyroid tissue under low power to view its two major cell types, the parathyroid cells and the oxyphil cells. Compare your observations to Figure 21.6b. The **parathyroid cells,** which synthesize parathyroid hormone (PTH), are small and abundant. The function of the scattered, much larger **oxyphil cells** is unknown.

Pancreas

1. Observe pancreas tissue under low power to identify the roughly circular **pancreatic islets** (also called *islets of Langerhans*), the endocrine portions of the pancreas. The islets are scattered amid the more numerous acinar cells and stain differently (usually lighter), which makes it possible to identify them. (See Figure 27.14.) The acinar cells produce the enzymatic exocrine product of the pancreas.

2. Focus on islet cells under high power. Notice that they are densely packed and have no definite arrangement. In contrast, the cuboidal acinar cells are arranged around secretory ducts. In Figure 21.6c, it is possible to distinguish the **alpha (α) cells,** which tend to cluster at the periphery of the islets and produce glucagon, from the **beta (β) cells,** which stain lighter and synthesize insulin.

Pituitary Gland

1. Observe the general structure of the pituitary gland under low power to differentiate the glandular anterior lobe from the neural posterior lobe.

2. Using the high-power lens, focus on the nests of cells of the anterior lobe. It is possible to identify the specialized cell types that secrete the specific hormones when differential stains are used. Using Figure 21.6d as a guide, locate the reddish pink–stained **acidophil cells,** which produce growth hormone and prolactin, and the **basophil cells,** stained blue to purple in color, which produce the tropic hormones (TSH, ACTH, FSH, and LH). **Chromophobes,** the third cellular population, do not take up the stain and appear colorless. The role of the chromophobes is controversial, but they do not appear to be involved in hormone production.

3. Now focus on the posterior lobe. Observe the nerve fibers (axons of hypophyseal neurons) that make up most of this portion of the pituitary. Also note the **pituicytes,** neuroglia that are randomly distributed among the nerve fibers. Refer to Figure 21.6e as you scan the slide.

Which two hormones are stored here?

_____ and _____

What is their source? _____

WHY THIS **MATTERS** | **Growth Hormone Uses and Abuses**

Growth hormone is an anabolic hormone that contributes to tissue building while mobilizing fat stores. Its primary targets include cartilage, bone, and muscle. Growth hormone was first approved by the U.S. Food and Drug Administration (FDA) for the treatment of children with growth disorders. Since then it has been approved to treat muscle-wasting disease associated with HIV/AIDS and a small number of other disorders. Ironically, growth hormone is more famous for its abuse than for its therapeutic applications. For example, athletes inject growth hormone to improve strength, speed, and endurance; actors use growth hormone as a chemical "fountain of youth" to smooth wrinkles and decrease body fat. ■

Adrenal Gland

1. Hold the slide of the adrenal gland up to the light to distinguish the cortex and medulla areas. Then scan the cortex under low power to distinguish the three cortical areas. Refer to Figure 21.6f and the following descriptions to identify the cortical areas:

- Connective tissue capsule of the adrenal gland.

- The outermost **zona glomerulosa,** where most mineralocorticoid production occurs and where the tightly packed cells are arranged in spherical clusters.

- The deeper intermediate **zona fasciculata,** which produces glucocorticoids. This is the thickest part of the cortex. Its cells are arranged in parallel cords.

- The innermost cortical zone, the **zona reticularis** abutting the medulla, which produces sex hormones and some glucocorticoids. The cells here stain intensely and form a branching network.

2. Switch to higher power to view the large, lightly stained cells of the adrenal medulla, which produce epinephrine and norepinephrine. Notice their clumped arrangement. ■

Endocrine Disorders

Many endocrine disorders are a result of either hyposecretion (underproduction) or hypersecretion (overproduction) of a given hormone. The characteristics of select endocrine disorders are summarized in **Table 21.5.** As you read through the table, recall the targets for the hormones and the effects of normal secretion levels.

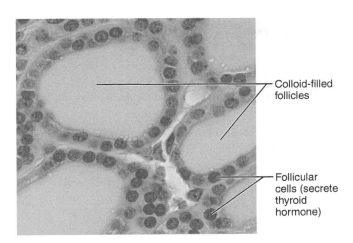

Colloid-filled follicles

Follicular cells (secrete thyroid hormone)

(a) Thyroid gland (480×)

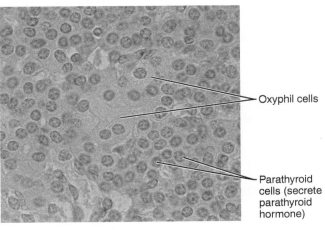

Oxyphil cells

Parathyroid cells (secrete parathyroid hormone)

(b) Parathyroid gland (420×)

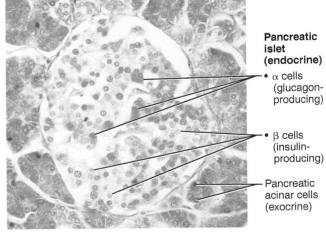

Pancreatic islet (endocrine)
- α cells (glucagon-producing)
- β cells (insulin-producing)

Pancreatic acinar cells (exocrine)

(c) Pancreatic islet (210×)

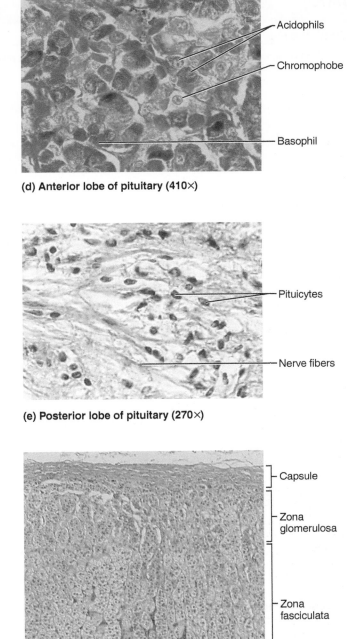

Acidophils

Chromophobe

Basophil

(d) Anterior lobe of pituitary (410×)

Pituicytes

Nerve fibers

(e) Posterior lobe of pituitary (270×)

Capsule

Zona glomerulosa

Zona fasciculata

Zona reticularis

Adrenal medulla

(f) Adrenal gland (105×)

Figure 21.6 Microscopic anatomy of selected endocrine organs.

Table 21.5   Summary of Select Endocrine Homeostatic Imbalances

Hormone	Effects of hyposecretion	Effects of hypersecretion
Growth hormone	*In children:* **pituitary dwarfism,** which results in short stature with normal proportions	*In children:* **gigantism,** abnormally tall *In adults:* **acromegaly,** abnormally large bones of the face, feet, and hands
Antidiuretic hormone	**Diabetes insipidus,** a condition characterized by thirst and excessive urine output	**Syndrome of inappropriate ADH secretion,** a condition characterized by fluid retention, headache, and disorientation
Thyroid hormone	*In children:* **cretinism,** mental retardation with a disproportionately short-sized body *In adults:* **myxedema,** low metabolic rate, edema, physical and mental sluggishness	**Graves' disease,** elevated metabolic rate, sweating, irregular heart rate, weight loss, protrusion of the eyeballs, and nervousness
Parathyroid hormone	**Hypoparathyroidism,** neural excitability with tetany (muscle spasms) and convulsions	**Hyperparathyroidism,** loss of calcium from bones, causing deformation, and spontaneous fractures
Insulin	**Diabetes mellitus,** which results in an inability of cells to take up and utilize glucose and in loss of glucose in the urine (may be due to hyposecretion or hypoactivity of insulin)	**Hypoglycemia,** which results in low blood sugar and is characterized by anxiety, nervousness, tremors, and weakness

GROUP CHALLENGE

Odd Hormone Out

Each box below contains four hormones. One of the listed hormones does *not* share a characteristic that the other three do. Work in groups of three, and discuss the characteristics of the four hormones in each group. On a separate piece of paper, one student will record the characteristics for each hormone for the group. For each set of four hormones, discuss the possible candidates for the "odd hormone" and which characteristic it lacks, based upon your recorded notes. Once you have come to a consensus among your group, circle the hormone that doesn't belong with the others, and explain why it is singled out. Sometimes there may be multiple reasons why the hormone doesn't belong with the others.

1. Which is the "odd hormone"?		Why is it the odd one out?
ACTH	LH	
Oxytocin	FSH	

2. Which is the "odd hormone"?		Why is it the odd one out?
Aldosterone	Epinephrine	
Cortisol	ADH	

3. Which is the "odd hormone"?		Why is it the odd one out?
PTH	LH	
Testosterone	FSH	

4. Which is the "odd hormone"?		Why is it the odd one out?
Insulin	Calcitonin	
Cortisol	Glucagon	

Functional Anatomy of the Endocrine Glands

Gross Anatomy and Basic Function of the Endocrine Glands

1. Both the endocrine and nervous systems are major regulating systems of the body; however, the nervous system has been compared to a text message and the endocrine system to mailing a letter. Briefly explain this comparison.

   _____

   _____

   _____

2. Define *hormone*. _____

   _____

3. Chemically, hormones belong chiefly to two molecular groups: _____ and _____.

4. Define *target cell*. _____

   _____

5. Given that hormones travel in the bloodstream, why don't all tissues respond to all hormones? _____

   _____

6. Identify the endocrine organ described by each of the following statements.

   _____ 1. located in the anterior neck; produces key hormones for metabolism

   _____ 2. produces the hormones that are stored in the posterior pituitary

   _____ 3. a mixed gland, located behind the stomach and close to the small intestine

   _____ 4. paired glands suspended in the scrotum

   _____ 5. bilobed gland located in the sella turcica

   _____ 6. found in the pelvic cavity of the female, concerned with production of ova and female hormones

   _____ 7. found in the upper thorax overlying the heart; large during youth

   _____ 8. found in the roof of the third ventricle of the brain

7. The table below lists the functions of many of the hormones you have studied. From the keys below, fill in the hormones responsible for each function, and the endocrine glands that produce each hormone. Glands may be used more than once.

Hormones Key:

ACTH	FSH	prolactin
ADH	glucagon	PTH
aldosterone	insulin	T_3 / T_4
cortisol	LH	testosterone
epinephrine	oxytocin	TSH
estrogens	progesterone	

Glands Key:

adrenal cortex	pancreas
adrenal medulla	parathyroid gland
anterior lobe of pituitary	posterior lobe of pituitary
hypothalamus	testes
ovaries	thyroid gland

Function	Hormone(s)	Synthesizing gland(s)
Regulate the function of another endocrine gland (tropic)	1.	
	2.	
	3.	
	4.	
Maintain salt and water balance in the extracellular fluid	1.	
	2.	
Directly involved in milk production and ejection	1.	
	2.	
Controls the rate of body metabolism and cellular oxidation	1.	
Regulates blood calcium levels	1.	
Regulate blood glucose levels; produced by the same "mixed" gland	1.	
	2.	
Released in response to stressors	1.	
	2.	
Drives development of secondary sex characteristics in males	1.	
Directly responsible for regulating the menstrual cycle	1.	
	2.	

8. Although the pituitary gland is sometimes referred to as the master gland of the body, the hypothalamus exerts control over the pituitary gland. How does the hypothalamus control the functioning of both the anterior and posterior lobes?

_____

_____

9. Name the hormone(s) produced in *inadequate* amounts that directly result in the following conditions.

_____   1. tetany

_____   2. excessive urination without high blood glucose levels

_____   3. loss of glucose in urine

_____   4. abnormally small stature, normal proportions

_____   5. low metabolic rate, mental and physical sluggishness

10. Name the hormone(s) produced in *excessive* amounts that directly result in the following conditions.

_____   1. in the adult: large bones of the hands, feet, and face

_____   2. bulging eyeballs, nervousness, increased pulse rate, sweating

_____   3. demineralization of bones, spontaneous fractures

WHY THIS MATTERS ┃ 11. Explain why growth hormone is an anabolic hormone. _____

_____

12. Considering the primary target organs of growth hormone, explain why growth hormone is not a tropic

hormone. _____

_____

Dissection and Identification: Selected Endocrine Organs of the Cat

13. How do the locations of the endocrine organs in the cat compare with those in the human?

_____

_____

14. Name two endocrine organs located in the throat region: _____ and _____

15. Name three endocrine organs located in the abdominal cavity.

_____

16. Given the assumption (not necessarily true) that human beings have more stress than cats, which endocrine organs would you expect to be relatively larger in humans?

_____

17. Cats are smaller animals than humans. Which would you expect to have a (relatively speaking) more active thyroid gland—

cats or humans? _____ Why?

_____

_____

Microscopic Anatomy of Selected Endocrine Glands

18. Choose a response from the key below to name the hormone(s) produced by the cell types listed.

Key: a. calcitonin   d. glucocorticoids   g. PTH
         b. GH, prolactin   e. insulin   h. T_4 / T_3
         c. glucagon   f. mineralocorticoids   i. TSH, ACTH, FSH, LH

_____ 1. parafollicular cells of the thyroid   _____ 6. zona fasciculata cells

_____ 2. follicular cells of the thyroid   _____ 7. zona glomerulosa cells

_____ 3. beta cells of the pancreatic islets   _____ 8. parathyroid cells

_____ 4. alpha cells of the pancreatic islets   _____ 9. acidophil cells of the anterior
                                                                               pituitary

_____ 5. basophil cells of the anterior pituitary

19. Identify the endocrine glands, and name all structures indicated by a leader line.

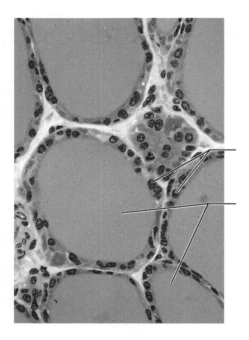
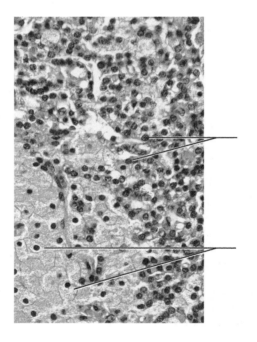

_____   _____

20. ✚ Pituitary gland tumors can secrete excess amounts of growth hormone. Describe the signs and symptoms that these tumors

cause in an adult experiencing hypersecretion of the growth hormone. _____

_____

21. ✚ Tumors of the adrenal medulla, called pheochromocytomas, cause hypersecretion of catecholamines. Describe the expected

signs and symptoms of this tumor. _____

_____

Blood

MATERIALS

General Supply Area*

- Disposable gloves
- Safety glasses
- Bucket or large beaker containing 10% household bleach solution for slide and glassware disposal
- Spray bottles containing 10% bleach solution
- Autoclave bag
- Designated lancet (sharps) disposal container
- Plasma (obtained from an animal hospital or prepared by centrifuging animal blood [for example, cattle or sheep blood] obtained from a biological supply house)
- Test tubes and test tube racks
- Wide-range pH paper
- Stained smears of human blood from a biological supply house or, if desired by the instructor, heparinized animal blood obtained from a biological supply house or an animal hospital (for example, dog blood), or EDTA-treated red cells (reference cells†) with blood type labels obscured (available from Immucor, Inc.)
- Clean microscope slides
- Glass stirring rods
- Wright's stain in a dropper bottle
- Distilled water in a dropper bottle
- Sterile lancets
- Absorbent cotton balls
- Alcohol swabs (wipes)
- Paper towels

*Note to the Instructor: See directions for handling of soiled glassware and disposable items (page 27).
†The blood in these kits (each containing four blood cell types—A1, A2, B, and O—individually supplied in 10-ml vials) is used to calibrate cell counters and other automated clinical laboratory equipment. This blood has been carefully screened and can be safely used by students for blood typing and determining hematocrits. It is not usable for hemoglobin determinations or coagulation studies.

Text continues on next page.

LEARNING OUTCOMES

- ☐ Name the two major components of blood, and state their average percentages in whole blood.
- ☐ Describe the composition and functional importance of plasma.
- ☐ Define *formed elements*. Name the cell types that comprise the formed elements, state their relative percentages, and describe their major functions.
- ☐ Identify erythrocytes, basophils, eosinophils, monocytes, lymphocytes, and neutrophils on a microscopic preparation or an appropriate image.
- ☐ Provide the normal values for a total white blood cell count and a total red blood cell count, and explain why these tests are important.
- ☐ Conduct the following blood tests in the laboratory, state their norms, and explain the importance of each: differential white blood cell count, hematocrit, hemoglobin determination, clotting time, and plasma cholesterol concentration.
- ☐ Perform an ABO and Rh blood typing test in the laboratory, and discuss why administering mismatched blood causes transfusion reactions.
- ☐ Define *leukocytosis, leukopenia, leukemia, polycythemia,* and *anemia;* and cite a possible cause for each condition.

PRE-LAB QUIZ

1. Circle True or False. There are no special precautions that I need to observe when performing today's lab.
2. Three types of formed elements found in blood include erythrocytes, leukocytes, and:
   a. electrolytes   b. fibers   c. platelets   d. sodium salts
3. Circle the correct underlined term. Mature <u>erythrocytes</u> / <u>leukocytes</u> are the most numerous blood cells and do not have a nucleus.
4. The least numerous but largest of all agranulocytes is the:
   a. basophil   b. lymphocyte   c. monocyte   d. neutrophil
5. _____ are the leukocytes responsible for releasing histamine and other mediators of inflammation.
   a. Basophils   b. Eosinophils   c. Monocytes   d. Neutrophils
6. _____ are essential for blood clotting.
7. Circle the correct underlined term. When determining the <u>hematocrit</u> / <u>hemoglobin</u>, you will centrifuge whole blood in order to allow the formed elements to sink to the bottom of the sample.
8. Circle the correct underlined term. The normal hematocrit value for <u>females</u> / <u>males</u> is generally higher than that of the opposite sex.
9. Circle the correct underlined term. Blood typing is based on the presence of proteins known as <u>antigens</u> / <u>antibodies</u> on the outer surface of the red blood cell plasma membrane.
10. Circle True or False. If an individual is transfused with the wrong type blood, the recipient's antibodies react with the donor's blood antigens, eventually clumping and hemolyzing the donated RBCs.

- ☐ Compound microscope
- ☐ Immersion oil
- ☐ Assorted slides of white blood count pathologies labeled "Unknown Sample _____"
- ☐ Timer

Because many blood tests are to be conducted in this exercise, it is advisable to set up a number of appropriately labeled supply areas for the various tests, as designated below. Some needed supplies are located in the general supply area.
Note: Artificial blood prepared by Ward's Natural Science can be used for differential counts, hematocrit, and blood typing.

Activity 4: Hematocrit
- ☐ Heparinized capillary tubes
- ☐ Microhematocrit centrifuge and reading gauge (if the reading gauge is not available, a millimeter ruler may be used)
- ☐ Capillary tube sealer or modeling clay

Activity 5: Hemoglobin Determination
- ☐ Hemoglobinometer, hemolysis applicator, and lens paper, or Tallquist hemoglobin scale and test paper

Activity 6: Coagulation Time
- ☐ Capillary tubes (nonheparinized)
- ☐ Fine triangular file

Activity 7: Blood Typing
- ☐ Blood typing sera (anti-A, anti-B, and anti-Rh [anti-D])

- ☐ Rh typing box
- ☐ Wax marking pencil
- ☐ Toothpicks
- ☐ Blood test cards or microscope slides
- ☐ Medicine dropper

Activity 8: Demonstration
- ☐ Microscopes set up with prepared slides demonstrating the following bone (or bone marrow) conditions: macrocytic hypochromic anemia, microcytic hypochromic anemia, sickle cell disease, lymphocytic leukemia (chronic), and eosinophilia

Activity 9: Cholesterol Measurement
- ☐ Cholesterol test cards and color scale

I n this exercise you will study plasma and formed elements of blood and conduct various hematologic tests. These tests are useful diagnostic tools for physicians because blood composition (number and types of blood cells, and chemical composition) reflects the status of many body functions and malfunctions.

⚠ **ALERT: Special precautions when handling blood.** This exercise provides information about blood from several sources: human, animal, human treated, and artificial blood. The instructor will decide whether to use animal blood for testing or to have students test their own blood, in accordance with the educational goals of the student group. For example, for students in the nursing or laboratory technician curricula, learning how to safely handle human blood and human wastes is essential. Whenever blood is being handled, pay special attention to safety precautions. Use these precautions regardless of the source of the blood. This will both teach good technique and ensure the safety of the students.

Follow exactly the safety precautions listed below.

1. Wear safety gloves at all times. Discard appropriately.

2. Wear safety glasses throughout the exercise.

3. Handle only your own, freshly drawn (human) blood.

4. Be sure you understand the instructions and have all supplies on hand before you begin any part of the exercise.

5. Do not reuse supplies and equipment once they have been exposed to blood.

6. Keep the lab area clean. Do not let anything that has come in contact with blood touch surfaces or other individuals in the lab. Keep track of the location of any supplies and equipment that come into contact with blood.

7. Immediately after use, dispose of lancets in a designated disposal container. Do not put them down on the lab bench, even temporarily.

8. Dispose of all used cotton balls, alcohol swabs, blotting paper, and so forth in autoclave bags and place all soiled glassware in containers of 10% bleach solution.

9. Wipe down the lab bench with 10% bleach solution when you are finished.

Composition of Blood

Circulating blood is a rather viscous substance that varies from bright red to a dull brick-red, depending on the amount of oxygen it is carrying. Oxygen-rich blood is bright red. The average volume of blood in the body is about 5–6 liters in adult males and 4–5 liters in adult females.

Blood is classified as a type of connective tissue because it consists of cells within a matrix. The nonliving fluid matrix is the **plasma** and the cells, and cell fragments are the **formed elements**. The fibers typical of a connective tissue matrix become visible in blood only when clotting occurs. They then appear as fibrin threads, which form the structural basis for clot formation.

More than 100 different substances are dissolved or suspended in plasma (**Figure 22.1**), which is over 90% water. These include nutrients, gases, hormones, various wastes and metabolites, many types of proteins, and electrolytes. The composition of plasma varies continuously as cells remove or add substances to the blood.

Three main types of formed elements are present in blood (**Table 22.1**). Most numerous are **erythrocytes,** or **red blood cells (RBCs),** which are literally sacs of hemoglobin molecules that transport the bulk of the oxygen carried in the blood (and a small percentage of the carbon dioxide). **Leukocytes,** or **white blood cells (WBCs),** are part of the body's nonspecific defenses and the immune system, and **platelets** function in hemostasis (blood clot formation) together; they make up <1% of whole blood. Formed elements normally constitute about 45% of whole blood; plasma accounts for the remaining 55%.

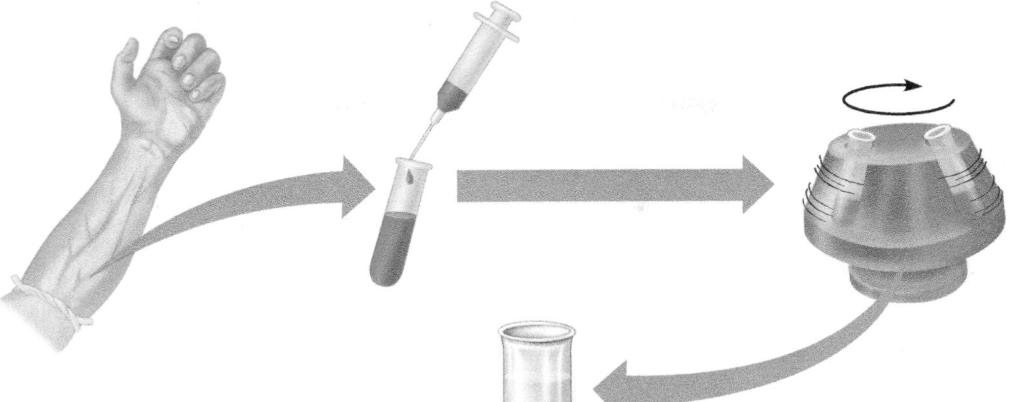

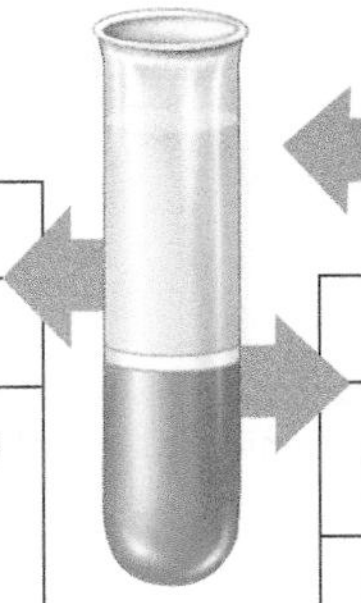

Figure 22.1 The composition of blood. Note that leukocytes and platelets are found in the band between plasma (above) and erythrocytes (below). This band is known as the buffy coat.

Determining the Physical Characteristics of Plasma

Go to the general supply area, and carefully pour a few milliliters of plasma into a test tube. Also obtain some wide-range pH paper, and then return to your laboratory bench to make the following observations.

pH of Plasma

Test the pH of the plasma with wide-range pH paper. Record

the pH observed. _____

Color and Clarity of Plasma

Hold the test tube up to a source of natural light. Note and record its color and degree of transparency. Is it clear, translucent, or opaque?

Color _____

Degree of transparency _____

Consistency

Dip your finger and thumb into plasma, and then press them firmly together for a few seconds. Gently pull them apart.

Table 22.1   Summary of Formed Elements of Blood

Cell type	Illustration	Description*	Cells/mm³ (μl) of blood	Function
Erythrocytes (red blood cells, RBCs)		Biconcave, anucleate disc; orange-pink color; diameter 7–8 μm	4–6 million	Transport oxygen and carbon dioxide
Leukocytes (white blood cells, WBCs)		Spherical, nucleated cells	4800–10,800	
Granulocytes Neutrophil		Nucleus multilobed; pale red and blue cytoplasmic granules; diameter 10–12 μm	3000–7000 Differential count: 50–70%	Phagocytize pathogens or debris
Eosinophil		Nucleus bilobed; red cytoplasmic granules; diameter 10–14 μm	100–400 Differential count: 2–4%	Kill parasitic worms; slightly phagocytic; complex role in allergy and asthma
Basophil		Nucleus lobed; large blue-purple cytoplasmic granules; diameter 10–14 μm	20–50 Differential count: <1%	Release histamine and other mediators of inflammation; contain heparin, an anticoagulant
Agranulocytes Lymphocyte		Nucleus spherical or indented; pale blue cytoplasm; diameter 5–17 μm	1500–3000 Differential count: 25–45%	Mount immune response by direct cell attack or via antibody production
Monocyte		Nucleus U- or kidneyshaped; gray-blue cytoplasm; diameter 14–24 μm	100–700 Differential count: 3–8%	In tissues, develop into macrophages that phagocytize pathogens or debris
Platelets		Cytoplasmic fragments containing granules; stain deep purple; diameter 2–4 μm	150,000–400,000	Seal small tears in blood vessels; instrumental in blood clotting

*Appearance when stained with Wright's stain.

How would you describe the consistency of plasma (slippery, watery, sticky, granular)? Record your observations.

_____ ▬

ACTIVITY 2

Examining the Formed Elements of Blood Microscopically

In this section, you will observe blood cells on an already prepared (purchased) blood slide or on a slide prepared from your own blood or blood provided by your instructor.

• If you are using a purchased blood slide, obtain a slide and begin your observations at step 6.

• If you are testing blood provided by a biological supply source or an animal hospital, obtain a tube of the supplied blood, disposable gloves, and the supplies listed in step 1, except for the lancets and alcohol swabs. After donning gloves, go to step 3b to begin your observations.

• If you are examining your own blood, you will perform all the steps that follow *except* step 3b.

1. Obtain two glass slides, a glass stirring rod, dropper bottles of Wright's stain and distilled water, two or three lancets, cotton balls, and alcohol swabs. Bring this equipment to the laboratory bench. Clean the slides thoroughly, and dry them.

2. Open the alcohol swab packet, and scrub your third or fourth finger with the swab. (Because the pricked finger may be a little sore later, it is better to prepare a finger on the hand used less often.) Swing your hand in a cone-shaped path for 10 to 15 seconds. This will dry the alcohol and cause your fingers to become filled with blood. Then, open the lancet packet and grasp the lancet by its blunt end. Quickly jab the pointed end into the prepared finger to produce a free flow of blood. It is *not* a good idea to squeeze or "milk" the finger, because

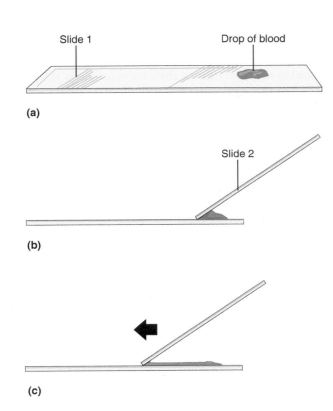

Slide 1   Drop of blood

(a)

Slide 2

(b)

(c)

Figure 22.2 Procedure for making a blood smear.
(a) Place a drop of blood on slide 1 approximately
½ inch from one end. **(b)** Hold slide 2 at a 30° to 40°
angle to slide 1 (it should touch the drop of blood) and
allow blood to spread along entire bottom edge of
angled slide. **(c)** Smoothly advance slide 2 to end of
slide 1 (blood should run out before reaching the end
of slide 1). Then lift slide 2 away from slide 1, and place
it on a paper towel.

this forces out tissue fluid as well as blood. If the blood is not
flowing freely, make another puncture.

⚠ _Under no circumstances are you to use a lancet for_
more than one puncture. Dispose of the lancets in the
designated disposal container immediately after use.

3a. With a cotton ball, wipe away the first drop of blood;
then allow another large drop of blood to form. Touch the
blood to one of the cleaned slides approximately 1.3 cm, or
½ inch, from the end. Then quickly (to prevent clotting) use
the second slide to form a blood smear **(Figure 22.2)**. When
properly prepared, the blood smear is uniformly thin. If the
blood smear appears streaked, the blood probably began to
clot or coagulate before the smear was made, and another slide
should be prepared. Continue at step 4.

3b. Dip a glass rod in the blood provided, and transfer a gen-
erous drop of blood to the end of a cleaned microscope slide.
For the time being, lay the glass rod on a paper towel on the
bench. Then, as described in step 3a (Figure 22.2), use the
second slide to make your blood smear.

4. Allow the blood smear slide to air-dry. When it is completely dry, it will look dull. Place it on a paper towel, and
add 5–10 drops of Wright's stain. Count the number of drops
of stain used. Allow the stain to remain on the slide for 3 to

4 minutes, and then flood the slide with an equal number of
drops of distilled water. Allow the water and Wright's stain
mixture to remain on the slide for 4 or 5 minutes or until a
metallic green film or scum is apparent on the fluid surface.

5. Rinse the slide with a stream of distilled water. Then flood
it with distilled water, and allow it to lie flat until the slide
becomes translucent and takes on a pink cast. Then stand
the slide on its long edge on the paper towel, and allow it
to dry completely. Once the slide is dry, you can begin your
observations.

6. Obtain a microscope and scan the slide under low power
to find the area where the blood smear is the thinnest. After
scanning the slide under low power to find the areas with the
largest numbers of WBCs, read the following descriptions
of cell types, and find each one in Figure 22.1, Table 22.1,
Figure 22.3, and **Figure 22.4**. Then, switch to the oil
immersion lens, and observe the slide carefully to identify
each cell type.

7. Set your prepared slide aside for use in Activity 3.

Erythrocytes

Erythrocytes, or red blood cells, which average 7.5 µm in
diameter, vary in color from an orange-pink color to pale
pink, depending on the effectiveness of the stain. They have a
distinctive biconcave disk shape and appear paler in the center
than at the edge (Figure 22.3).

As you observe the slide, notice that the red blood
cells are by far the most numerous blood cells seen in the
field. Their number averages 4.5 million to 5.5 million
cells per cubic millimeter of blood (for women and men,
respectively).

Red blood cells differ from the other blood cells because
they are anucleate (lacking a nucleus) when mature and circulating in the blood. As a result, they are unable to reproduce

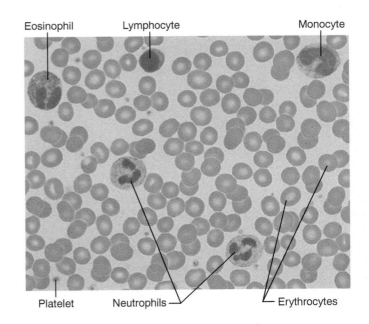

Eosinophil   Lymphocyte   Monocyte

Platelet   Neutrophils   Erythrocytes

**Figure 22.3 Photomicrograph of a human blood smear
stained with Wright's stain (715×).**

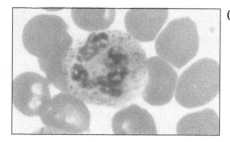

(a) Neutrophil:
multilobed nucleus,
pale red and blue
cytoplasmic
granules

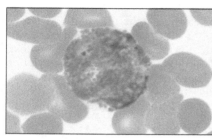

(b) Eosinophil:
bilobed nucleus,
red cytoplasmic
granules

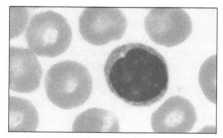

(c) Basophil:
bilobed nucleus,
purplish black
cytoplasmic
granules

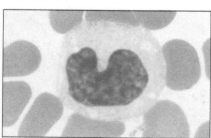

**(d) Lymphocyte
(small):**
large spherical
nucleus, thin rim of
pale blue cytoplasm

(e) Monocyte:
kidney-shaped
nucleus, abundant
pale blue cytoplasm

Figure 22.4 Leukocytes. In each case, the leukocytes are surrounded by erythrocytes (all 1750×, Wright's stain).

or repair damage and have a limited life span of 100 to 120 days, after which they begin to fragment and are destroyed, mainly in the spleen.

In various anemias, the red blood cells may appear pale (an indication of decreased hemoglobin content) or may be nucleated (an indication that the bone marrow is turning out cells prematurely). +

Leukocytes

Leukocytes, or white blood cells, are nucleated cells that are formed in the bone marrow from the same blood stem cells *(hematopoietic stem cells)* as red blood cells. They are much less numerous than the red blood cells, averaging from 4800 to 10,800 cells per cubic millimeter. The life span of leukocytes varies. They can survive for minutes or decades, depending on the type of leukocyte and tissue activity. Basically, white blood cells are protective, pathogen-destroying cells that are transported to all parts of the body in the blood or lymph. Important to their protective function is their ability to move in and out of blood vessels, a process called **diapedesis,** and to wander through body tissues by **amoeboid motion** to reach sites of inflammation or tissue destruction. They are classified into two major groups, depending on whether or not they contain conspicuous granules in their cytoplasm.

Granulocytes make up the first group. The granules in their cytoplasm stain differentially with Wright's stain, and they have peculiarly lobed nuclei, which often consist of expanded nuclear regions connected by thin strands of nucleoplasm. There are three types of granulocytes: **neutrophils, eosinophils,** and **basophils.**

The second group, **agranulocytes,** contains no *visible* cytoplasmic granules. Although found in the bloodstream, they are much more abundant in lymphoid tissues. There are two types of agranulocytes: **lymphocytes** and **monocytes.** The specific characteristics of leukocytes are described in Table 22.1. Photomicrographs of the leukocytes illustrate their different appearances (Figure 22.4).

Students are often asked to list the leukocytes in order from the most abundant to the least abundant. The following silly phrase may help you with this task: *N*ever *l*et *m*onkeys *e*at *b*ananas (neutrophils, lymphocytes, monocytes, eosinophils, basophils).

Platelets

Platelets are cell fragments of large multinucleate cells **(megakaryocytes)** formed in the bone marrow. They appear as darkly staining, irregularly shaped bodies interspersed among the blood cells (see Figure 22.3). The normal platelet count in blood ranges from 150,000 to 400,000 per cubic millimeter. Platelets are instrumental in the clotting process that occurs in plasma when blood vessels are ruptured.

After you have identified these cell types on your slide, observe charts and three-dimensional models of blood cells if these are available. Do not dispose of your slide, because you will use it later for the differential white blood cell count. ■

Hematologic Tests

When someone enters a hospital as a patient, several hematologic tests are routinely done to determine general level of health as well as the presence of pathological conditions. You will be conducting the most common of these tests in this exercise.

⚠ Materials such as cotton balls, lancets, and alcohol swabs are used in nearly all of the following diagnostic tests. These supplies are at the general supply area and should be properly disposed of (glassware to the bleach bucket, lancets in a designated disposal container, and disposable items to the autoclave bag) immediately after use.

Other necessary supplies and equipment are at specific supply areas marked according to the test with which they are used. Because nearly all of the tests require a finger stick, if you will be using your own blood it might be wise to quickly read through the tests to determine in which instances more than one preparation can be done from the same finger stick. For example, the hematocrit capillary tubes and sedimentation rate samples might be prepared at the same time. A little planning will save you the discomfort of multiple finger sticks.

An alternative to using blood obtained from the finger stick technique is to use heparinized blood samples supplied by your instructor. The purpose of using heparinized tubes is to prevent the blood from clotting. Thus blood collected and stored in such tubes will be suitable for all tests except coagulation time testing.

Total White and Red Blood Cell Counts

A **total WBC count** or **total RBC count** determines the total number of that cell type per unit volume of blood. Total WBC and RBC counts are a routine part of any physical exam. Most clinical agencies use computers to conduct these counts. Total RBC and WBC counts will not be done here, but the importance of such counts (both normal and abnormal values) is briefly described below.

Total White Blood Cell Count

Because white blood cells are an important part of the body's defense system, it is essential to note any abnormalities in them.

Leukocytosis, a WBC count over 11,000 cells/mm^3 may indicate bacterial or viral infection, metabolic disease, hemorrhage, or poisoning by drugs or chemicals. A decrease in the white cell number below 4000/mm^3 (**leukopenia**) is usually due to exposure to certain chemicals or toxins, including anticancer agents. A person with leukopenia lacks the usual protective mechanisms. **Leukemia,** a malignant disorder of the lymphoid tissues characterized by uncontrolled proliferation of abnormal WBCs accompanied by a reduction in the number of RBCs and platelets, is detectable not only by a total WBC count, but also by a differential WBC count. ✛

Total Red Blood Cell Count

Because RBCs are absolutely necessary for oxygen transport, a doctor typically investigates any excessive change in their number immediately.

An increase in the number of RBCs (**polycythemia**) may result from bone marrow cancer or from living at high altitudes where less oxygen is available. A decrease in the number of RBCs results in anemia. The term **anemia** simply indicates a decreased oxygen-carrying capacity of blood that may result from a decrease in RBC number or size, or a decreased hemoglobin content of the RBCs. A decrease in RBCs may result suddenly from hemorrhage or more gradually from conditions that destroy RBCs or hinder RBC production. ✛

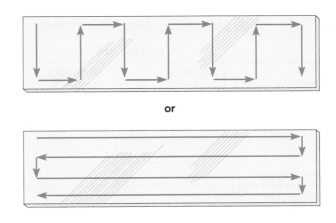

Figure 22.5 Alternative methods of moving the slide for a differential WBC count.

Differential White Blood Cell Count

To perform a **differential white blood cell count,** 100 WBCs are counted and classified according to type. Such a count is routine in a physical examination and in diagnosing illness, because any abnormality in percentages of WBC types may indicate a problem and the source of pathology.

ACTIVITY 3

Conducting a Differential White Blood Cell Count

1. Use the slide prepared for identification of the blood cells in Activity 2 or a prepared slide provided by your instructor. Begin at the edge of the smear, and move the slide in a systematic manner on the microscope stage—either up and down or from side to side as indicated in **Figure 22.5**.

2. Record each type of white blood cell you observe by making a count on the **Activity 3 chart** (for example, ||||| || = 7 cells) until you have observed and recorded a total of 100 WBCs. Using the following equation, compute the percentage of each WBC type counted, and record the percentages on the **Hematologic Test Data Sheet** (page 370).

$$\text{Percent (\%)} = \frac{\# \text{ observed}}{\text{Total } \# \text{ counted (100)}} \times 100$$

3. Select a slide marked "Unknown sample," record the slide number, and again use the Activity 3 chart to conduct a differential count. Record the percentages on the Hematologic Test Data Sheet (page 370).

How does the differential count from the unknown sample slide compare to the normal percentages given for each type in Table 22.1?

_____

_____

Activity 3: Count of 100 WBCs

Cell type	Number observed	
	Student smear	Unknown sample # ____
Neutrophils		
Eosinophils		
Basophils		
Lymphocytes		
Monocytes		

Using the text and other references, try to determine the blood pathology on the unknown slide. Defend your answer.

_____

_____

4. How does your differential white blood cell count compare to the percentages given in Table 22.1?

_____ ▬

Hematocrit

The **hematocrit** is routinely determined when anemia is suspected. Centrifuging whole blood spins the formed elements to the bottom of the tube, with plasma forming the top layer (see Figure 22.1). Because the blood cell population is primarily RBCs, the hematocrit is generally considered equivalent to the RBC volume, and this is the only value reported. However,

the relative percentage of WBCs can be differentiated, and both WBC and plasma volume will be reported here. Normal hematocrit values for the male and female, respectively, are 47.0 ± 5 and 42.0 ± 5.

ACTIVITY 4

Determining the Hematocrit

The hematocrit is determined by the micromethod, so only a drop of blood is needed. If possible (and the centrifuge allows), all members of the class should prepare their capillary tubes at the same time so the centrifuge can be run only once.

1. Obtain two heparinized capillary tubes, capillary tube sealer or modeling clay, a lancet, alcohol swabs, and some cotton balls.

2. If you are using your own blood, use an alcohol swab to cleanse a finger, prick the finger with a lancet, and allow the blood to flow freely. Wipe away the first few drops and, holding the red-line-marked end of the capillary tube to the blood drop, allow the tube to fill at least three-fourths full by capillary action **(Figure 22.6a)**. If the blood is not flowing freely, the end of the capillary tube will not be completely submerged in the blood during filling, air will enter, and you will have to prepare another sample.

If you are using instructor-provided blood, simply immerse the red-marked end of the capillary tube in the blood sample, and fill it three-quarters full as just described.

3. Plug the blood-containing end by pressing it into the capillary tube sealer or clay (Figure 22.6b). Prepare a second tube in the same manner.

4. Place the prepared tubes opposite one another in the head radial grooves of the microhematocrit centrifuge with the

---

Hematologic Test Data Sheet

Differential WBC count:

WBC	Student smear	Unknown sample # _____
% neutrophils	_____	_____
% eosinophils	_____	_____
% basophils	_____	_____
% lymphocytes	_____	_____
% monocytes	_____	_____

Hematocrit:

RBC _____ % of blood volume

WBC _____ % of blood volume ⎫ not

Plasma _____ % of blood ⎬ generally reported ⎭

Hemoglobin (Hb) content:

Hemoglobinometer (type: _____)

_____ g/100 ml of blood; _____ % Hb

Tallquist method _____ g/100 ml of blood; _____ % Hb

Ratio (hematocrit to grams of Hb per 100 ml of blood): _____

Coagulation time _____

Blood typing:

ABO group _____ Rh factor _____

Total cholesterol level _____ mg/dl of blood

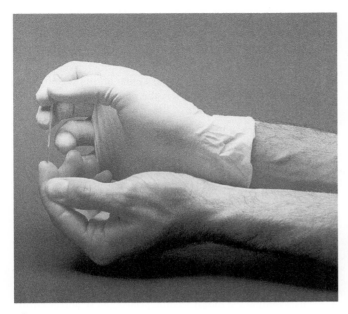

(a)

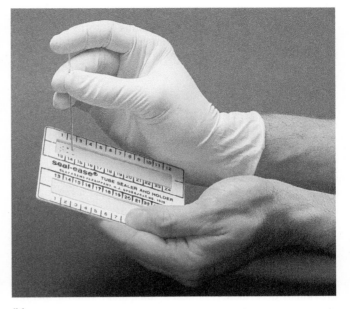

(b)

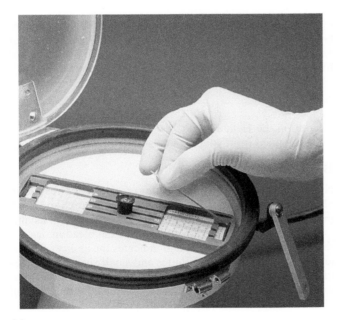

(c)

Figure 22.6 Steps in a hematocrit determination.
(a) Fill a heparinized capillary tube with blood. **(b)** Plug
the blood-containing end of the tube with clay. **(c)** Place
the tube in a microhematocrit centrifuge. (Centrifuge
must be balanced.)

sealed ends abutting the rubber gasket at the centrifuge
periphery (Figure 22.6c). This loading procedure balances the
centrifuge and prevents blood from spraying everywhere by
centrifugal force. *Make a note of the numbers of the grooves
your tubes are in.* When all the tubes have been loaded, make
sure the centrifuge is properly balanced, and secure the centrifuge cover. Turn the centrifuge on, and set the timer for 4
or 5 minutes.

5. Determine the percentage of RBCs, WBCs, and plasma by
using the microhematocrit reader. The RBCs are the bottom
layer, the plasma is the top layer, and the WBCs are the buff-
colored layer between the two. If the reader is not available,
use a millimeter ruler to measure the length of the filled capillary tube occupied by each element, and compute its percentage by using the following formula:

$$\frac{\text{Height of the column composed of the element (mm)}}{\text{Height of the original column of whole blood (mm)}} \times 100$$

Record your calculations below and on the Hematologic Test
Data Sheet (page 370).

% RBC _____ % WBC _____ % plasma _____

Usually WBCs constitute 1% of the total blood volume. How
do your blood values compare to this figure and to the normal
percentages for RBCs and plasma? (See Figure 22.1.)

_____

As a rule, a hematocrit is considered a more accurate test than
the total RBC count for determining the RBC composition of
the blood. A hematocrit within the normal range generally
indicates a normal RBC number, whereas an abnormally high
or low hematocrit is cause for concern. ▬

Hemoglobin Concentration

As noted earlier, a person can be anemic even with a normal
RBC count. Because hemoglobin (Hb) is the RBC protein
responsible for oxygen transport, perhaps the most accurate
way of measuring the oxygen-carrying capacity of the blood
is to determine its hemoglobin content. Oxygen, which combines reversibly with the heme (iron-containing portion) of the
hemoglobin molecule, is picked up by the blood cells in the
lungs and unloaded in the tissues. Thus, the more hemoglobin

molecules the RBCs contain, the more oxygen they will be able to transport. Normal blood contains 12 to 18 g of hemoglobin per 100 ml of blood. Hemoglobin content in men is slightly higher (13 to 18 g) than in women (12 to 16 g).

ACTIVITY 5

Determining Hemoglobin Concentration

Several techniques have been developed to estimate the hemoglobin content of blood, ranging from the old, rather inaccurate Tallquist method to expensive hemoglobinometers, which are precisely calibrated and yield highly accurate results. Directions for both the Tallquist method and a hemoglobinometer are provided here.

Tallquist Method

1. Obtain a Tallquist hemoglobin scale, test paper, lancets, alcohol swabs, and cotton balls.

2. Use instructor-provided blood or prepare the finger as previously described. (For best results, make sure the alcohol evaporates before puncturing your finger.) Place one good-sized drop of blood on the special absorbent paper provided with the color scale. The blood stain should be larger than the holes on the color scale.

3. As soon as the blood has dried and loses its glossy appearance, match its color, under natural light, with the color standards by moving the specimen under the comparison scale so that the blood stain appears at all the various apertures. Do not allow the blood to dry to a brown color, as this will result in an inaccurate reading. Because the colors on the scale represent 1% variations in hemoglobin content, it may be necessary to estimate the percentage if the color of your blood sample is intermediate between two color standards.

4. On the Hematologic Test Data Sheet (page 370), record your results as the percentage of hemoglobin concentration and as grams per 100 ml of blood.

Hemoglobinometer Determination

1. Obtain a hemoglobinometer, hemolysis applicator, alcohol swab, and lens paper, and bring them to your bench. Test the hemoglobinometer light source to make sure it is working; if not, request new batteries before proceeding and test it again.

2. Remove the blood chamber from the slot in the side of the hemoglobinometer, and disassemble the blood chamber by separating the glass plates from the metal clip. Notice as you do this that the larger glass plate has an H-shaped depression cut into it that acts as a moat to hold the blood, whereas the smaller glass piece is flat and serves as a coverslip.

3. Clean the glass plates with an alcohol swab, and then wipe them dry with lens paper. Hold the plates by their sides to prevent smearing during the wiping process.

4. Reassemble the blood chamber (remember: larger glass piece on the bottom with the moat up), but leave the moat plate about halfway out to provide adequate exposed surface to charge it with blood.

5. Obtain a drop of blood (from the provided sample or from your fingertip as before), and place it on the depressed area of the moat plate that is closest to you **(Figure 22.7a)**.

6. Using the wooden hemolysis applicator, stir or agitate the blood to rupture (lyse) the RBCs (Figure 22.7b). This usually takes 35 to 45 seconds. Hemolysis is complete when the blood appears transparent rather than cloudy.

7. Push the blood-containing glass plate all the way into the metal clip and then firmly insert the charged blood chamber back into the slot on the side of the instrument (Figure 22.7c).

8. Hold the hemoglobinometer in your left hand, with your left thumb resting on the light switch located on the underside of the instrument. Look into the eyepiece, and notice that there is a green area divided into two halves (a split field).

9. With the index finger of your right hand, slowly move the slide on the right side of the hemoglobinometer back and forth until the two halves of the green field match (Figure 22.7d).

10. Note and record on the Hematologic Test Data Sheet (page 370) the grams of Hb (hemoglobin)/100 ml of blood indicated on the uppermost scale by the index mark on the slide. Also record % Hb, indicated by one of the lower scales.

11. Disassemble the blood chamber once again, and carefully place its parts (glass plates and clip) into a bleach-containing beaker.

Generally speaking, the relationship between the hematocrit and grams of hemoglobin per 100 ml of blood is 3:1—for example, a hematocrit of 36% with 12 g of Hb per 100 ml of blood is a ratio of 3:1. How do your values compare?

Record on the Hematologic Test Data Sheet (page 370) the value obtained from your data.■

Bleeding Time

Normally a sharp prick of the finger or earlobe results in bleeding that lasts from 2 to 7 minutes (Ivy method) or 0 to 5 minutes (Duke method), although other factors such as altitude affect the time. The amount of time the bleeding lasts is referred to as **bleeding time** and indicates the ability of platelets to stop bleeding in capillaries and small vessels. Absence of some clotting factors may affect bleeding time, but prolonged bleeding time is most often associated with deficient or abnormal platelets.

Coagulation Time

Hemostasis is a protective mechanism that is set into motion when a blood vessel breaks. Hemostasis responds rapidly to stop bleeding. During hemostasis, three events occur in the following order: vascular spasm, platelet plug formation, and coagulation (blood clotting). Platelet plug formation and coagulation are illustrated in **Figure 22.8a**.

Blood clotting, or **coagulation,** is a process that requires the interaction of many substances normally present in the plasma as well as some released by platelets and injured tissues. The injured tissues and platelets release **tissue factor (TF)** and **phosphatidylserine** (formerly known as **platelet factor 3**) respectively, which trigger the clotting mechanism, or cascade. Tissue factor and phosphatidylserine interact with other blood protein clotting factors and calcium ions to form **prothrombin activator,** which in turn converts **prothrombin** (present in plasma) to **thrombin.** Thrombin then

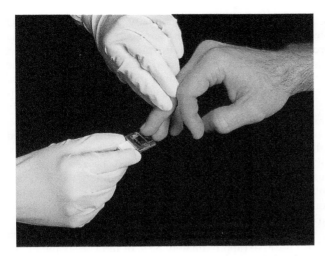

(a) Add a drop of blood to the moat plate of the blood chamber. The blood must flow freely.

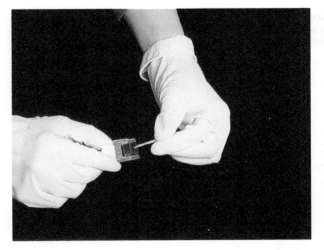

(b) The blood sample is hemolyzed with a wooden hemolysis applicator. Complete hemolysis requires 35 to 45 seconds.

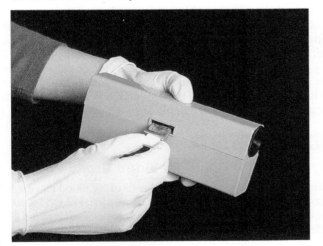

(c) Insert the charged blood chamber into the slot on the side of the hemoglobinometer.

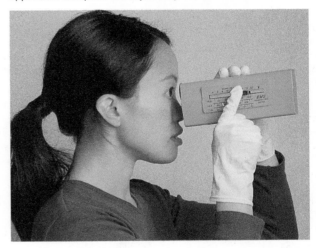

(d) Find the colors of the green split screen by moving the slide with the right index finger. When the two colors match in density, read the grams/100 ml and % Hb on the scale.

Figure 22.7 Hemoglobin determination using a hemoglobinometer.

acts enzymatically to polymerize (combine) the soluble **fibrinogen** proteins (present in plasma) into insoluble **fibrin,** which forms a meshwork of strands that traps the RBCs and forms the basis of the clot (Figure 22.8b). Normally, blood removed from the body clots within 2 to 6 minutes.

ACTIVITY 6

Determining Coagulation Time

1. Obtain a *nonheparinized* capillary tube, a timer (or watch), a lancet, cotton balls, a triangular file, and alcohol swabs.

2. Clean and prick the finger to produce a free flow of blood. Discard the lancet in the disposal container.

3. Place one end of the capillary tube in the blood drop, and hold the opposite end at a lower level to collect the sample.

4. Lay the capillary tube on a paper towel after collecting the sample.

Record the time._____

5. At 30-second intervals, make a small nick on the tube close to one end with the triangular file, and then carefully break the tube. Slowly separate the ends to see if a gel-like thread of fibrin spans the gap. When this occurs, record below and on the Hematologic Test Data Sheet (page 370) the time for coagulation to occur. Are your results within the normal time range?

_____

6. Put used supplies in the autoclave bag and broken capillary tubes into the sharps container. ■■

Blood Typing

Blood typing is a system of blood classification based on the presence of specific glycoproteins on the outer surface of the RBC plasma membrane. Such proteins are called **antigens**

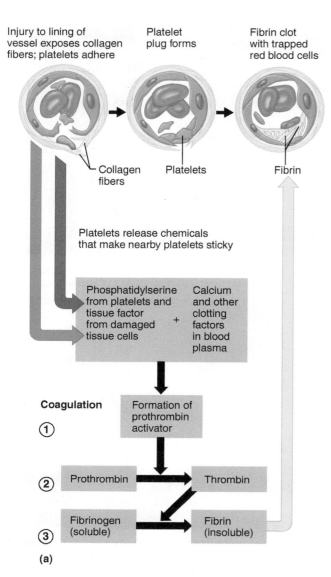

Platelets release chemicals that make nearby platelets sticky

Coagulation

(a)

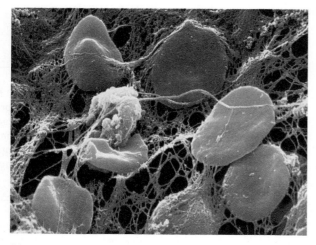

(b)

Figure 22.8 Events of platelet plug formation and coagulation. (a) Simple schematic of events. Steps numbered 1–3 represent the major events of coagulation. **(b)** Photomicrograph of RBCs trapped in a fibrin mesh (2500×).

or **agglutinogens** and are genetically determined. For ABO blood groups, these antigens are accompanied by plasma proteins, **antibodies,** or **agglutinins.** These antibodies act against RBCs carrying antigens that are not present on the person's own RBCs. If the donor blood type doesn't match, the recipient's antibodies react with the donor's blood antigens, causing the RBCs to clump, agglutinate, and eventually hemolyze. It is because of this phenomenon that a person's blood must be carefully typed before a whole blood or packed cell transfusion.

Several blood typing systems exist, based on the various possible antigens, but the factors routinely typed for are antigens of the ABO and Rh blood groups, which are most commonly involved in transfusion reactions. The basis of the ABO typing is shown in **Table 22.2**.

Individuals whose red blood cells carry the Rh antigen are Rh positive (approximately 85% of the U.S. population); those lacking the antigen are Rh negative. Unlike ABO blood groups, neither the blood of the Rh-positive (Rh^+) nor the blood of Rh-negative (Rh^-) individuals carries preformed anti-Rh antibodies. This is understandable in the case of the Rh-positive individual. However, Rh-negative persons who receive transfusions of Rh-positive blood become sensitized by the Rh antigens of the donor RBCs, and their systems begin to produce anti-Rh antibodies. On subsequent exposures to Rh-positive blood, typical transfusion reactions occur, resulting in the clumping and hemolysis of the donor blood cells.

ACTIVITY 7

Typing for ABO and Rh Blood Groups

Blood may be typed on glass slides or using blood test cards. Both methods are described next. The artificial blood kit does not use any body fluids and produces results similar to but not identical to results for human blood.

Typing Blood Using Glass Slides

1. Obtain two clean microscope slides, a wax marking pencil, anti-A, anti-B, and anti-Rh typing sera, toothpicks, lancets, alcohol swabs, a medicine dropper, and the Rh typing box.

2. Divide slide 1 into halves with the wax marking pencil. Label the lower left-hand corner "anti-A" and the lower right-hand corner "anti-B." Mark the bottom of slide 2 "anti-Rh."

3. Place one drop of anti-A serum on the *left* side of slide 1. Place one drop of anti-B serum on the *right* side of slide 1. Place one drop of anti-Rh serum in the center of slide 2.

4. If you are using your own blood, cleanse your finger with an alcohol swab, pierce the finger with a lancet, and wipe away the first drop of blood. Obtain 3 drops of freely flowing blood, placing one drop on each side of slide 1 and a drop on slide 2. Immediately dispose of the lancet in a designated disposal container.

If you are using instructor-provided animal blood or red blood cells treated with EDTA (an anticoagulant), use a medicine dropper to place one drop of blood on each side of slide 1 and a drop of blood on slide 2.

 5. Quickly mix each blood-antiserum sample with a *fresh* toothpick. Then dispose of the toothpicks and used alcohol swab in the autoclave bag.

Table 22.2   ABO Blood Typing

ABO blood type	Antigens present on RBC membranes	Antibodies present in plasma	% of U.S. population		
			White	Black	Asian
A	A	Anti-B	40	27	28
B	B	Anti-A	11	20	27
AB	A and B	None	4	4	5
O	Neither	Anti-A and anti-B	45	49	40

6. Place slide 2 on the Rh typing box, and rock it gently back and forth. (A slightly higher temperature is required for precise Rh typing than for ABO typing.)

7. After 2 minutes, observe all three blood samples for evidence of clumping. The agglutination that occurs in the positive test for the Rh factor is very fine and difficult to interpret; thus if there is any question, observe the slide under the microscope. Record your observations in the **Activity 7 chart.**

8. Interpret your ABO results (see the examples of each type in **Figure 22.9**). If you observed clumping on slide 2, you are Rh positive. If not, you are Rh negative.

9. Record your blood type in the Hematologic Test Data Sheet (page 370).

10. Put the used slides in the bleach-containing bucket at the general supply area; put disposable supplies in the autoclave bag.

Using Blood Typing Cards

1. Obtain a blood typing card marked A, B, and Rh; dropper bottles of anti-A serum, anti-B serum, and anti-Rh serum; toothpicks; lancets; and alcohol swabs.

2. Place a drop of anti-A serum in the spot marked anti-A, place a drop of anti-B serum on the spot marked anti-B, and place a drop of anti-Rh serum on the spot marked anti-Rh (or anti-D).

3. Carefully add a drop of blood to each of the spots marked "Blood" on the card. If you are using your own blood, refer to step 4 in the alternative instructions, "Typing Blood Using Glass Slides." Immediately discard the lancet in the designated disposal container.

4. Using a new toothpick for each test, mix the blood sample with the antibody. Dispose of the toothpicks appropriately.

5. Gently rock the card to allow the blood and antibodies to mix.

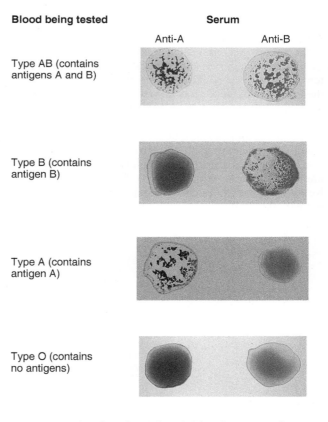

Activity 7: Blood Typing		
Result	Observed (+)	Not observed (−)
Presence of clumping with anti-A		
Presence of clumping with anti-B		
Presence of clumping with anti-Rh		

Figure 22.9 Blood typing of ABO blood types. When serum containing anti-A or anti-B antibodies (agglutinins) is added to a blood sample, agglutination will occur between the antibody and the corresponding antigen (agglutinogen A or B). As illustrated, agglutination occurs with both sera in blood group AB, with anti-B serum in blood group B, with anti-A serum in blood group A, and with neither serum in blood group O.

6. After 2 minutes, observe the card for evidence of clumping. The Rh clumping is very fine and may be difficult to observe. Record your observations in the Activity 7 chart. Use Figure 22.9 to interpret your results.

7. Record your blood type in the Hematologic Test Data Sheet (page 370), and discard the card in an autoclave bag. ▬

ACTIVITY 8

Observing Demonstration Slides

Look at the slides of *macrocytic hypochromic anemia, microcytic hypochromic anemia, sickle cell disease, lymphocytic leukemia* (chronic), and *eosinophilia* that have been put on demonstration by your instructor. Record your observations in the appropriate section of the Review Sheet for this exercise. You can refer to your notes, the text, and other references later to respond to questions about the blood pathologies represented on the slides. ▬

Cholesterol Concentration in Plasma

Atherosclerosis is the disease process in which the body's blood vessels become increasingly blocked by plaques. By narrowing the arteries, the plaques can contribute to hypertensive heart disease. They also serve as starting points for the formation of blood clots (thrombi), which may break away and block smaller vessels farther downstream in the circulatory pathway causing heart attacks or strokes.

Cholesterol is a major component of the smooth muscle plaques formed during atherosclerosis. No physical examination of an adult is considered complete until cholesterol levels are assessed along with other risk factors. A normal value for total plasma cholesterol in adults ranges from 130 to 200 mg per 100 ml of plasma; you will use blood to make such a determination.

Although the total plasma cholesterol concentration is valuable information, it may be misleading, particularly if a person's high-density lipoprotein (HDL) level is high and low-density lipoprotein (LDL) level is relatively low. Cholesterol, being water insoluble, is transported in the blood complexed to lipoproteins. In general, cholesterol bound into HDLs is destined to be degraded by the liver and then eliminated from the body, whereas that forming part of the LDLs is "traveling" to the body's tissue cells. When LDL levels are excessive, cholesterol is deposited in the blood vessel walls; hence, LDLs are considered to carry the "bad" cholesterol.

ACTIVITY 9

Measuring Plasma Cholesterol Concentration

1. Go to the appropriate supply area, and obtain a cholesterol test card and color scale, a lancet, and an alcohol swab.

2. Clean your fingertip with the alcohol swab, allow it to dry, then prick it with a lancet. Place a drop of blood on the test area of the card. Put the lancet in the designated disposal container.

3. After 3 minutes, remove the blood sample strip from the card and discard in the autoclave bag.

4. Analyze the underlying test spot, using the included color scale. Record the cholesterol level below and on the Hematologic Test Data Sheet (page 370).

Total cholesterol level _____ mg/dl

5. Before leaving the laboratory, use the spray bottle of bleach solution to saturate a paper towel, and thoroughly wash down your laboratory bench. ▬

Blood

Composition of Blood

1. What is the blood volume of an average-sized adult male? _____ liters

Of an average-sized adult female? _____ liters

2. What determines whether blood is bright red or a dull brick-red? _____

_____

3. Use the key to identify the cell type(s) or blood elements that fit the following descriptive statements. (Some terms will be used more than once.)

Key: a. red blood cell   d. basophil   g. lymphocyte
          b. megakaryocyte   e. monocyte   h. platelets
          c. eosinophil   f. neutrophil   i. plasma

_____ 1. most numerous leukocyte

_____, _____, and_____ 2. granulocytes (3)

_____ 3. also called an erythrocyte; anucleate formed element

_____, _____, _____ 4. phagocytic leukocytes (3)

_____ _____ 5. agranulocytes

_____ 6. precursor cell of platelets

_____ 7. cell fragments

_____ 8. involved in destroying parasitic worms

_____ 9. releases histamine; promotes inflammation

_____ 10. produces antibodies

_____ 11. transports oxygen

_____ 12. primarily water, noncellular; the fluid matrix of blood

_____ 13. exits a blood vessel to develop into a macrophage

_____, _____, _____,

_____, _____ 14. the five types of white blood cells

4. Define *formed elements*. _____

_____

5. Describe the consistency and color of the plasma you observed in the laboratory. _____

_____

6. What is the average life span of a red blood cell? How does its anucleate condition affect this life span?

_____

_____

7. Identify the leukocytes shown in the photomicrographs below.

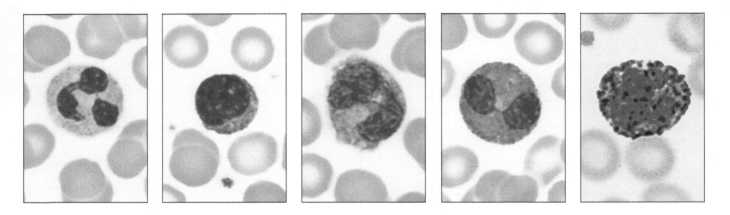

_____   _____   _____   _____   _____

8. Correctly identify the blood disorders described in column A by matching them with selections from column B.

Column A		Column B
_____ 1.	abnormal increase in the number of WBCs	a. anemia
_____ 2.	abnormal increase in the number of RBCs	b. leukocytosis
_____ 3.	condition of too few RBCs or of RBCs with hemoglobin deficiencies	c. leukopenia.
		d. polycythemia
_____ 4.	abnormal decrease in the number of WBCs	

Hematologic Tests

9. In the chart below, record information from the blood tests you read about or conducted. Complete the chart by recording values for healthy male adults and indicating the significance of high or low values for each test.

Test	Student test results	Normal values (healthy male adults)	Significance	
			High values	**Low values**
Total WBC count	No data			
Total RBC count	No data			
Hematocrit				
Hemoglobin determination				
Bleeding time	No data			
Coagulation time				

10. Why is a differential WBC count more valuable than a total WBC count when one is trying to determine the specific source

of pathology? _____

_____

11. Define *hematocrit.* _____

12. If you had a high hematocrit, would you expect your hemoglobin determination to be high or low? _____

Why? _____

_____

13. What is an anticoagulant? _____

Name two anticoagulants used in conducting the hematologic tests. _____

and _____

What is the body's natural anticoagulant? _____

14. If your blood agglutinates with anti-A but not anti-B sera, your ABO blood type would be _____

To which ABO blood groups could you donate blood? _____

From which ABO donor types could you receive blood? _____

Which ABO blood type is most common? _____ Least common? _____

15. Assume the blood of two patients has been typed for ABO blood type.

Typing results
Mr. Adams:

Blood drop and
anti-A serum

Blood drop and
anti-B serum

Typing results
Mr. Calhoon:

Blood drop and
anti-A serum

Blood drop and
anti-B serum

On the basis of these results, Mr. Adams has type _____ blood, and Mr. Calhoon has type _____ blood.

16. Record your observations of the five demonstration slides viewed.

a. Macrocytic hypochromic anemia: _____

b. Microcytic hypochromic anemia: _____

_____

c. Sickle cell disease: _____

d. Lymphocytic leukemia (chronic): _____

_____

e. Eosinophilia: _____

Which of the slides (a through e) above corresponds with the following conditions? (Consult your textbook or other reference source.)

_____ 1. iron-deficient diet   _____ 4. lack of vitamin B_{12}

_____ 2. a type of bone marrow cancer   _____ 5. a tapeworm infestation in the body

_____ 3. genetic defect that causes hemoglobin   _____ 6. a bleeding ulcer
         to become sharp/spiky

17. Plasmapheresis is a procedure in which blood is removed, its plasma is separated from the formed elements, and the formed elements are returned to the patient or donor. Kidney transplants usually require that the donor and recipient have the same blood type. If plasmapheresis is administered to the patient before and after the transplant surgery, rejection of the kidney is unlikely to occur. Explain why.

_____

_____

18. ✚ Bleeding disorders are usually a result of thrombocytopenia, a deficiency of platelets. Considering the mechanism of

hemostasis, explain why thrombocytopenia could lead to abnormal bleeding. _____

_____

_____

Anatomy of the Heart

MATERIALS

- X-ray image of the human thorax; X-ray viewing box
- Three-dimensional heart model and torso model or laboratory chart showing heart anatomy
- Red and blue pencils
- Highlighter
- Three-dimensional models of cardiac and skeletal muscle
- Compound microscope
- Prepared slides of cardiac muscle (l.s.)
- Preserved sheep heart, pericardial sacs intact or fresh hearts, pericardial sacs intact (if possible)
- Dissecting instruments and tray
- Pointed glass rods or blunt probes
- Disposable gloves
- Small plastic metric rulers
- Container for disposal of organic debris
- Laboratory detergent
- Spray bottle with 10% household bleach solution

LEARNING OUTCOMES

- ☐ Describe the location of the heart.
- ☐ Name and describe the covering and lining tissues of the heart.
- ☐ Name and locate the major anatomical areas and structures of the heart when provided with an appropriate model, image, or dissected sheep heart, and describe the function of each.
- ☐ Explain how the atrioventricular and semilunar valves operate.
- ☐ Distinguish blood vessels carrying oxygen-rich blood from those carrying carbon dioxide–rich blood, and describe the system used to color-code them in images.
- ☐ Explain why the heart is called a double pump, and compare the pulmonary and systemic circuits.
- ☐ Trace the pathway of blood through the heart.
- ☐ Trace the functional blood supply of the heart, and name the associated blood vessels.
- ☐ Describe the histology of cardiac muscle, and state the importance of its intercalated discs and the spiral arrangement of its cells.

PRE-LAB QUIZ

1. The heart is enclosed in a double-walled sac called the:
   a. apex   b. mediastinum   c. pericardium   d. thorax
2. The heart is divided into _____ chambers.
   a. two   b. three   c. four   d. five
3. What is the name of the two receiving chambers of the heart?
   _____
4. The left ventricle discharges blood into the _____, from which all systemic arteries of the body diverge to supply the body tissues.
   a. aorta   c. pulmonary vein
   b. pulmonary artery   d. vena cava
5. Circle True or False. Blood flows through the heart in one direction—from the atria to the ventricles.
6. Circle the correct underlined term. The right atrioventricular valve, or tricuspid valve / mitral valve, prevents backflow into the right atrium when the right ventricle is contracting.
7. Circle the correct underlined term. The heart serves as a double pump. The right / left side serves as the pulmonary circuit pump, shunting carbon dioxide–rich blood to the lungs.
8. The blood vessels that supply blood to the heart itself are the:
   a. aortas   c. coronary arteries
   b. carotid arteries   d. pulmonary trunks
9. Two microscopic features of cardiac cells that help distinguish them from other types of muscle cells are branching of the cells and:
   a. intercalated discs   c. sarcolemma
   b. myosin fibers   d. striations
10. Circle the correct underlined term. In the heart, the left / right ventricle has thicker walls and a basically circular cavity shape.

The major function of the **cardiovascular system** is transportation. Using blood as the transport vehicle, the system carries oxygen, nutrients, cell wastes, electrolytes, and many other substances vital to the body's homeostasis to and from the body cells. The system's propulsive force is the contracting heart, which can be compared to a muscular pump equipped with one-way valves. As the heart contracts, it forces blood into a closed system of large and small plumbing tubes (blood vessels) within which the blood is confined and circulated.

Gross Anatomy of the Human Heart

The **heart,** a cone-shaped organ approximately the size of a fist, is located within the mediastinum of the thorax. It is flanked laterally by the lungs, posteriorly by the vertebral column, and anteriorly by the sternum **(Figure 23.1)**. Its more pointed **apex** extends slightly to the left and rests on the diaphragm, approximately at the level of the fifth intercostal space. Its broader **base,** from which the great vessels emerge, lies beneath the second rib and points toward the right shoulder.

☐ If an X- ray image of a human thorax is available, verify the relationships described above (otherwise, use Figure 23.1).

Check the box when you have completed this task.

Figure 23.2 shows three views of the heart—anterior and posterior views and a frontal section.

The heart is enclosed within a double-walled sac called the *pericardium*. The thin **visceral pericardium,** or **epicardium,** is closely applied to the heart muscle. It reflects downward at the base of the heart to form its companion serous membrane, the outer, loosely applied **parietal pericardium,** which is attached at the heart apex to the diaphragm. Serous fluid produced by these membranes allows the heart to beat in a relatively frictionless environment. The serous parietal pericardium, in turn, lines the loosely fitting superficial **fibrous pericardium** composed of dense connective tissue.

The walls of the heart are composed of three layers:

- **Epicardium:** The outer layer, which is also the visceral pericardium.
- **Myocardium:** The middle layer and thickest layer, which is composed mainly of cardiac muscle. It is reinforced with dense fibrous connective tissue, the *cardiac skeleton,* which is thicker around the heart valves and at the base of the great vessels leaving the heart.
- **Endocardium:** The inner lining of the heart, which covers the heart valves and is continuous with the inner lining of the great vessels. It is composed of simple squamous epithelium resting on areolar connective tissue.

Heart Chambers

The heart is divided into four chambers: two superior **atria** (singular: *atrium*) and two inferior **ventricles.** The septum that divides the heart longitudinally is referred to as the **interatrial septum** where it separates the atria and the **interventricular septum** where it separates the ventricles. Functionally, the atria are receiving chambers and are relatively ineffective as pumps.

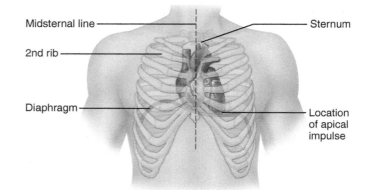

Figure 23.1 Location of the heart in the thorax. The apical impulse occurs where the heart's apex touches the chest wall.

The inferior thick-walled ventricles, which form the bulk of the heart, are the discharging chambers. They force blood out of the heart into the large arteries that emerge from its base.

Heart Valves

Four valves enforce a one-way blood flow through the heart chambers. The **atrioventricular (AV) valves** are located between the atrium and the ventricle on the left and right side of the heart. The **semilunar (SL) valves** are located between a ventricle and a great vessel **(Figure 23.3)**.

- **Tricuspid valve:** The right AV valve has three flaplike cusps anchored to the **papillary muscles** of the ventricular wall by tiny, white collagenic cords called **chordae tendineae** (literally, "heart strings")
- **Mitral valve** (*bicuspid valve*): The left AV valve has two flaplike cusps anchored to the papillary muscles by chordae tendineae.

The AV valves are open and hang into the ventricles when blood is flowing into the atria and the ventricles are relaxed. When the ventricles contract, the blood in the ventricles is compressed, causing the AV valves to move superiorly and close the opening between the atrium and the ventricle. The chordae tendineae, pulled tight by the contracting papillary muscles, anchor the cusps in the closed position and prevent the backflow of blood from the ventricles into the atria like an umbrella being turned inside out by a strong wind.

- **Pulmonary (SL) valve:** Has three pocketlike cusps located between the right ventricle and the pulmonary trunk.
- **Aortic (SL) valve:** Has three pocketlike cusps located between the left ventricle and the aorta.

The SL valves are open and flattened against the wall of the vessel when the contraction of the ventricles pushes blood into the great vessels. When the ventricles relax, blood flows backward toward the ventricle and the cusps fill with blood, closing the SL valves. This prevents the backflow of blood from the great vessels into the ventricles.

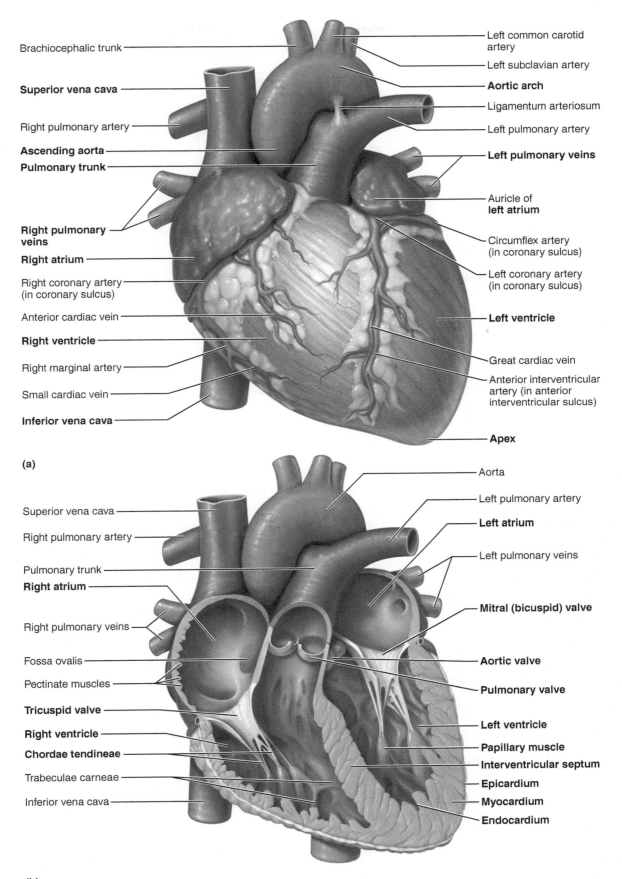

Brachiocephalic trunk

Superior vena cava

Right pulmonary artery

Ascending aorta

Pulmonary trunk

Right pulmonary veins

Right atrium

Right coronary artery (in coronary sulcus)

Anterior cardiac vein

Right ventricle

Right marginal artery

Small cardiac vein

Inferior vena cava

Left common carotid artery

Left subclavian artery

Aortic arch

Ligamentum arteriosum

Left pulmonary artery

Left pulmonary veins

Auricle of left atrium

Circumflex artery (in coronary sulcus)

Left coronary artery (in coronary sulcus)

Left ventricle

Great cardiac vein

Anterior interventricular artery (in anterior interventricular sulcus)

Apex

(a)

Superior vena cava

Right pulmonary artery

Pulmonary trunk

Right atrium

Right pulmonary veins

Fossa ovalis

Pectinate muscles

Tricuspid valve

Right ventricle

Chordae tendineae

Trabeculae carneae

Inferior vena cava

Aorta

Left pulmonary artery

Left atrium

Left pulmonary veins

Mitral (bicuspid) valve

Aortic valve

Pulmonary valve

Left ventricle

Papillary muscle

Interventricular septum

Epicardium

Myocardium

Endocardium

(b)

Figure 23.2 Gross anatomy of the human heart. (a) Anterior view. **(b)** Frontal section.

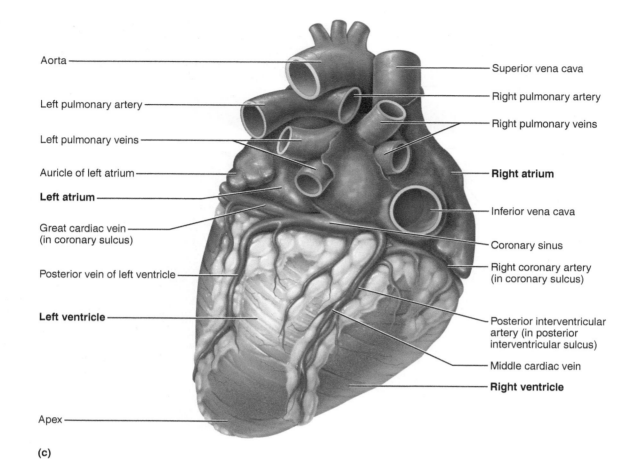

Aorta

Left pulmonary artery

Left pulmonary veins

Auricle of left atrium

Left atrium

Great cardiac vein
(in coronary sulcus)

Posterior vein of left ventricle

Left ventricle

Apex

Superior vena cava

Right pulmonary artery

Right pulmonary veins

Right atrium

Inferior vena cava

Coronary sinus

Right coronary artery
(in coronary sulcus)

Posterior interventricular
artery (in posterior
interventricular sulcus)

Middle cardiac vein

Right ventricle

(c)

Figure 23.2 *(continued)* **Gross anatomy of the human heart. (c)** Posterior view.

ACTIVITY 1

Using the Heart Model to Study Heart Anatomy

Locate in Figure 23.2 all the structures described above. Then, observe the human heart model and laboratory charts, and identify the same structures without referring to the figure. ▬

Pulmonary, Systemic, and Coronary Circulations

Pulmonary and Systemic Circuits

The heart functions as a double pump:

- The right side of the heart pumps oxygen-poor blood entering its chambers to the lungs to unload carbon dioxide and to pick up oxygen. The blood vessels that carry blood to and from the lungs form the **pulmonary circuit.** The function of the pulmonary circuit is strictly to provide for gas exchange.

- The left side of the heart pumps oxygenated blood returning from the lungs to the body tissues. The blood vessels that carry blood to and from all body tissues form the **systemic circuit. (Figure 23.4**, page 386)

The following steps describe the blood flow through the right side of the heart (pulmonary circuit):

1. The right atrium receives oxygen-poor blood from the body via the venae cavae (**superior vena cava** and **inferior vena cava**) and the coronary sinus.

2. From the right atrium, blood flows through the tricuspid valve to the right ventricle.

3. From the right ventricle, blood flows through the pulmonary valve into the **pulmonary trunk.**

4. The pulmonary trunk branches into left and right **pulmonary arteries,** which carry blood to the lungs, where the blood unloads carbon dioxide and picks up oxygen.

5. Oxygen-rich blood returns to the heart via four pulmonary veins.

The remaining steps describe the blood flow through the left side of the heart (systemic circuit):

6. Oxygen-rich blood enters the left atrium via four pulmonary veins.

7. From the left atrium, blood flows through the mitral valve to the left ventricle.

8. From the left ventricle, blood flows through the aortic valve to the **aorta.**

9. Oxygen-rich blood is delivered to the body tissues by the systemic arteries.

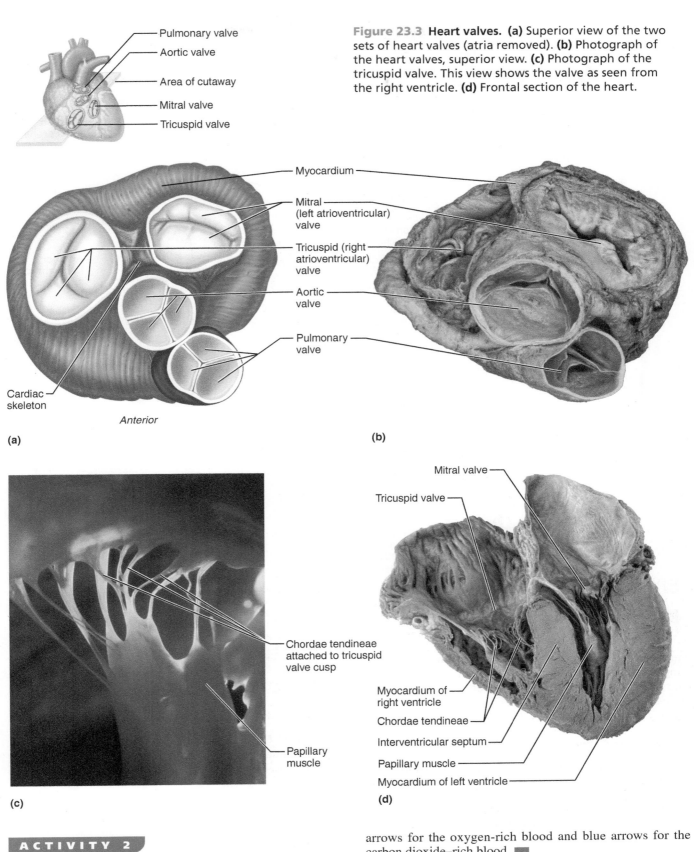

Pulmonary valve

Aortic valve

Area of cutaway

Mitral valve

Tricuspid valve

Figure 23.3 Heart valves. (a) Superior view of the two sets of heart valves (atria removed). **(b)** Photograph of the heart valves, superior view. **(c)** Photograph of the tricuspid valve. This view shows the valve as seen from the right ventricle. **(d)** Frontal section of the heart.

Myocardium

Mitral (left atrioventricular) valve

Tricuspid (right atrioventricular) valve

Aortic valve

Pulmonary valve

Cardiac skeleton

Anterior

(a)

(b)

Mitral valve

Tricuspid valve

Chordae tendineae attached to tricuspid valve cusp

Papillary muscle

Myocardium of right ventricle

Chordae tendineae

Interventricular septum

Papillary muscle

Myocardium of left ventricle

(c)

(d)

Tracing the Path of Blood Through the Heart

Use colored pencils to trace the pathway of a red blood cell through the heart by adding arrows to Figure 23.2b. Use red arrows for the oxygen-rich blood and blue arrows for the carbon dioxide–rich blood. ◼

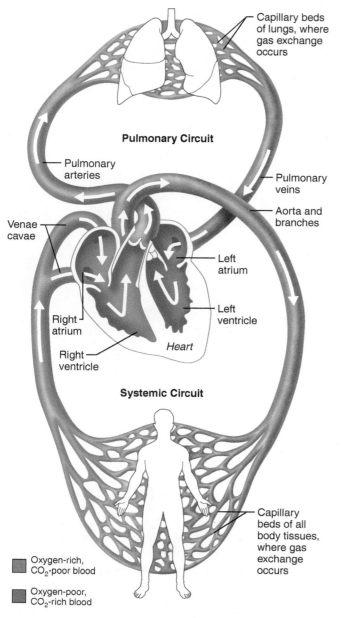

Figure 23.4 The systemic and pulmonary circuits. For simplicity, the actual number of two pulmonary arteries and four pulmonary veins has been reduced to one each in this art.

Coronary Circulation

Even though the heart chambers are almost continually bathed with blood, this contained blood does not nourish the myocardium. The functional blood supply of the heart is provided by the right and left coronary arteries (Figure 23.2 and **Figure 23.5**). **Table 23.1** summarizes the arteries that supply blood to the heart and the veins that drain the blood.

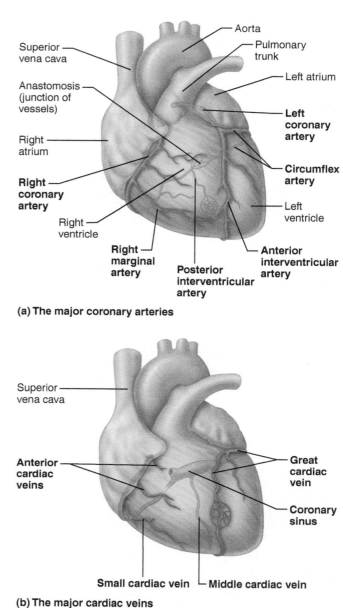

(a) The major coronary arteries

(b) The major cardiac veins

Figure 23.5 Coronary circulation.

ACTIVITY 3

Using the Heart Model to Study Coronary Circulation

1. Obtain a highlighter, and highlight all the names of the blood vessels in Figure 23.2a and c. Note how arteries and veins travel together.

2. On a model of the heart, locate all the blood vessels shown in Figure 23.5. Use your finger to trace the pathway of blood from the right coronary artery to the lateral aspect of the right side of the heart and back to the right atrium. Name the arteries and veins along the pathway. Trace the pathway of blood from the left coronary artery to the anterior ventricular walls and back to the right atrium. Name the arteries and veins along the pathway. Note that there are multiple different pathways to distribute blood to these parts of the heart. ■

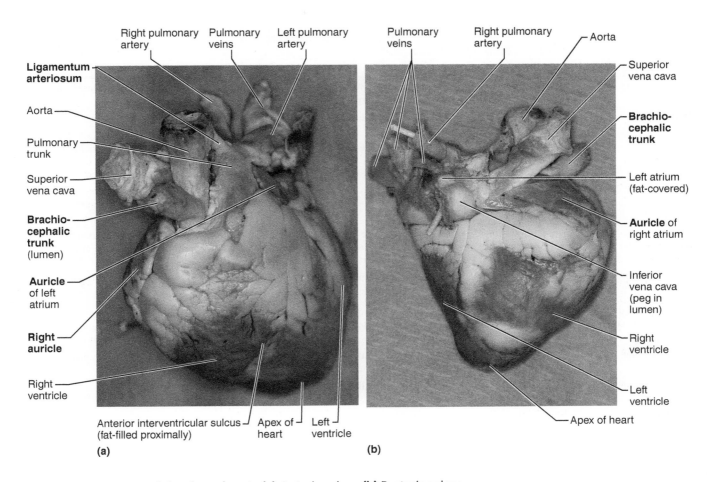

Figure 23.6 **Anatomy of the sheep heart. (a)** Anterior view. **(b)** Posterior view.

DISSECTION
The Sheep Heart

Dissecting a sheep heart is a valuable exercise because it is similar in size and structure to the human heart. Refer to **Figure 23.6** as you proceed with the dissection.

1. Obtain a preserved sheep heart, a dissecting tray, dissecting instruments, a glass probe, a plastic ruler, and gloves.

2. Observe the texture of the fibrous pericardium. Also, note its point of attachment to the heart. Where is it attached?

_____

_____

3. If the fibrous pericardial sac is still intact, slit it open and cut it from its attachments. Observe the slippery parietal pericardium that lines the sac and the visceral pericardium (epicardium) that covers the heart wall. Using a sharp scalpel, carefully pull a little of the epicardium away from the myocardium. How do its position, thickness, and apposition to the heart differ from those of the parietal pericardium?

_____

_____

_____

4. Examine the external surface of the heart. Notice the accumulation of adipose tissue, which in many cases marks the separation of the chambers and the location of the coronary arteries. Carefully scrape away some of the fat with a scalpel to expose the coronary blood vessels.

5. Identify the base and apex of the heart, and then identify the two wrinkled **auricles,** earlike flaps of tissue projecting from the atria. The balance of the heart muscle is ventricular tissue. To identify the left ventricle, compress the ventricles on each side of the longitudinal fissures carrying the coronary blood vessels. The side that feels thicker and more solid is the left ventricle. The right ventricle is much thinner and feels somewhat flabby when compressed. This difference reflects the greater demand placed on the left ventricle, which must pump blood through the much longer systemic circuit, a pathway with much higher resistance than the pulmonary circuit served by the right ventricle. Hold the heart in its anatomical position (Figure 23.6a), with the anterior surface uppermost. In this position, the left ventricle composes the entire apex and the left side of the heart.

6. Identify the pulmonary trunk and the aorta extending from the superior aspect of the heart. The pulmonary trunk is more anterior, and you may see its division into the right and left pulmonary arteries if it has not been cut too closely to the heart. The thicker-walled aorta, which branches almost immediately, is located just beneath the pulmonary trunk. The first observable branch of the sheep aorta, the **brachiocephalic trunk,** is identifiable unless the aorta has been cut immediately as it leaves the heart.

Table 23.1   Coronary Circulation (Figure 23.5)

Arteries	Description	Areas supplied/branches
Right coronary artery (RCA)	Branches from the ascending aorta just above the aortic valve and encircles the heart in the coronary sulcus.	Its branches include the right marginal artery and the posterior interventricular artery.
Right marginal artery	Branches off the RCA and is located in the lateral portion of the right ventricle.	Supplies the lateral right side of the heart.
Posterior interventricular artery	Branches off the RCA and is located in the posterior interventricular sulcus.	Supplies the posterior walls of the ventricles and the posterior portion of the interventricular septum. Near the apex of the heart it merges (anastomoses) with the anterior interventricular artery.
Left coronary artery (LCA)	Branches from the ascending aorta and passes posterior to the pulmonary trunk.	Its branches include the anterior interventricular artery and the circumflex artery.
Anterior interventricular artery	Branches off the LCA and is located in the anterior interventricular sulcus. This artery is referred to clinically as the left anterior descending artery (LAD).	Supplies the anterior portion of the interventricular septum and the anterior walls of both ventricles.
Circumflex artery	Branches off the LCA; located in the coronary sulcus.	Supplies the left atrium and the posterior portion of the left ventricle.
Veins	**Description**	**Areas drained**
Great cardiac vein	Located in the anterior interventricular sulcus, parallel to the anterior interventricular artery.	Anterior portions of the right and left ventricles.
Middle cardiac vein	Located in the posterior interventricular sulcus, parallel to the posterior interventricular artery.	Posterior portions of the right and left ventricles.
Small cardiac vein	Located on the lateral right ventricle, parallel to the right marginal artery.	Lateral right ventricle.
Coronary sinus	Located in the coronary sulcus on the posterior surface of the heart; drains into the right atrium.	The entire heart; the great, middle and small cardiac veins all drain into the coronary sinus.
Anterior cardiac veins	Located on the anterior surface of the right atrium.	They drain directly into the right atrium.

Carefully clear away some of the fat between the pulmonary trunk and the aorta to expose the **ligamentum arteriosum,** a cordlike remnant of the **ductus arteriosus.** (In the fetus, the ductus arteriosus allows blood to pass directly from the pulmonary trunk to the aorta, thus bypassing the nonfunctional fetal lungs.)

7. Cut through the wall of the aorta until you see the aortic valve. Identify the two openings to the coronary arteries just above the valve. Insert a probe into one of these holes to see if you can follow the course of a coronary artery.

8. Turn the heart to view its posterior surface (compare it to the view in Figure 23.6b). Notice that the right and left ventricles appear equal-sized in this view. Try to identify the four thin-walled pulmonary veins entering the left atrium. Identify the superior and inferior venae cavae entering the right atrium. Because of the way the heart is trimmed, the pulmonary veins and superior vena cava may be very short or missing. If possible, compare the approximate diameter of the superior vena cava with the diameter of the aorta.

Which is larger?_____

Which has thicker walls?_____

Why do you suppose these differences exist?

_____

_____

_____

9. Insert a probe into the superior vena cava, through the right atrium, and out the inferior vena cava. Use scissors to cut along the probe so that you can view the interior of the right atrium. Observe the tricuspid valve.

How many cusps does it have?_____

10. Return to the pulmonary trunk, and cut through its anterior wall until you can see the pulmonary (semilunar) valve **(Figure 23.7).** How does its action differ from that of the tricuspid valve?

_____

_____

_____

Extend the cut through the pulmonary trunk into the right ventricle. Cut down, around, and up through the tricuspid valve to make the cut continuous with the cut across the right atrium (see Figure 23.7).

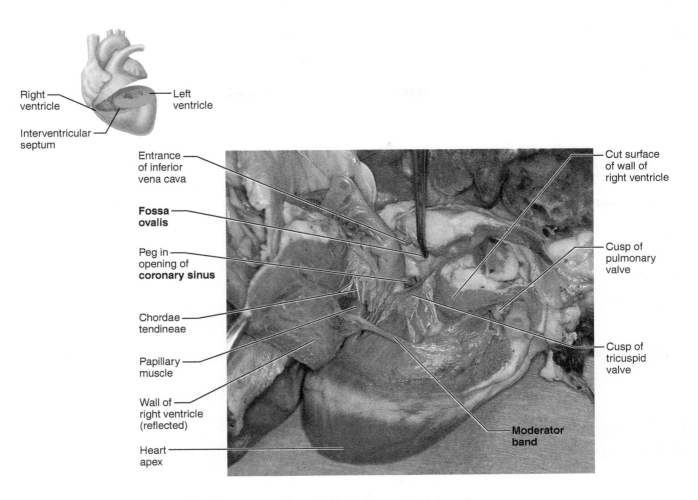

Figure 23.7 Right side of the sheep heart opened and reflected to reveal internal structures. Overview diagram illustrates the anatomical differences between the right and left ventricles. The left ventricle has thicker walls, and its cavity is basically circular. By contrast, the right ventricle cavity is crescent-shaped and wraps around the left ventricle.

11. Reflect the cut edges of the superior vena cava, right atrium, and right ventricle to obtain the view seen in Figure 23.7. Observe the comblike ridges of muscle throughout most of the right atrium. This is called **pectinate muscle** (*pectin* means "comb"). Identify, on the posterior atrial wall, the large opening of the inferior vena cava, and follow it to its external opening with a probe. Notice that the atrial walls in the vicinity of the venae cavae are smooth and lack the roughened appearance (pectinate musculature) of the other regions of the atrial walls. Just below the inferior vena caval opening, identify the opening of the **coronary sinus.** Nearby, locate an oval depression, the **fossa ovalis,** in the interatrial septum. This depression marks the site of an opening in the fetal heart, the **foramen ovale,** which allows blood to pass from the right to the left atrium, thus bypassing the fetal lungs.

12. Identify the papillary muscles in the right ventricle, and follow their attached chordae tendineae to the cusps of the tricuspid valve. Notice the pitted and ridged appearance **(trabeculae carneae)** of the inner ventricular muscle.

13. Identify the **moderator band** (septomarginal band), a bundle of cardiac muscle fibers connecting the interventricular

septum to anterior papillary muscles. It contains a branch of the atrioventricular bundle and helps coordinate contraction of the ventricle.

14. Make a longitudinal incision through the left atrium, and continue it into the left ventricle. Notice how much thicker the myocardium of the left ventricle is than that of the right ventricle. Measure the thickness of the left and right ventricular walls in millimeters, and record the numbers:

left _____

right_____

Compare the *shape* of the left ventricular cavity to the shape of the right ventricular cavity (see the overview diagram in Figure 23.7).

_____

_____

_____

Are the papillary muscles and chordae tendineae observed in the right ventricle also present in the left ventricle? _____

Count the number of cusps in the mitral valve. How does this compare with the number seen in the tricuspid valve?

_____

_____

How do the sheep valves compare with their counterparts in humans?

_____

15. Reflect the cut edges of the atrial wall, and attempt to locate the entry points of the pulmonary veins into the left atrium. Follow the pulmonary veins, if present, to the heart exterior with a probe. Notice how thin walled these vessels are.

16. Dispose of the organic debris in the designated container, clean the dissecting tray and instruments with detergent and water, and wash the lab bench with bleach solution before continuing on to Activity 4. ■

Microscopic Anatomy of Cardiac Muscle

The cardiac cells, crisscrossed by connective tissue fibers for strength, are arranged in spiral or figure-8-shaped bundles (**Figure 23.8**, overview diagram). When the heart contracts, its internal chambers become smaller, forcing the blood into the large arteries leaving the heart.

ACTIVITY 4

Examining Cardiac Muscle Tissue Anatomy

1. Observe the three-dimensional model of cardiac muscle, examining its branching cells and the areas where the cells interlock, the **intercalated discs.**

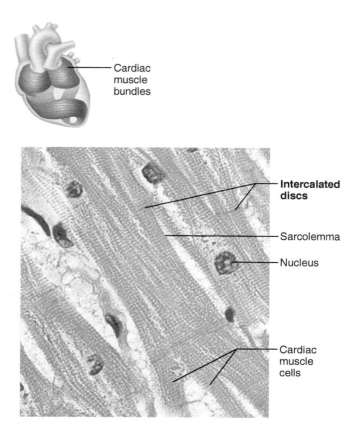

Figure 23.8 Photomicrograph of cardiac muscle (875×). The overview diagram illustrates the circular and spiral arrangement of cardiac muscle bundles in the myocardium of the heart.

2. Compare the model of cardiac muscle to the model of skeletal muscle. Note the similarities and differences between the two kinds of muscle tissue.

3. Observe a longitudinal section of cardiac muscle under high power. Identify the nucleus, striations, intercalated discs, and sarcolemma of the individual cells, and then compare your observations to the photomicrograph in Figure 23.8. ■

Anatomy of the Heart

Gross Anatomy of the Human Heart

1. An anterior view of the heart is shown here. Match each structure listed with the correct letter in the figure.

_____ 1. right atrium

_____ 2. right ventricle

_____ 3. left atrium

_____ 4. left ventricle

_____ 5. superior vena cava

_____ 6. ascending aorta

_____ 7. aortic arch

_____ 8. brachiocephalic trunk

_____ 9. left common carotid artery

_____ 10. left subclavian artery

_____ 11. pulmonary trunk

_____ 12. left pulmonary arteries

_____ 13. ligamentum arteriosum

_____ 14. left pulmonary veins

_____ 15. right coronary artery

_____ 16. circumflex artery

_____ 17. anterior interventricular artery

_____ 18. apex of heart

_____ 19. great cardiac vein

2. What is the function of the fluid that fills the pericardial sac? _____

_____

3. Match the terms in the key to the descriptions provided below.

_____ 1.   location of the heart in the thorax

_____ 2.   tricuspid and mitral valves

_____ 3.   discharging chambers of the heart

_____ 4.   visceral pericardium

_____ 5.   receiving chambers of the heart

_____ 6.   layer composed of cardiac muscle

_____ 7.   provide nutrient blood to the heart muscle

_____ 8.   lining of the heart chambers

_____ 9.   pulmonary and aortic valves

_____ 10.   drains blood into the right atrium

Key:

a.   atria

b.   atrioventricular valves

c.   coronary arteries

d.   coronary sinus

e.   endocardium

f.   epicardium

g.   mediastinum

h.   myocardium

i.   semilunar valves

j.   ventricles

4. Which valves are anchored by chordae tendineae? _____

_____

5. Which valves close when the cusps fill with blood? _____

_____

Pulmonary, Systemic, and Coronary Circulations

6. Describe the role of the pulmonary circuit. _____

_____

_____

7. Describe the role of the systemic circuit. _____

_____

_____

8. Name the three vessels that deliver oxygen-poor blood to the right atrium.

_____

9. Starting with the right atrium, trace a drop of blood through the heart and lungs, naming the following structures: aorta, aortic valve, left atrium, left ventricle, mitral valve, pulmonary arteries, pulmonary capillaries, pulmonary valve, pulmonary trunk, pulmonary veins, right atrium, right ventricle, and tricuspid valve.

1. _____   8. _____

2. _____   9. _____

3. _____   10. _____

4. _____   11. _____

5. _____   12. _____

6. _____   13. _____

7. _____

Dissection of the Sheep Heart

10. During the sheep heart dissection, you were asked initially to identify the right and left ventricles without cutting into the heart. During this procedure, what differences did you observe between the two chambers?

_____

_____

Knowing that structure and function are related, how would you say this structural difference reflects the relative functions

of these two heart chambers? _____

_____

11. Semilunar valves prevent backflow into the _____; atrioventricular valves prevent

backflow into the _____. Using your own observations, explain how the operation

of the semilunar valves differs from that of the mitral and tricuspid valves. _____

_____

_____

_____

12. Compare and contrast the structure of the right and left atrioventricular valves. _____

_____

Microscopic Anatomy of Cardiac Muscle

13. How would you distinguish the structure of cardiac muscle from the structure of skeletal muscle? _____

_____

_____

14. Add the following terms to the photograph of cardiac muscle below.

    a.   intercalated discs   b.   nucleus   c.   cardiac muscle cell

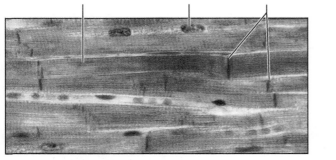

Describe the unique anatomical features of cardiac muscle. What role does the unique structure of cardiac muscle play in its function?

_____

_____

_____

_____

_____

_____

15. ✚ A proximal LAD lesion is a blockage in the left anterior descending artery, also known as the anterior interventricular artery. Explain why a heart attack caused by an obstruction of this artery is sometimes referred to as the "widow maker"

heart attack. _____

_____

_____

16. ✚ Congestive heart failure (CHF) is an inability of the heart to pump sufficient blood to the body. One sign of CHF is excess fluid in the tissue spaces, known as edema. Describe the location of the edema if the left side of the heart fails, compared to

the location of edema if the right side of the heart fails. _____

_____

_____

Anatomy of Blood Vessels

MATERIALS

- Compound microscope
- Prepared microscope slides showing cross sections of an artery and a vein
- Anatomical charts of human arteries and veins (or a three-dimensional model of the human circulatory system)
- Anatomical charts of the following specialized circulations: pulmonary circulation, hepatic portal circulation, fetal circulation, arterial supply and cerebral arterial circle of the brain (or a brain model showing this circulation)
- Dissection animal, tray, and instruments
- Bone cutters
- Scissors
- Disposable gloves
- Safety glasses
- Embalming fluid
- Organic debris container

LEARNING OUTCOMES

- ☐ Describe the tunics of blood vessel walls, and state the function of each layer.
- ☐ Correlate differences in artery, vein, and capillary structure with the functions of these vessels.
- ☐ Recognize a cross-sectional view of an artery and vein when provided with a microscopic view or an appropriate image.
- ☐ List and identify the major arteries arising from the aorta, and indicate the body region supplied by each.
- ☐ Describe the cerebral arterial circle, and discuss its importance in the body.
- ☐ List and identify the major veins draining into the superior and inferior venae cavae, and indicate the body regions drained.
- ☐ Describe these special circulations in the body: pulmonary circulation, hepatic portal system, and fetal circulation, and discuss the important features of each.
- ☐ Identify important blood vessels of the cat, and describe anatomical differences between the vascular system of the human and the cat.

PRE-LAB QUIZ

1. Circle the correct underlined term. <u>Arteries</u> / <u>Veins</u> drain tissues and return blood to the heart.
2. Circle True or False. Gas exchange takes place between tissue cells and blood through capillary walls.
3. The _____ is the largest artery of the body.
   a. aorta   c. femoral artery
   b. carotid artery   d. subclavian artery
4. Circle the correct underlined term. The largest branch of the abdominal aorta, the <u>renal</u> / <u>superior mesenteric</u> artery, supplies most of the small intestine and the first half of the large intestine.
5. The anterior tibial artery terminates with the _____ artery, which is often palpated in patients with circulatory problems to determine the circulatory efficiency of the lower limb.
   a. dorsalis pedis   c. obturator
   b. external iliac   d. tibial
6. Circle the correct underlined term. Veins draining the head and upper extremities empty into the <u>superior</u> / <u>inferior</u> vena cava.
7. Located in the lower limb, the _____ is the longest vein in the body.
   a. external iliac   c. great saphenous
   b. fibular   d. internal iliac
8. Circle the correct underlined term. The <u>renal</u> / <u>hepatic</u> veins drain the liver.
9. The function of the _____ is to drain the digestive viscera and carry dissolved nutrients to the liver for processing.
   a. fetal circulation
   b. hepatic portal circulation
   c. pulmonary circulation system
10. Circle the correct underlined term. In the developing fetus, the umbilical <u>artery</u> / <u>vein</u> carries blood rich in nutrients and oxygen to the fetus.

rteries, which carry blood away from the heart, and veins, which drain the tissues and return blood to the heart, function simply as conducting vessels or conduits. Only the tiny capillaries that connect the arterioles and venules and branch throughout the tissues directly serve the needs of the body's cells. It is through the capillary walls that exchanges are made between tissue cells and blood.

In this exercise you will examine the microscopic structure of blood vessels and identify the major arteries and veins of the systemic circulation and other special circulations.

Microscopic Structure of the Blood Vessels

Except for the microscopic capillaries, the walls of blood vessels are constructed of three coats, or *tunics* **(Figure 24.1)**.

- **Tunica intima:** Lines the lumen of a vessel and is composed of the *endothelium,* subendothelial layer, and internal elastic membrane. The simple squamous cells of the endothelium fit closely together, forming an extremely smooth blood vessel lining that helps decrease resistance to blood flow.

- **Tunica media:** Middle coat, composed primarily of smooth muscle and elastin. The smooth muscle plays an active role in regulating the diameter of blood vessels, which in turn alters blood flow and blood pressure.

- **Tunica externa:** Outermost tunic, composed of areolar or fibrous connective tissue. Its function is basically supportive and protective. In larger vessels, the tunica externa contains a system of tiny blood vessels, the **vasa vasorum.**

In general, the walls of arteries are thicker than those of veins. The tunica media in particular tends to be much heavier and contains substantially more smooth muscle and elastic tissue. Arteries, which are closer to the pumping action of the heart, must be able to expand as an increased volume of blood is propelled into them during systole and then recoil passively as the blood flows off into the circulation during diastole. The anatomical differences between the different types of vessels reflect their functional differences. **Table 24.1** summarizes the structure and function of various blood vessels.

Valves in veins act to prevent backflow of blood in much the same manner as the semilunar valves of the heart. The skeletal muscle "pump" also promotes venous return; as the skeletal muscles surrounding the veins contract and relax, the blood is milked through the veins toward the heart. Anyone who has been standing relatively still for an extended time has experienced swelling in the ankles, caused by blood pooling in the feet during the period of muscle inactivity! Pressure changes

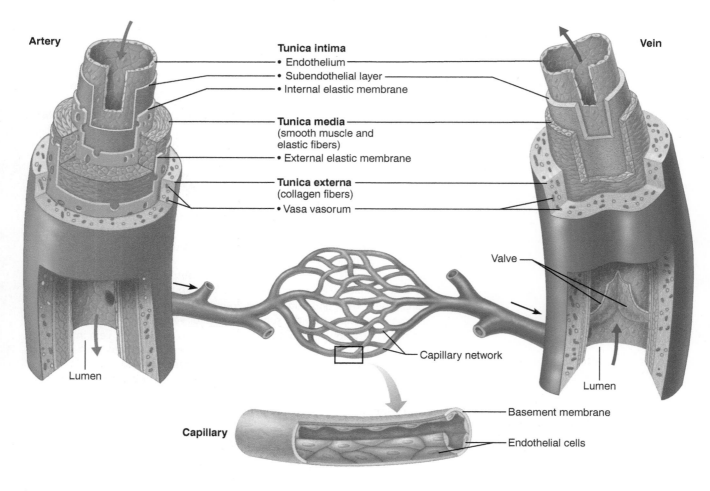

Figure 24.1 **Generalized structure of arteries, veins, and capillaries.**

Table 24.1   Summary of Blood Vessel Anatomy and Physiology

Type of vessel	Description	Average lumen diameter	Average wall thickness	Function
Elastic (conducting) arteries	Largest, most elastic arteries. Contain more elastic tissue than other arteries.	1.5 cm	1.0 mm	Act as a pressure reservoir, expanding and recoiling for continuous blood flow. Examples: aorta, brachiocephalic artery, and common carotid artery.
Muscular (distributing) arteries	Medium-sized arteries, accounting for most arteries found in the body. They have less elastic tissue and more smooth muscle than other arteries.	0.6 cm	1.0 mm	Better ability to constrict and less stretchable than elastic arteries. They distribute blood to specific areas of the body. Examples: brachial artery and radial artery.
Arterioles	Smallest arteries, with a very thin tunica externa and only a few layers of smooth muscle in the tunica media.	37 μm	6 μm	Blood flows from arterioles into a capillary bed. They play a role in regulating the blood flow to specific areas of the body.
Capillaries	Contain only a tunica intima.	9 μm	0.5 μm	Provide for the exchange of materials (gases, nutrients, etc.) between the blood and tissue cells.
Venules	Smallest veins. All tunics are very thin, with at most two layers of smooth muscle and no elastic tissue.	20 μm	1 μm	Drain capillary beds and merge to form veins.
Veins	Contain more fibrous tissue in the tunica externa than corresponding arteries. The tunica media is thinner, with a larger lumen than the corresponding artery.	0.5 cm	0.5 mm	Low-pressure vessels; return blood to the heart. Valves prevent the backflow of the blood.

that occur in the thorax during breathing also aid the return of blood to the heart.

☐ To demonstrate how efficiently venous valves prevent backflow of blood, perform the following simple experiment. Allow one hand to hang by your side until the blood vessels on the dorsal aspect become distended. Place two fingertips against one of the distended veins, and pressing firmly, move the superior finger proximally along the vein, and then release this finger. The vein will remain flattened and collapsed despite gravity. Then remove the distal fingertip, and observe the rapid filling of the vein.

Check the box when you have completed this task.

ACTIVITY 1

Examining the Microscopic Structure of Arteries and Veins

1. Obtain a slide showing a cross-sectional view of blood vessels and a microscope.

2. Scan the section to identify a thick-walled artery Use **Figure 24.2** as a guide. Very often, but not always, an arterial lumen will appear scalloped because its walls are constricted by the elastic tissue of the tunica media.

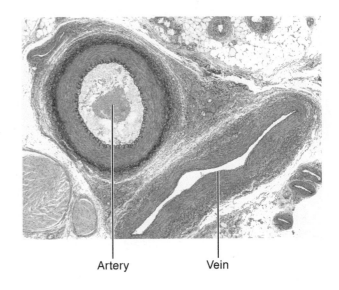

Artery   Vein

Figure 24.2 Photomicrograph of a muscular artery and corresponding vein in cross section (40×).

3.   Identify a vein. Its lumen may be elongated or irregularly shaped and collapsed, and its walls will be considerably thinner. Notice the difference in the relative amount of elastic fibers in the tunica media of the two vessels. Also, note the thinness of the intima layer, which is composed of flat squamous cells. ■

Major Systemic Arteries of the Body

The **aorta** is the largest artery of the body. It has three major regions: ascending aorta, aortic arch, and descending aorta. Extending upward as the **ascending aorta** from the left ventricle, it arches posteriorly and to the left as the **aortic arch** and then courses downward as the **descending aorta** through the thoracic cavity. Called the **thoracic aorta** from T_5 to T_{12}, the descending aorta penetrates the diaphragm to enter the abdominal cavity just anterior to the vertebral column. As it enters the abdominal cavity, it becomes the **abdominal aorta.** The branches of the ascending aorta and aortic arch are summarized in **Table 24.2**. Branches of the thoracic and abdominal aorta are summarized in Tables 24.3 and 24.4.

As you locate the arteries on **Figure 24.3**, notice that in many cases, the name of the artery reflects the body region traversed (axillary, subclavian, brachial, popliteal), the organ served (renal, hepatic), or the bone followed (tibial, femoral, radial, ulnar).

Aortic Arch

The **brachiocephalic** (literally, "arm-head") **trunk** is the first branch of the aortic arch **(Figure 24.4)**. The other two major arteries branching off the aortic arch are the **left common carotid artery** and the **left subclavian artery.** The brachiocephalic trunk persists briefly before dividing into the **right common carotid artery** and the **right subclavian artery.**

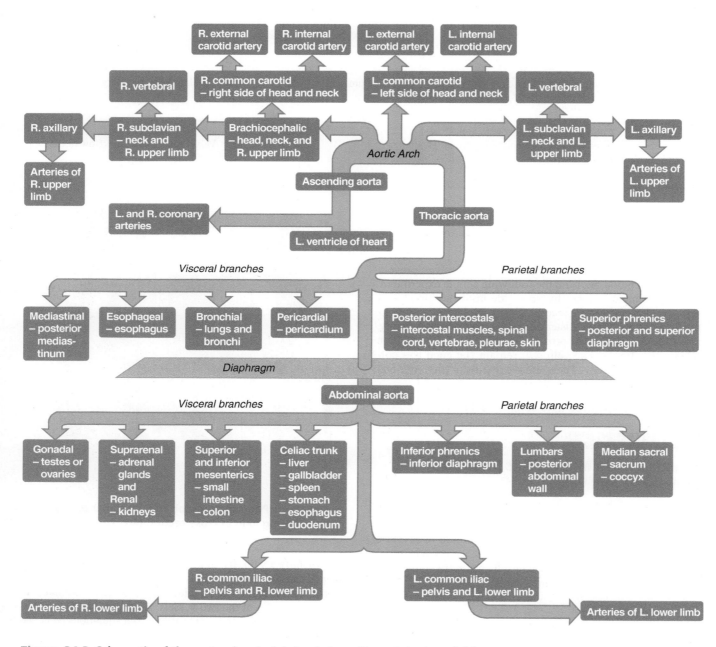

Figure 24.3 Schematic of the systemic arterial circulation. (R. = right, L. = left)

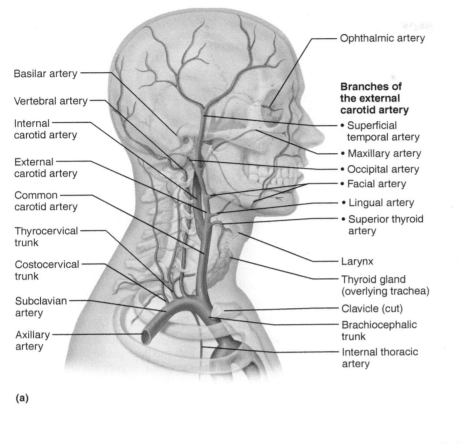

(a)

Ophthalmic artery

Basilar artery

Vertebral artery

Internal carotid artery

External carotid artery

Common carotid artery

Thyrocervical trunk

Costocervical trunk

Subclavian artery

Axillary artery

Branches of the external carotid artery
- Superficial temporal artery
- Maxillary artery
- Occipital artery
- Facial artery
- Lingual artery
- Superior thyroid artery

Larynx

Thyroid gland (overlying trachea)

Clavicle (cut)

Brachiocephalic trunk

Internal thoracic artery

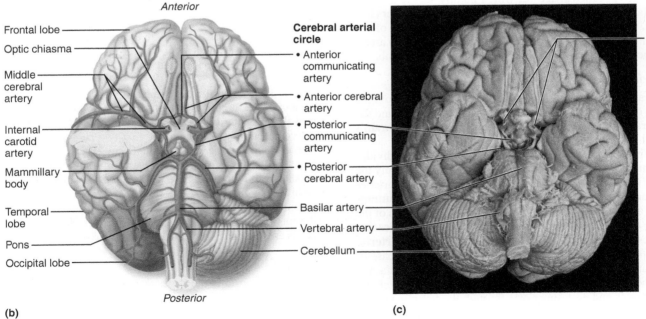

(b)

Anterior

Frontal lobe

Optic chiasma

Middle cerebral artery

Internal carotid artery

Mammillary body

Temporal lobe

Pons

Occipital lobe

Posterior

Cerebral arterial circle
- Anterior communicating artery
- Anterior cerebral artery
- Posterior communicating artery
- Posterior cerebral artery

Basilar artery

Vertebral artery

Cerebellum

(c)

Internal carotid arteries

Figure 24.4 Arteries of the head, neck, and brain. (a) Right aspect. **(b)** Drawing of the cerebral arteries, inferior view. Right side of cerebellum and part of right temporal lobe have been removed. **(c)** Cerebral arterial circle (circle of Willis), inferior view of brain.

Table 24.2   The Aorta: Ascending Aorta and Aortic Arch (Figure 24.3)

Ascending aorta branches	Structures served
Right coronary artery	The myocardium of the heart (see Exercise 23)
Left coronary artery	The myocardium of the heart (see Exercise 23)
Aortic arch branches	**Structures served**
Brachiocephalic trunk (branches into right common carotid and right subclavian arteries)	Right common carotid artery – right side of the head and neck Right subclavian artery – right upper limb
Left common carotid artery	Left side of the head and neck
Left subclavian artery	Left upper limb

Arteries Serving the Head and Neck

The common carotid artery on each side divides to form an internal and external carotid artery. The **internal carotid artery** serves the brain and gives rise to the **ophthalmic artery,** which supplies orbital structures. The **external carotid artery** supplies the tissues external to the skull, largely via its **superficial temporal, maxillary, facial,** and **occipital** arterial branches.

The right and left subclavian arteries each give off several branches to the head and neck. The first of these is the **vertebral artery,** which runs up the posterior neck to supply the cerebellum, part of the brain stem, and the posterior cerebral hemispheres. Issuing just lateral to the vertebral artery are the **thyrocervical trunk,** which mainly serves the thyroid gland and some scapular muscles, and the **costocervical trunk,** which supplies deep neck muscles and some of the upper intercostal muscles. In the armpit, the subclavian artery becomes the axillary artery, which serves the upper limb.

Arteries Serving the Brain

The brain is supplied by two pairs of arteries arising from the region of the aortic arch—the internal carotid arteries and the vertebral arteries. (Figure 24.4b is a diagram of the brain's arterial supply).

Within the cranium, each internal carotid artery divides into **anterior** and **middle cerebral arteries,** which supply the bulk of the cerebrum. The right and left anterior cerebral arteries are connected by a short shunt called the **anterior communicating artery.** This shunt—along with shunts from each of the middle cerebral arteries, called the **posterior communicating arteries**—contributes to the formation of the **cerebral arterial circle** *(circle of Willis),* an arterial anastomosis at the base of the brain surrounding the pituitary gland and the optic chiasma.

The paired **vertebral arteries** diverge from the subclavian arteries and pass superiorly through the foramina of the transverse process of the cervical vertebrae to enter the skull through the foramen magnum. Within the skull, the vertebral arteries unite to form a single **basilar artery,** which continues superiorly along the anterior aspect of the brain stem, giving off branches to the pons, cerebellum, and inner ear. At the pons-midbrain border, the basilar artery divides to form the **posterior cerebral arteries,** which supply portions of the temporal and occipital lobes of the cerebrum and complete the cerebral arterial circle posteriorly.

The uniting of the blood supply of the internal carotid arteries and the vertebral arteries via the cerebral arterial circle is a protective device that provides alternate pathways for blood to reach the brain tissue if an artery should become blocked or if blood flow is impaired anywhere in the system.

Arteries Serving the Thorax and Upper Limbs

As the **axillary artery** runs through the axilla, it gives off several branches to the chest wall and shoulder girdle **(Figure 24.5).** These include the **thoracoacromial artery** (to shoulder and pectoral region), the **lateral thoracic artery** (lateral chest wall), the **subscapular artery** (to scapula and dorsal thorax), and the **anterior** and **posterior circumflex humeral arteries** (to the shoulder and the deltoid muscle). At the inferior edge of the teres major muscle, the axillary artery becomes the **brachial artery** as it enters the arm. The brachial artery gives off a deep branch, the **deep artery of the arm,** and as it nears the elbow it gives off several small branches. At the elbow, the brachial artery divides into the **radial** and **ulnar arteries,** which follow the same-named bones to supply the forearm and hand.

The **internal thoracic arteries** that arise from the subclavian arteries supply the mammary glands, most of the thorax wall, and anterior intercostal structures via their **anterior intercostal artery** branches. The first two pairs of **posterior intercostal arteries** arise from the costocervical trunk, noted above. The more inferior pairs arise from the thoracic aorta. Not shown in Figure 24.5 are the small arteries that serve the diaphragm *(phrenic arteries),* esophagus *(esophageal arteries),* bronchi *(bronchial arteries),* and other structures of the mediastinum *(mediastinal* and *pericardial arteries).* However, these vessels are listed in Figure 24.3 and summarized in Table 24.3.

Thoracic Aorta

The thoracic aorta is the superior portion of the descending aorta (Figure 24.5). It begins where the aortic arch ends and ends just as it pierces the diaphragm. The main branches of the thoracic aorta are summarized in **Table 24.3**.

Abdominal Aorta

Although several small branches of the descending aorta serve the thorax, the more major branches of the descending aorta are those serving the abdominal organs and ultimately the lower limbs **(Figure 24.6).**

Most branches of the abdominal aorta serve the abdominal organs. The major branches of the abdominal aorta are summarized in **Table 24.4**.

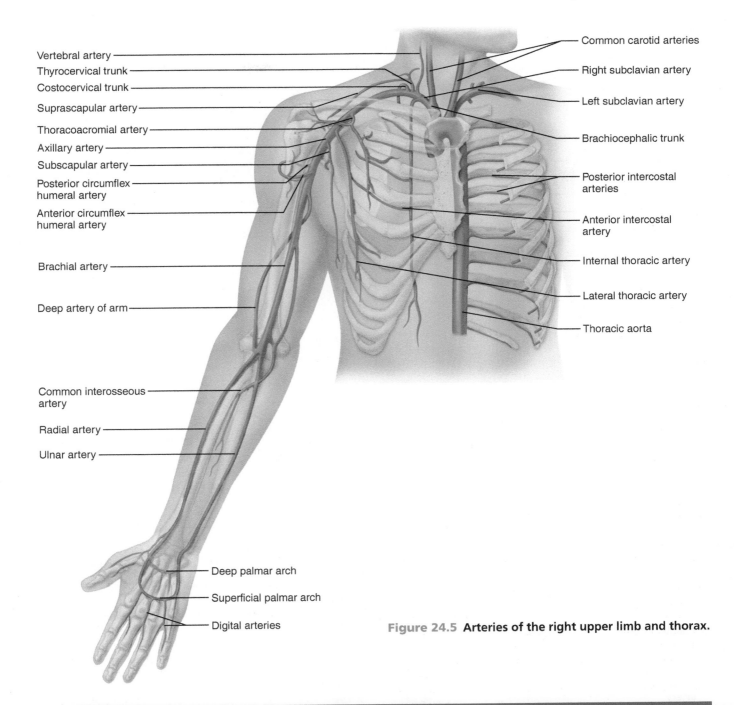

Vertebral artery

Thyrocervical trunk

Costocervical trunk

Suprascapular artery

Thoracoacromial artery

Axillary artery

Subscapular artery

Posterior circumflex humeral artery

Anterior circumflex humeral artery

Brachial artery

Deep artery of arm

Common interosseous artery

Radial artery

Ulnar artery

Deep palmar arch

Superficial palmar arch

Digital arteries

Common carotid arteries

Right subclavian artery

Left subclavian artery

Brachiocephalic trunk

Posterior intercostal arteries

Anterior intercostal artery

Internal thoracic artery

Lateral thoracic artery

Thoracic aorta

Figure 24.5 Arteries of the right upper limb and thorax.

Table 24.3 The Aorta: Thoracic Aorta (Figure 24.3) (Note that many of these arteries vary in number from person to person)	
Visceral thoracic aorta branches	**Structures served**
Pericardial arteries	Pericardium, the serous membrane of the heart
Bronchial arteries	Bronchi, bronchioles, and lungs
Esophageal arteries	Esophagus
Mediastinal arteries	Posterior mediastinum
Parietal thoracic aorta branches	**Structures served**
Posterior intercostal arteries (inferior pairs)	Intercostal muscles, spinal cord, vertebrae, and skin
Subcostal arteries	Intercostal muscles, spinal cord, vertebrae, and skin
Superior phrenic arteries	Posterior, superior part of the diaphragm

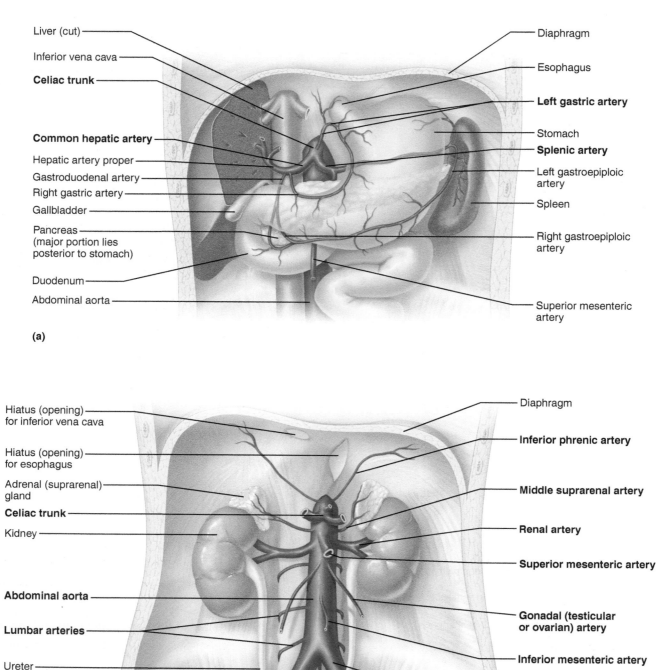

Liver (cut)

Inferior vena cava

Celiac trunk

Common hepatic artery

Hepatic artery proper

Gastroduodenal artery

Right gastric artery

Gallbladder

Pancreas (major portion lies posterior to stomach)

Duodenum

Abdominal aorta

Diaphragm

Esophagus

Left gastric artery

Stomach

Splenic artery

Left gastroepiploic artery

Spleen

Right gastroepiploic artery

Superior mesenteric artery

(a)

Hiatus (opening) for inferior vena cava

Hiatus (opening) for esophagus

Adrenal (suprarenal) gland

Celiac trunk

Kidney

Abdominal aorta

Lumbar arteries

Ureter

Median sacral artery

Diaphragm

Inferior phrenic artery

Middle suprarenal artery

Renal artery

Superior mesenteric artery

Gonadal (testicular or ovarian) artery

Inferior mesenteric artery

Common iliac artery

(b)

Figure 24.6 Arteries of the abdomen. (a) The celiac trunk and its major branches. **(b)** Major branches of the abdominal aorta.

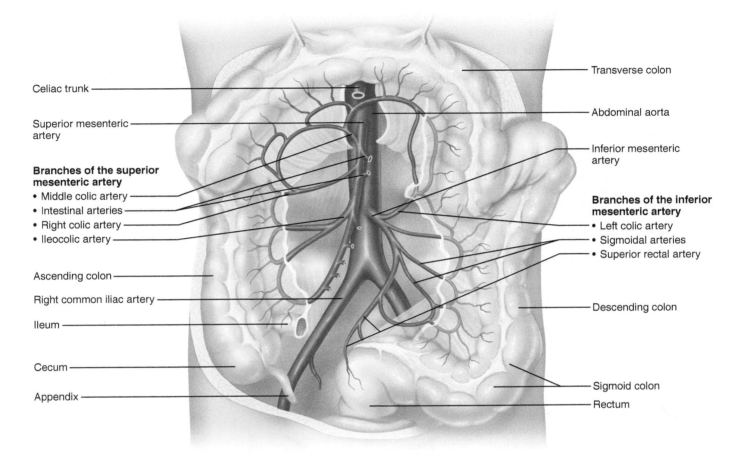

Celiac trunk

Superior mesenteric artery

Branches of the superior mesenteric artery
• Middle colic artery
• Intestinal arteries
• Right colic artery
• Ileocolic artery

Ascending colon

Right common iliac artery

Ileum

Cecum

Appendix

Transverse colon

Abdominal aorta

Inferior mesenteric artery

Branches of the inferior mesenteric artery
• Left colic artery
• Sigmoidal arteries
• Superior rectal artery

Descending colon

Sigmoid colon

Rectum

(c)

Figure 24.6 *(continued)* **(c)** Distribution of the superior and inferior mesenteric arteries, transverse colon pulled superiorly.

Table 24.4 The Aorta: Abdominal Aorta (Figure 24.6)	
Branches	**Structures served**
Inferior phrenic arteries	Inferior surface of the diaphragm
Celiac trunk: left gastric artery	Stomach and esophagus
Celiac trunk: splenic artery	Branches to the spleen; short gastric arteries branch to the stomach; and the left gastroepiploic artery branches to the stomach
Celiac trunk: common hepatic artery	Branches into the hepatic artery proper (its branches serve the liver, gallbladder, and stomach) and the gastroduodenal artery (its branches serve the stomach, pancreas, and duodenum)
Superior mesenteric artery	Most of the small intestine and the first part of the large intestine
Middle suprarenal arteries	Adrenal glands that sit on top of the kidneys
Renal arteries	Kidneys
Gonadal arteries	Ovarian arteries (female) –ovaries Testicular arteries (male) –testes
Inferior mesenteric artery	Distal portion of the large intestine
Lumbar arteries	Posterior abdominal wall
Median sacral artery	Sacrum and coccyx
Common iliac arteries	The distal abdominal aorta splits to form the left and right common iliac arteries, which serve the pelvic organs, lower abdominal wall, and the lower limbs

Arteries Serving the Lower Limbs

Each of the common iliac arteries extends for about 5 cm (2 inches) into the pelvis before it divides into the internal and external iliac arteries **(Figure 24.7)**. The **internal iliac artery** supplies the gluteal muscles via the **superior** and **inferior gluteal arteries** and the adductor muscles of the medial thigh via the **obturator artery,** as well as the external genitalia and perineum (via the *internal pudendal artery,* not illustrated).

The **external iliac artery** supplies the anterior abdominal wall and the lower limb. As it continues into the thigh, its name changes to **femoral artery.** Proximal branches of the femoral artery, the **circumflex femoral arteries,** supply the head and neck of the femur and the hamstring muscles. The femoral artery gives off a deep branch, the **deep artery of the thigh** (also called the *deep femoral artery*), which is the main supply to the thigh muscles (hamstrings, quadriceps, and adductors). In the knee region, the femoral artery briefly becomes the **popliteal artery;** its subdivisions—the **anterior** and **posterior tibial arteries**—supply the leg, ankle, and foot. The posterior tibial, which supplies flexor muscles, gives off one main branch, the **fibular artery,** that serves the lateral calf (fibular muscles). It then divides into the **lateral** and **medial plantar arteries** that supply blood to the plantar surface of the foot. The anterior tibial artery supplies the extensor muscles and terminates with the **dorsalis pedis artery.** The dorsalis pedis supplies the dorsum of the foot and continues on as the **arcuate artery,** which issues the **dorsal metatarsal arteries** to the metatarsus of the foot. The dorsalis pedis is often palpated in patients with circulation problems of the leg to determine the circulatory efficiency to the limb as a whole.

☐ Palpate your own dorsalis pedis artery.

Check the box when you have completed this task.

ACTIVITY 2

Locating Arteries on an Anatomical Chart or Model

Now that you have identified the arteries in Figures 24.3–24.7, attempt to locate and name them (without a reference) on a large anatomical chart or three-dimensional model of the vascular system. ▬

Major Systemic Veins of the Body

Arteries are generally located in deep, well-protected body areas. However, many veins follow a more superficial course and are often easy to see and palpate on the body surface. Most deep veins parallel the course of the major arteries, and in many cases the naming of the veins and arteries is identical except for the designation of the vessels as veins. Whereas the major systemic arteries branch off the aorta, the veins tend to converge on the venae cavae, which enter the right atrium of the heart. Veins draining the head and upper extremities empty into the **superior vena cava,** and those draining the lower body empty into the **inferior vena cava.** A schematic **(Figure 24.8)** of the systemic veins and their relationship to the venae cavae will get you started.

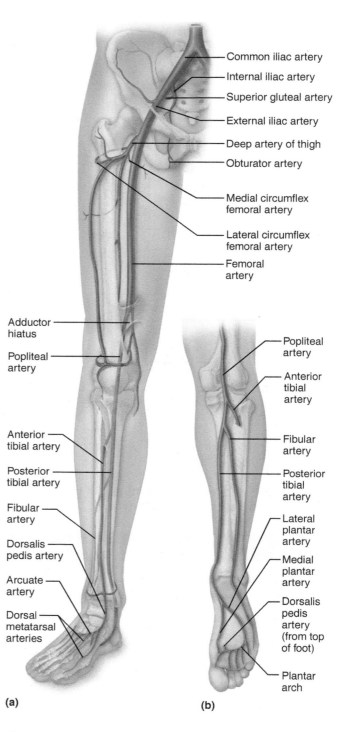

(a)   (b)

Figure 24.7 Arteries of the right pelvis and lower limb. (a) Anterior view. **(b)** Posterior view.

Veins Draining into the Inferior Vena Cava

The inferior vena cava, a much longer vessel than the superior vena cava, returns blood to the heart from all body regions below the diaphragm (see Figure 24.8). It begins in the lower abdominal region with the union of the paired **common iliac veins,** which drain venous blood from the legs and pelvis.

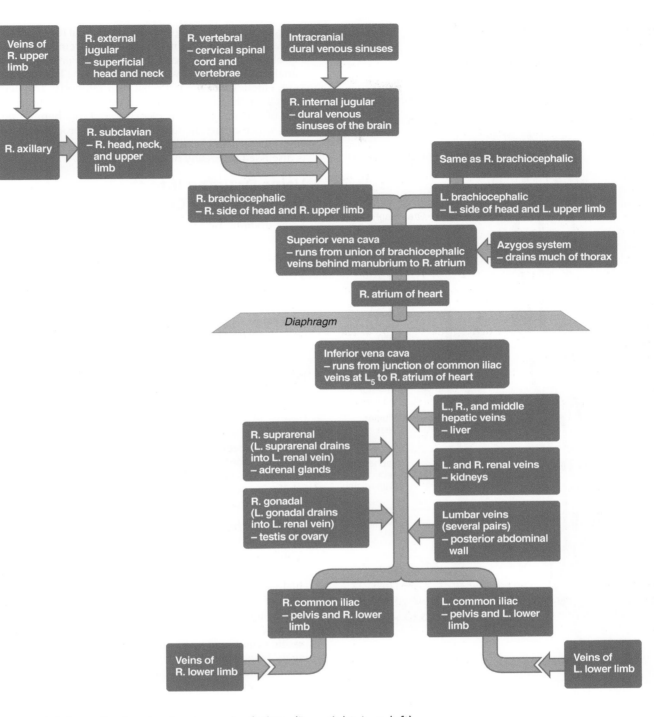

Figure 24.8 Schematic of systemic venous circulation. (R. = right, L. = left)

Veins of the Lower Limbs

Each common iliac vein is formed by the union of the **internal iliac vein,** draining the pelvis, and the **external iliac vein,** which receives venous blood from the lower limb **(Figure 24.9).** Veins of the leg include the **anterior** and **posterior tibial veins,** which drain the calf and foot. The anterior tibial vein is a superior continuation of the **dorsalis pedis vein** of the foot. The posterior tibial vein is formed by the union of the **medial** and **lateral plantar veins** and ascends deep in the calf muscles. It receives the **fibular vein** in the calf and then joins with the anterior tibial vein at the knee to produce the **popliteal vein,** which crosses the back of the knee. The popliteal vein becomes the **femoral vein** in the thigh; the femoral vein in turn becomes the external iliac vein in the inguinal region.

The **great saphenous vein,** a superficial vein, is the longest vein in the body. Beginning in common with the **small saphenous vein** from the **dorsal venous arch,** it extends up the medial side of the leg, knee, and thigh to empty into the femoral vein. The small saphenous vein runs along the

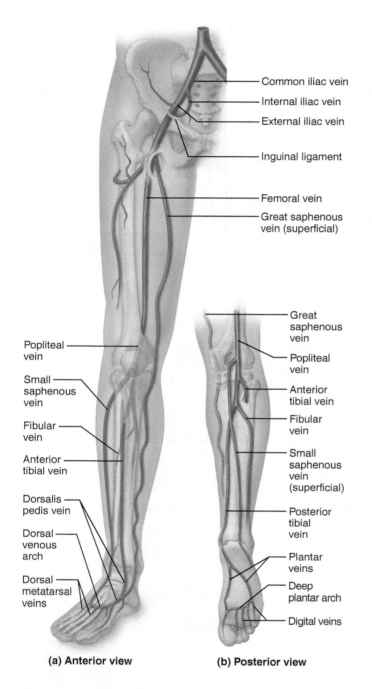

Common iliac vein

Internal iliac vein

External iliac vein

Inguinal ligament

Femoral vein

Great saphenous vein (superficial)

Great saphenous vein

Popliteal vein

Popliteal vein

Small saphenous vein

Anterior tibial vein

Fibular vein

Fibular vein

Anterior tibial vein

Small saphenous vein (superficial)

Dorsalis pedis vein

Posterior tibial vein

Dorsal venous arch

Plantar veins

Dorsal metatarsal veins

Deep plantar arch

Digital veins

(a) Anterior view   **(b) Posterior view**

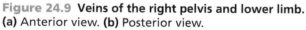

**Figure 24.9 Veins of the right pelvis and lower limb.
(a)** Anterior view. **(b)** Posterior view.

lateral aspect of the foot and through the calf muscle, which it drains, and then empties into the popliteal vein at the knee (Figure 24.9b).

Veins of the Abdomen

Moving superiorly in the abdominal cavity **(Figure 24.10)**, the inferior vena cava receives blood from the posterior abdominal wall via several pairs of **lumbar veins,** and from the right ovary or testis via the **right gonadal vein.** (The **left gonadal** [**ovarian** or **testicular**] **vein** drains into the left renal

vein superiorly.) The paired **renal veins** drain the kidneys. Just above the right renal vein, the **right suprarenal vein** (receiving blood from the adrenal gland on the same side) drains into the inferior vena cava, but its partner, the **left suprarenal vein,** empties into the left renal vein inferiorly. The **hepatic veins** drain the liver. The unpaired veins draining the digestive tract organs empty into a special vessel, the **hepatic portal vein,** which carries blood to the liver to be processed before it enters the systemic venous system. (The hepatic portal system is discussed separately on pages 409–410.)

Veins Draining into the Superior Vena Cava

Veins draining into the superior vena cava are named from the superior vena cava distally, _but remember that the flow of blood is in the opposite direction._

Veins of the Head and Neck

The **right** and **left brachiocephalic veins** drain the head, neck, and upper extremities and unite to form the superior vena cava **(Figure 24.11)**. Notice that although there is only one brachiocephalic artery, there are two brachiocephalic veins.

Branches of the brachiocephalic veins include the internal jugular, vertebral, and subclavian veins. The **internal jugular veins** are large veins that drain the superior sagittal sinus and other **dural venous sinuses.** As they move inferiorly, they receive blood from the head and neck via the **superficial temporal** and **facial veins.** The **vertebral veins** drain the posterior aspect of the head, including the cervical vertebrae and spinal cord. The **subclavian veins** receive venous blood from the upper extremity. The **external jugular vein** joins the subclavian vein near its origin to return the venous drainage of the extracranial (superficial) tissues of the head and neck.

Veins of the Upper Limb and Thorax

As the subclavian vein traverses the axilla, it becomes the **axillary vein** and then the **brachial vein** as it courses along the posterior aspect of the humerus **(Figure 24.12)**. The brachial vein is formed by the union of the deep **radial** and **ulnar veins** of the forearm. The superficial venous drainage of the arm includes the **cephalic vein,** which courses along the lateral aspect of the arm and empties into the axillary vein; the **basilic vein,** found on the medial aspect of the arm and entering the brachial vein; and the **median cubital vein,** which runs between the cephalic and basilic veins in the anterior aspect of the elbow (this vein is often the site of choice for removing blood for testing purposes). The **median antebrachial vein** lies between the radial and ulnar veins, and terminates variably by entering the cephalic or basilic vein at the elbow.

The **azygos system** (Figure 24.12) drains the intercostal muscles of the thorax and provides an accessory venous system to drain the abdominal wall. The **azygos vein,** which drains the right side of the thorax, enters the dorsal aspect of the superior vena cava immediately before that vessel enters the right atrium. Also part of the azygos system are the **hemiazygos** (a continuation of the **left ascending lumbar vein** of the abdomen) and the **accessory hemiazygos veins,** which together drain the left side of the thorax and empty into the azygos vein.

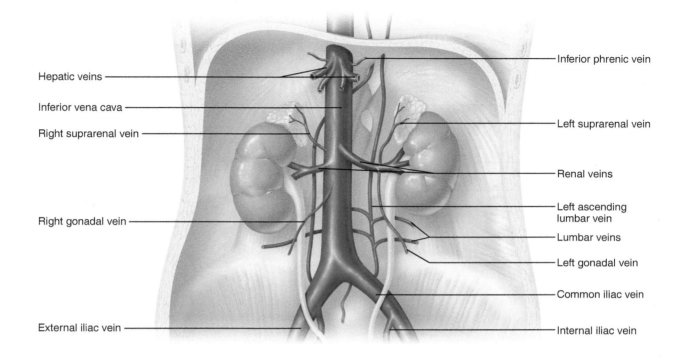

Hepatic veins
Inferior vena cava
Right suprarenal vein
Right gonadal vein
External iliac vein

Inferior phrenic vein
Left suprarenal vein
Renal veins
Left ascending lumbar vein
Lumbar veins
Left gonadal vein
Common iliac vein
Internal iliac vein

Figure 24.10 **Venous drainage of abdominal organs not drained by the hepatic portal vein.**

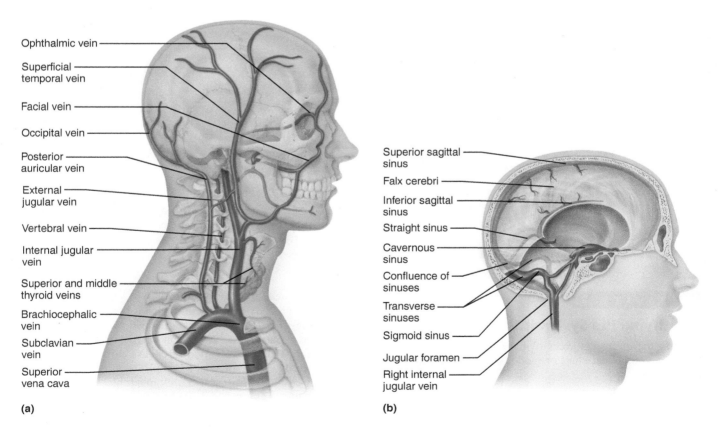

Ophthalmic vein
Superficial temporal vein
Facial vein
Occipital vein
Posterior auricular vein
External jugular vein
Vertebral vein
Internal jugular vein
Superior and middle thyroid veins
Brachiocephalic vein
Subclavian vein
Superior vena cava

Superior sagittal sinus
Falx cerebri
Inferior sagittal sinus
Straight sinus
Cavernous sinus
Confluence of sinuses
Transverse sinuses
Sigmoid sinus
Jugular foramen
Right internal jugular vein

(a)

(b)

Figure 24.11 **Venous drainage of the head, neck, and brain.** **(a)** Veins of the head and neck, right superficial aspect. **(b)** Dural venous sinuses, right aspect.

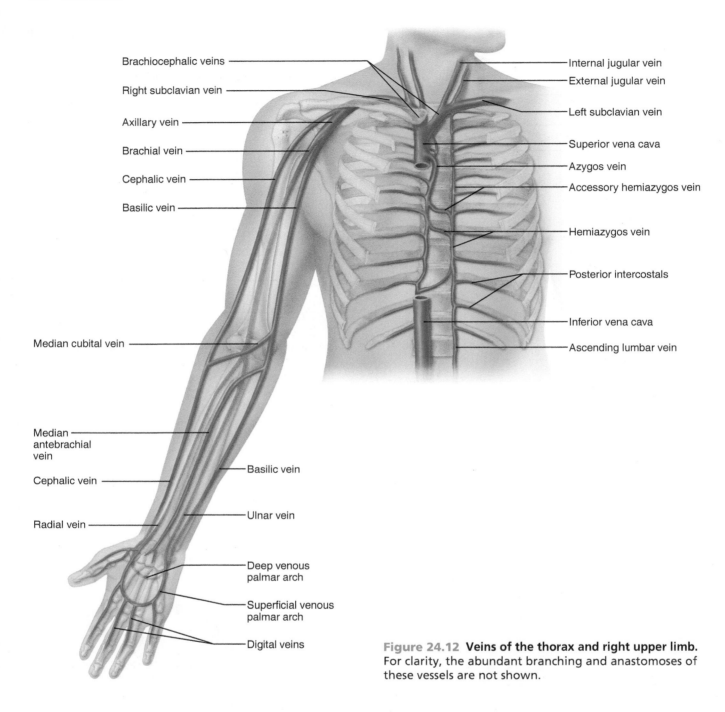

Brachiocephalic veins
Right subclavian vein
Axillary vein
Brachial vein
Cephalic vein
Basilic vein

Internal jugular vein
External jugular vein
Left subclavian vein
Superior vena cava
Azygos vein
Accessory hemiazygos vein
Hemiazygos vein
Posterior intercostals
Inferior vena cava
Ascending lumbar vein

Median cubital vein

Median antebrachial vein
Cephalic vein
Radial vein

Basilic vein
Ulnar vein
Deep venous palmar arch
Superficial venous palmar arch
Digital veins

Figure 24.12 Veins of the thorax and right upper limb. For clarity, the abundant branching and anastomoses of these vessels are not shown.

ACTIVITY 3

Identifying the Systemic Veins

Identify the important veins of the systemic circulation on the large anatomical chart or model without referring to the figures. ■

Special Circulations

Pulmonary Circulation

The pulmonary circulation (discussed previously in relation to heart anatomy on page 384) differs in many ways from systemic circulation because it does not serve the metabolic needs of the body tissues with which it is associated (in this case, lung tissue). It functions instead to bring the blood into close contact with the alveoli of the lungs to permit gas exchanges that rid the blood of excess carbon dioxide and replenish its supply of vital oxygen. The arteries of the pulmonary circulation are structurally much like veins, and they create a low-pressure bed in the lungs. (If the arterial pressure in the systemic circulation is 120/80, the pressure in the pulmonary artery is likely to be approximately 24/8.) The functional blood supply of the lungs is provided by the **bronchial arteries** (not shown), which diverge from the thoracic portion of the descending aorta.

ACTIVITY 4

Identifying Vessels of the Pulmonary Circulation

Find the vessels of the pulmonary circulation in **Figure 24.13** and on an anatomical chart (if one is available). ▬

Pulmonary circulation begins with the large **pulmonary trunk,** which leaves the right ventricle and divides into the **right** and **left pulmonary arteries** about 5 cm (2 inches) above its origin. The right and left pulmonary arteries plunge into the lungs, where they subdivide into **lobar arteries** (three on the right and two on the left). The lobar arteries accompany the main bronchi into the lobes of the lungs and branch extensively within the lungs to form arterioles, which finally terminate in the capillary networks surrounding the alveolar sacs of the lungs. The respiratory gases diffuse across the walls of the alveoli and **pulmonary capillaries.** The pulmonary capillary beds are drained by venules, which converge to form sequentially larger veins and finally the four **pulmonary veins** (two leaving each lung), which return the blood to the left atrium of the heart.

Hepatic Portal Circulation

Blood vessels of the hepatic portal circulation drain the digestive viscera, spleen, and pancreas and deliver this blood to the liver for processing via the **hepatic portal vein** (formed by the union of the splenic and superior mesenteric veins). If a meal has recently been eaten, the hepatic portal blood will be nutrient-rich. The liver is the key body organ involved in maintaining proper sugar, fatty acid, and amino acid concentrations in the blood, and the hepatic portal system ensures that these substances pass through the liver before entering the systemic circulation. As blood travels through the liver sinusoids, some of the nutrients are removed to be stored or processed in various ways for release to the general circulation. At the same time, the hepatocytes are detoxifying alcohol and other possibly harmful chemicals present in the blood, and the liver's macrophages are removing bacteria and other debris from the passing blood. The liver in turn is drained by the hepatic veins that enter the inferior vena cava.

ACTIVITY 5

Tracing the Hepatic Portal Circulation

Locate on **Figure 24.14** and on an anatomical chart of the hepatic portal circulation (if available), the vessels named below in bold type. ▬

The **splenic vein** carries blood from the spleen, parts of the pancreas, and the stomach. The splenic vein unites with the **superior mesenteric vein** to form the hepatic portal vein. The superior mesenteric vein drains the small intestine, part of the large intestine, and the stomach. The **inferior mesenteric vein,** which drains the distal portion of the large intestine and rectum, empties into the splenic vein just before the splenic vein merges with the superior mesenteric vein.

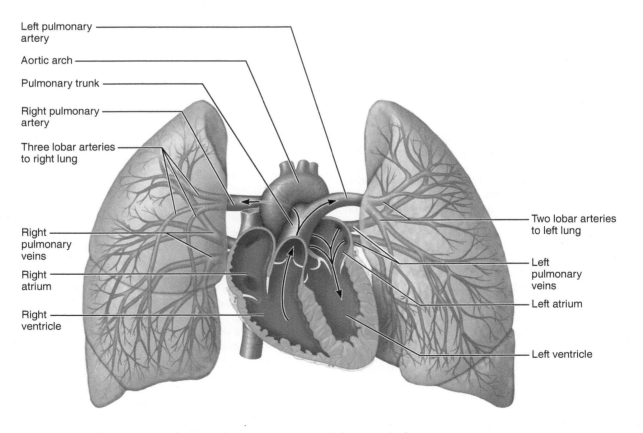

Left pulmonary artery

Aortic arch

Pulmonary trunk

Right pulmonary artery

Three lobar arteries to right lung

Right pulmonary veins

Right atrium

Right ventricle

Two lobar arteries to left lung

Left pulmonary veins

Left atrium

Left ventricle

Figure 24.13 The pulmonary circulation. The pulmonary arterial system is shown in blue to indicate that the blood it carries is relatively oxygen-poor. The pulmonary venous drainage is shown in red to indicate that the blood it transports is oxygen-rich.

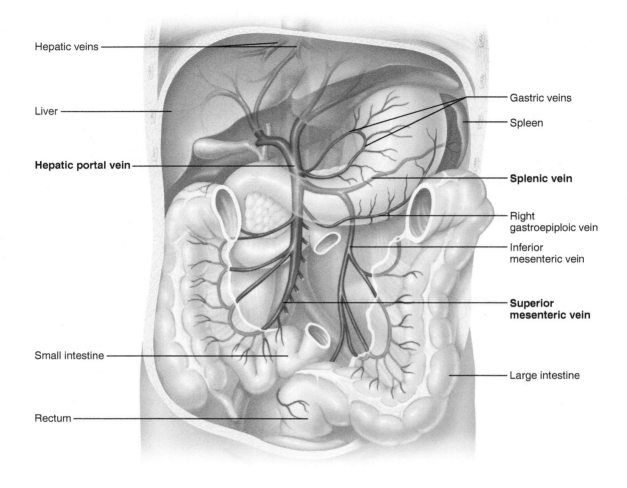

Figure 24.14 Hepatic portal circulation.

Fetal Circulation

In a developing fetus, the lungs and digestive system are not yet functional, and all nutrient, excretory, and gaseous exchanges occur through the placenta **(Figure 24.15a)**. Nutrients and oxygen move across placental barriers from the mother's blood into fetal blood, and carbon dioxide and other metabolic wastes move from the fetal blood supply to the mother's blood.

ACTIVITY 6

Tracing the Pathway of Fetal Blood Flow

The pathway of fetal blood flow is indicated with arrows on Figure 24.15a. Follow the flow of blood from the placenta to the fetal heart and back, noting the names of all specialized fetal circulatory structures. ▪

Fetal blood travels through the umbilical cord, which contains three blood vessels: one large umbilical vein and two smaller umbilical arteries. The **umbilical vein** carries blood rich in nutrients and oxygen to the fetus; the **umbilical arteries** carry blood laden with carbon dioxide and waste from the fetus to the placenta. The umbilical arteries, which transport blood away from the fetal heart, meet the umbilical vein at the *umbilicus* and wrap around the vein within the

cord en route to their placental attachments. Newly oxygenated blood flows in the umbilical vein superiorly toward the fetal heart. Some of this blood perfuses the liver, but the larger proportion is ducted through the relatively nonfunctional liver to the inferior vena cava via a shunt vessel called the **ductus venosus,** which carries the blood to the right atrium of the heart.

Because fetal lungs are nonfunctional and collapsed, two shunting mechanisms ensure that blood almost entirely bypasses the lungs. Much of the blood entering the right atrium is shunted into the left atrium through the **foramen ovale,** a flaplike opening in the interatrial septum. The left ventricle then pumps the blood out the aorta to the systemic circulation. Blood that does enter the right ventricle and is pumped out of the pulmonary trunk encounters a second shunt, the **ductus arteriosus,** a short vessel connecting the pulmonary trunk and the aorta. Because the collapsed lungs present an extremely high-resistance pathway, blood more readily enters the systemic circulation through the ductus arteriosus.

The aorta carries blood to the tissues of the body; this blood ultimately finds its way back to the placenta via the umbilical arteries. The only fetal vessel that carries highly oxygenated blood is the umbilical vein. All other vessels contain varying degrees of oxygen-rich and oxygen-poor blood.

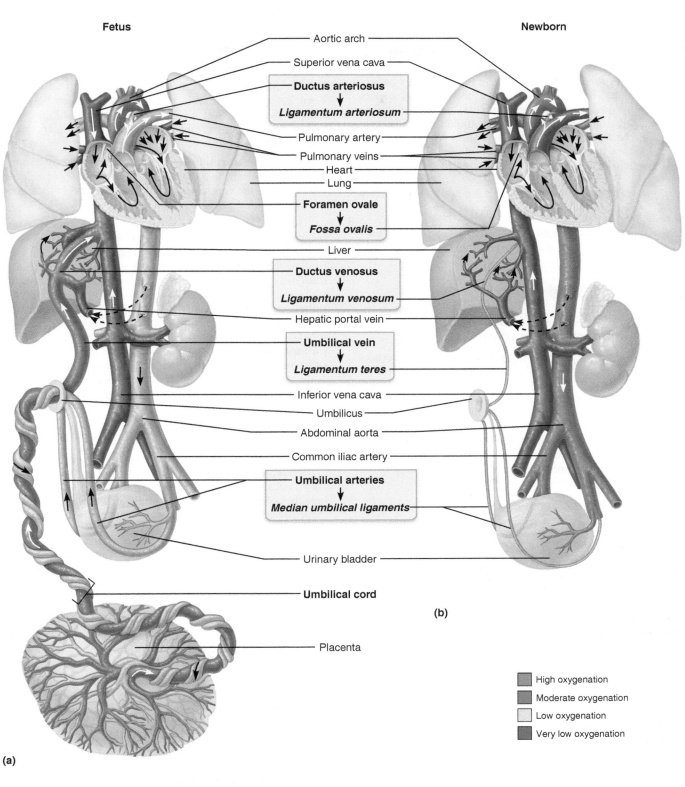

Fetus

Newborn

Aortic arch

Superior vena cava

Ductus arteriosus
↓
Ligamentum arteriosum

Pulmonary artery

Pulmonary veins

Heart

Lung

Foramen ovale
↓
Fossa ovalis

Liver

Ductus venosus
↓
Ligamentum venosum

Hepatic portal vein

Umbilical vein
↓
Ligamentum teres

Inferior vena cava

Umbilicus

Abdominal aorta

Common iliac artery

Umbilical arteries
↓
Median umbilical ligaments

Urinary bladder

Umbilical cord

(b)

Placenta

High oxygenation

Moderate oxygenation

Low oxygenation

Very low oxygenation

(a)

Figure 24.15 Circulation in the fetus and newborn. Arrows on blood vessels indicate direction of blood flow. **(a)** Special adaptations for embryonic and fetal life. Arrows in the blue boxes go from the fetal structure to what it becomes after birth (the postnatal structure). The umbilical vein (red) carries oxygen- and nutrient-rich blood from the placenta to the fetus. The umbilical arteries (pink) carry waste-laden blood from the fetus to the placenta. **(b)** Changes in the cardiovascular system at birth. The umbilical vessels are occluded, as are the liver and lung bypasses (ductus venosus and ductus arteriosus, and the foramen ovale).

At birth, or shortly after, the foramen ovale closes and becomes the **fossa ovalis,** and the ductus arteriosus collapses and is converted to the fibrous **ligamentum arteriosum** (Figure 24.15b). Lack of blood flow through the umbilical vessels leads to their eventual obliteration, and the circulatory pattern becomes that of the adult. Remnants of the umbilical arteries persist as the **median umbilical ligaments** on the inner surface of the anterior abdominal wall; the occluded umbilical vein becomes the **ligamentum teres** (or **round ligament** of the liver); and the ductus venosus becomes a fibrous band called the **ligamentum venosum** on the inferior surface of the liver.

DISSECTION AND IDENTIFICATION
The Blood Vessels of the Cat

If you have already opened your animal's ventral body cavity and identified many of its organs, begin this exercise with Activity 8. ■

ACTIVITY 7

Opening the Ventral Body Cavity and Preliminary Organ Identification

If the ventral body cavity has not yet been opened, do so now by following your instructor's instructions (see Exercise 21 on page 353).

Before you identify and trace the blood supply of the various organs of the cat, it can be helpful to do a preliminary identification of ventral body cavity organs shown in **Figure 24.16.** You will study the organ systems contained in the ventral cavity in later exercises, so the objective here is simply to identify the most important organs. Locate and identify the following body cavity organs (refer to Figure 24.16).

Thoracic Cavity Organs

Heart: In the mediastinum enclosed by the pericardium.

Lungs: Flanking the heart.

Thymus: Superior to and partially covering the heart (see Figure 21.5, page 355). The thymus is quite large in young cats but is largely replaced by fat as cats age.

Abdominal Cavity Organs

Liver: Posterior to the diaphragm.

- Lift the large, drapelike, fat-infiltrated greater omentum covering the abdominal organs to expose the following:

Stomach: Dorsally located and to the left side of the liver.

Spleen: A flattened, brown organ curving around the lateral aspect of the stomach.

Small intestine: Continuing posteriorly from the stomach.

Large intestine: Taking a U-shaped course around the small intestine and terminating in the rectum. ■

ACTIVITY 8

Identifying the Blood Vessels

1. Carefully clear away any thymus tissue or fat obscuring the heart and the large vessels associated with the heart. Before identifying the blood vessels, try to locate the *phrenic nerve* (from the cervical plexus), which innervates the diaphragm.

The phrenic nerves lie ventral to the root of the lung on each side, as they pass to the diaphragm. Also attempt to locate the *vagus nerve* (cranial nerve X) passing laterally along the trachea and dorsal to the root of the lung.

2. Slit the parietal pericardium, and reflect it superiorly. Then, cut it away from its heart attachments. Review the structures of the heart. Notice its pointed inferior end (apex) and its broader superior base. Identify the two *atria,* which appear darker than the inferior *ventricles.*

3. Identify the **aorta,** the largest artery in the body, issuing from the left ventricle. Also identify the *coronary arteries* in the sulcus on the ventral surface of the heart; these should be injected with red latex.

4. Identify the two large venae cavae—the **superior** and **inferior venae cavae**—entering the right atrium. The superior vena cava is the largest dark-colored vessel entering the base of the heart. These vessels are called the *precava* and *postcava,* respectively, in the cat. The caval veins drain the same relative body areas as in humans. Also identify the **pulmonary trunk** (usually injected with blue latex) extending anteriorly from the right ventricle. The right and left pulmonary arteries branch off the pulmonary trunk. Trace the **pulmonary arteries** until they enter the lungs. Locate the **pulmonary veins** entering the left atrium and the ascending aorta arising from the left ventricle and running dorsal to the precava and to the left of the body midline.

Arteries of the Cat

Begin your dissection of the arterial system of the cat. Refer to **Figure 24.17** and Figure 24.20.

1. Reidentify the aorta as it emerges from the left ventricle. As you observed in the dissection of the sheep heart, the first branches of the aorta are the **coronary arteries,** which supply the myocardium. The coronary arteries emerge from the base of the aorta and can be seen on the surface of the heart. Follow the aorta as it arches (aortic arch), and identify its major branches. In the cat, the aortic arch gives off two large vessels, the **brachiocephalic artery** and the **left subclavian artery.** The brachiocephalic artery has three major branches, the right subclavian artery and the right and left common carotid arteries. Note that in humans, the left common carotid artery directly branches off the aortic arch.

2. Follow the **right common carotid artery** along the right side of the trachea as it moves anteriorly, giving off branches to the neck muscles, thyroid gland, and trachea. At the level of the larynx, it branches to form the **external** and **internal carotid arteries.** The internal carotid is quite small in the cat and it may be difficult to locate. It may even be absent. The distribution of the carotid arteries parallels that in humans.

3. Follow the **right subclavian artery** laterally. It gives off four branches, the first being the tiny **vertebral artery,** which along with the internal carotid artery provides the arterial circulation of the brain. Other branches of the subclavian artery include the **costocervical trunk** (to the costal and cervical regions), the **thyrocervical trunk** (to the shoulder), and the **internal thoracic (mammary) artery** (serving the ventral thoracic wall). As the subclavian passes in front of the first rib, it becomes the **axillary artery.** Its branches, which supply the trunk and shoulder muscles, are the **ventral thoracic**

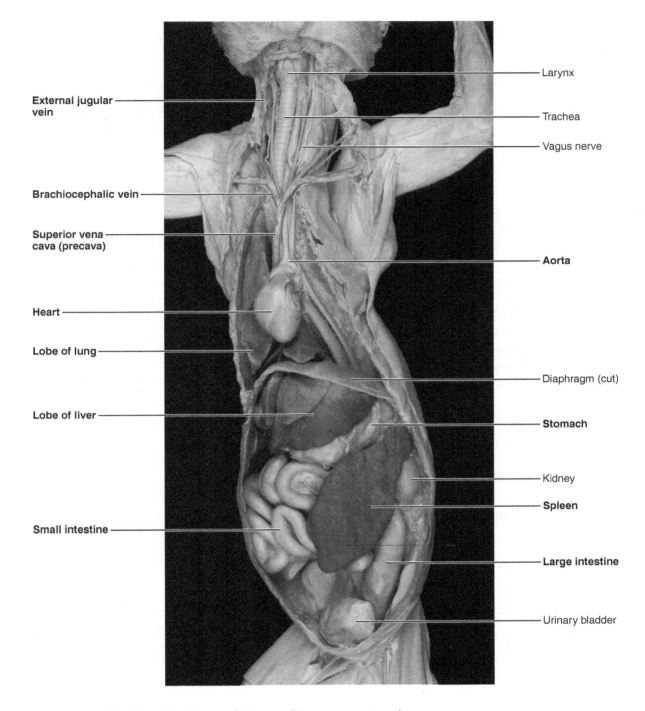

External jugular vein

Brachiocephalic vein

Superior vena cava (precava)

Heart

Lobe of lung

Lobe of liver

Small intestine

Larynx

Trachea

Vagus nerve

Aorta

Diaphragm (cut)

Stomach

Kidney

Spleen

Large intestine

Urinary bladder

Figure 24.16 Ventral body cavity organs of the cat. (Greater omentum has been removed.)

artery (the pectoral muscles), the **long thoracic artery** (pectoral muscles and latissimus dorsi), and the subscapular artery (the trunk muscles). As the axillary artery enters the arm, it is called the **brachial artery,** and it travels with the median nerve down the length of the humerus. At the elbow, the brachial artery branches to produce the two major arteries serving the forearm and hand, the **radial** and **ulnar arteries.**

4. Return to the thorax, lift the left lung, and follow the course of the **descending aorta** through the thoracic cavity. The esophagus overlies it along its course. Notice the paired

intercostal arteries that branch laterally from the aorta in the thoracic region.

5. Follow the aorta through the diaphragm into the abdominal cavity. Carefully pull the peritoneum away from its ventral surface, and identify the following vessels:

Celiac trunk: The first branch diverging from the aorta immediately as it enters the abdominal cavity; supplies the stomach, liver, gallbladder, pancreas, and spleen. (Trace as many of its branches to these organs as possible.)

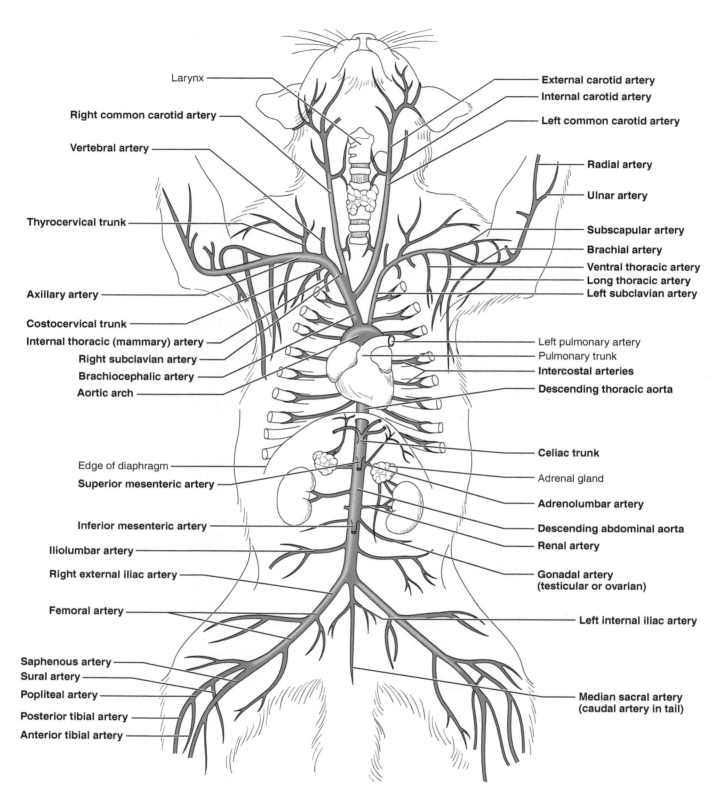

Figure 24.17 Arterial system of the cat. See also Figure 24.20 on page 419.

Superior mesenteric artery: Immediately posterior to the celiac trunk; supplies the small intestine and most of the large intestine. (Spread the mesentery of the small intestine to observe the branches of this artery as they run to supply the small intestine.)

Adrenolumbar arteries: Paired arteries diverging from the aorta slightly posterior to the superior mesenteric artery; supply the muscles of the body wall and adrenal glands.

Renal arteries: Paired arteries supplying the kidneys.

Gonadal arteries (testicular or ovarian): Paired arteries supplying the gonads.

Inferior mesenteric artery: An unpaired thin vessel arising from the ventral surface of the aorta posterior to the gonadal arteries; supplies the second half of the large intestine.

Iliolumbar arteries: Paired, rather large arteries that supply the body musculature in the iliolumbar region.

External iliac arteries: Paired arteries that continue through the body wall and pass under the inguinal ligament to the hindlimb.

6. After giving off the external iliac arteries, the aorta persists briefly and then divides into three arteries: the two **internal iliac arteries,** which supply the pelvic viscera, and the **median sacral artery.** As the median sacral artery enters the tail, it comes to be called the **caudal artery.** Note that there is no common iliac artery in the cat.

7. Trace the external iliac artery into the thigh, where it becomes the **femoral artery.** The femoral artery is most easily identified in the *femoral triangle* at the medial surface of the upper thigh. Follow the femoral artery as it courses through the thigh (along with the femoral vein and nerve) and gives off branches to the thigh muscles. (These various branches are indicated on Figure 24.17.) As you approach the knee, the **saphenous artery** branches off the femoral artery to supply the medial portion of the leg. The femoral artery then descends deep to the knee to become the **popliteal artery** in the popliteal region. The popliteal artery in turn gives off two main branches, the **sural artery** and the **posterior tibial artery,** and continues as the **anterior tibial artery.** These branches supply the leg and foot.

Veins of the Cat

Begin your dissection of the venous system of the cat. Refer to **Figure 24.18** and Figure 24.20. Keep in mind that the vessels are named for the region drained, not for the point of union with other veins.

1. Reidentify the **superior vena cava (precava)** as it enters the right atrium. Trace it anteriorly to identify veins that enter it:

Azygos vein: Passing directly into its dorsal surface; drains the thoracic intercostal muscles.

Internal thoracic (mammary) veins: Drain the chest and the abdominal walls.

Right vertebral vein: Drains the spinal cord and brain; usually enters right side of precava approximately at the level of the internal thoracic veins but may enter the brachiocephalic vein in your specimen.

Right and left brachiocephalic veins: Form the precava by their union.

2. Reflect the pectoral muscles, and trace the brachiocephalic vein laterally. Identify the two large veins that unite to form it—the external jugular vein and the subclavian vein. Notice that this differs from what is seen in humans, where the brachiocephalic veins are formed by the union of the internal jugular and subclavian veins.

3. Follow the **external jugular vein** as it travels anteriorly along the side of the neck to the point where it is joined on its medial surface by the **internal jugular vein.** The internal jugular veins are small and may be difficult to identify in the cat. Notice the difference in cat and human jugular veins. The internal jugular is considerably larger in humans and drains into the subclavian vein. In the cat, the external jugular is larger, and the interior jugular vein drains into it. Identify the *common carotid artery,* since it accompanies the internal jugular vein in this region. Also attempt to find the *sympathetic trunk,* which is located in the same area running lateral to the trachea. Several other vessels drain into the external jugular vein (transverse scapular vein, facial veins, and others). These are not discussed here but are shown on the figure and may be traced if time allows.

4. Return to the shoulder region, and follow the path of the **subclavian vein** as it moves laterally toward the forelimb. It becomes the **axillary vein** as it passes in front of the first rib and runs through the brachial plexus, giving off several branches, the first of which is the **subscapular vein.** The subscapular vein drains the proximal part of the arm and shoulder. The four other branches that receive drainage from the shoulder, pectoral, and latissimus dorsi muscles are shown in the figure but need not be identified in this dissection.

5. Follow the axillary vein into the arm, where it becomes the **brachial vein.** You can locate this vein on the medial side of the arm accompanying the brachial artery and nerve. Trace it to the point where it receives the **radial** and **ulnar veins** (which drain the forelimb) at the inner bend of the elbow. Also locate the superficial **cephalic vein** on the dorsal side of the arm. It communicates with the brachial vein via the median cubital vein in the elbow region and then enters the transverse scapular vein in the shoulder.

6. Reidentify the **inferior vena cava (postcava),** and trace it to its passage through the diaphragm. Notice again as you follow its course that the intercostal veins drain into a much smaller vein lying dorsal to the postcava, the **azygos vein.**

7. Attempt to identify the **hepatic veins** entering the postcava from the liver. You may be able to see these if some of the anterior liver tissue is scraped away where the postcava enters the liver.

8. Displace the intestines to the left side of the body cavity, and proceed posteriorly to identify the following veins in order. All of these veins empty into the postcava and drain the organs served by the same-named arteries. In the cat, variations in the connections of the veins to be located are common, and in some cases the postcaval vein may be below the level of the renal veins. If you observe deviations, call them to the attention of your instructor.

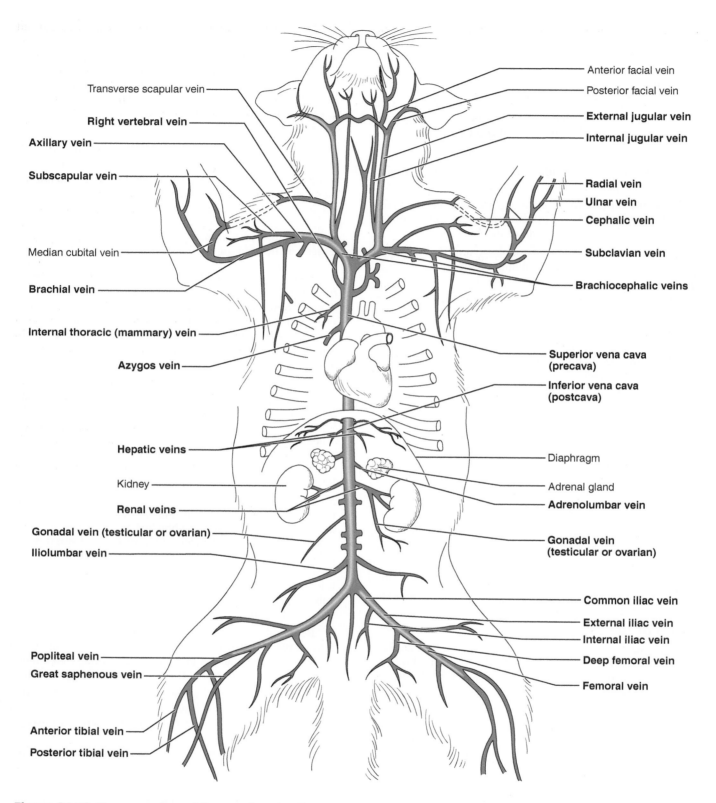

Figure 24.18 Venous system of the cat. See also Figure 24.20 on page 419.

Adrenolumbar veins: From the adrenal glands and body wall.

Renal veins: From the kidneys (it is common to find two renal veins on the right side).

Gonadal veins (testicular or ovarian veins): The left vein of this venous pair enters the left renal vein anteriorly.

Iliolumbar veins: Drain muscles of the back.

Common iliac veins: Unite to form the postcava.

The common iliac veins are formed in turn by the union of the **internal iliac** and **external iliac veins.** The more medial internal iliac veins receive branches from the pelvic organs and gluteal region, whereas the external iliac vein receives venous drainage from the lower limb. As the external iliac vein enters the thigh by running beneath the inguinal ligament, it receives the **deep femoral vein,** which drains the thigh and the external genital region. Just inferior to that point, the external iliac vein becomes the **femoral vein,** which receives blood from the thigh, leg, and foot. Follow the femoral vein down the thigh to identify the **great saphenous vein,** a superficial vein that travels up the inner aspect of the calf and across the inferior portion of the gracilis muscle (accompanied by the great saphenous artery and nerve) to enter the femoral vein. The femoral vein is formed by the union of this vein and the popliteal vein. The **popliteal vein** is located deep in the thigh beneath the semimembranosus and semitendinosus muscles in the popliteal space accompanying the popliteal artery. Trace the popliteal vein to its point of division into the **posterior** and **anterior tibial veins,** which drain the leg.

9. Trace the hepatic portal drainage system in your cat (refer to **Figure 24.19**). Locate the **hepatic portal vein** by removing the peritoneum between the first portion of the small intestine and the liver. It appears brown because of coagulated blood, and it is unlikely that it or any of the vessels of this circulation contain latex. In the cat, the hepatic portal vein is formed by the union of the **gastrosplenic** and **superior mesenteric veins.** (In humans, the hepatic portal vein is formed by the union of the splenic and superior mesenteric veins.) If possible, locate the following vessels, which empty into the hepatic portal vein:

Gastrosplenic vein: Carries blood from the spleen and stomach; located dorsal to the stomach.

Superior (cranial) mesenteric vein: A large vein draining the small and large intestines and the pancreas.

Inferior (caudal) mesenteric vein (not shown): Parallels the course of the inferior mesenteric artery and empties into the superior mesenteric vein. In humans, this vessel merges with the splenic vein.

Coronary vein: Drains the lesser curvature of the stomach (not shown).

Pancreaticoduodenal veins (anterior and posterior): The anterior branch empties into the hepatic portal vein; the posterior branch empties into the superior mesenteric vein. (In humans, both of these are branches of the superior mesenteric vein.)

If the structures of the lymphatic system of the cat are to be studied during this laboratory session, turn to the lymphatic system dissection (Exercise 25) for instructions to conduct the study. Otherwise, follow the boxed instructions to prepare your cat for storage and clean the area (page 219). ▄

View **Figure 24.20** showing the major blood vessels of the dissected cat.

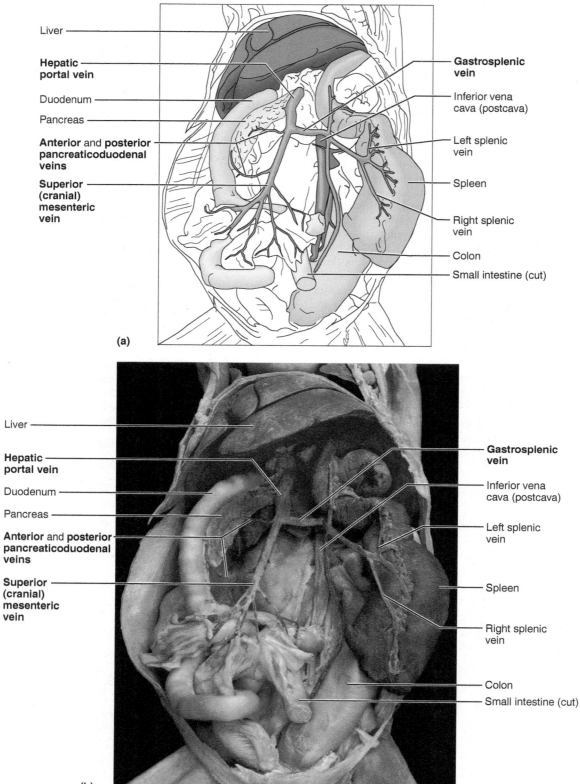

Figure 24.19 Hepatic portal circulation of the cat. (a) Diagram. **(b)** Photograph of hepatic portal system of the cat. Intestines have been partially removed. The mesentery of the small intestine has been partially dissected to show the veins of the portal system.

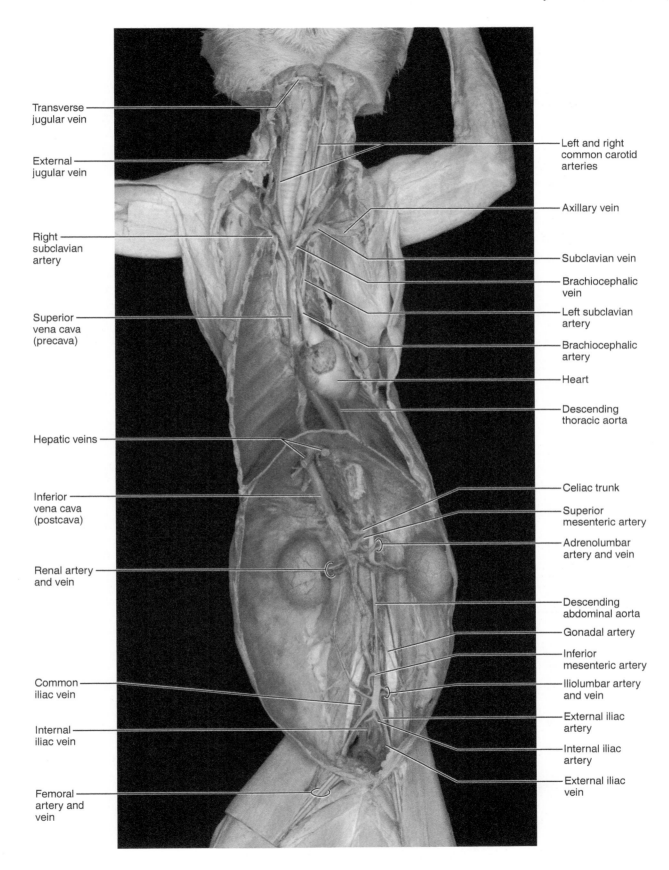

Transverse jugular vein

External jugular vein

Right subclavian artery

Superior vena cava (precava)

Hepatic veins

Inferior vena cava (postcava)

Renal artery and vein

Common iliac vein

Internal iliac vein

Femoral artery and vein

Left and right common carotid arteries

Axillary vein

Subclavian vein

Brachiocephalic vein

Left subclavian artery

Brachiocephalic artery

Heart

Descending thoracic aorta

Celiac trunk

Superior mesenteric artery

Adrenolumbar artery and vein

Descending abdominal aorta

Gonadal artery

Inferior mesenteric artery

Iliolumbar artery and vein

External iliac artery

Internal iliac artery

External iliac vein

Figure 24.20 Cat dissected to reveal major blood vessels (summary figure).

Anatomy of Blood Vessels

Microscopic Structure of the Blood Vessels

1. Use the key choices to identify the blood vessel tunics described. (Responses may be used more than once.)

   Key: a.   tunica intima   b.   tunica media   c.   tunica externa

   _____ 1.   innermost tunic

   _____ 2.   bulky middle tunic; contains smooth muscle and elastin

   _____ 3.   its smooth surface decreases resistance to blood flow

   _____ 4.   tunic of capillaries

   _____, _____, _____ 5.   tunic(s) of arteries and veins

   _____ 6.   regulates blood vessel diameter

   _____ 7.   most superficial tunic

2. Describe the essential function that the capillaries provide.

   _____

   _____

   _____

3. Cross-sectional views of an artery and of a vein are shown here. Identify each; and on the lines to the sides, note the structural details that enabled you to make these identifications:

   (vessel type)   (vessel type)

   (a) _____   (a) _____

   (b) _____   (b) _____

4. Describe the role that valves play in returning blood to the heart. _____

   _____

   _____

5. Name two events *occurring within the body* that aid in venous return.

   _____ and _____

6. Why are the walls of arteries proportionately thicker than those of the corresponding veins?_____

_____

_____

Major Systemic Arteries and Veins of the Body

7. Use the key on the right to identify the arteries or veins described on the left.

_____ 1. vessel that is paired in the venous system, but only a single vessel is present in the arterial system

_____ 2. these arteries supply the myocardium

_____, _____ 3. two paired arteries serving the brain

_____ 4. vein that runs between the cephalic and basilic veins

_____ 5. artery on the dorsum of the foot

_____ 6. main artery that serves the thigh muscles

_____ 7. supplies the diaphragm

_____ 8. formed by the union of the radial and ulnar veins

_____, _____ 9. two superficial veins of the arm

_____ 10. artery serving the kidney

_____ 11. veins draining the liver

_____ 12. artery that supplies the distal half of the large intestine

_____ 13. divides into the external and internal carotid arteries

_____ 14. what the external iliac artery becomes on entry into the thigh

_____ 15. drains blood from the spleen, pancreas, and part of the stomach

_____ 16. supplies most of the small intestine

_____ 17. join to form the inferior vena cava

_____ 18. an arterial trunk that has three major branches, which run to the liver, spleen, and stomach

_____ 19. major artery serving the tissues external to the skull

_____, _____, _____, _____ 20. four veins serving the leg

_____ 21. artery generally used to take the pulse at the wrist

Key:
a. anterior tibial
b. basilic
c. brachial
d. brachiocephalic
e. celiac trunk
f. cephalic
g. common carotid
h. common iliac
i. coronary
j. deep artery of the thigh
k. dorsalis pedis
l. external carotid
m. femoral
n. fibular
o. great saphenous
p. hepatic
q. inferior mesenteric
r. internal carotid
s. median cubital
t. phrenic
u. posterior tibial
v. radial
w. renal
x. splenic
y. superior mesenteric
z. vertebral

8. What two paired arteries enter the skull to supply the brain?

_____ and _____

9. Branches of the paired arteries just named cooperate to form a ring of blood vessels encircling the pituitary gland, at the base

of the brain. _____

What name is given to this communication network?_____

What is its function? _____

_____

10. What portion of the brain is served by the anterior and middle cerebral arteries? _____

_____

Both the anterior and middle cerebral arteries arise from the _____ arteries.

11. Trace the pathway of a drop of blood from the aorta to the left occipital lobe of the brain, noting all structures through which

it flows. _____

_____

_____

_____

12. Trace the blood flow for each of the following situations.

a.   From the mitral valve to the tricuspid valve by way of the great toe: _____

_____

_____

_____

_____

b.   From the pulmonary vein to the pulmonary artery by way of the right side of the brain: _____

_____

_____

_____

_____

_____

_____

13. The human arterial and venous systems are diagrammed on this page and the next. Identify all indicated blood vessels.

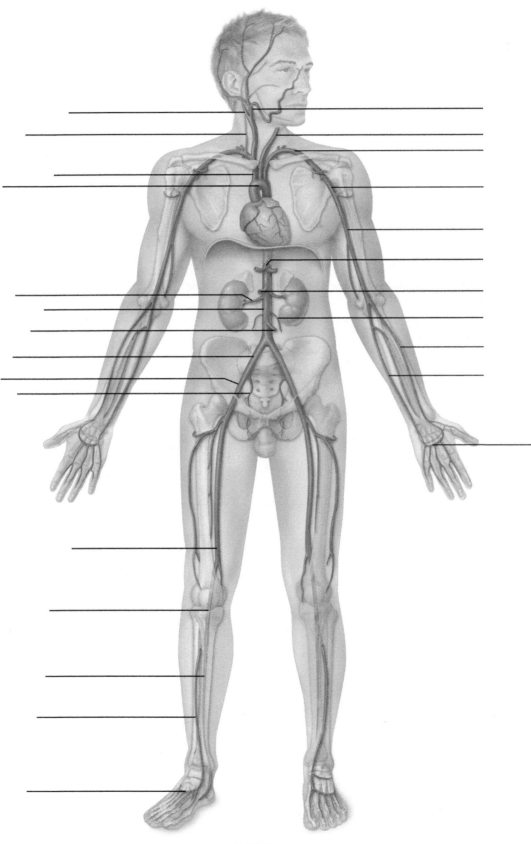

Arteries

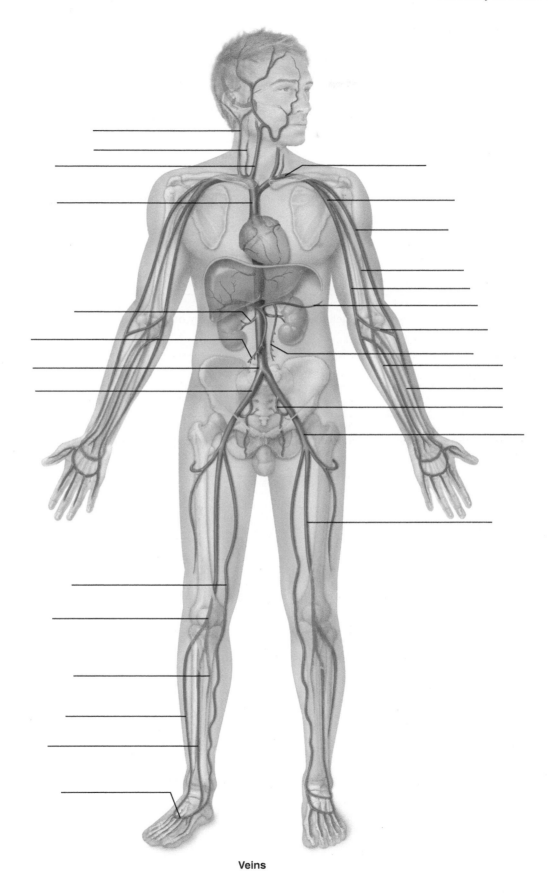

Veins

Pulmonary Circulation

14. Trace the pathway of a carbon dioxide gas molecule in the blood from the inferior vena cava until it leaves the bloodstream. Name all structures (vessels, heart chambers, and others) passed through en route.

_____

_____

_____

15. Trace the pathway of an oxygen gas molecule from an alveolus of the lung to the right ventricle of the heart. Name all

structures through which it passes. Circle the areas of gas exchange. _____

_____

_____

16. Most arteries of the adult body carry oxygen-rich blood, and the veins carry oxygen-poor blood. How does this differ in the

pulmonary arteries and veins? _____

_____

_____

17. How do the arteries of the pulmonary circulation differ structurally from the systemic arteries? What condition is indicated

by this anatomical difference? _____

_____

_____

Hepatic Portal Circulation

18. Why is the blood that drains into the hepatic portal circulation nutrient-rich? _____

_____

19. Why is this blood carried to the liver before it enters the systemic circulation? _____

_____

_____

20. The hepatic portal vein is formed by the union of the _____ and the

_____. The _____ vein carries blood from the _____,

_____, and _____. The _____ vein drains

the _____, _____, and _____. The _____

vein empties into the splenic vein and drains the _____ and _____.

21. Trace the flow of a drop of blood from the small intestine to the right atrium of the heart, noting all structures encountered or passed through on the way. _____

_____

_____

_____

Fetal Circulation

22. The failure of two of the fetal bypass structures to become obliterated after birth can cause congenital heart disease, in which the infant's blood is improperly oxygenated. Which two structures are these?

_____ and _____

23. For each of the following structures, first indicate its function in the fetus; then, note its fate (what happens to it or what it is converted to after birth). **Circle the blood vessel that carries the most oxygen-rich blood.**

Structure	Function in fetus	Fate and postnatal structure
Umbilical artery		
Umbilical vein		
Ductus venosus		
Ductus arteriosus		
Foramen ovale		

24. What organ serves as a respiratory/digestive/excretory organ for the fetus? _____

Dissection and Identification: The Blood Vessels of the Cat

25. What differences did you observe between the origin of the left common carotid arteries in the cat and in the human?

_____

_____

Between the origin of the internal and external iliac arteries? _____

_____

26. How do the relative sizes of the external and internal jugular veins differ in the human and the cat? _____

_____

_____

27. In the cat, the inferior vena cava is called the _____ ,

and the superior vena cava is referred to as the _____ .

28. Describe the location of the following blood vessels:

ascending aorta: _____

aortic arch: _____

descending thoracic aorta: _____

descending abdominal aorta: _____

29. ✚ A peripherally inserted central catheter (PICC) line involves the use of a long, thin tube to deliver medications or nutrients to a patient. For adult patients, it is usually inserted into the right cephalic vein. Trace the route that a medication would take

to reach the right atrium. List all vessels on the route. _____

_____

_____

30. ✚ A patient with iliofemoral deep vein thrombosis (IFDVT) has a blood clot located in the femoral or external iliac vein. Such a patient is at risk of the clot traveling to the lungs, resulting in a pulmonary embolism. Trace the route of the clot from

the femoral vein to the pulmonary artery. List all vessels on the route. _____

_____

_____

The Lymphatic System and Immune Response

MATERIALS

- Large anatomical chart of the human lymphatic system
- Prepared slides of lymph node, spleen, and tonsil
- Compound microscope
- Dissection tray and instruments
- Animal specimen from previous dissections
- Embalming fluid
- Disposable gloves
- Safety glasses
- Organic debris container

LEARNING OUTCOMES

☐ State the function of the lymphatic system, name its components, and compare its function to that of the blood vascular system.

☐ Describe the formation and composition of lymph, and discuss how it is transported through the lymphatic vessels.

☐ Relate immunological memory, specificity, and differentiation of self from nonself to immune function.

☐ Differentiate between the roles of B cells and T cells in the immune response.

☐ Describe the structure and function of lymph nodes, and indicate the location of T cells, B cells, and macrophages in a typical lymph node.

☐ Describe the major microanatomical features of the spleen and tonsils.

☐ Compare and contrast the lymphatic structures of the cat to those of the human.

PRE-LAB QUIZ

1. Circle True or False. The lymphatic system protects the body by removing foreign material such as bacteria from the lymphatic stream.
2. Lymph is excess:
   a. blood that has escaped from veins
   b. tissue fluid that has leaked out of capillaries
   c. tissue fluid that has escaped from arteries
3. Circle True or False. Collecting lymphatic vessels have three tunics and are equipped with valves like veins.
4. _____, which serve as filters for the lymphatic system, occur at various points along the lymphatic vessels.
   a. Glands
   b. Lymph nodes
   c. Valves
5. Circle True or False. The immune response is a systemic response that occurs when the body recognizes a substance as foreign and acts to destroy or neutralize it.
6. Three characteristics of the immune response are the ability to distinguish self from nonself, memory, and:
   a. autoimmunity   b. specificity   c. susceptibility
7. Circle the correct underlined term. B cells / T cells differentiate in the thymus.
8. Circle the correct underlined term. T cells mediate humoral / cellular immunity because they destroy cells infected with viruses and certain bacteria and parasites.
9. Circle True or False. Antibodies are produced by plasma cells in response to antigens and are found in all body secretions.
10. Circle True or False. The thymus contains both T and B cell–dependent regions.

The lymphatic system has two chief functions: (1) it transports tissue fluid that has leaked out of the vascular system and returns it to the blood vessels, and (2) it protects the body by removing foreign material such as bacteria from the lymphatic stream and by serving as a site for lymphocyte "policing" of body fluids and for lymphocyte multiplication.

The Lymphatic System

The **lymphatic system** consists of a network of lymphatic vessels (lymphatics), lymphoid tissue, lymph nodes, and a number of other lymphoid organs, such as the tonsils, thymus, and spleen. We will focus on the lymphatic vessels and lymph nodes in this section. The white blood cells, which are the central actors in body immunity, are described later in this exercise.

Distribution and Function of Lymphatic Vessels and Lymph Nodes

As blood circulates through the body, the hydrostatic and osmotic pressures operating at the capillary beds result in fluid outflow at the arterial end of the bed and in fluid return at the venous end. However, not all of the lost fluid is returned to the bloodstream by this mechanism. It is the microscopic, blind-ended **lymphatic capillaries (Figure 25.1a)**, which branch through nearly all the tissues of the body, that pick up this leaked fluid and carry it through successively larger vessels—**collecting lymphatic vessels** to **lymphatic trunks**—until the lymph finally returns to the blood vascular system through one of the two large ducts in the thoracic region (Figure 25.1b). The **right lymphatic duct** drains lymph from the right upper extremity, head, and thorax delivered by the jugular, subclavian, and bronchomediastinal trunks. The large **thoracic duct** receives lymph from the rest of the body (see Figure 25.1c). The enlarged terminus of the thoracic duct is the **cisterna chyli,** which receives lymph from the digestive organs. In humans, both ducts empty the lymph into the venous circulation at the junction of the internal jugular vein and the subclavian vein, on their respective sides of the body. Notice that the lymphatic system, lacking both a contractile "heart" and arteries, is a one-way system; it carries lymph only toward the heart.

Like veins of the blood vascular system, the collecting lymphatic vessels have three tunics and are equipped with valves **(Figure 25.2)**. However, lymphatics tend to be thinner walled, to have *more* valves, and to anastomose (form branching networks) more than veins. Because the lymphatic system is a pumpless system, lymph transport depends largely on the milking action of the skeletal muscles and on pressure changes within the thorax that occur during breathing.

As lymph is transported, it filters through bean-shaped **lymph nodes,** which cluster along the lymphatic vessels of the body. There are hundreds of lymph nodes, but because they are usually embedded in connective tissue, they are not ordinarily seen. Particularly large collections of lymph nodes are found in the inguinal, axillary, and cervical regions of the body.

ACTIVITY 1

Identifying the Organs of the Lymphatic System

Study the large anatomical chart to observe the general plan of the lymphatic system. Notice the distribution of lymph nodes, various lymphatics, the lymphatic trunks, the location of the right lymphatic duct, the thoracic duct, and the cisterna chyli. ■

The Immune Response

The collection of body defenses that protect us against microbes, cancer cells, and other foreign bodies is collectively known as the *immune response*. It consists of two main types of defense systems: the *innate (nonspecific) defenses* and the *adaptive (specific) defenses*.

• The innate defenses include surface barriers, such as the skin and mucous membranes, as well as internal defenses that include phagocytes, inflammation, and fever. We are born with these protective mechanisms.

• The adaptive defenses are referred to as specific defenses because the key players in this system have a "lock-and-key" recognition for foreign molecules.

Major Characteristics of the Adaptive Immune Response

Three characteristics of the adaptive immune response are (1) **memory,** (2) **specificity,** and (3) **self-tolerance.** The adaptive immune system's "memory" for previously encountered foreign antigens (the chicken pox virus, for example) is remarkably accurate and highly specific.

An almost limitless variety of things are *antigens,* which are capable of provoking an immune response and reacting with the products of the response. Nearly all foreign proteins, many polysaccharides, bacteria and their toxins, viruses, mismatched RBCs, cancer cells, and many small molecules (haptens) can be antigenic. The cells that recognize antigens and initiate the adaptive immune response are lymphocytes, the second most numerous members of the leukocyte, or white blood cell (WBC), population. Each immunocompetent lymphocyte has receptors on its surface allowing it to bind with only one or a few very similar antigens, thus providing specificity.

As a rule, our own proteins are tolerated, a fact that reflects the ability of the immune system to distinguish our own tissues (self) from foreign antigens (nonself). Nevertheless, an inability to recognize self can and does occasionally happen, and the immune system then attacks the body's own tissues. This phenomenon is called *autoimmunity*. Autoimmune diseases include multiple sclerosis (MS), myasthenia gravis, Graves' disease, rheumatoid arthritis (RA), and diabetes mellitus.

Organs, Cells, and Cell Interactions of the Immune Response

The immune system utilizes as part of its arsenal the **lymphoid organs** and **lymphoid tissues,** which include the thymus, lymph nodes, spleen, tonsils, appendix, and red bone marrow. Of these, the thymus and red bone marrow are

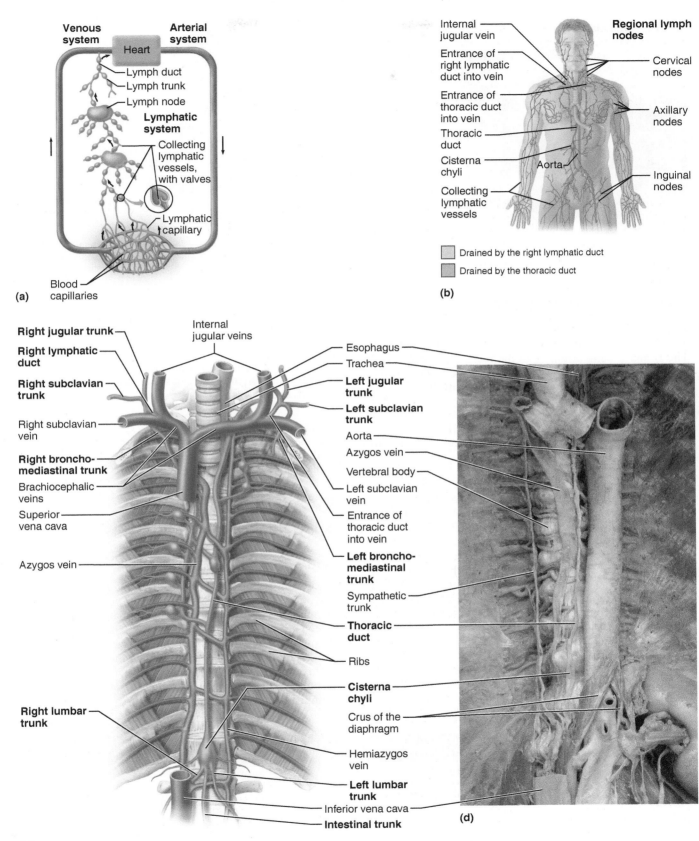

(a) Venous system / Arterial system

Heart

Lymph duct
Lymph trunk
Lymph node

Lymphatic system

Collecting lymphatic vessels, with valves

Lymphatic capillary

Blood capillaries

(b)

Internal jugular vein
Entrance of right lymphatic duct into vein
Entrance of thoracic duct into vein
Thoracic duct
Cisterna chyli
Collecting lymphatic vessels

Regional lymph nodes

Cervical nodes
Axillary nodes
Aorta
Inguinal nodes

☐ Drained by the right lymphatic duct
☐ Drained by the thoracic duct

(c)

Right jugular trunk
Right lymphatic duct
Right subclavian trunk
Right subclavian vein
Right broncho-mediastinal trunk
Brachiocephalic veins
Superior vena cava
Azygos vein
Right lumbar trunk

Internal jugular veins

Esophagus
Trachea
Left jugular trunk
Left subclavian trunk
Aorta
Azygos vein
Vertebral body
Left subclavian vein
Entrance of thoracic duct into vein
Left broncho-mediastinal trunk
Sympathetic trunk
Thoracic duct
Ribs
Cisterna chyli
Crus of the diaphragm
Hemiazygos vein
Left lumbar trunk
Inferior vena cava
Intestinal trunk

(d)

Figure 25.1 Lymphatic system. (a) Simplified scheme of the relationship of lymphatic vessels to blood vessels of the cardiovascular system. **(b)** Distribution of lymphatic vessels and lymph nodes. The green-shaded area represents body area drained by the right lymphatic duct. **(c)** Relationship of the major lymphatic trunks to the thoracic and right lymphatic ducts, and the entry points of the ducts into the subclavian veins. **(d)** Photograph showing the course of the thoracic duct.

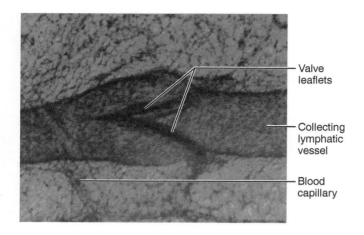

Figure 25.2 Collecting lymphatic vessel (120×).

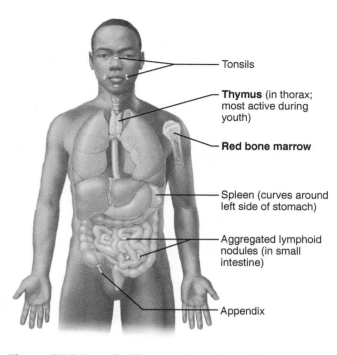

Figure 25.3 Lymphoid organs. Location of the primary and secondary lymphoid organs. Primary lymphoid organs are in bold.

considered to be the *primary lymphoid organs*. The others are *secondary lymphoid organs* and *tissues* (**Figure 25.3**). The primary cells that provide for the adaptive immune response are the **B** and **T lymphocytes,** also known as **B** and **T cells.** The B and T cells originate in the red bone marrow. Each cell type must go through a maturation process whereby they become **immunocompetent** and **self-tolerant.** Immunocompetence involves the addition of receptors on the cell surface that recognize and bind to a specific antigen. Self-tolerance involves the cell's ability to distinguish self from nonself. B cells mature in the red bone marrow. T cells travel to the thymus for their maturation process.

After maturation, the B and T cells leave the bone marrow and thymus, respectively; enter the bloodstream; and travel to peripheral (secondary) lymphoid organs, where clonal selection occurs. **Clonal selection** is triggered when an antigen binds to the specific cell-surface receptors of a T or B cell. This event causes the lymphocyte to proliferate rapidly, forming a clone of like cells, all bearing the same antigen-specific receptors. Then, in the presence of certain regulatory signals, the progeny of the clone specialize, or differentiate—some forming memory cells and others becoming effector cells. Upon subsequent meetings with the same antigen, the immune response proceeds considerably faster because the troops are already mobilized and awaiting further orders, so to speak. Additional characteristics of B and T cells are compared in **Table 25.1.**

If T lymphocytes fail to differentiate in the thymus, both antibody and cell-mediated immune functions will be significantly depressed. Additionally, the observation

that the thymus naturally shrinks with age has been correlated with the decline in immune function that occurs in elderly individuals. ✚

ACTIVITY 2

Studying the Microscopic Anatomy of a Lymph Node, the Spleen, and a Tonsil

1. Obtain a compound microscope and prepared slides of a lymph node, the spleen, and a tonsil. As you examine the lymph node slide, notice the following anatomical features, which are depicted in **Figure 25.4b**. The node is enclosed within a fibrous **capsule,** from which connective tissue septa (**trabeculae**) extend inward to divide the node into several compartments. Very fine strands of reticular connective tissue issue from the trabeculae, forming the stroma of the gland within which cells are found.

Table 25.1	Comparison of B and T Lymphocytes	
Properties	**B lymphocytes**	**T lymphocytes**
Type of Immune Response	Humoral immunity (antibody-mediated immunity)	Cellular immunity (cell-mediated immunity)
Site of Maturation	Red bone marrow	Thymus
Effector Cells	Plasma cells (antibody-secreting cells)	Cytotoxic T cells, helper T cells, regulatory T cells
Memory Cell Formation	Yes	Yes
Functions	Plasma cells produce antibodies that inactivate antigen and tag antigen for destruction.	Cytotoxic T cells attack infected cells and tumor cells. Helper T cells activate B cells and other T cells.

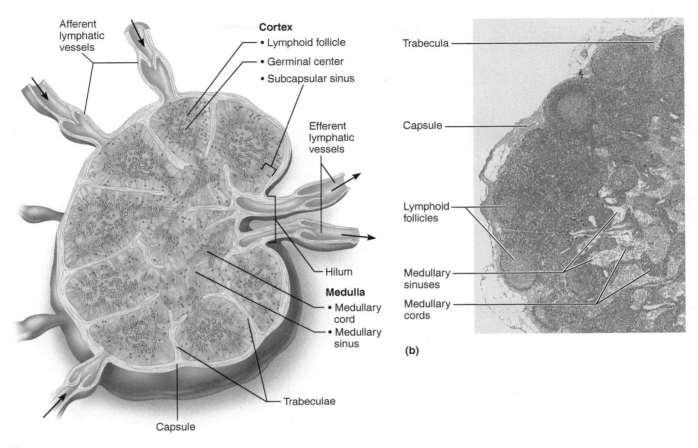

Figure 25.4 Structure of lymph node. (a) Longitudinal section of a lymph node. The arrows indicate the direction of the lymph flow. **(b)** Photomicrograph of a part of a lymph node (20×).

In the outer region of the node, the **cortex,** some of the cells are arranged in globular masses, referred to as germinal centers. The **germinal centers** contain rapidly dividing B cells. The rest of the cortical cells are primarily T cells that circulate continuously, moving from the blood into the node and then exiting from the node in the lymphatic stream.

In the internal portion of the node, the **medulla,** the cells are arranged in cordlike fashion. Most of the medullary cells are macrophages. Macrophages are important not only for their phagocytic function but also because they play an essential role in "presenting" the antigens to the T cells.

Lymph enters the node through a number of *afferent vessels,* circulates through *lymph sinuses* within the node, and leaves the node through *efferent vessels* at the **hilum.** Because each node has fewer efferent than afferent vessels, the lymph flow stagnates somewhat within the node. This allows time for generating an immune response and for the macrophages to remove debris from the lymph before it reenters the blood vascular system.

2. As you observe the slide of the spleen, look for the areas of lymphocytes suspended in reticular fibers, the **white pulp,** clustered around central arteries **(Figure 25.5)**. The remaining tissue in the spleen is the **red pulp,** which is composed of splenic sinusoids and areas of reticular tissue and macrophages called the **splenic cords.** The white pulp, composed primarily of lymphocytes, is responsible for the immune functions of the spleen. Macrophages remove worn-out red blood cells, debris, bacteria, viruses, and toxins from blood flowing through the sinuses of the red pulp.

3. As you examine the tonsil slide, notice the **lymphoid follicles** containing **germinal centers** surrounded by scattered lymphocytes. The characteristic **tonsillar crypts** (invaginations of the mucosal epithelium) of the tonsils trap bacteria and other foreign material **(Figure 25.6)**. Eventually the bacteria work their way into the lymphoid tissue and are destroyed. ▬

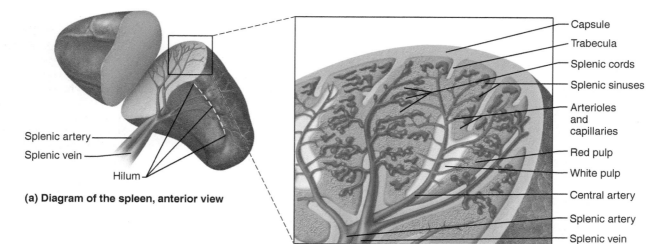

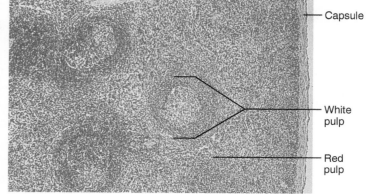

(a) Diagram of the spleen, anterior view

Splenic artery
Splenic vein
Hilum

Capsule
Trabecula
Splenic cords
Splenic sinuses
Arterioles and capillaries
Red pulp
White pulp
Central artery
Splenic artery
Splenic vein

(b) Diagram of spleen histology

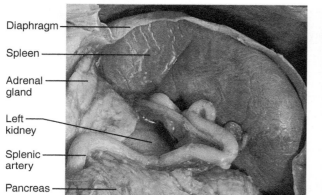

Diaphragm
Spleen
Adrenal gland
Left kidney
Splenic artery
Pancreas

(c) Photograph of the spleen in its normal position in the abdominal cavity, anterior view

Capsule

White pulp

Red pulp

(d) Photomicrograph of spleen tissue (75×). The white pulp, a lymphoid tissue with many lymphocytes, is surrounded by red pulp containing abundant erythrocytes.

Figure 25.5 The spleen.

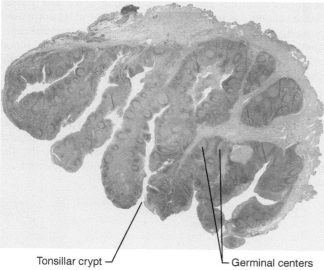

Tonsillar crypt

Germinal centers in lymphoid follicles

Figure 25.6 Histology of a palatine tonsil. The luminal surface is covered with epithelium that invaginates deeply to form crypts (10×).

DISSECTION AND IDENTIFICATION:
The Main Lymphatic Ducts of the Cat

ACTIVITY 3

Identifying the Main Lymphatic Ducts of the Cat

1. Don disposable gloves and safety glasses. Obtain your cat and a dissecting tray and instruments. Because lymphatic vessels are extremely thin-walled, it is difficult to locate them in a dissection unless the animal has been triply injected (with yellow or green latex for the lymphatic system). However, the large thoracic duct can be localized and identified.

2. Move the thoracic organs to the side to locate the **thoracic duct.** Typically it lies just to the left of the mid-dorsal line, abutting the dorsal aspect of the descending aorta. It is usually about the size of pencil lead and red-brown, with a segmented or beaded appearance caused by the valves within it. Trace it anteriorly to the site where it passes behind the left brachiocephalic vein and then bends and enters the venous system at the junction of the left subclavian and external jugular veins. If the veins are well injected, some of the blue latex may have slipped past the valves and entered the first portion of the thoracic duct.

GROUP CHALLENGE

Compare and Contrast Lymphoid Organs and Tissues

Work in groups of three to discuss the characteristics of each lymphoid structure listed in the **Group Challenge chart** below. On a separate piece of paper, one student will record the characteristics for each structure for the group. The group will then consider each pair of structures listed in the chart, discuss the similarities and differences for each pair, and complete the chart based on consensus answers. Use your text or another appropriate reference as needed for comparing aggregated lymphoid nodules and the thymus to the other organs and tissues.

Lymphoid Pair	Similarities	Differences
Lymph node Spleen		
Lymph node Tonsil		
Aggregated lymphoid nodules Tonsils		
Tonsil Spleen		
Thymus Spleen		

3. While in this region, also attempt to identify the short **right lymphatic duct** draining into the right subclavian vein, and notice the collection of lymph nodes in the axillary region.

4. If the cat is triply injected, trace the thoracic duct posteriorly to identify the **cisterna chyli,** a saclike enlargement of its distal end. This structure, which receives fat-rich lymph from the intestine, begins at the level of the diaphragm and can be located posterior to the left kidney.

5. When you finish identifying these lymphatic structures, clean the dissecting instruments and tray, and follow the boxed instructions to prepare your cat for storage and clean the area (page 219). ▬

The Lymphatic System and Immune Response

The Lymphatic System

1. Match the terms below with the correct letters on the diagram.

_____ 1. aggregated lymphoid nodules (in small intestine)

_____ 2. appendix

_____ 3. axillary lymph nodes

_____ 4. cervical lymph nodes

_____ 5. cisterna chyli

_____ 6. inguinal lymph nodes

_____ 7. lymphatic vessels

_____ 8. red bone marrow

_____ 9. right lymphatic duct

_____ 10. spleen

_____ 11. thoracic duct

_____ 12. thymus gland

_____ 13. tonsils

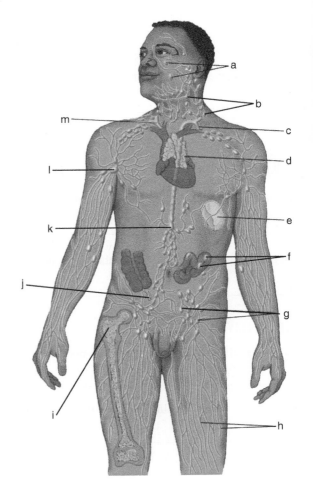

2. Explain why the lymphatic system is a one-way system, whereas the blood vascular system is a two-way system.

_____

_____

3. How do collecting lymphatic vessels resemble veins? _____

   How do lymphatic capillaries differ from blood capillaries? _____

   _____

4. What is the function of the lymphatic vessels? _____

   _____

5. What is lymph? _____

6. What factors are involved in the flow of lymphatic fluid? _____

   _____

7. What name is given to the terminal duct draining lymph from most of the body? _____

8. What is the cisterna chyli? _____

   _____

9. Which portion of the body is drained by the right lymphatic duct? _____

10. Note three areas where lymph nodes are densely clustered: _____

    _____, and _____ _____

11. What are the two major functions of the lymph nodes? _____

    and _____

The Immune Response

12. Describe the effector cells involved in humoral immunity. _____

    _____

13. Describe the effector cells involved in cell-mediated immunity. What is the function of T cells in the immune response?

    _____

    _____

Studying the Microscopic Anatomy of a Lymph Node, the Spleen, and a Tonsil

14. In the diagram of a lymph node below, label the following: afferent lymphatic vessel, efferent lymphatic vessel, lymphoid follicle, trabeculae, subcapsular sinus, capsule, and hilum.

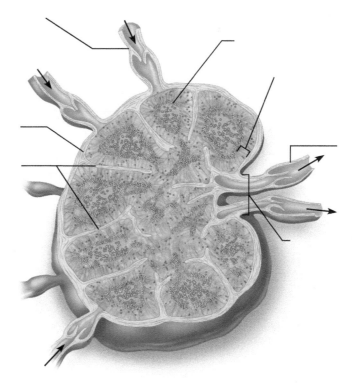

15. What structural characteristic ensures a *slow* flow of lymph through a lymph node? _____

_____

Why is this desirable? _____

_____

16. What similarities in structure and function are found in the lymph nodes, spleen, and tonsils? _____

_____

_____

Dissection and Identification: The Main Lymphatic Ducts of the Cat

17. How does the cat's lymphatic drainage pattern compare to that of humans? _____

_____

18. What is the role of each of the following?

   a.   thoracic duct: _____

_____

   b.   right lymphatic duct: _____

_____

   c.   cisterna chyli: _____

_____

19. ➕ Lymphedema is a condition characterized by insufficient movement of lymph in the lymphatic vessels. Fluid builds up in the tissues and in the lymphatic vessels of the limbs. Explain why exercise would have a positive effect on this condition.

_____

_____

20. ➕ Buboes are a key sign for diagnosing bubonic plague. Buboes are swollen lymph nodes that can become necrotic and turn black. Predict where on the body buboes would be most likely to develop in cases of the bubonic plague.

_____

_____

Anatomy of the Respiratory System

EXERCISE

MATERIALS

- Resin cast of the bronchial tree (if available)
- Human torso model
- Thoracic cavity structures model and/or chart of the respiratory system
- Larynx model (if available)
- Preserved inflatable lung preparation (obtained from a biological supply house) or fresh sheep pluck
- Source of compressed air
- 0.6 m (2-foot) length of laboratory rubber tubing
- Dissecting tray and instruments
- Disposable gloves
- Safety glasses
- Disposable autoclave bag
- Animal specimen from previous dissections
- Prepared slides of the following (if available): trachea (x.s.), lung tissue, both normal and pathological specimens (for example, sections taken from lung tissues exhibiting bronchitis, pneumonia, emphysema, or lung cancer)
- Compound and dissecting microscopes
- Organic debris container

LEARNING OUTCOMES

- ☐ State the major functions of the respiratory system.
- ☐ Define the following terms: *pulmonary ventilation, external respiration,* and *internal respiration.*
- ☐ Identify the major respiratory system structures on models or appropriate images, and describe the function of each.
- ☐ Describe the difference between the conducting and respiratory zones, and indicate which is referred to as *anatomical dead space.*
- ☐ Name the serous membrane that encloses each lung, and describe its structure.
- ☐ Demonstrate lung inflation in a fresh sheep pluck or preserved tissue specimen.
- ☐ Recognize the histologic structure of the trachea and lung tissue microscopically or in an image, and describe the functions served by the observed structures.
- ☐ Identify the major respiratory system organs in a dissected animal.

PRE-LAB QUIZ

1. The major role of the respiratory system is to:
   a. dispose of waste products in a solid form
   b. permit the flow of nutrients through the body
   c. supply the body with carbon dioxide and dispose of oxygen
   d. supply the body with oxygen and dispose of carbon dioxide
2. Circle True or False. Four processes—pulmonary ventilation, external respiration, transport of respiratory gases, and internal respiration—must all occur in order for the respiratory system to function fully.
3. The upper respiratory structures include the nose, the larynx, and the:
   a. epiglottis   b. lungs   c. pharynx   d. trachea
4. Circle the correct underlined term. The <u>thyroid cartilage</u> / <u>arytenoid cartilage</u> is the largest and most prominent of the laryngeal cartilages.
5. Circle True or False. The epiglottis forms a lid over the larynx when we swallow food; it closes off the respiratory passageway to incoming food or drink.
6. Air flows from the larynx to the trachea, from which it enters the:
   a. left and right lungs   c. pharynx
   b. left and right main bronchi   d. segmental bronchi
7. Circle the correct underlined term. The lining of the trachea is pseudostratified ciliated <u>columnar epithelium</u> / <u>transitional epithelium</u>, which propels dust particles, bacteria, and other debris away from the lungs.
8. Circle True or False. All but the smallest branches of the bronchial tree have cartilaginous reinforcements in their walls.
9. _____, tiny balloonlike expansions of the alveolar sacs, are composed of a single thin layer of squamous epithelium. They are the main structural and functional units of the lung and the actual sites of gas exchange.
10. Circle the correct underlined term. Fissures divide the lungs into lobes, three on the right and <u>two</u> / <u>three</u> on the left.

Body cells require an abundant and continuous supply of oxygen. As the cells use oxygen, they release carbon dioxide, a waste product that the body must get rid of. The major role of the **respiratory system,** our focus in this exercise, is to supply the body with oxygen and dispose of carbon dioxide. To fulfill this role, at least four distinct processes, collectively referred to as **respiration,** must occur:

Pulmonary ventilation: The tidelike movement of air into and out of the lungs so that the gases are continuously changed and refreshed. Also more simply called *breathing*.

External respiration: The gas exchange between the blood and the air-filled chambers of the lungs.

Transport of respiratory gases: The transport of respiratory gases between the lungs and tissue cells of the body using blood as the transport vehicle.

Internal respiration: Exchange of gases between systemic blood and tissue cells.

Only the first two processes are exclusive to the respiratory system, but all four must occur for the respiratory system to function completely. Hence, the respiratory and circulatory systems are irreversibly linked.

Upper Respiratory System Structures

The upper respiratory system structures—the external nose, nasal cavity, pharynx, and paranasal sinuses—are summarized in Table 26.1 and illustrated in Figure 26.1. As you read through the descriptions in the table, identify each structure in the figures. Note that different sources divide the upper and lower respiratory systems slightly differently.

Table 26.1 Structures of the Upper Respiratory System (Figure 26.1)		
Structure	**Description**	**Function**
External Nose	Externally visible, its inferior surface has nostrils (nares). Supported by bone and cartilage and covered with skin.	The nostrils provide an entrance for air into the respiratory system.
Nasal Cavity (includes the structures listed below)	Lined with respiratory mucosa composed of pseudostratified ciliated columnar epithelium. The floor of the cavity is formed by the hard and soft palates.	Functions to filter, warm, and moisten incoming air; resonance chambers for voice production.
Nasal vestibule	Anterior portion of the nasal cavity; contains sebaceous and sweat glands and numerous hair follicles.	Filters coarse particles from the air.
Nasal septum	Formed by the vomer, perpendicular plate of the ethmoid bone, and septal cartilage.	Divides the nasal cavity into left and right sides.
Superior, middle, and inferior nasal conchae	Turbinates that project medially from the lateral walls of the cavity. Each concha has a corresponding meatus beneath it.	Increase the surface area of the mucosa, which enhances air turbulence and aids in trapping large particles in the mucus.
Posterior nasal apertures	Posterior openings of the nasal cavity.	Provide an exit for the air into the nasopharynx.
Pharynx (3 subdivisions: nasopharynx, oropharynx, and laryngopharynx—listed below)		
Nasopharynx	Superior portion of the pharynx located posterior to the nasal cavity; lined with pseudostratified ciliated columnar epithelium. The pharyngeal tonsil and openings of the pharyngotympanic tubes (surrounded by tubal tonsils) are located in this region.	Provides for the passage of air from the nasal cavity. Tonsils in the region provide protection against pathogens.
Oropharynx	Located posterior to the oral cavity and extends from the soft palate to the epiglottis; lined with stratified squamous epithelium. Its lateral walls contain the palatine tonsils. The lingual tonsils are in the anterior oropharynx at the base of the tongue.	Provides for the passage of air and swallowed food. Tonsils provide protection against pathogens.
Laryngopharynx	Extends from the epiglottis to the larynx; lined with stratified squamous epithelium. It diverges into respiratory and digestive branches.	Provides for the passage of air and swallowed food.
Pharyngotympanic Tube	Tube that opens into the lateral walls of the nasopharynx and connects the nasopharynx to the middle ear.	Allows the middle ear pressure to equalize with the atmospheric pressure.
Paranasal Sinuses	Surround the nasal cavity and are named for the bones in which they are located. Lined with pseudostratified ciliated columnar epithelium.	Act as resonance chambers for speech; warm and moisten incoming air.

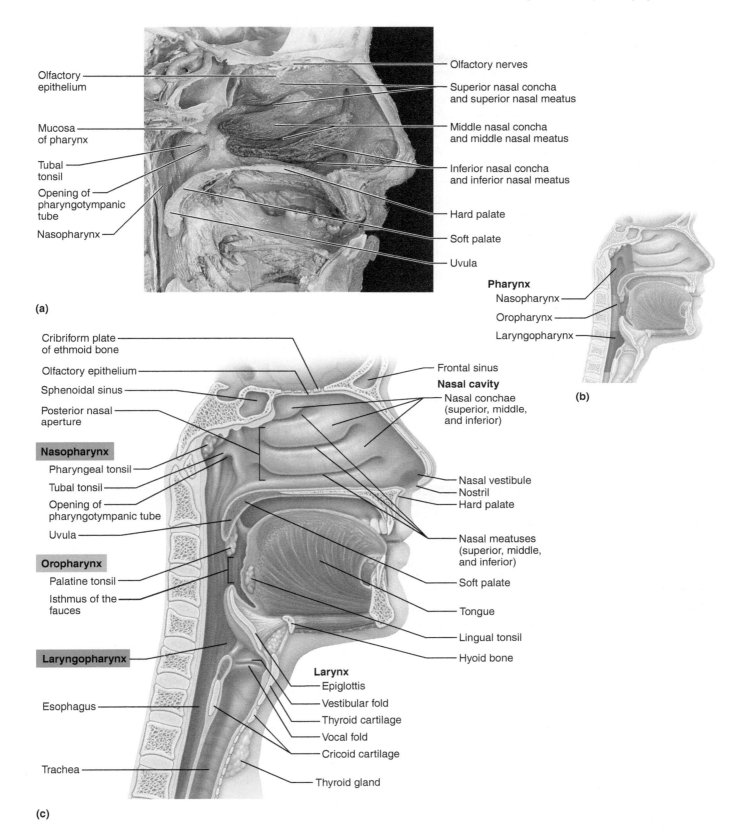

Figure 26.1 Structures of the upper respiratory tract (midsagittal section).
(a) Photograph. **(b)** Regions of the pharynx. **(c)** Diagram.

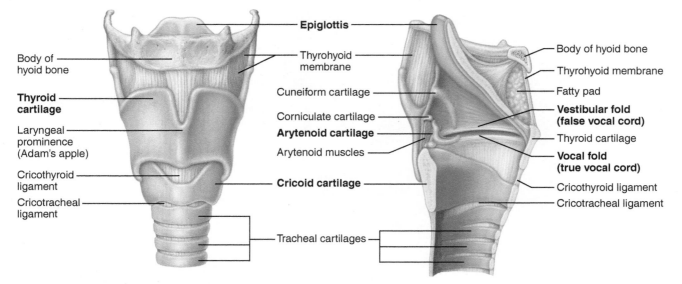

(a) Anterior superficial view   **(b) Sagittal view; anterior surface to the right**

Figure 26.2 The larynx. Attached tracheal cartilages are also shown.

Lower Respiratory System Structures

Anatomically, the lower respiratory structures include the larynx, trachea, bronchi, and lungs. The **larynx,** or voice box, attaches superiorly to the hyoid bone and contains a number of important structures summarized in **Table 26.2** and illustrated in **Figure 26.2.** From the larynx, air enters the **trachea,** or windpipe, and travels down the trachea to the level of the *sternal angle.* There the passageway divides into the right and left **main (primary) bronchi (Figure 26.3),** which plunge into their respective lungs at an indented area called the **hilum** (see Figure 26.5c). The right main bronchus is wider, shorter, and more vertical than the left, and foreign objects that enter the respiratory passageways are more likely to become stuck there.

Table 26.2	Structures of the Larynx (Figure 26.2)	
Structure	**Description**	**Function**
Larynx (includes the structures listed below)	Tube connecting the laryngopharynx and the trachea. Nine cartilages are present. Epithelium superior to the vocal folds is stratified squamous. Epithelium inferior to the vocal folds is pseudostratified ciliated columnar.	Air passageway; prevents food from entering the lower respiratory tract. Responsible for voice production.
Thyroid cartilage	Large cartilage made up of hyaline cartilage. Its laryngeal prominence is commonly referred to as the Adam's apple.	Forms the framework of the larynx.
Cricoid cartilage	Single ring of hyaline cartilage located inferior to the thyroid cartilage and superior to the trachea.	Attaches the larynx to the trachea via the cricotracheal ligament.
Arytenoid cartilage	Paired pyramid-shaped hyaline cartilages.	Anchor the vocal folds (true vocal cords).
Corniculate cartilage	Paired small horn-shaped hyaline cartilages located atop the arytenoid cartilages.	Form part of the posterior wall of the larynx.
Cuneiform cartilage	Paired wedge-shaped hyaline cartilages.	Form the lateral aspect of the laryngeal wall.
Epiglottis	Single flap of elastic cartilage anchored to the inner rim of the thyroid cartilage.	"Guardian of the airways" forms a lid over the larynx during swallowing.
Vocal folds (true vocal cords)	Mucosal folds composed of mostly elastic fibers covered with mucous membrane; attached to the arytenoid cartilage.	Vibrate with expired air for sound production.
Vestibular folds (false vocal cords)	Located superior to the vocal folds. Mucosal folds similar in composition to the vocal folds.	Protect the vocal folds and help to close the glottis when we swallow.
Glottis	The vocal folds and the slitlike passageway between the vocal folds.	Plays a role in the Valsalva maneuver.

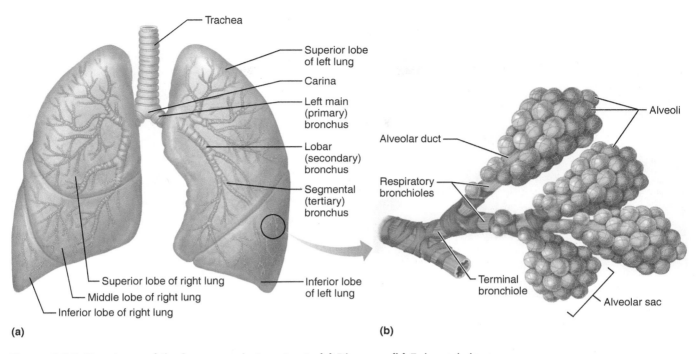

Figure 26.3 **Structures of the lower respiratory tract. (a)** Diagram. **(b)** Enlarged view of alveoli.

The trachea is lined with a ciliated mucus-secreting, pseudostratified columnar epithelium. The cilia propel mucus (produced by goblet cells) laden with dust particles, bacteria, and other debris away from the lungs and toward the throat, where it can be expectorated or swallowed. The walls of the trachea are reinforced with C-shaped cartilaginous rings, the incomplete portion located posteriorly. These C-shaped cartilages serve a double function. The incomplete parts allow the esophagus to expand anteriorly when a large food bolus is swallowed. The solid portions reinforce the trachea walls to maintain its open passageway regardless of the pressure changes that occur during breathing.

The main bronchi further divide into smaller and smaller branches—the lobar (secondary), segmental (tertiary), and on down—finally becoming the **bronchioles.** Each bronchiole divides into many **terminal bronchioles** which are less than 0.5 mm in diameter. Each terminal bronchiole branches into two or more **respiratory bronchioles** (Figure 26.3b). All but the smallest branches have cartilaginous reinforcements in their walls, usually in the form of small plates of hyaline cartilage rather than cartilaginous rings. As the respiratory tubes get smaller and smaller, the relative amount of smooth muscle in their walls increases as the amount of cartilage declines and finally disappears. Additionally, the epithelium of the bronchioles changes from pseudostratified columnar to columnar and then to cuboidal in the terminal bronchioles and respiratory bronchioles. The continuous branching of the respiratory passageways in the lungs is often referred to as the **bronchial tree.**

• Observe a resin cast of respiratory passages if one is available for observation in the laboratory.

The respiratory bronchioles subdivide into several **alveolar ducts,** which terminate in alveolar sacs that resemble clusters of grapes. **Alveoli,** tiny balloonlike expansions that each represent a single grape in the alveolar sac, are composed of a single thin layer of squamous epithelium overlying a basement membrane. The external surfaces of the alveoli are covered with a network of pulmonary capillaries **(Figure 26.4).** Together, the alveolar and capillary walls and their fused basement membranes form the **respiratory membrane,** also called the *blood air barrier.*

Because gas exchanges occur by simple diffusion across the respiratory membrane, the alveoli, alveolar ducts, and respiratory bronchioles are referred to collectively as **respiratory zone structures.** All other respiratory passageways (from the nasal cavity to the terminal bronchioles) simply serve as access or exit routes to and from these gas exchange chambers and are called **conducting zone structures** or *anatomical dead space.*

The Lungs and Pleurae

The paired lungs are soft, spongy organs that occupy the entire thoracic cavity except for the *mediastinum,* which houses the heart, bronchi, esophagus, and other organs **(Figure 26.5).** Each lung is connected to the mediastinum by a **root** containing its vascular and bronchial attachments. The structures of the root enter (or leave) the lung via a medial indentation called the *hilum.* All structures distal to the main bronchi are found within the lung substance. A lung's **apex,** the narrower superior aspect, lies just deep to the clavicle, and its **base,** the inferior concave surface, rests on the diaphragm. Anterior, lateral, and posterior lung surfaces are in close contact with the ribs and, hence, are collectively called the **costal surface.** The medial surface of the left lung exhibits a concavity called the **cardiac notch,** which accommodates the heart where it extends left from the body midline. Fissures divide the lungs into a number of **lobes**—two in the left lung and three in the right. Other than the respiratory passageways and air spaces that make up the bulk of their volume, the lungs are mostly elastic connective tissue, which allows them to recoil passively during expiration.

Each lung is enclosed in a double-layered sac of serous membrane called the **pleura.** The outer layer, the **parietal**

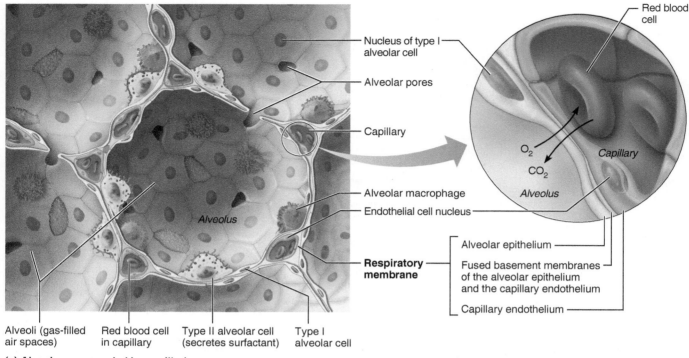

Alveoli (gas-filled air spaces)	Red blood cell in capillary	Type II alveolar cell (secretes surfactant)	Type I alveolar cell

(a) Alveolus surrounded by capillaries

(b) Respiratory membrane

Figure 26.4 Relationship between the alveoli and pulmonary capillaries involved in gas exchange.

pleura, is attached to the thoracic walls and the **diaphragm;** the inner layer, covering the lung tissue, is the **visceral pleura.** The two pleural layers are separated by the **pleural cavity,** which is filled with a thin film of *pleural fluid.* Produced by the pleurae, this fluid allows the lungs to glide without friction over the thoracic wall during breathing.

ACTIVITY 1

Identifying Respiratory System Organs

Before proceeding, be sure to locate on the torso model, thoracic cavity structures model, larynx model, or an anatomical chart all the respiratory structures described—both upper and lower respiratory system organs (see Figure 26.5). ▬

ACTIVITY 2

Examining Prepared Slides of Trachea and Lung Tissue

1. Obtain a compound microscope and a slide of a cross section of the tracheal wall. Identify the smooth muscle layer, the hyaline cartilage supporting rings, and the pseudostratified ciliated epithelium. Use **Figure 26.6** as a guide; also try to identify a few goblet cells in the epithelium (see Figure 5.3d, page 57).

2. Obtain a slide of lung tissue for examination. The alveolus is the main structural and functional unit of the lung and is the actual site of gas exchange. Identify a bronchiole **(Figure 26.7a)** and the simple squamous epithelium of the alveolar walls (Figure 26.7b).

3. Examine slides of pathological lung tissues, and compare them to the normal lung specimens. Record your observations in the Review Sheet at the end of this exercise. ▬

ACTIVITY 3

Demonstrating Lung Inflation in a Sheep Pluck

A *sheep pluck* includes the larynx, trachea with attached lungs, the heart and pericardium, and portions of the major blood vessels found in the mediastinum.

⚠ Don disposable gloves, obtain a dissecting tray and a fresh sheep pluck (or a preserved pluck of another animal), and identify the lower respiratory system organs. Once you have completed your observations, insert a hose from an air compressor (vacuum pump) into the trachea and alternately allow air to flow into and out of the lungs. Notice how the lungs inflate. This observation is educational in a preserved pluck, but it is a spectacular sight in a fresh one. When a fresh pluck is used, the lung pluck changes color (becomes redder) as hemoglobin in trapped RBCs becomes loaded with oxygen.

⚠ Dispose of the gloves in the autoclave bag immediately after use. ▬

 DISSECTION AND IDENTIFICATION
The Respiratory System of the Cat

In this dissection exercise, you will be examining both the gross and microscopic structure of respiratory system organs. Don disposable gloves and safety glasses, and then obtain your dissection animal, dissecting tray, and instruments. ▬

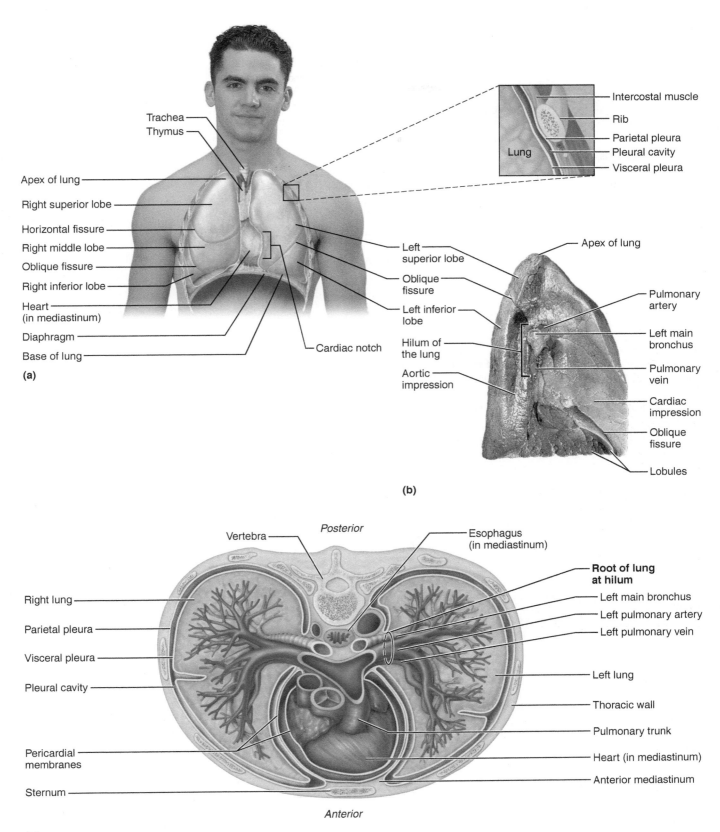

Figure 26.5 Anatomical relationships of organs in the thoracic cavity. (a) Anterior view of the thoracic organs. The lungs flank the central mediastinum. The inset at upper right depicts the pleura and the pleural cavity. **(b)** Photograph of medial aspect of left lung. **(c)** Transverse section through the superior part of the thorax, showing the lungs and the main organs in the mediastinum.

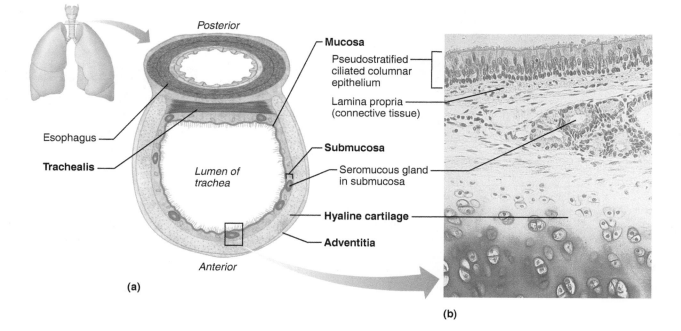

Posterior

Mucosa
Pseudostratified ciliated columnar epithelium
Lamina propria (connective tissue)

Esophagus

Submucosa
Seromucous gland in submucosa

Trachealis

Lumen of trachea

Hyaline cartilage

Adventitia

Anterior

(a)

(b)

Figure 26.6 Microscopic structure of the trachea. (a) Cross-sectional view of the trachea. **(b)** Photomicrograph of a portion of the tracheal wall (125×).

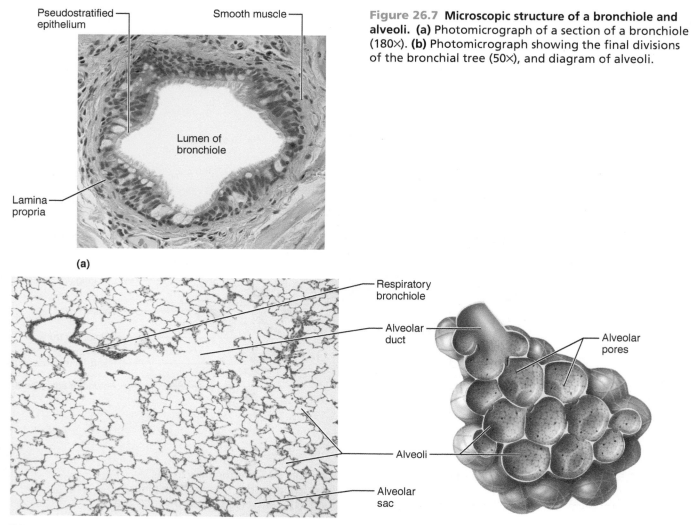

Pseudostratified epithelium

Smooth muscle

Lumen of bronchiole

Lamina propria

(a)

Figure 26.7 Microscopic structure of a bronchiole and alveoli. (a) Photomicrograph of a section of a bronchiole (180×). **(b)** Photomicrograph showing the final divisions of the bronchial tree (50×), and diagram of alveoli.

Respiratory bronchiole

Alveolar duct

Alveolar pores

Alveoli

Alveolar sac

(b)

ACTIVITY 4

Identifying Organs of the Respiratory System of the Cat

1. Examine the external nares, oral cavity, and oropharynx (**Figure 26.8**). Use a probe to demonstrate the continuity between the oropharynx and the nasopharynx above. The pharynx continues as the laryngopharynx, which lies immediately dorsal to the larynx (not shown).

2. After securing the animal to the dissecting tray, dorsal surface down, expose the more distal respiratory structures by retracting the cut muscle and rib cage. Do not sever nerves and blood vessels located on either side of the trachea if these have not been studied. If you have not previously opened the thoracic cavity, make a medial longitudinal incision through the neck muscles and thoracic musculature to expose and view the thoracic organs (see Figure 21.3, page 354).

3. Identify the structures named in steps 3 through 5 (refer to **Figures 26.9, 26.10,** and **26.11**). Examine the **trachea,** and

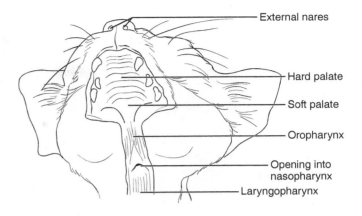

Figure 26.8 External nares, oral cavity, and pharynx of the cat. The larynx has been dissected free and reflected towards the thorax.

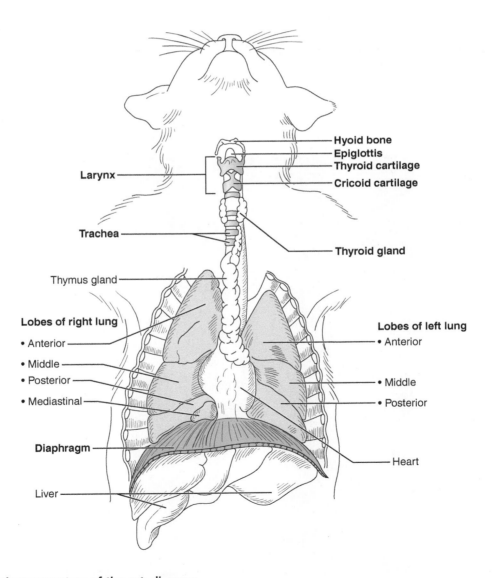

Figure 26.9 Respiratory system of the cat, diagram.

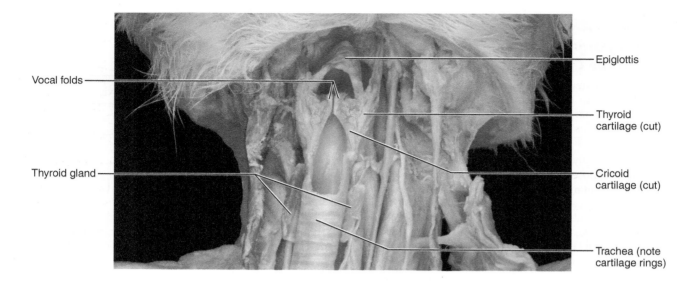

Vocal folds

Thyroid gland

Epiglottis

Thyroid cartilage (cut)

Cricoid cartilage (cut)

Trachea (note cartilage rings)

Figure 26.10 **Anterior view of larynx (opened), trachea, and thyroid gland.**

determine by finger examination whether the cartilage rings are complete or incomplete posteriorly. Locate the **thyroid gland** inferior to the **larynx** on the trachea. Free the larynx from the attached muscle tissue for ease of examination. Identify the **thyroid** and **cricoid cartilages** and the flaplike **epiglottis.** Find the *hyoid bone,* located anterior to the larynx. Make a longitudinal incision through the ventral wall of the larynx, and locate the *vocal folds* (Figure 26.10) and *vestibular folds* on the inner wall.

4. Locate the large *right* and *left common carotid arteries* (Figure 26.11) and the *internal jugular* veins on either side of the trachea. Also locate a conspicuous white band, the *vagus nerve,* which lies alongside the trachea, adjacent to the common carotid artery.

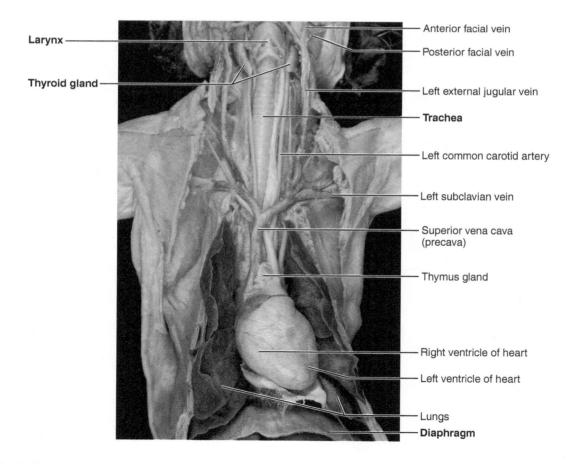

Larynx

Thyroid gland

Anterior facial vein

Posterior facial vein

Left external jugular vein

Trachea

Left common carotid artery

Left subclavian vein

Superior vena cava (precava)

Thymus gland

Right ventricle of heart

Left ventricle of heart

Lungs

Diaphragm

Figure 26.11 **Photograph of the respiratory system of the cat.**

5. Examine the contents of the thoracic cavity. Follow the trachea as it bifurcates into two **main (primary) bronchi,** which plunge into the **lungs.** Note that there are two *pleural cavities* containing the lungs and that each lung is composed of many lobes. In humans there are three lobes in the right lung and two in the left. How does this compare to what you see in the cat?

_____

In the mediastinum, identify the pericardial sac (if it is still present) containing the heart. Examine the pleura, and note its exceptionally smooth texture.

6. Locate the **diaphragm** and the **phrenic nerve.** The phrenic nerve, clearly visible as a white "thread" running along the pericardium to the diaphragm, controls the activity of the diaphragm in breathing. Lift one lung, and find the esophagus beneath the parietal pleura. Follow it through the diaphragm to the stomach. ▄▄

ACTIVITY 5

Observing Lung Tissue Microscopically

Make a longitudinal incision in the outer tissue of one lung lobe beginning at a main bronchus. Attempt to follow part of the bronchial tree from this point down into the smaller subdivisions. Carefully observe the cut lung tissue (under a dissecting microscope, if one is available), noting the richness of the vascular supply and the irregular or spongy texture of the lung. Before you leave the laboratory, follow the boxed instructions to prepare your cat for storage and clean the area (page 219). ▄▄

Anatomy of the Respiratory System

Upper and Lower Respiratory System Structures

1. Complete the labeling of the model of the respiratory structures (sagittal section) shown below.

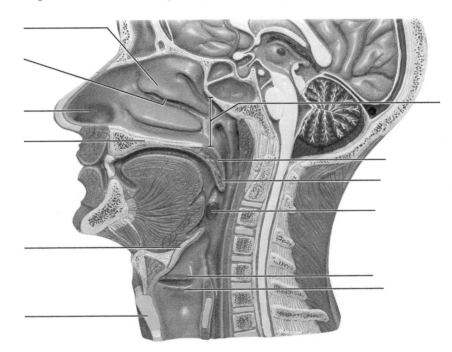

2. Two pairs of mucosal folds are found in the larynx. Which pair are the true vocal cords (superior or inferior)?

_____

3. Name the specific cartilages in the larynx that correspond to the following descriptions.

forms the Adam's apple: _____   shaped like a ring: _____

a "lid" for the larynx: _____   vocal cord attachment: _____

4. Why is it important that the human trachea is reinforced with cartilaginous rings?

_____

Why is it important that the rings are incomplete posteriorly? _____

_____

5. What is the function of the pleural fluid? _____

_____

_____

6. Name two functions of the nasal conchae. _____

and _____

7. The following questions refer to the main bronchi.

Which is longer? _____   Larger in diameter? _____   More horizontal? _____

Which is more likely to trap a foreign object that has entered the respiratory passageways? _____

8. Appropriately label all structures provided with leader lines on the model shown below.

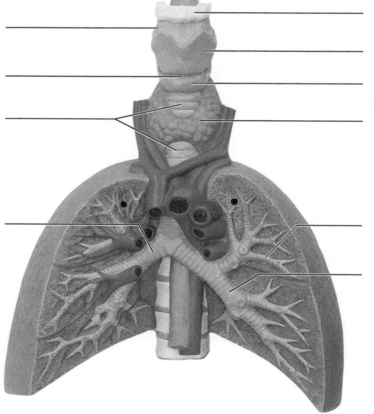

9. Trace a molecule of oxygen from the nostrils to the pulmonary capillaries of the lungs: Nostrils →

_____

_____

_____

10. Match the terms in column B to the descriptions in column A. (Not all terms will be used.)

Column A

_____ 1. connects the larynx to the main bronchi

_____ 2. includes terminal and respiratory as subtypes

_____ 3. food passageway posterior to the trachea

_____ 4. covers the glottis during swallowing of food

_____ 5. contains the vocal cords

_____ 6. indentation on the lung where the lung root structures enter and exit

_____ 7. pleural layer lining the walls of the thorax

_____ 8. site from which oxygen enters the pulmonary blood

_____ 9. connects the middle ear to the nasopharynx

_____ 10. pleural layer in contact with the surface of the lung

_____ 11. increase air turbulence in the nasal cavity

_____ 12. separates the oral cavity from the nasal cavity

Column B

a. alveolus

b. bronchiole

c. conchae

d. epiglottis

e. esophagus

f. hilum

g. larynx

h. palate

i. parietal pleura

j. pharyngotympanic tube

k. trachea

l. visceral pleura

11. Which portions of the respiratory system are referred to as _anatomical dead space_?_____

_____

_____

Why? _____

12. Define the following terms.

external respiration: _____

_____

internal respiration: _____

_____

13. On the diagram below, identify the alveolar duct, respiratory bronchioles, terminal bronchiole, alveoli, and alveolar sac.

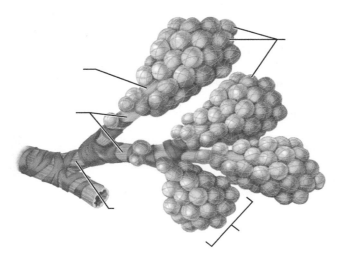

Examining Prepared Slides of Tracheal and Lung Tissue

14. The tracheal epithelium is ciliated and has goblet cells. What is the function of each of these modifications?

cilia: _____

goblet cells: _____

15. The tracheal epithelium is said to be pseudostratified. Why? _____

_____

_____

16. What structural characteristics of the alveoli make them an ideal site for the diffusion of gases?

_____

Why does oxygen move from the alveoli into the pulmonary capillary blood? _____

_____

_____

17. If you observed pathological lung sections, record your observations. Also record how the tissue differed from normal lung tissue. Complete the table below using your answers.

Slide type	Observations	Comparison to normal lung tissue

Demonstrating Lung Inflation in a Sheep Pluck

18. Does the lung inflate part by part or as a whole, like a balloon?_____

19. What happened when the pressure was released? _____

20. What type of tissue ensures this phenomenon?_____

Dissection and Identification: The Respiratory System of the Cat

21. Are the cartilaginous rings in the cat trachea complete or incomplete? _____

22. How does the number of lung lobes in the cat compare with the number in humans? _____

_____

23. Describe the appearance of the bronchial tree in the cat lung. _____

_____

_____

24. Describe the appearance of lung tissue under the dissecting microscope. _____

_____

_____

25. ⊞ Epiglottitis is a condition in which the epiglottis is inflamed. It is most often caused by a bacterial infection. Explain

why this type of inflammation is life-threatening. _____

_____

26. ⊞ Pneumonia is an infectious disease in which fluid accumulates in the alveoli. Patients who are diagnosed with pneumonia are monitored for their oxygen saturation levels. Describe how pneumonia could affect the amount of oxygen in the

blood. _____

_____

Anatomy of the Digestive System

MATERIALS

- Dissectible torso model
- Anatomical chart of the human digestive system
- Prepared slides of the liver, pancreas, and mixed salivary glands; of longitudinal sections of the esophagus-stomach junction and a tooth; and of cross sections of the stomach, duodenum, ileum, and large intestine
- Compound microscope
- Three-dimensional model of a villus (if available)
- Jaw model or human skull
- Three-dimensional model of liver lobules (if available)
- *Human Digestive System* video
- Dissection animal, tray, and instruments
- Bone cutters
- Disposable gloves
- Safety glasses
- Embalming fluid
- Hand lens
- Organic debris container

LEARNING OUTCOMES

- ☐ State the overall function of the digestive system.
- ☐ Describe the general histologic structure of the alimentary canal wall, and identify the following structures on an appropriate image of the wall: mucosa, submucosa, muscularis externa, and serosa or adventitia.
- ☐ Identify on a model or image the organs of the alimentary canal, and name their subdivisions, if any.
- ☐ Describe the general function of each of the digestive system organs and structures.
- ☐ List and explain the specializations of the structure of the stomach and small intestine that contribute to their functional roles.
- ☐ Name and identify the accessory digestive organs, listing a function for each.
- ☐ Describe the anatomy of the generalized tooth, and name the human deciduous and permanent teeth.
- ☐ List the major enzymes or enzyme groups produced by the salivary glands, stomach, small intestine, and pancreas.
- ☐ Recognize microscopically or in an image the histologic structure of the following organs: small intestine, tooth, liver, salivary glands, and stomach.
- ☐ Identify on a dissected animal the organs that make up the alimentary canal and the accessory organs of the digestive system.

PRE-LAB QUIZ

1. The digestive system:
   a. eliminates undigested food
   b. provides the body with nutrients
   c. provides the body with water
   d. all of the above
2. Circle the correct underlined term. <u>Digestion</u> / <u>Absorption</u> occurs when small molecules pass through epithelial cells into the blood for distribution to the body cells.
3. The _____ abuts the lumen of the alimentary canal and consists of epithelium, lamina propria, and muscularis mucosae.
   a. mucosa
   b. serosa
   c. submucosa
4. Circle the correct underlined term. Approximately 25 cm long, the <u>esophagus</u> / <u>alimentary canal</u> conducts food from the pharynx to the stomach.
5. Wavelike contractions of the digestive tract that propel food along are called:
   a. digestion   c. ingestion
   b. elimination   d. peristalsis

Text continues on next page.

6. The _____ is located on the left side of the abdominal cavity and is hidden by the liver and diaphragm.
   a. gallbladder   c. small intestine
   b. large intestine   d. stomach

7. Circle True or False. Nearly all nutrient absorption occurs in the small intestine.

8. Circle the correct underlined term. The <u>ascending colon</u> / <u>descending colon</u> traverses down the left side of the abdominal cavity and becomes the sigmoid colon.

9. A tooth consists of two major regions, the crown and the:
   a. dentin   c. gingiva
   b. enamel   d. root

10. Located inferior to the diaphragm, the _____ is the largest gland in the body.
    a. gallbladder   c. pancreas
    b. liver   d. thymus

The **digestive system** provides the body with the nutrients, water, and electrolytes essential for health. The organs of this system ingest, digest, and absorb food and eliminate the undigested remains as feces.

The digestive system consists of a hollow tube extending from the mouth to the anus, into which various accessory organs empty their secretions (**Figure 27.1**). For ingested food to become available to the body cells, it must first be broken down into its smaller diffusible molecules—a process called **digestion.** The digested end products can then pass through the epithelial cells lining the tract into the blood for distribution to the body cells—a process called **absorption.**

The organs of the digestive system are traditionally separated into two major groups: the **alimentary canal, or gastrointestinal (GI) tract,** and the **accessory digestive organs.** The alimentary canal consists of the mouth, pharynx, esophagus, stomach, and small and large intestines. The accessory structures include the teeth, which physically break down foods, and the salivary glands, gallbladder, liver, and pancreas, which secrete their products into the alimentary canal.

General Histologic Plan of the Alimentary Canal

From the esophagus to the anal canal, the basic structure of the alimentary canal is similar. So, it makes sense to begin our study by learning the features of this structure. As we study individual parts of the alimentary canal, we will note how this basic plan is modified to enable each subsequent organ to perform its unique digestive functions.

Essentially, the alimentary canal wall has four basic layers, or **tunics.** From the lumen outward, these are the *mucosa, the submucosa,* the *muscularis externa,* and either a *serosa or adventitia* (**Figure 27.2**). Each of these tunics has a predominant tissue type and a specific function in the digestive process.

Table 27.1 summarizes the characteristics of the layers of the wall of the alimentary canal.

Table 27.1   Alimentary Canal Wall Layers (Figure 27.2)

Layer	Subdivision of the layer	Tissue type	Major functions (generalized for the layer)
Mucosa	Epithelium	Stratified squamous epithelium in the mouth, oropharynx, laryngopharynx, esophagus, and anus; simple columnar epithelium in the remainder of the canal	Secretion of mucus, digestive enzymes, and hormones; absorption of end products into the blood; protection against infectious disease.
	Lamina propria	Areolar connective tissue with blood vessels; many lymphoid follicles, especially as tonsils and mucosa-associated lymphoid tissue (MALT)	
	Muscularis mucosae	A thin layer of smooth muscle	
Submucosa	N/A	Areolar and dense irregular connective tissue containing blood vessels, lymphatic vessels, and nerve fibers (submucosal nerve plexus)	Blood vessels absorb and transport nutrients. Elastic fibers help maintain the shape of each organ.
Muscularis externa	Circular layer	Inner layer of smooth muscle	Segmentation and peristalsis of digested food along the tract are regulated by the myenteric nerve plexus.
	Longitudinal layer	Outer layer of smooth muscle	
Serosa* (visceral peritoneum)	Connective tissue	Areolar connective tissue	Reduces friction as the digestive system organs slide across one another.
	Epithelium (mesothelium)	Simple squamous epithelium	

*Since the esophagus is outside the peritoneal cavity, the serosa is replaced by an adventitia made of aerolar connective tissue that binds the esophagus to surrounding tissues.

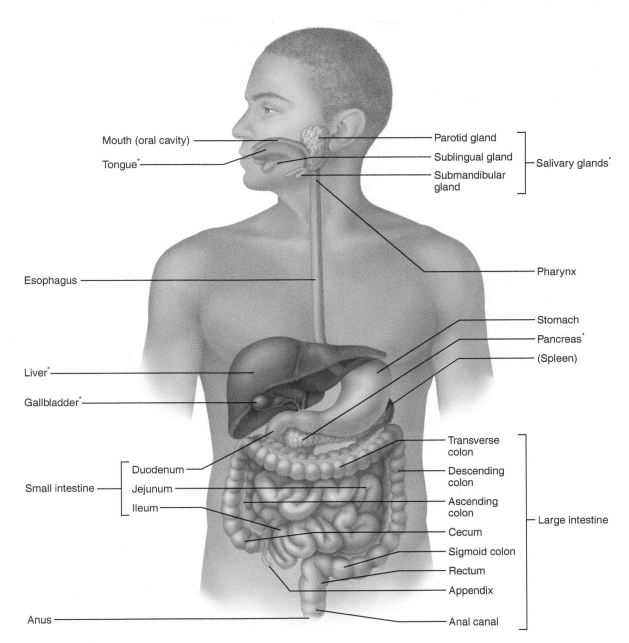

Figure 27.1 The human digestive system: alimentary canal and accessory organs.
Organs with asterisks are accessory organs. Those without asterisks are alimentary
canal organs (except the spleen, which is an organ of the lymphatic system).

Organs of the Alimentary Canal

Identifying Organs of the Alimentary Canal

The sequential pathway and fate of food as it passes through
the alimentary canal are described in the next sections. Identify each structure in Figure 27.1 and on the torso model or
anatomical chart of the digestive system as you work. ▬

Oral Cavity, or Mouth

Food enters the digestive tract through the **oral cavity, or
mouth (Figure 27.3)**. Within this mucous membrane–lined

cavity are the gums, teeth, tongue, and openings of the ducts
of the salivary glands. The **lips (labia)** protect the anterior
opening of the **oral orifice.** The **cheeks** form the mouth's lateral walls, and the **palate,** its roof. The anterior portion of
the palate is referred to as the **hard palate** because the palatine processes of the maxillae and the palatine bones underlie
it. The posterior **soft palate** is a fibromuscular structure that is
unsupported by bone. The **uvula,** a fingerlike projection of the
soft palate, extends inferiorly from its posterior margin. The
floor of the oral cavity is occupied by the muscular **tongue
(Figure 27.4)**, which is largely supported by the *mylohyoid muscle* and attaches to the hyoid bone, mandible, styloid processes,
and pharynx. A membrane called the **lingual frenulum** secures
the inferior midline of the tongue to the floor of the mouth.

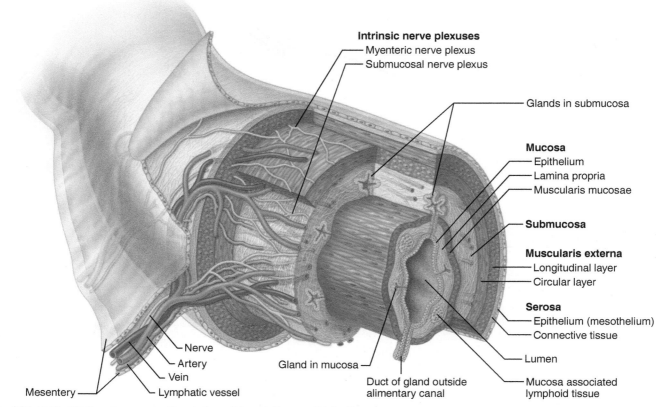

Intrinsic nerve plexuses
— Myenteric nerve plexus
— Submucosal nerve plexus

Glands in submucosa

Mucosa
— Epithelium
— Lamina propria
— Muscularis mucosae

Submucosa

Muscularis externa
— Longitudinal layer
— Circular layer

Serosa
— Epithelium (mesothelium)
— Connective tissue

Nerve
Artery
Vein
Lymphatic vessel

Mesentery

Gland in mucosa

Duct of gland outside alimentary canal

Lumen

Mucosa associated lymphoid tissue

(a) Longitudinal and cross-sectional views through the small intestine

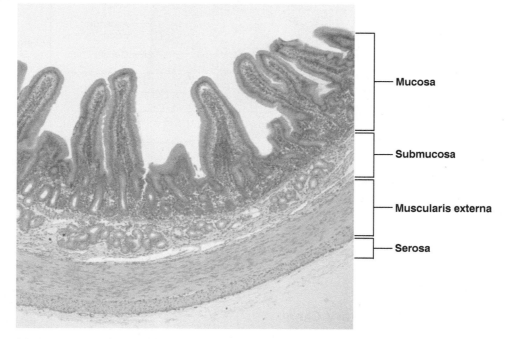

Mucosa

Submucosa

Muscularis externa

Serosa

(b) Light micrograph cross section through the small intestine (90×)

Figure 27.2 Basic structural pattern of the alimentary canal wall.

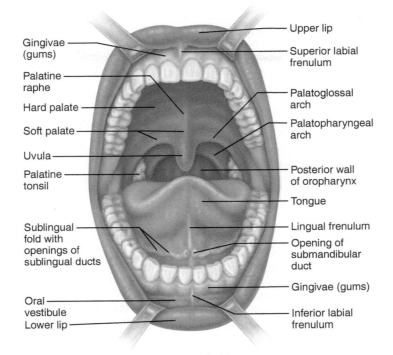

Gingivae (gums)
Palatine raphe
Hard palate
Soft palate
Uvula
Palatine tonsil
Sublingual fold with openings of sublingual ducts
Oral vestibule
Lower lip

Upper lip
Superior labial frenulum
Palatoglossal arch
Palatopharyngeal arch
Posterior wall of oropharynx
Tongue
Lingual frenulum
Opening of submandibular duct
Gingivae (gums)
Inferior labial frenulum

Figure 27.3 Anterior view of the oral cavity.

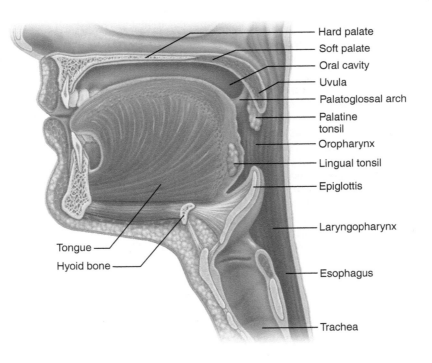

Hard palate
Soft palate
Oral cavity
Uvula
Palatoglossal arch
Palatine tonsil
Oropharynx
Lingual tonsil
Epiglottis
Laryngopharynx
Esophagus
Trachea

Tongue
Hyoid bone

Figure 27.4 Sagittal view of the head showing oral cavity and pharynx.

The space between the lips and cheeks and the teeth and gums is the **oral vestibule;** the area that lies within the teeth and gums is the *oral cavity* proper. (The teeth and gums are discussed in more detail on pages 471–472.)

On each side of the mouth at its posterior end are masses of lymphoid tissue, the **palatine tonsils** (see Figure 27.3).

Each lies in a concave area bounded anteriorly and posteriorly by membranes, the **palatoglossal arch** and the **palatopharyngeal arch,** respectively. Another mass of lymphoid tissue, the **lingual tonsil** (see Figure 27.4), covers the base of the tongue, posterior to the oral cavity proper. The tonsils, in common with other lymphoid tissues, are part of the body's defense system.

Structure	Description
Cardia (cardial part)	The area surrounding the cardial orifice through which food enters the stomach
Fundus	The dome-shaped area that is located superior and lateral to the cardia
Body	Midportion of the stomach and largest region
Pyloric part:	Funnel-shaped pouch that forms the distal stomach
Pyloric antrum	Wide superior portion of the pyloric part
Pyloric canal	Narrow tubelike portion of the pyloric part
Pylorus	Distal end of the pyloric part that is continuous with the small intestine
Pyloric sphincter	Valve that controls the emptying of the stomach into the small intestine

Table 27.2 Parts of the Stomach (Figure 27.5)

Very often in young children, the palatine tonsils become inflamed and enlarge, partially blocking the entrance to the pharynx posteriorly and making swallowing difficult and painful. This condition is called **tonsillitis.** ✚

Three pairs of salivary glands duct their secretion, saliva, into the oral cavity. One component of saliva, salivary amylase, begins the digestion of starchy foods within the oral cavity. (The salivary glands are discussed in more detail on page 472.)

As food enters the mouth, it is mixed with saliva and masticated (chewed). The cheeks and lips help hold the food between the teeth during mastication, and the highly mobile tongue manipulates the food during chewing and initiates swallowing. Thus the mechanical and chemical breakdown of food begins before the food has left the oral cavity.

Pharynx

When the tongue initiates swallowing, the food passes posteriorly into the pharynx, a common passageway for food, fluid, and air (see Figure 27.4). The pharynx is subdivided anatomically into three parts—the **nasopharynx** (behind the nasal cavity), the **oropharynx** (behind the oral cavity extending from the soft palate to the epiglottis), and the **laryngopharynx** (behind the larynx, extending from the epiglottis to the larynx). The walls of the pharynx consist largely of two layers of skeletal muscles: an inner layer of longitudinal muscle and an outer layer of circular constrictor muscles, which initiate wavelike contractions that propel the food inferiorly into the esophagus. Its mucosa, like that of the oral cavity, contains a protective stratified squamous epithelium.

Esophagus

The **esophagus** extends from the laryngopharynx through the diaphragm to the gastroesophageal sphincter in the superior aspect of the stomach. Approximately 25 cm long in humans, it is essentially a food passageway that conducts food to the stomach in a wavelike peristaltic motion. The esophagus has no digestive or absorptive function. The walls at its superior end contain skeletal muscle, which is replaced by smooth muscle in the area nearing the stomach. The **gastroesophageal sphincter,** a slight thickening of the smooth muscle layer at the esophagus-stomach junction, controls food passage into the stomach (**Figure 27.5**).

Stomach

The **stomach** (Figures 27.1 and 27.5) is primarily located in the upper left quadrant of the abdominopelvic cavity and is nearly hidden by the liver and diaphragm. The stomach is made up of several regions, summarized in **Table 27.2**.

Mesentery is the general term that refers a double layer of peritoneum—a sheet of two serous membranes fused together—that extends from the digestive organs to the body wall. There are two mesenteries, the **greater omentum** and **lesser omentum,** that connect to the stomach. The lesser omentum extends from the liver to the **lesser curvature** of the stomach. The greater omentum extends from the **greater curvature** of the stomach, reflects downward, and covers most of the abdominal organs in an apronlike fashion. Figure 27.7 on page 467 illustrates the omenta as well as the other peritoneal attachments of the abdominal organs.

The stomach is a temporary storage region for food as well as a site for mechanical and chemical breakdown of food. It contains a third (innermost) *obliquely* oriented layer of smooth muscle in its muscularis externa that allows it to churn, mix, and pummel the food, physically reducing it to smaller fragments. **Gastric glands** of the mucosa secrete hydrochloric acid (HCl) and hydrolytic enzymes. The *mucosal glands* also secrete a viscous mucus that helps prevent the stomach itself from being digested by the proteolytic enzymes. Most digestive activity occurs in the pyloric part of the stomach. After the food is processed in the stomach, it resembles a creamy mass called **chyme,** which enters the small intestine through the pyloric sphincter.

ACTIVITY 2

Studying the Histologic Structure of Selected Digestive System Organs

To prepare for the histologic studies you will be conducting in the lab, obtain a microscope and the following slides: salivary glands (submandibular or sublingual); pancreas; liver; cross sections of the duodenum, ileum, stomach, and large intestine; and longitudinal sections of a tooth and the esophagus-stomach junction.

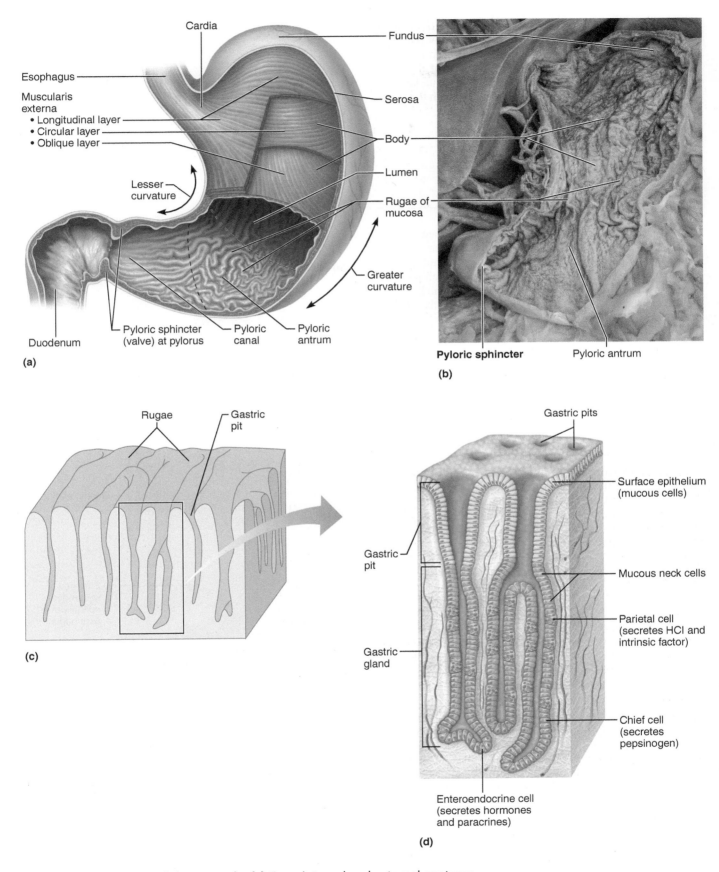

(a)

Cardia

Esophagus

Muscularis externa
• Longitudinal layer
• Circular layer
• Oblique layer

Lesser curvature

Duodenum

Pyloric sphincter (valve) at pylorus

Pyloric canal

Pyloric antrum

Fundus

Serosa

Body

Lumen

Rugae of mucosa

Greater curvature

(b)

Pyloric sphincter

Pyloric antrum

(c)

Rugae

Gastric pit

(d)

Gastric pits

Surface epithelium (mucous cells)

Gastric pit

Mucous neck cells

Gastric gland

Parietal cell (secretes HCl and intrinsic factor)

Chief cell (secretes pepsinogen)

Enteroendocrine cell (secretes hormones and paracrines)

Figure 27.5 Anatomy of the stomach. (a) Gross internal and external anatomy.
(b) Cadaver photograph of internal aspect of stomach. **(c)** Section of the stomach wall
showing rugae and gastric pits. **(d)** Detailed structure of the gastric pits and glands.

1. **Stomach:** The stomach slide will be viewed first. Refer to Figure 27.6a as you scan the tissue under low power to locate the muscularis externa; then move to high power to more closely examine this layer. Try to pick out the three smooth muscle layers. How does the extra oblique layer of smooth muscle found in the stomach correlate with the stomach's churning movements?

_____

_____

_____

Identify the gastric glands and the gastric pits (see Figures 27.5 and 27.6b). If the section is taken from the stomach fundus and is appropriately stained, you can identify, in the gastric glands, the blue-staining **chief cells,** which produce pepsinogen, and the red-staining **parietal cells,** which secrete HCl and intrinsic factor. The enteroendocrine cells that release hormones and paracrines are indistinguishable. Draw a small section of the stomach wall, and label it appropriately.

2. **Esophagus-stomach junction:** Scan the slide under low power to locate the mucosal junction between the end of the esophagus and the beginning of the stomach, the gastroesophageal junction. Compare your observations to Figure 27.6c. What is the functional importance of the epithelial differences seen in the two organs?

_____

_____

_____

Small Intestine

The **small intestine** is a convoluted tube, 6 to 7 m (about 20 feet) long in a cadaver but only about 2 m (6 feet) long during life because of its muscle tone. It extends from the pyloric sphincter to the ileocecal valve. The small intestine is suspended by a double layer of peritoneum, the fan-shaped **mesentery,** from the posterior abdominal wall **(Figure 27.7)**, and it lies, framed laterally and superiorly by the large intestine, in the abdominal cavity. The small intestine has three subdivisions (Figure 27.1):

1. The **duodenum** extends from the pyloric sphincter for about 25 cm (10 inches) and curves around the head of the pancreas; most of the duodenum lies in a retroperitoneal position.

2. The **jejunum,** continuous with the duodenum, extends for 2.5 m (about 8 feet). Most of the jejunum occupies the umbilical region of the abdominal cavity.

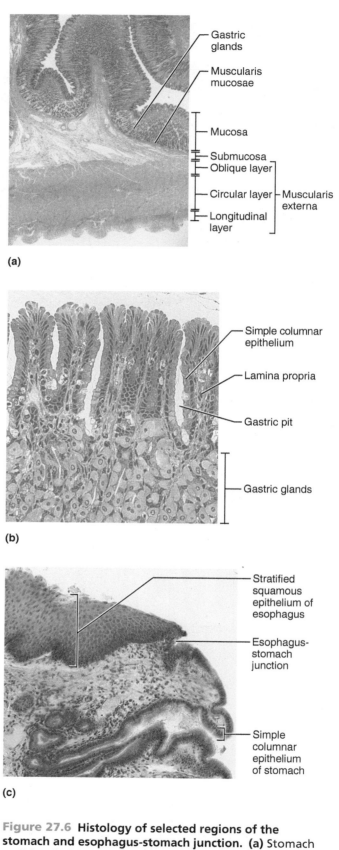

(a)

(b)

(c)

Figure 27.6 Histology of selected regions of the stomach and esophagus-stomach junction. (a) Stomach wall (85×). **(b)** Gastric pits and glands (170×). **(c)** Esophagus-stomach junction, longitudinal section (130×).

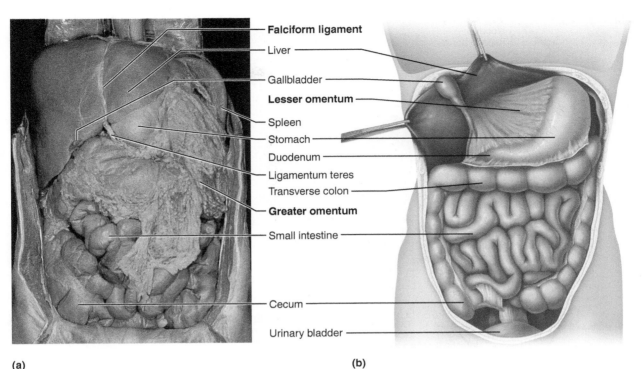

(a)

(b)

— Falciform ligament
Liver —
Gallbladder —
Lesser omentum —
Spleen —
Stomach —
Duodenum —
Ligamentum teres —
Transverse colon —
Greater omentum —
Small intestine —
Cecum —
Urinary bladder —

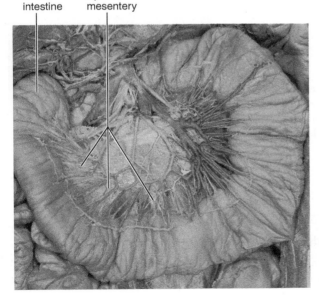

Small   Spread
intestine mesentery

(c)

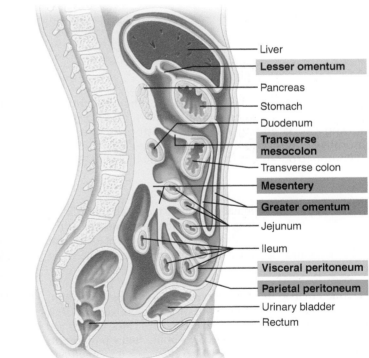

Liver
Lesser omentum
Pancreas
Stomach
Duodenum
Transverse mesocolon
Transverse colon
Mesentery
Greater omentum
Jejunum
Ileum
Visceral peritoneum
Parietal peritoneum
Urinary bladder
Rectum

(d)

Figure 27.7 Peritoneal attachments of the abdominal organs. Superficial anterior views of the abdominal cavity. **(a)** Cadaver photograph with the greater omentum in place. **(b)** Diagram showing greater omentum removed and liver and gallbladder reflected superiorly. **(c)** Mesentery of the small intestine. **(d)** Sagittal view of a male torso. Mesentery labels appear in colored boxes.

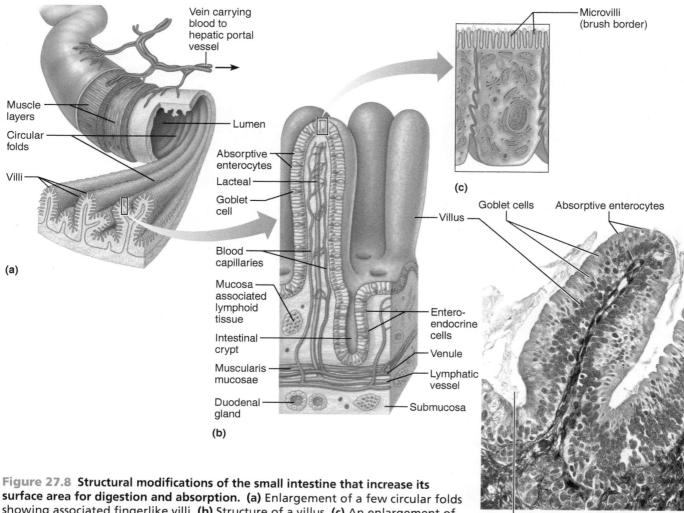

Figure 27.8 Structural modifications of the small intestine that increase its surface area for digestion and absorption. **(a)** Enlargement of a few circular folds showing associated fingerlike villi. **(b)** Structure of a villus. **(c)** An enlargement of absorptive enterocytes that exhibit microvilli on their free (luminal) surface. **(d)** Photomicrograph of the mucosa showing villi (250×).

3. The **ileum**, the terminal portion of the small intestine, is about 3.6 m (12 feet) long and joins the large intestine at the **ileocecal valve.** It is located inferiorly and somewhat to the right in the abdominal cavity, but its major portion lies in the pubic region.

Two types of enzymes complete digestion in the small intestine: **brush border enzymes,** which are hydrolytic enzymes bound to the microvilli of the columnar epithelial cells; and, more important, pancreatic enzymes, which are ducted into the duodenum largely via the **main pancreatic duct.** Bile (formed in the liver) also enters the duodenum via the **bile duct** in the same area. At the duodenum, the ducts join to form the bulblike **hepatopancreatic ampulla** and empty their products into the duodenal lumen through the **major duodenal papilla,** an orifice controlled by a muscular valve called the **hepatopancreatic sphincter.**

Nearly all nutrient absorption occurs in the small intestine, where three structural modifications increase the absorptive surface of the mucosa: the microvilli, villi, and circular folds **(Figure 27.8)**.

• **Microvilli** are microscopic projections of the surface plasma membrane of the columnar epithelial lining cells of the mucosa.

• **Villi** are the fingerlike projections of the mucosa tunic that give it a velvety appearance and texture.

• The **circular folds** are deep folds of the mucosa and submucosa layers that force chyme to spiral through the intestine, mixing it and slowing its progress.

These structural modifications decrease in frequency and size toward the end of the small intestine. Any residue remaining undigested and unabsorbed at the terminus of the small intestine enters the large intestine through the ileocecal valve. In contrast, the amount of lymphoid tissue in the submucosa of the small intestine (especially the **aggregated lymphoid nodules,** also called *Peyer's patches;* see Figure 27.9b) increases along the length of the small intestine and is very apparent in the ileum.

ACTIVITY 3

Observing the Histologic Structure of the Small Intestine

1. **Duodenum:** Secure the slide of the duodenum to the microscope stage. Observe the tissue under low power to identify the four basic tunics of the intestinal wall—that is, the **mucosa** and its three sublayers, the **submucosa,** the **muscularis externa,** and the **serosa,** or visceral peritoneum. Consult **Figure 27.9a** to help you identify the scattered mucus-producing **duodenal glands** in the submucosa.

What type of epithelium do you see here? _____

_____

Examine the large leaflike *villi,* which increase the surface area for absorption. Notice the scattered mucus-producing goblet cells in the epithelium of the villi. Note also the **intestinal crypts** (see also Figure 27.8), invaginated areas of the mucosa between the villi containing the cells that produce intestinal juice, a watery mucus-containing mixture that serves as a carrier fluid for absorption of nutrients from the chyme.

2. **Ileum:** The structure of the ileum resembles that of the duodenum, except that the villi are less elaborate because most of the absorption has occurred by the time that chyme reaches the ileum. Secure a slide of the ileum to the microscope stage for viewing. Observe the villi, and identify the four layers of the wall and the large, generally spherical aggregated lymphoid nodules (Figure 27.9b). What tissue type are aggregated lymphoid nodules?

_____

3. If a villus model is available, identify the following cells or regions before continuing: absorptive epithelium, goblet cells, lamina propria, muscularis mucosae, capillary bed, and lacteal. If possible, also identify the intestinal crypts. ▬

Large Intestine

The **large intestine** (**Figure 27.10**) is about 1.5 m (5 feet) long and extends from the ileocecal valve to the anus. It encircles the small intestine on three sides and consists of the following subdivisions: **cecum, appendix, colon, rectum,** and **anal canal.**

The blind wormlike appendix, which hangs from the cecum, is a trouble spot in the large intestine. Because it is generally twisted, it provides an ideal location for bacteria to accumulate and multiply. Inflammation of the appendix, or appendicitis, is the result. ✚

The colon is divided into several distinct regions. The **ascending colon** travels up the right side of the abdominal

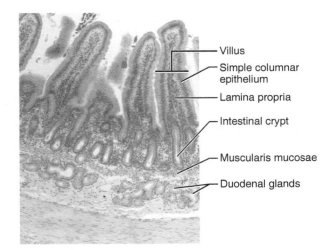

(a)

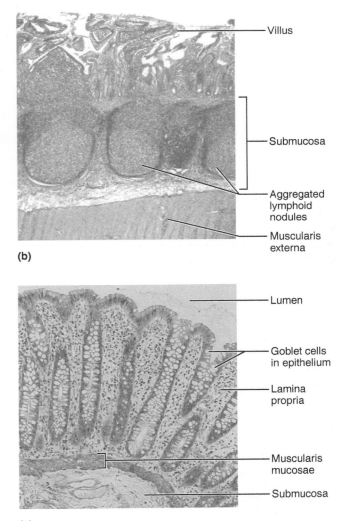

(b)

(c)

Figure 27.9 Histology of selected regions of the small and large intestines. Cross-sectional views. **(a)** Duodenum of the small intestine (180×). **(b)** Ileum of the small intestine (35×). **(c)** Large intestine (120×).

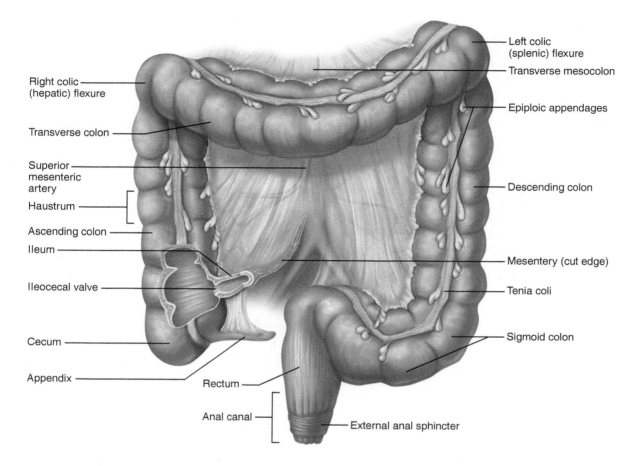

Figure 27.10 The large intestine. (Section of the cecum removed to show the ileocecal valve.)

cavity and makes a right-angle turn at the **right colic (hepatic) flexure** to cross the abdominal cavity as the **transverse colon.** It then turns at the **left colic (splenic) flexure** and continues down the left side of the abdominal cavity as the **descending colon,** where it takes an S-shaped course as the **sigmoid colon.** The sigmoid colon, rectum, and the anal canal lie in the pelvis anterior to the sacrum and thus are not considered abdominal cavity structures. Except for the transverse and sigmoid colons, the colon is retroperitoneal.

The anal canal terminates in the **anus,** the opening to the exterior of the body. The anal canal has two sphincters, a voluntary *external anal sphincter* composed of skeletal muscle, and an involuntary *internal anal sphincter* composed of smooth muscle. The sphincters are normally closed except during defecation, when undigested food and bacteria are eliminated from the body as feces.

In the large intestine, the longitudinal layer of the muscularis externa is reduced to three longitudinal bands called the **teniae coli.** Because these bands are shorter than the rest of the wall of the large intestine, they cause the wall to pucker into small pocketlike sacs called **haustra.** Fat-filled pouches of visceral peritoneum, called *epiploic appendages,* hang from the colon's surface.

The major function of the large intestine is to consolidate and propel the unusable fecal matter toward the anus and eliminate it from the body. While it does this task, it

(1) provides a site where intestinal bacteria manufacture vitamins B and K; and (2) reclaims most of the remaining water from undigested food, thus conserving body water.

Watery stools, or **diarrhea,** result from any condition that rushes undigested food residue through the large intestine before it has had sufficient time to absorb the water. Conversely, when food residue remains in the large intestine for extended periods, water is absorbed, and the stool becomes hard and difficult to pass, causing **constipation. ✚**

> **ACTIVITY 4**

Examining the Histologic Structure of the Large Intestine

Secure a slide of the large intestine to the microscope stage for viewing. Observe the villi, and note the numerous goblet cells in the epithelium (Figure 27.9c). Why do you think the large intestine produces so much mucus?

_____

_____

Accessory Digestive Organs

Teeth

By the age of 21, two sets of teeth have developed **(Figure 27.11)**. The initial set, called the **deciduous** (or **milk**) **teeth,** normally appears between the ages of 6 months and 2½ years. The first of these to erupt are the lower central incisors. The child begins to shed the deciduous teeth around the age of 6, and a second set of teeth, the **permanent teeth,** gradually replaces them. As the deeper permanent teeth progressively enlarge and develop, the roots of the deciduous teeth are resorbed, leading to their final shedding.

Teeth are classified as **incisors, canines** *(eyeteeth, cuspids),* **premolars** *(bicuspids),* and **molars.** Teeth names reflect differences in relative structure and function. The incisors are chisel-shaped and exert a shearing action used in biting. Canines are cone-shaped teeth used for the tearing of food. The premolars have two *cusps* (grinding surfaces); the molars have broad crowns with rounded cusps specialized for the fine grinding of food.

Dentition is described by means of a **dental formula,** which designates the numbers, types, and position of the teeth in one side of the jaw. Because tooth arrangement is bilaterally symmetrical, it is necessary to designate only one side of the jaw. The complete dental formula for the deciduous teeth from the medial aspect of each jaw and proceeding posteriorly is as follows:

$$\frac{\text{Upper teeth: 2 incisors, 1 canine, 0 premolars, 2 molars}}{\text{Lower teeth: 2 incisors, 1 canine, 0 premolars, 2 molars}} \times 2$$

This formula is generally abbreviated to read as follows:

$$\frac{2,1,0,2}{2,1,0,2} \times 2 \text{ (20 deciduous teeth)}$$

The permanent teeth are then described by the following dental formula:

$$\frac{2,1,2,3}{2,1,2,3} \times 2 \text{ (32 permanent teeth)}$$

Although 32 is designated as the normal number of permanent teeth, not everyone develops a full set. In many people, the third molars, commonly called *wisdom teeth,* never erupt.

ACTIVITY 5

Identifying Types of Teeth

Identify the four types of teeth (incisors, canines, premolars, and molars) on the jaw model or human skull. ■

A tooth consists of two major regions, the *crown* and the *root.* These two regions meet at the **neck** near the gum line. A longitudinal section made through a tooth shows the following basic anatomical plan **(Figure 27.12)**. The **crown** is the superior portion of the tooth visible above the **gingiva, or gum,** which surrounds the tooth. The surface of the crown is covered by **enamel.** Enamel is the hardest substance in the body and is fairly brittle. It consists of 95% to 97% inorganic calcium salts (chiefly $CaPO_4$) and thus is heavily mineralized. The crevice between the end of the crown and the upper margin of the gingiva is referred to as the *gingival sulcus.*

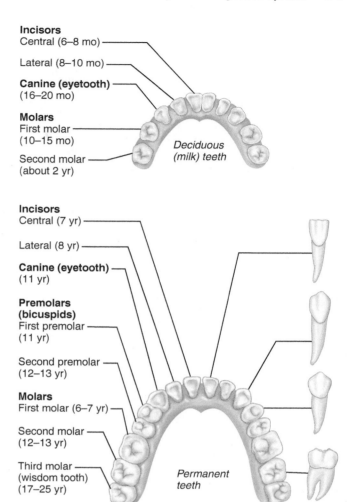

Incisors
Central (6–8 mo)
Lateral (8–10 mo)
Canine (eyetooth)
(16–20 mo)
Molars
First molar
(10–15 mo)
Second molar
(about 2 yr)

Deciduous (milk) teeth

Incisors
Central (7 yr)
Lateral (8 yr)
Canine (eyetooth)
(11 yr)
Premolars (bicuspids)
First premolar
(11 yr)
Second premolar
(12–13 yr)
Molars
First molar (6–7 yr)
Second molar
(12–13 yr)
Third molar
(wisdom tooth)
(17–25 yr)

Permanent teeth

Figure 27.11 Human dentition. (Approximate time of teeth eruption shown in parentheses.)

That portion of the tooth embedded in the bone is the **root.** The outermost surface of the root is covered by **cement,** which is similar to bone in composition and less brittle than enamel. The cement attaches the tooth to the **periodontal ligament,** which holds the tooth in the tooth socket and exerts a cushioning effect. **Dentin,** which makes up the bulk of the tooth, is the bonelike material interior to the enamel and cement.

Dentin surrounds the **pulp cavity** which is filled with pulp. **Pulp** is composed of connective tissue liberally supplied with blood vessels, nerves, and lymphatics that provide for tooth sensation and supply nutrients to the tooth tissues. **Odontoblasts,** specialized cells in the outer margins of the pulp cavity, produce and maintain the dentin. Odontoblasts have slender processes that extend into the *dentinal tubules* of the dentin. The pulp cavity extends into distal portions of the root and becomes the **root canal.** An opening at the root apex, the **apical foramen,** provides a route of entry into the tooth for blood vessels, nerves, and other structures from the tissues beneath.

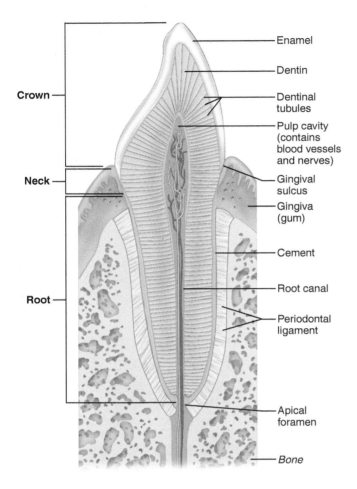

Figure 27.12 Longitudinal section of human canine tooth within its bony socket.

Studying Microscopic Anatomy of the Tooth

Observe a slide of a longitudinal section of a tooth, and compare your observations with the structures detailed in Figure 27.12. Identify as many of these structures as possible. ◼

Salivary Glands

Three pairs of major **salivary glands** (Figure 27.13) empty their secretions into the oral cavity.

Parotid glands: Large glands located anterior to the ear and ducting into the mouth over the second upper molar through the parotid duct.

Submandibular glands: Located along the medial aspect of the mandibular body in the floor of the mouth, and ducting under the tongue to the base of the lingual frenulum.

Sublingual glands: Small glands located most anteriorly in the floor of the mouth and emptying under the tongue via several small ducts.

Food in the mouth and mechanical pressure (even chewing rubber bands or wax) stimulate the salivary glands to secrete saliva. Saliva consists primarily of a viscous glycoprotein called *mucin,* which moistens the food and helps to bind it together into a mass called a **bolus,** and a clear serous fluid containing the enzyme *salivary amylase.* Salivary amylase begins the digestion of starch, breaking it down into oligosaccharides and disaccharides. Parotid gland secretion is mainly serous, whereas the submandibular is a mixed gland that produces both mucin and serous components. The sublingual gland produces mostly mucin.

Examining Salivary Gland Tissue

Examine salivary gland tissue under low power and then high power to become familiar with the appearance of glandular tissue. Notice the clustered arrangement of the cells around their ducts. The cells are basically triangular, with their pointed ends facing the duct opening. If possible, differentiate mucus-producing cells, which look hollow or have a clear cytoplasm, from serous cells, which produce the clear, enzyme-containing fluid and have granules in their cytoplasm. The serous cells often form *demilunes* (caps) around the more central mucous cells. (Figure 27.13b may be helpful in this task.) ◼

Pancreas

The **pancreas** is a soft, triangular gland that extends horizontally across the posterior abdominal wall from the spleen to the duodenum (Figure 27.14). Like the duodenum, it is a retroperitoneal organ (see Figure 27.7). The pancreas has both an endocrine function, producing the hormones insulin and glucagon, and an exocrine function. Its exocrine secretion, which includes many hydrolytic enzymes produced by the acinar cells, is secreted into the duodenum through the pancreatic ducts. Pancreatic juice is very alkaline. Its high concentration of bicarbonate ion (HCO_3^-) neutralizes the acidic chyme entering the duodenum from the stomach, enabling the pancreatic and intestinal enzymes to operate at their optimal pH, which is slightly alkaline.

Examining the Histology of the Pancreas

Observe pancreatic tissue under low power and then high power to distinguish between the lighter-staining, endocrine-producing clusters of cells called **pancreatic islets** and the deeper-staining **acinar cells,** which produce the hydrolytic enzymes and form the major portion of the pancreatic tissue (Figure 27.14). Notice the arrangement of the exocrine cells around their central ducts. If the tissue is differentially stained, you will also be able to identify specifically the lavender blue–stained insulin-secreting beta cells and the red-stained glucagon-secreting alpha cells of the pancreas. ◼

Liver and Gallbladder

The **liver** (Figure 27.1 and Figure 27.15), the largest gland in the body, is located inferior to the diaphragm, more to the right than the left side of the body. As noted earlier, it hides the stomach from view in a superficial observation of abdominal contents. The human liver has four lobes and is suspended from the diaphragm and anterior abdominal wall by the **falciform ligament** (Figures 27.7a and 27.15).

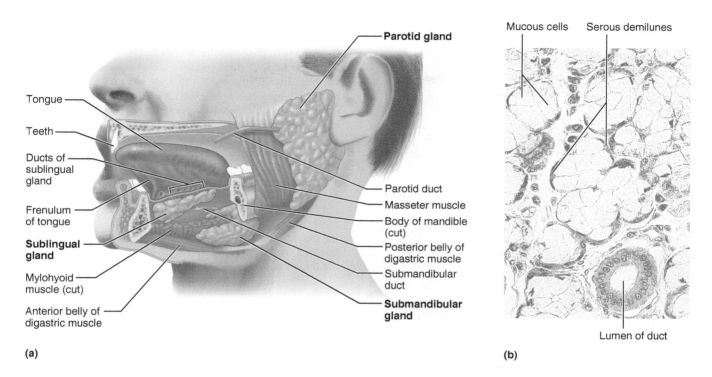

Parotid gland

Tongue

Teeth

Ducts of sublingual gland

Frenulum of tongue

Sublingual gland

Mylohyoid muscle (cut)

Anterior belly of digastric muscle

Parotid duct

Masseter muscle

Body of mandible (cut)

Posterior belly of digastric muscle

Submandibular duct

Submandibular gland

(a)

Mucous cells   Serous demilunes

Lumen of duct

(b)

Figure 27.13 The salivary glands. (a) The parotid, submandibular, and sublingual salivary glands associated with the left aspect of the oral cavity. **(b)** Photomicrograph of the sublingual salivary gland (215×), which is a mixed salivary gland.

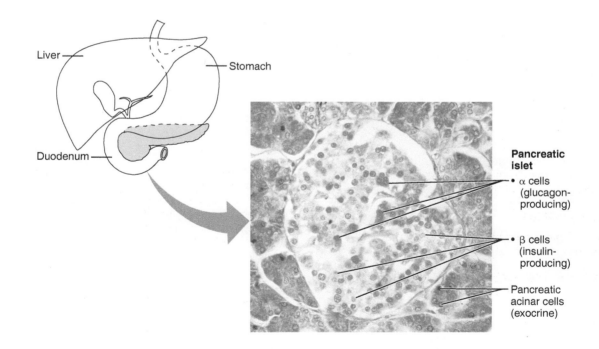

Liver

Stomach

Duodenum

Pancreatic islet

• α cells (glucagon-producing)

• β cells (insulin-producing)

Pancreatic acinar cells (exocrine)

Figure 27.14 Histology of the pancreas. The pancreatic islet cells produce insulin and glucagon (hormones). The acinar cells synthesize digestive enzymes for "export" to the duodenum.

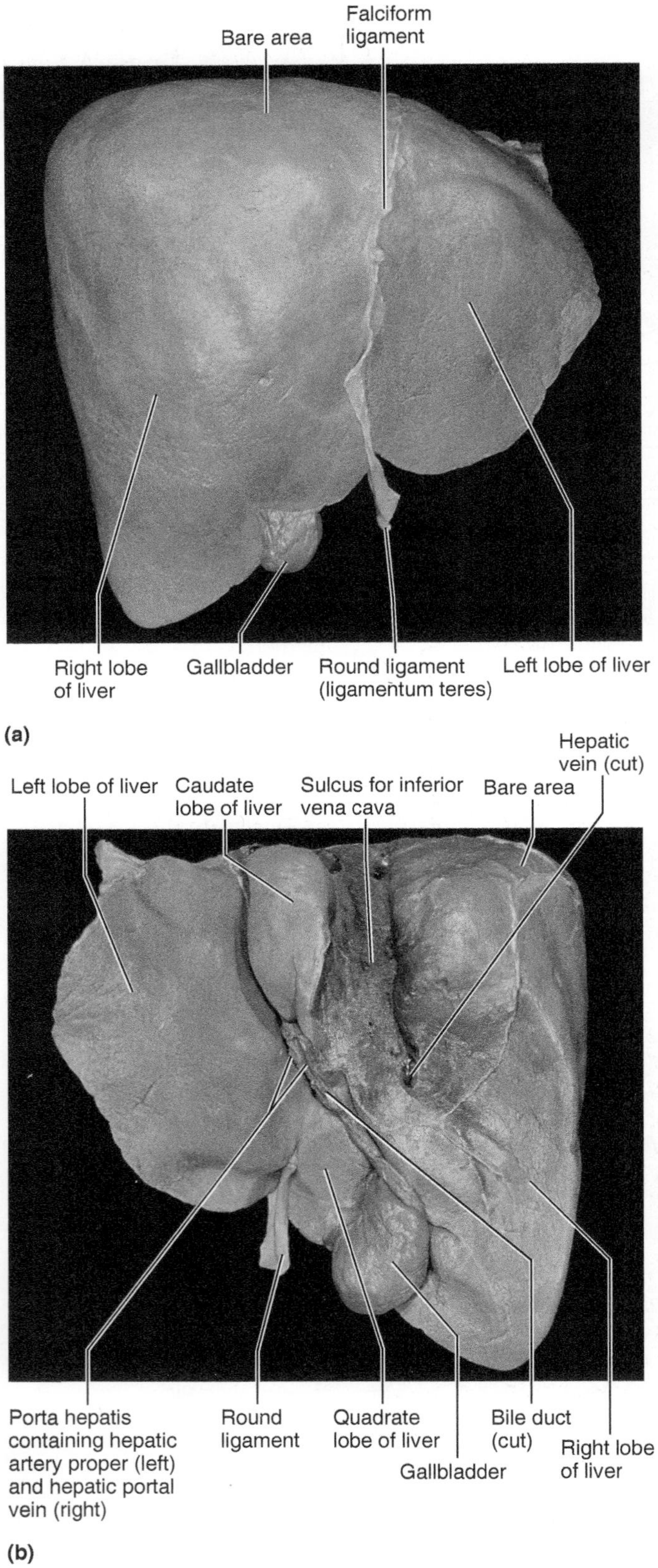

Bare area

Falciform
ligament

Right lobe
of liver

Gallbladder

Round ligament
(ligamentum teres)

Left lobe of liver

(a)

Left lobe of liver

Caudate
lobe of liver

Sulcus for inferior
vena cava

Bare area

Hepatic
vein (cut)

Porta hepatis
containing hepatic
artery proper (left)
and hepatic portal
vein (right)

Round
ligament

Quadrate
lobe of liver

Gallbladder

Bile duct
(cut)

Right lobe
of liver

(b)

Figure 27.15 Gross anatomy of the human liver.
(a) Anterior view. **(b)** Posteroinferior aspect. The four liver
lobes are separated by a group of fissures in this view.

The liver is one of the body's most important organs, and it performs many metabolic roles. However, its digestive function is to produce bile, which leaves the liver through the **common hepatic duct** and then enters the duodenum through the **bile duct** (Figure 27.16). Bile has no enzymatic action but emulsifies fats by breaking up fat globules into small droplets, which creates a larger surface area to improve the efficiency of the enzyme lipase. Without bile, very little fat digestion or absorption occurs.

When digestive activity is not occurring in the digestive tract, bile backs up into the **cystic duct** and enters the **gallbladder,** a small green sac on the inferior surface of the liver. Bile is stored there until needed for the digestive process. While in the gallbladder, bile is concentrated by the removal of water and some ions. When fat-rich food enters the duodenum, a hormonal stimulus causes the gallbladder to contract, releasing the stored bile and making it available to the duodenum.

If the common hepatic or bile duct is blocked (for example, by wedged gallstones), bile is prevented from entering the small intestine, accumulates, and eventually backs up into the liver. This exerts pressure on the liver cells, and bile begins to enter the bloodstream. As the bile circulates through the body, the tissues become yellow, or jaundiced.

Blockage of the ducts is just one cause of jaundice. More often it results from actual liver problems such as **hepatitis,** which is any inflammation of the liver, or **cirrhosis,** a condition in which the liver is severely damaged and becomes hard and fibrous. Cirrhosis is prevalent in people who drink excessive alcohol for many years. ✚

As demonstrated by its highly organized anatomy, the liver (Figure 27.17) is very important in the initial processing of the nutrient-rich blood draining the digestive organs. Its structural and functional units are called **lobules.** Each

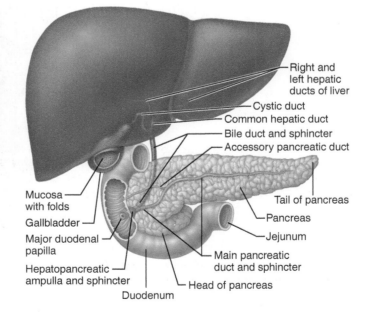

Right and
left hepatic
ducts of liver

Cystic duct

Common hepatic duct

Bile duct and sphincter

Accessory pancreatic duct

Mucosa
with folds

Gallbladder

Major duodenal
papilla

Hepatopancreatic
ampulla and sphincter

Duodenum

Tail of pancreas

Pancreas

Jejunum

Main pancreatic
duct and sphincter

Head of pancreas

Figure 27.16 Ducts of accessory digestive organs.

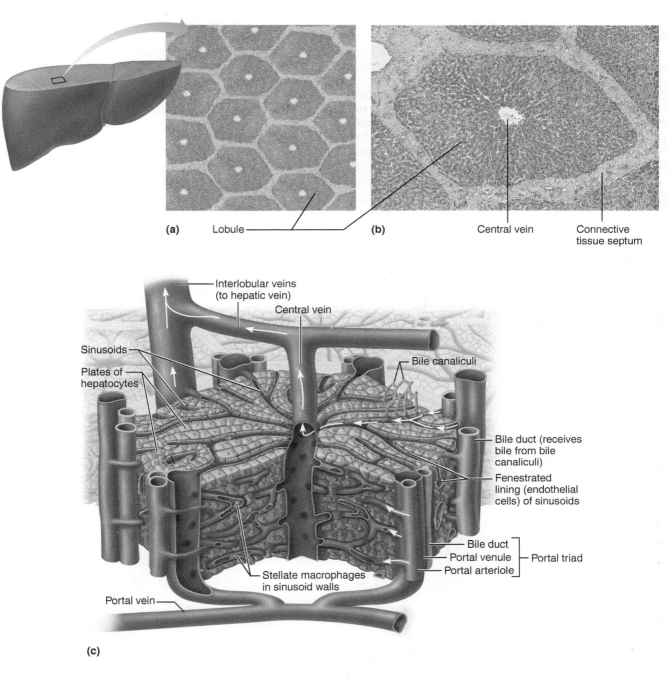

(a) Lobule

(b) Central vein   Connective tissue septum

Interlobular veins (to hepatic vein)

Central vein

Sinusoids

Plates of hepatocytes

Bile canaliculi

Bile duct (receives bile from bile canaliculi)

Fenestrated lining (endothelial cells) of sinusoids

Bile duct
Portal venule — Portal triad
Portal arteriole

Stellate macrophages in sinusoid walls

Portal vein

(c)

Figure 27.17 Microscopic anatomy of the liver. (a) Schematic view of the cut surface of the liver showing the hexagonal nature of its lobules. **(b)** Photomicrograph of one liver lobule (50×). **(c)** Enlarged three-dimensional diagram of one liver lobule. Arrows show direction of blood flow. Bile flows in the opposite direction toward the bile ducts.

lobule is a basically hexagonal structure consisting of cord-like arrays of **hepatocytes, or** liver cells, which radiate outward from a central vein running upward in the longitudinal axis of the lobule. At each of the six corners of the lobule is a **portal triad,** so named because three basic structures are always present there: a *portal arteriole* (a branch of the *hepatic artery,* the functional blood supply of the liver), a *portal venule* (a branch of the *hepatic portal vein* carrying

nutrient-rich blood from the digestive viscera), and a *bile duct.* Between the hepatocytes are blood-filled spaces, or **sinusoids,** through which blood from the hepatic portal vein and hepatic artery percolates. **Stellate macrophages,** special phagocytic cells also called **hepatic macrophages,** line the sinusoids and remove debris such as bacteria from the blood as it flows past while the hepatocytes pick up oxygen and nutrients. Much of the glucose transported to the liver from the digestive system is stored as glycogen in the liver for later use, and amino acids are taken from the blood by the liver cells and utilized to make plasma proteins. The sinusoids empty into the central vein, and the blood ultimately drains from the liver via the *hepatic veins.*

Bile is continuously being made by the hepatocytes. It flows through tiny canals, the **bile canaliculi,** which run between adjacent hepatocytes toward the bile duct branches in the triad regions, where the bile eventually leaves the liver. Notice that the directions of blood and bile flow in the liver lobule are exactly opposite.

ACTIVITY 9

Examining the Histology of the Liver

Examine a slide of liver tissue, and identify as many structural features as possible (see Figure 27.17). Also examine a three-dimensional model of liver lobules if this is available. ■

DISSECTION AND IDENTIFICATION
The Digestive System of the Cat

Don gloves and safety glasses. Obtain your cat and secure it to the dissecting tray, dorsal surface down. Obtain all necessary dissecting instruments. If you have completed the dissection of the circulatory and respiratory systems, the abdominal cavity is already exposed, and many of the digestive system structures have been identified. However, duplication of effort generally provides a good learning experience, so you will trace and identify all of the digestive system structures in this exercise.

If the abdominal cavity has not been previously opened, make a midline incision from the rib cage to the pubic symphysis. Then make four lateral cuts—two parallel to the rib cage and two at the inferior margin of the abdominal

cavity—so that the abdominal wall can be reflected back while you examine the abdominal contents.

Observe the shiny membrane lining the inner surface of the abdominal wall, which is the **parietal peritoneum.** We will return to identify digestive organs in the abdominal cavity shortly. But first we will expose and identify the salivary glands. ■

ACTIVITY 10

Exposing and Viewing the Salivary Glands and Oral Cavity Structures

1. To expose and identify the **salivary glands** remove the skin from one side of the head, and clear the connective tissue away from the angle of the jaw, below the ear, and superior to the masseter muscle. Many lymph nodes are in this area, and you should remove them if they obscure the salivary glands, which are lighter tan and lobular in structure. The cat has five pairs of salivary glands, but only those glands described in humans are easily localized and identified (Figure 27.18). Locate the **parotid gland** on the cheek just inferior to the ear. Follow its duct over the surface of the masseter muscle to the angle of the mouth. The **submandibular gland** is posterior to the parotid, near the angle of the jaw, and the **sublingual gland** (not shown in the figure) is just anterior to the submandibular gland within the lower jaw. The ducts of the submandibular and sublingual glands run deep and parallel to each other and empty on the side of the frenulum of the tongue. These need not be identified on the cat.

2. To expose and identify the structures of the oral cavity, cut through the mandible with bone cutters just anterior to the angle of the jaw to free the lower jaw from the maxilla.

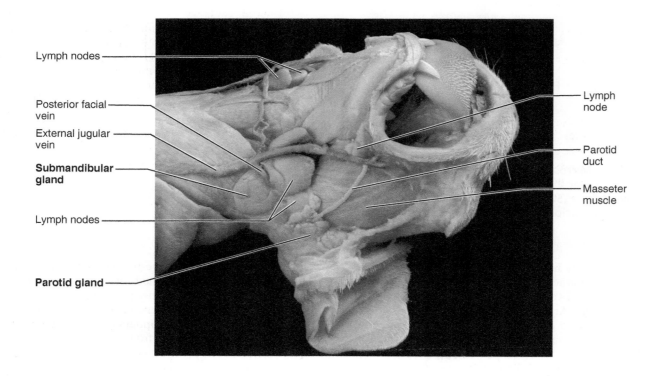

Lymph nodes

Posterior facial vein

External jugular vein

Submandibular gland

Lymph nodes

Parotid gland

Lymph node

Parotid duct

Masseter muscle

Figure 27.18 Photograph of salivary glands of the cat.

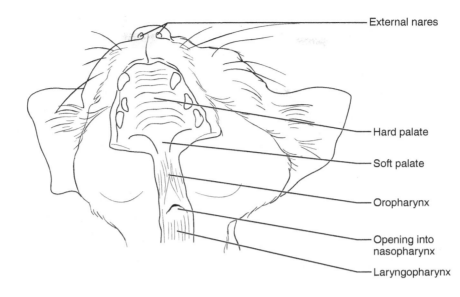

External nares

Hard palate

Soft palate

Oropharynx

Opening into
nasopharynx

Laryngopharynx

Figure 27.19 Oral cavity of the cat.

Identify the **hard** and **soft palates,** and use a probe to trace the hard palate to its posterior limits **(Figure 27.19).** Note the transverse ridges, or *rugae,* on the hard palate, which play a role in holding food in place while chewing.

Do these appear in humans? _____

Does the cat have a uvula? _____

Identify the **oropharynx** at the rear of the oral cavity and the palatine tonsils on the posterior walls at the junction between the oral cavity and oropharynx. Identify the **tongue,** and rub your finger across its surface to feel the papillae. Some of the papillae, especially at the anterior end of the tongue, should feel sharp and bristly. These are the filiform papillae. What do you think their function is?

_____

Locate the **lingual frenulum** attaching the tongue to the floor of the mouth. Trace the tongue posteriorly until you locate the **epiglottis,** the flap of tissue that covers the entrance to the respiratory passageway when swallowing occurs. Identify the **esophageal opening** posterior to the epiglottis.

Observe the **teeth** of the cat. The dental formula for the adult cat is as follows:

$$\frac{3,1,3,1}{3,1,2,1} \times 2 \text{ (30 teeth)}$$

ACTIVITY 11

Identifying Alimentary Canal Organs

1. Locate the abdominal alimentary canal structures (refer to **Figure 27.20**).

2. Identify the large reddish brown **liver** just beneath the diaphragm and the greater omentum, an apron of mesentery

riddled with fat (not shown) that covers the abdominal contents. The greater omentum is attached to the greater curvature of the stomach; its immune cells and macrophages help to protect the abdominal cavity. Lift the greater omentum, noting its two-layered structure and attachments, and lay it to the side or remove it to make subsequent organ identifications easier. Does the liver of the cat have the same number of lobes as the human liver?

_____

3. Lift the liver and examine its inferior surface to locate the **gallbladder,** a dark greenish sac embedded in the liver's ventral surface. Identify the **falciform ligament,** a delicate layer of mesentery separating the main lobes of the liver (right and left median lobes) and attaching the liver superiorly to the abdominal wall. Also identify the thickened area along the posterior edge of the falciform ligament, the *round ligament,* or *ligamentum teres,* a remnant of the umbilical vein of the fetus.

4. Displace the left lobes of the liver to expose the **stomach.** Identify the cardial part, fundus, body, and pyloric part of the stomach. What is the general shape of the stomach?

_____

Locate the **lesser omentum,** the serous membrane attaching the lesser curvature of the stomach to the liver, and identify the large spleen curving around the greater curvature of the stomach. Make an incision through the stomach wall to expose its inner surface. When the stomach is empty, its mucosa has large folds called **rugae.** As the stomach fills, the rugae gradually disappear and are no longer visible. Can you see any rugae? Identify the **pyloric sphincter** at the distal end of the stomach.

5. Lift the stomach and locate the **pancreas,** which appears as a grayish or brownish diffuse glandular mass in the mesentery. It extends from the vicinity of the spleen and greater

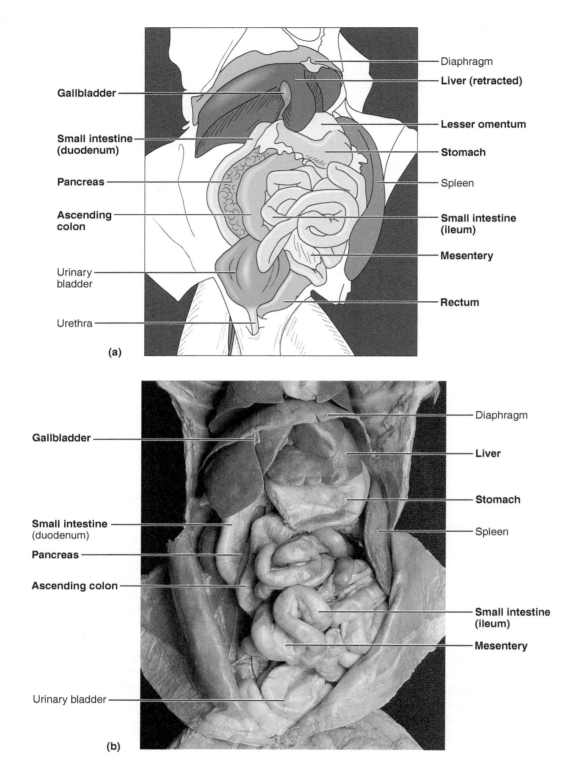

Figure 27.20 Digestive organs of the cat. (a) Diagram. **(b)** Photograph. The greater omentum has been cut from its attachment to the stomach.

curvature of the stomach and wraps around the duodenum. Try to find the **pancreatic duct** as it empties into the duodenum at a bulbous area referred to as the **hepatopancreatic ampulla.** Close to the pancreatic duct, locate the **bile duct,** and trace its path superiorly to the point where it diverges into the **cystic duct** (gallbladder duct) and the **common hepatic duct** (duct from the liver). Notice that the duodenum assumes a looped position.

6. Lift the **small intestine** to investigate how it is attached to the posterior body wall by the **mesentery.** Observe the mesentery closely. What types of structures do you see in this double peritoneal fold?

_____

Other than providing support for the intestine, what other functions does the mesentery have?

_____

_____

Trace the path of the small intestine from its proximal, (duodenal) end to its distal (ileal) end. Can you see any obvious differences in the external anatomy of the small intestine from one end to the other?

_____

With a scalpel, slice open the distal portion of the ileum, and flush out the inner surface with water. Feel the inner surface with your fingertip. How does it feel?

_____

Use a hand lens to see whether you can see any **villi** and to locate the areas of lymphoid tissue called **aggregated lymphoid nodules,** which appear as scattered white patches on the inner intestinal surface. See also Figure 27.9b.

Return to the duodenal end of the small intestine. Make an incision into the duodenum. As before, flush the surface with water, and feel the inner surface. Does it feel any different than the ileal mucosa?

_____ If so, describe the difference. _____

_____

Use the hand lens to observe the villi. What differences do you see in the villi in the two areas of the small intestine?

_____

_____

7. Make an incision into the junction between the ileum and cecum to locate the **ileocecal valve (Figure 27.21)**. Observe

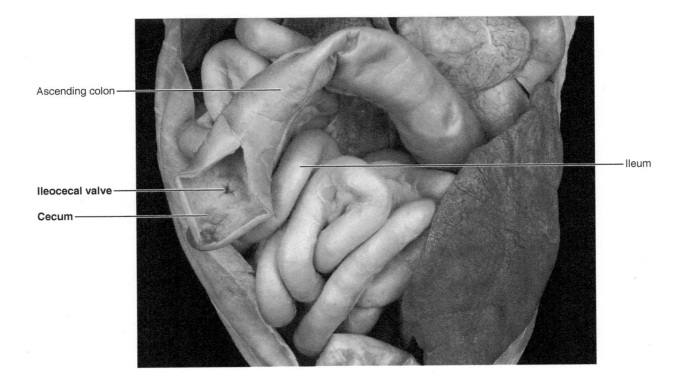

Figure 27.21 Ileocecal valve.

the **cecum,** the initial expanded part of the large intestine. You may have to remove lymph nodes from this area to observe it clearly. Does the cat have an appendix?

_____

8. Identify the short ascending, transverse, and descending portions of the **colon** and the **mesocolon,** a membrane that attaches the colon to the posterior body wall. Trace the descending colon to the **rectum,** which penetrates the body wall, and identify the **anus.**

Identify the two portions of the peritoneum, the parietal peritoneum lining the abdominal wall (identified previously) and the visceral peritoneum, which is the outermost layer of the wall of the abdominal organs.

9. Before you leave the laboratory, follow the boxed instructions to prepare your cat for storage and clean the area (page 219). ■■

Anatomy of the Digestive System

General Histologic Plan of the Alimentary Canal

1. The general anatomical features of the alimentary canal are listed below. Fill in the table to complete the information.

Wall layer	Subdivisions of the layer	Major functions
Mucosa		
Submucosa	(Not applicable)	
Muscularis externa		
Serosa or adventitia	(Not applicable)	

Organs of the Alimentary Canal

2. The tubelike digestive system canal that extends from the mouth to the anus is known as the _____ canal

   or the _____ tract.

3. How is the muscularis externa of the stomach modified? _____

   _____

   How does this modification relate to the function of the stomach? _____

   _____

4. What transition in epithelial type exists at the esophagus-stomach junction? _____

   _____

   How do the epithelia of these two organs relate to their specific functions? _____

   _____

5. Differentiate the colon from the large intestine. _____

   _____

6. Match the items in column B with the descriptive statements in column A.

Column A

_____ 1. structure that suspends the small intestine from the posterior body wall

_____ 2. fingerlike extensions of the intestinal mucosa that increase the surface area for absorption

_____ 3. large collections of lymphoid tissue found in the submucosa of the small intestine

_____ 4. deep folds of the mucosa and submucosa that extend completely or partially around the circumference of the small intestine

_____ 5. mobile organ that manipulates food in the mouth and initiates swallowing

_____ 6. conduit for both air and food

_____ 7. food passage that has no digestive or absorptive function

_____ 8. folds of the gastric mucosa

_____ 9. pocketlike sacs of the large intestine

_____ 10. projections of the plasma membrane of a mucosal epithelial cell

_____ 11. valve at the junction of the small and large intestines

_____ 12. primary region of nutrient absorption

_____ 13. membrane securing the tongue to the floor of the mouth

_____ 14. absorbs water and forms feces

_____ 15. area between the teeth and lips/cheeks

_____ 16. wormlike sac that outpockets from the cecum

_____ 17. initiates protein digestion

_____ 18. structure attached to the lesser curvature of the stomach

_____ 19. covers most of the abdominal organs like an apron

_____ 20. valve controlling food movement from the stomach into the duodenum

_____ 21. posterosuperior boundary of the oral cavity

_____ 22. region containing two sphincters through which feces are expelled from the body

_____ 23. bone-supported anterosuperior boundary of the oral cavity

Column B

a. aggregated lymphoid nodules

b. anus

c. appendix

d. circular folds

e. esophagus

f. frenulum

g. greater omentum

h. hard palate

i. haustra

j. ileocecal valve

k. large intestine

l. lesser omentum

m. mesentery

n. microvilli

o. oral vestibule

p. pharynx

q. pyloric sphincter

r. rugae

s. small intestine

t. soft palate

u. stomach

v. tongue

w. villi

7. Correctly identify all organs depicted in the diagram below.

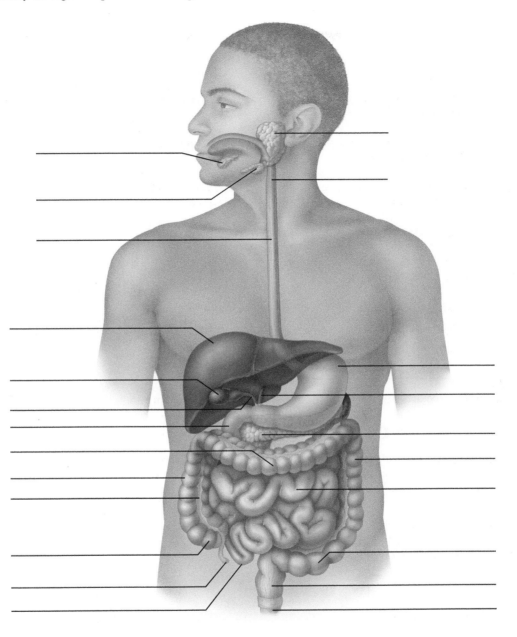

8. You have studied the histologic structure of a number of organs in this exercise. The stomach and duodenum are diagrammed below. Label the structures indicated by leader lines.

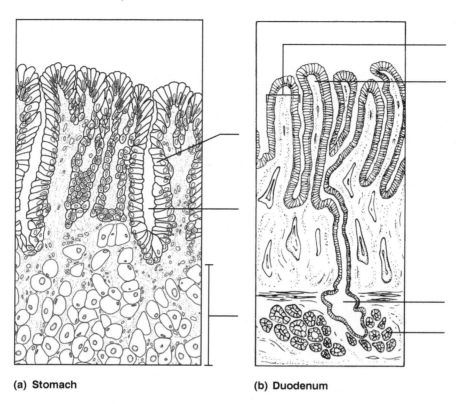

(a) **Stomach**   (b) **Duodenum**

Accessory Digestive Organs

9. Correctly label all structures provided with leader lines in the diagram of a molar below. (Note: Some of the terms in the key for question 10 may be helpful in this task.)

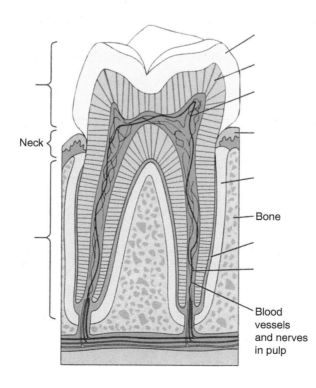

Neck

Bone

Blood
vessels
and nerves
in pulp

10. Use the key to identify each tooth area described below.

_____ 1. visible portion of the tooth

_____ 2. material covering the tooth root

_____ 3. hardest substance in the body

_____ 4. attaches the tooth to the tooth socket

_____ 5. portion of the tooth embedded in bone

_____ 6. forms the major portion of tooth structure; similar to bone

_____ 7. produces the dentin

_____ 8. site of blood vessels, nerves, and lymphatics

_____ 9. narrow gap between the crown and the gum

Key:

a.   cement

b.   crown

c.   dentin

d.   enamel

e.   gingival sulcus

f.   odontoblast

g.   periodontal ligament

h.   pulp

i.   root

11. In the human, the number of deciduous teeth is _____; the number of permanent teeth is _____.

12. The dental formula for permanent teeth is $\dfrac{2,1,2,3}{2,1,2,3} \times 2$. Explain what this means. _____

_____

_____

What is the dental formula for the deciduous teeth? _____ $\times$ _____ (_____ deciduous teeth)

13. Which teeth are the "wisdom teeth"? _____

_____

14. Various types of glands form a part of the alimentary canal wall or duct their secretions into it. Match the glands listed in column B with the function/locations described in column A.

Column A

_____ 1.   produce(s) mucus; found in the submucosa of the small intestine

_____ 2.   produce(s) a product containing amylase that begins starch breakdown in the mouth

_____ 3.   produce(s) many enzymes and an alkaline fluid that is secreted into the duodenum

_____ 4.   produce(s) bile that it secretes into the duodenum via the bile duct

_____ 5.   produce(s) HCl and pepsinogen

_____ 6.   found in the mucosa of the small intestine; produce(s) intestinal juice

Column B

a.   duodenal glands

b.   gastric glands

c.   intestinal crypts

d.   liver

e.   pancreas

f.   salivary glands

15. Which of the salivary glands produces a secretion that is mainly serous? _____

16. What is the role of the gallbladder? _____

17. Name three structures that form the portal triad of the liver. _____ ,

_____ , and _____

18. Where would you expect to find the stellate macrophages of the liver? _____

What is their function? _____

19. Why is the liver so dark red in living animals? _____

_____

20. The pancreas has two major populations of secretory cells—those in the islets and the acinar cells. Which population serves

the digestive process? _____

Dissection and Identification: The Digestive System of the Cat

21. Several differences between cat and human digestive anatomy should have become apparent during the dissection. Note the pertinent differences between the human and the cat relative to the following structures:

Structure	Cat	Human
Tongue papillae		
Number of liver lobes		
Appendix		

22. ➕ Pyloric stenosis is a type of gastric outlet obstruction caused by a narrowing of the pyloric part of the stomach. It is most

common in infants. Describe the clinical signs that you would expect to see with this condition. _____

_____

_____

23. ➕ Surgical removal of the gallbladder is called a *cholecystectomy*. The presence of gallstones that block any of the ducts that carry bile is the usual reason for the surgery. Explain why the gallbladder is not an essential organ, and predict possible

dietary changes that a patient might need to make post-cholecystectomy. _____

_____

_____

Anatomy of the Urinary System

MATERIALS

- Dissectible human torso model, three-dimensional model of the urinary system, and/or anatomical chart of the human urinary system
- Dissecting instruments and tray
- Pig or sheep kidney, doubly or triply injected
- Disposable gloves
- Safety glasses
- Animal specimen from previous dissections
- Embalming fluid
- Three-dimensional models of the cut kidney and of a nephron (if available)
- Compound microscope
- Prepared slides of section of kidney (l.s.) and of bladder (x.s.)
- Organic debris container
- Sticky notes

LEARNING OUTCOMES

☐ List the functions of the urinary system.

☐ Identify, on a model or image, the urinary system organs, and state the general function of each.

☐ Compare the course and length of the urethra in males and females.

☐ Identify these regions of the dissected kidney (longitudinal section): hilum, cortex, medulla, renal pyramids, major and minor calyces, pelvis, renal columns, and fibrous and perirenal fat capsules.

☐ Trace the blood supply of the kidney from the renal artery to the renal vein.

☐ Define *nephron,* and describe its anatomy.

☐ Define *glomerular filtration, tubular resorption,* and *tubular secretion,* and indicate the nephron areas involved in these processes.

☐ Define *micturition,* and explain the differences in the control of the internal and external urethral sphincters.

☐ Recognize the histologic structure of the kidney and ureter microscopically or in an image.

☐ Identify on a dissection specimen the urinary system organs, and describe the general function of each.

PRE-LAB QUIZ

1. Circle the correct underlined term. In its excretory role, the urinary system is primarily concerned with the removal of <u>carbon-containing</u> / <u>nitrogenous</u> wastes from the body.

2. Which structure(s) perform(s) the excretory and homeostatic functions of the urinary system?
   a. kidneys
   b. ureters
   c. urinary bladder
   d. all of the above

3. Circle the correct underlined term. The <u>cortex</u> / <u>medulla</u> of the kidney is segregated into triangular regions with a striped appearance.

4. Circle the correct underlined terms. As the renal artery approaches a kidney, it is divided into branches known as the <u>segmental arteries</u> / <u>afferent arterioles</u>.

5. What do we call the anatomical units responsible for the formation of urine? _____

6. This knot of coiled capillaries, found in the kidneys, forms the filtrate. It is the:
   a. arteriole
   b. glomerulus
   c. podocyte
   d. tubule

Text continues on next page.

7. The section of the renal tubule closest to the glomerular capsule is the:
   a. collecting duct
   b. distal convoluted tubule
   c. nephron loop
   d. proximal convoluted tubule

8. Circle the correct underlined term. The afferent / efferent arteriole drains the glomerular capillary bed.

9. Circle True or False. During tubular resorption, components of the filtrate move from the bloodstream into the tubule.

10. Circle the correct underlined term. The internal / external urethral sphincter consists of skeletal muscle and is voluntarily controlled.

Metabolism of nutrients produces wastes, including carbon dioxide, nitrogenous wastes, and ammonia, that must be eliminated if the body's normal function is to continue. Excretory processes involve multiple organ systems, with the **urinary system** primarily responsible for the removal of nitrogenous wastes from the body. In addition to this excretory function, the kidney maintains the electrolyte, acid-base, and fluid balances of the blood and is thus a major homeostatic organ of the body.

To perform its functions, the kidney acts first as a blood filter and then as a filtrate processor. It allows toxins, metabolic wastes, and excess ions to leave the body in the urine, while simultaneously retaining needed substances and returning them to the blood. Malfunction of the urinary system, particularly of the kidneys, leads to a failure in homeostasis which, unless corrected, is fatal.

Gross Anatomy of the Human Urinary System

The urinary system (Figure 28.1) consists of the paired kidneys and ureters and the single urinary bladder and urethra. The **kidneys** perform the functions described above and manufacture urine in the process. The remaining organs of the system provide temporary storage reservoirs or transportation channels for urine.

ACTIVITY 1

Identifying Urinary System Organs

Examine the human torso model, a large anatomical chart, or a three-dimensional model of the urinary system to locate and study the anatomy and relationships of the urinary organs.

1. Locate the paired kidneys on the dorsal body wall in the superior lumbar region. Notice that they are not positioned at exactly the same level. Because it is crowded by the liver, the right kidney is slightly lower than the left kidney. Three layers of support tissue surround each kidney. Beginning with the innermost layer, they are (1) a transparent *fibrous capsule*, (2) a *perirenal fat capsule*, and (3) the fibrous *renal fascia* that holds the kidneys in place in a retroperitoneal position.

In cases of rapid weight loss or in very thin individuals, the fat capsule may be reduced or scarce in amount. In such a case, the kidneys are less securely anchored, so they may drop to a more inferior position in the abdominal cavity. This phenomenon is called **ptosis.** +

2. Observe the **renal arteries** as they diverge from the descending aorta and plunge into the indented medial region, called the **hilum,** of each kidney. Note also the **renal veins,** which drain the kidneys, and the two **ureters,** which carry urine from the kidneys, moving it by peristalsis to the bladder for temporary storage.

3. Locate the **urinary bladder,** and observe the point of entry of the two ureters into this organ. Also locate the single **urethra,** which drains the bladder. The triangular region of the bladder that is delineated by the openings of the ureters and the urethra is referred to as the **trigone** (Figure 28.2).

4. Follow the path of the urethra to the body exterior. In the male, it is approximately 20 cm (8 inches) long, travels the length of the **penis,** and opens at its tip. The urethra is made up of three named regions—the *prostatic urethra,* the *intermediate part,* and *spongy urethra* (described in more detail in Exercise 29). The male urethra has a dual function. It carries urine to the body exterior, and it provides a passageway for semen ejaculation. Thus, in the male, the urethra is part of both the urinary and reproductive systems. In females, the urethra is short, approximately 4 cm (1½ inches) long (see Figure 28.2). There are no common urinary-reproductive pathways in the female, and the female's urethra serves only to transport urine to the body exterior. Its external opening, the **external urethral orifice,** lies anterior to the vaginal opening. ■

DISSECTION
Gross Internal Anatomy of the Pig or Sheep Kidney

1. In preparation for dissection, don gloves and safety glasses. Obtain a preserved sheep or pig kidney, dissecting tray, and instruments. Observe the kidney to identify the **fibrous capsule,** a smooth transparent membrane that adheres tightly to the external aspect of the kidney.

2. Find the ureter, renal vein, and renal artery at the hilum (indented) region. The renal vein has the thinnest wall and will be collapsed. The ureter is the largest of these structures and has the thickest wall.

3. Section the kidney in the frontal plane, through the longitudinal axis and locate the anatomical areas described below and depicted in Figure 28.3.

Renal cortex: The superficial kidney region, which is lighter in color. If the kidney is doubly injected with latex, you will see a predominance of red and blue latex specks in this region indicating its rich blood supply.

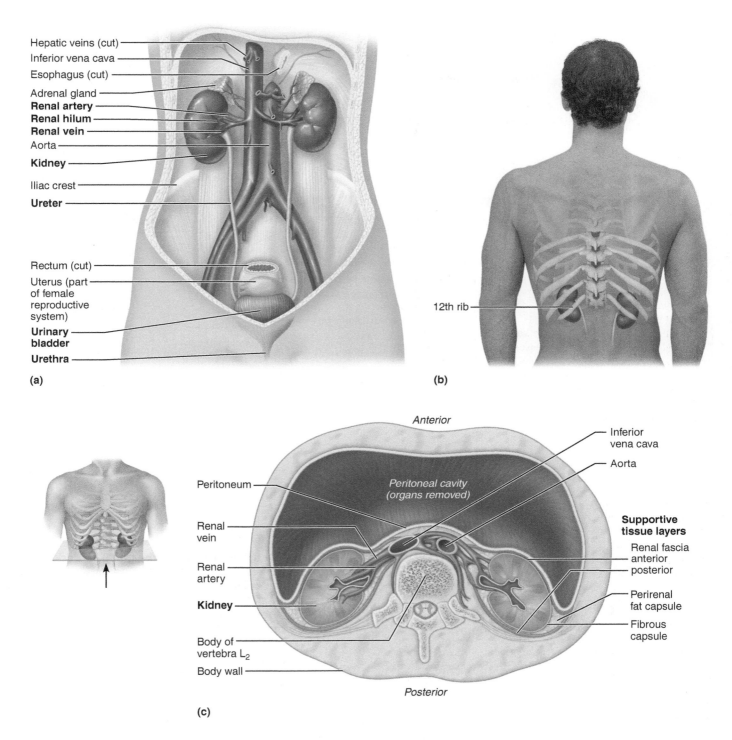

Hepatic veins (cut)
Inferior vena cava
Esophagus (cut)
Adrenal gland
Renal artery
Renal hilum
Renal vein
Aorta
Kidney
Iliac crest
Ureter
Rectum (cut)
Uterus (part of female reproductive system)
Urinary bladder
Urethra

(a)

12th rib

(b)

Anterior

Inferior vena cava
Aorta

Peritoneum

Peritoneal cavity (organs removed)

Supportive tissue layers
Renal fascia
  anterior
  posterior

Renal vein

Renal artery

Kidney

Perirenal fat capsule
Fibrous capsule

Body of vertebra L$_2$
Body wall

Posterior

(c)

Figure 28.1 Organs of the urinary system. (a) Anterior view of the female urinary organs. Most unrelated abdominal organs have been removed. **(b)** Posterior in situ view showing the position of the kidneys relative to the twelfth ribs. **(c)** Cross section of the abdomen viewed from inferior direction. Note the retroperitoneal position and supportive tissue layers of the kidney.

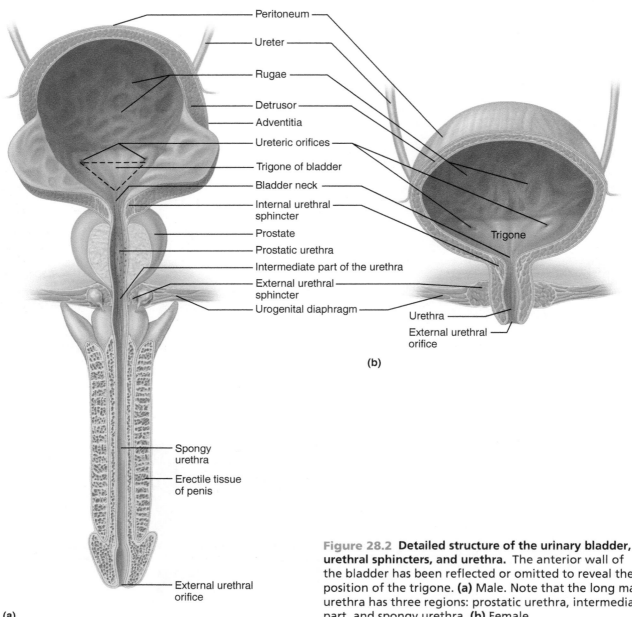

Peritoneum
Ureter
Rugae
Detrusor
Adventitia
Ureteric orifices
Trigone of bladder
Bladder neck
Internal urethral sphincter
Prostate
Prostatic urethra
Intermediate part of the urethra
External urethral sphincter
Urogenital diaphragm
Trigone
Urethra
External urethral orifice

(b)

Spongy urethra
Erectile tissue of penis
External urethral orifice

(a)

Figure 28.2 Detailed structure of the urinary bladder, urethral sphincters, and urethra. The anterior wall of the bladder has been reflected or omitted to reveal the position of the trigone. **(a)** Male. Note that the long male urethra has three regions: prostatic urethra, intermediate part, and spongy urethra. **(b)** Female.

Renal medulla: Deep to the cortex; a darker, reddish-brown color. The medulla is segregated into triangular regions that have a striped appearance—the **renal pyramids.** The base of each pyramid faces toward the cortex. Its more pointed *papilla,* or *apex,* points to the innermost kidney region.

Renal columns: Areas of tissue, more like the cortex in appearance, that dip inward between the pyramids separating them.

Renal pelvis: Extending inward from the hilum; a relatively flat, basinlike cavity that is continuous with the **ureter,** which exits from the hilum region. Fingerlike extensions of the pelvis should be visible. The larger, or primary, extensions are called the **major calyces** (singular: **calyx**); subdivisions of the major calyces are the **minor calyces.** Notice that the minor calyces terminate in cuplike areas that enclose the papillae and collect urine draining from the pyramidal tips into the pelvis.

Approximately a fourth of the total blood flow of the body is delivered to the kidneys each minute by the large **renal arteries.** As a renal artery approaches the kidney, it breaks up into branches called **segmental arteries,** which enter the hilum. Each segmental artery, in turn, divides into several **interlobar arteries,** which ascend in the renal columns. At the top of the cortx-medulla junction, the interlobar arteries branch into the **arcuate arteries,** which curve over the bases of the medullary pyramids. Small **cortical radiate arteries** branch off the arcuate arteries and ascend into the cortex, giving off the individual **afferent arterioles (Figure 28.4),** which lead to the **glomerulus,** a ball of capillaries found in the nephron. *Efferent arterioles* drain the glomerulus and feed into one of two capillary beds, either the *peritubular capillaries* or the *vasa recta.* Blood draining from the nephron capillary beds enters the **cortical radiate veins** and then drains

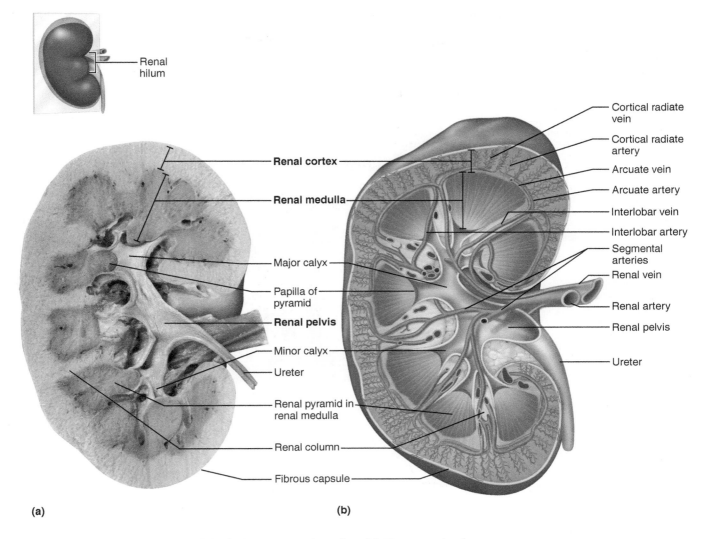

Renal hilum

Renal cortex

Renal medulla

Major calyx

Papilla of pyramid

Renal pelvis

Minor calyx

Ureter

Renal pyramid in renal medulla

Renal column

Fibrous capsule

Cortical radiate vein

Cortical radiate artery

Arcuate vein

Arcuate artery

Interlobar vein

Interlobar artery

Segmental arteries

Renal vein

Renal artery

Renal pelvis

Ureter

(a)   **(b)**

Figure 28.3 Internal anatomy of the kidney. Frontal section. **(a)** Photograph of a right kidney. **(b)** Diagram showing the larger blood vessels supplying the kidney tissue.

through the **arcuate veins** and the **interlobar veins** to finally enter the **renal vein** in the pelvis region. There are no segmental veins.

Dispose of the kidney specimen as your instructor specifies. ■

Functional Microscopic Anatomy of the Kidney and Bladder

Kidney

Each kidney contains over a million **nephrons,** which are the structural and functional units responsible for filtering the blood and forming urine. The detailed structure and the relative positioning of the nephrons in the kidney is depicted in Figure 28.4.

Each nephron consists of two major structures: a *renal corpuscle* and a *renal tubule.* The structures within the renal corpuscle and renal tubule are summarized individually in **Table 28.1** on page 494.

There are two kinds of nephrons, cortical and juxtamedullary (see Figure 28.4). **Cortical nephrons** are most numerous, making up about 85% of nephrons. They are located almost entirely within the renal cortex, except for small parts of their nephron loops that dip into the renal medulla. The renal corpuscles of **juxtamedullary nephrons** are located deep in the cortex at the border with the medulla; their long nephron loops penetrate deeply into the medulla. Juxtamedullary nephrons play an important role in concentrating urine.

The function of the nephron depends on several unique features of the renal circulation **(Figure 28.5)**. The capillary vascular supply consists of three distinct capillary beds, the *glomerulus,* the *peritubular capillary bed,* and the *vasa recta.* Vessels leading to and from the glomerulus, the first capillary bed, are both glomerular arterioles: the **afferent arteriole**

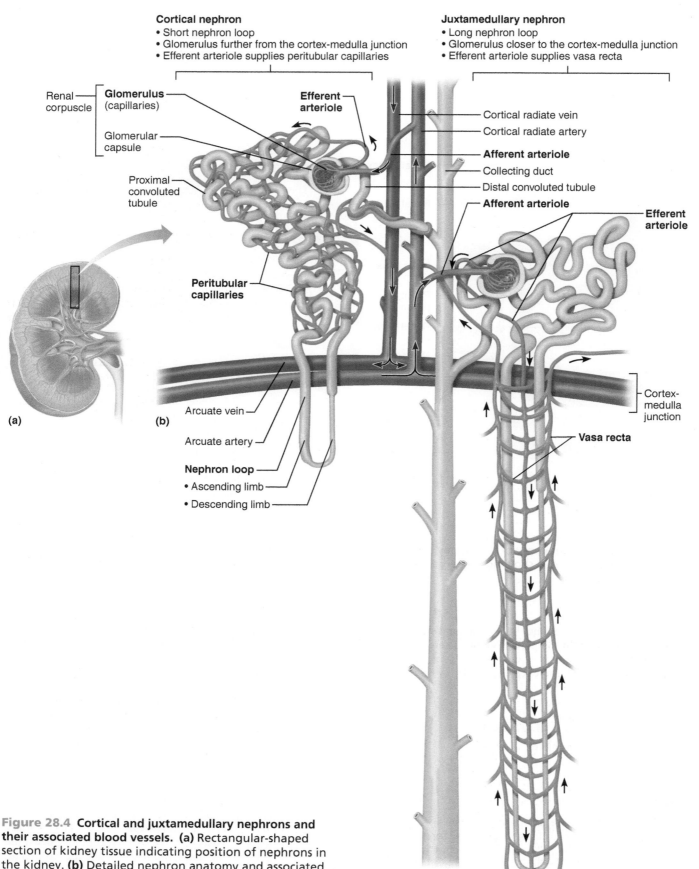

Cortical nephron
- Short nephron loop
- Glomerulus further from the cortex-medulla junction
- Efferent arteriole supplies peritubular capillaries

Juxtamedullary nephron
- Long nephron loop
- Glomerulus closer to the cortex-medulla junction
- Efferent arteriole supplies vasa recta

Renal corpuscle

Glomerulus (capillaries)

Glomerular capsule

Efferent arteriole

Cortical radiate vein
Cortical radiate artery
Afferent arteriole
Collecting duct
Distal convoluted tubule

Afferent arteriole

Efferent arteriole

Proximal convoluted tubule

Peritubular capillaries

(a)

(b)

Arcuate vein
Arcuate artery

Nephron loop
- Ascending limb
- Descending limb

Cortex-medulla junction

Vasa recta

Figure 28.4 Cortical and juxtamedullary nephrons and their associated blood vessels. (a) Rectangular-shaped section of kidney tissue indicating position of nephrons in the kidney. **(b)** Detailed nephron anatomy and associated blood supply. Arrows indicate direction of blood flow.

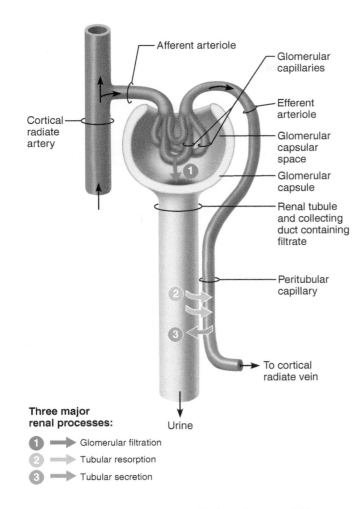

Three major
renal processes:

1 ➡ Glomerular filtration

2 ➡ Tubular resorption

3 ➡ Tubular secretion

Figure 28.5 A schematic uncoiled nephron. A kidney actually has millions of nephrons acting in parallel. The three major mechanisms by which the kidneys adjust the composition of plasma are **(a)** glomerular filtration, **(b)** tubular resorption, and **(c)** tubular secretion. Black arrows show the path of blood flow through the renal microcirculation.

feeds the bed while the **efferent arteriole** drains it. The glomerular capillary is unique in the body. It is a high-pressure bed along its entire length. Its high pressure is a result of two major factors: (1) the bed is *fed and drained* by arterioles, and (2) the afferent feeder arteriole is larger in diameter than the efferent arteriole draining the bed. The high hydrostatic pressure created by these two anatomical features forces fluid and blood components smaller than proteins out of the glomerulus into the glomerular capsule. That is, it forms the filtrate, which is processed by the nephron tubule.

The **peritubular capillary bed** arises from the efferent arteriole draining the glomerulus. This set of capillaries clings intimately to the renal tubule. The peritubular capillaries are *low-pressure* porous capillaries adapted for absorption rather than filtration and readily take up the solutes and water resorbed from the filtrate by the tubule cells. Efferent arterioles that supply juxtaglomerular nephrons tend not to form peritubular capillaries. Instead, they form long, straight, highly interconnected vessels called **vasa recta** that run

parallel and very close to the long nephron loops. This vasa recta is essential in concentrating urine.

Each nephron also has a **juxtaglomerular complex (JGC) (Figure 28.6)** located where the most distal portion of the ascending limb of the nephron loop touches the afferent arteriole. Helping to form the JGC are (1) *granular cells,* also called *juxtaglomerular (JG) cells,* in the arteriole walls that sense blood pressure in the afferent arteriole; (2) a group of columnar cells in the ascending limb of the nephron loop called the *macula densa* that monitors sodium chloride (NaCl) concentration in the filtrate; and (3) *extraglomerular mesangial cells* that may pass regulatory signals between the macula densa and granular cells. The role of the JGC is to regulate the rate of filtration and systemic blood pressure.

Urine forms as a result of three processes: *filtration, resorption,* and *secretion* (see Figure 28.5). **Filtration,** the role of the glomerulus, is largely a passive process in which a portion of the blood passes from the glomerular capillary into the glomerular capsule. During **tubular resorption,** many of the filtrate components move through the tubule cells and return to the blood in the peritubular capillaries. Some of this resorption is passive—such as that of water, which passes by

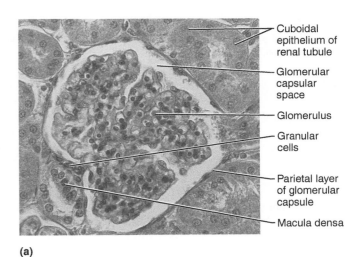

(a)

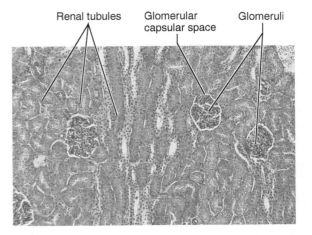

(b)

Figure 28.6 Microscopic structure of kidney tissue.
(a) Detailed structure of the nephron (300×).
(b) Low-power view of the renal cortex (75×).

Table 28.1 Structures of the Nephron (Figure 28.4)

Structure	Description	Epithelium	Function
Structures Within the Renal Corpuscle			
Glomerulus	A cluster of capillaries supplied by the afferent arteriole and drained by the efferent arteriole	Fenestrated endothelium (simple squamous)	Forms part of the filtration membrane
Visceral layer of the glomerular capsule	Podocytes that branch into foot processes	Simple squamous epithelium	Forms part of the filtration membrane. Spaces between the foot processes form filtration slits.
Parietal layer of the glomerular capsule	Outer impermeable wall of the glomerular capsule	Simple squamous epithelium	Forms the outside of the cuplike glomerular capsule. Plays no role in filtration.
Structures Within the Renal Tubule			
Proximal convoluted tubule (PCT)	Highly coiled first section of the renal tubule	Simple cuboidal with many microvilli and many mitochondria	Primary site of tubular reabsorption of water and solutes. Some secretion also occurs.
Descending limb of the nephron loop	First portion of the nephron loop	Simple cuboidal with some microvilli	Tubular reabsorption and secretion of water and solutes.
Descending thin limb of the nephron loop	A continuation of the descending limb	Simple squamous epithelium	Very permeable to water. Water is reabsorbed, but no solutes are reabsorbed.
Thick ascending limb of the nephron loop	In most nephrons the ascending limb is thick	Cuboidal or low columnar, with very few aquaporins	Not permeable to water. Solutes are reabsorbed actively and passively.
Distal convoluted tubule (DCT)	Coiled distal portion of the tubule	Simple cuboidal with few microvilli but many mitochondria	Some reabsorption of water and solutes and secretion, which are regulated to meet the body's needs
Collecting duct	Receives filtrate from the DCT of multiple nephrons	Simple cuboidal epithelium with two specialized cell types: principal cells and intercalated cells	Some reabsorption and secretion to conserve body fluids, maintain blood pH, and regulate solute concentrations.

osmosis—but the resorption of most substances depends on active transport processes and is highly selective. Substances that are almost entirely resorbed from the filtrate include water, glucose, and amino acids. Various ions are selectively resorbed or allowed to go out in the urine according to what is required to maintain appropriate blood pH and electrolyte composition. Waste products, including urea, creatinine, uric acid, and drug metabolites, are resorbed to a much lesser degree or not at all. Most (75% to 80%) of tubular resorption occurs in the proximal convoluted tubule.

Tubular secretion is essentially the reverse process of tubular resorption. Substances such as hydrogen and potassium ions and creatinine move from the blood of the peritubular capillaries through the tubular cells into the filtrate to be disposed of in the urine.

ACTIVITY 2

Studying Nephron Structure

1. Begin your study of nephron structure by identifying the glomerular capsule, proximal and distal convoluted tubule regions, and the nephron loop on a model of the nephron. Then, obtain a compound microscope and a prepared slide of kidney tissue to continue with the microscope study of the kidney.

2. Hold the longitudinal section of the kidney up to the light to identify the renal cortex and renal medulla. Then secure the slide on the microscope stage, and scan the slide under low power.

3. Move the slide so that you can see the cortical area. Identify a glomerulus, which appears as a ball of tightly packed material containing many small nuclei (Figure 28.6). It is usually surrounded by a vacant-appearing region (the *capsular space*) between the visceral and parietal layers of the glomerular capsule that surrounds it.

4. Notice that the renal tubules are cut at various angles. Try to differentiate the fuzzy cuboidal epithelium of the proximal convoluted tubule, which has dense microvilli, from the epithelium of the distal convoluted tubule with sparse microvilli. Also identify the thin-walled nephron loop. ■

Bladder

Although the kidney produces urine continuously, urine is usually removed from the body only when voiding is convenient. In the meantime, the **urinary bladder,** which receives urine via the ureters and discharges it via the urethra, stores it temporarily.

Voiding, or **micturition,** is the act of emptying the bladder. Two sphincter muscles, or valves (see Figure 28.2), the **internal urethral sphincter** and the **external urethral**

sphincter, control the outflow of urine from the bladder. Ordinarily, the bladder continues to collect urine until about 200 ml have accumulated, at which time the stretching of the bladder wall activates stretch receptors. Impulses transmitted to the central nervous system subsequently produce reflex contractions of the bladder wall through parasympathetic nervous system pathways via the pelvic splanchnic nerves. As contractions increase in force and frequency, stored urine is forced past the internal sphincter, which is a smooth muscle involuntary sphincter, into the superior part of the urethra. It is then that a person feels the urge to void. The inferior external sphincter consists of skeletal muscle and is voluntarily controlled. If it is not convenient to void, the opening of this sphincter can be inhibited. Conversely, if the time is convenient, the sphincter may be relaxed and the stored urine flushed from the body. If voiding is inhibited, the reflex contractions of the bladder cease temporarily, and urine continues to accumulate in the bladder. After another 200 to 300 ml of urine have been collected, the *micturition reflex* will again be initiated.

Lack of voluntary control over the external sphincter is referred to as **incontinence.** Incontinence is normal in children 2 years old or younger, because they have not yet gained control over the voluntary sphincter. In adults and older children, incontinence generally results from spinal cord injury, emotional problems, bladder irritability, or some other disorder of the urinary tract. ✚

ACTIVITY 3

Studying Bladder Structure

1. Return the kidney slide to the supply area, and obtain a slide of bladder tissue. Scan the bladder tissue. Identify its three layers: mucosa, muscular layer, and fibrous adventitia.

2. Study the highly specialized transitional epithelium of the mucosa. The plump, transitional epithelial cells have the ability to slide over one another, thus decreasing the thickness of the mucosa layer as the bladder fills and stretches to accommodate the increased urine volume. Depending on the degree of stretching of the bladder, the mucosa may be three to eight cell layers thick. (Compare the transitional epithelium of the mucosa to that shown in Figure 5.3h, page 59.)

3. Examine the heavy muscular wall (detrusor), which consists of three irregularly arranged muscular layers. The innermost and outermost muscle layers are arranged longitudinally; the middle layer is arranged circularly. Attempt to differentiate the three muscle layers.

4. Compare the structure of the bladder wall you are observing to the structure of the ureter wall shown in **Figure 28.7**. How are the two organs similar histologically?

_____

_____

What is/are the most obvious difference(s)?

_____

_____

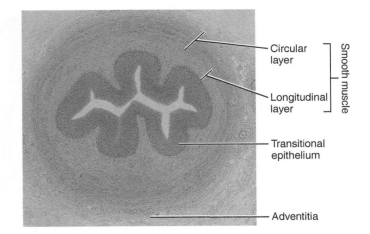

Figure 28.7 Structure of the ureter wall. Cross section of ureter (30×).

DISSECTION AND IDENTIFICATION
The Urinary System of the Cat

The structures of the reproductive and urinary systems are often considered together as the *urogenital system,* because they have common embryologic origins. However, the emphasis in this dissection is on identifying the structures of the urinary tract (**Figure 28.8** and **Figure 28.9**), with only a few references to contiguous reproductive structures. The anatomy of the reproductive system is studied in Exercise 29. ∎

ACTIVITY 4

Identifying Organs of the Urinary System

1. Don gloves and safety glasses. Obtain your dissection specimen, and place it ventral side up on the dissection tray. Reflect the abdominal viscera (in particular, the small intestine) to locate the kidneys high on the dorsal body wall. Note that the **kidneys** in the cat, as well as in the human, are retroperitoneal (behind the peritoneum).

2. Carefully remove the peritoneum, and clear away the bed of fat that invests the kidneys. Then locate the adrenal glands that lie superiorly and medial to the kidneys.

3. Identify the **renal artery** (red latex injected), the **renal vein** (blue latex injected), and the ureter at the hilum region of the kidney. You may find two renal veins leaving one kidney in the cat but not in humans.

4. To observe the gross internal anatomy of the kidney, slit the connective tissue *fibrous capsule* encasing a kidney, and peel it back. Make a midfrontal cut through the kidney, and examine one cut surface with a hand lens to identify the granular *cortex* and the central darker *medulla,* which will appear striated. Notice that the cat's renal medulla consists of just one pyramid, in contrast to the multipyramidal human kidney.

Text continues on page 498.

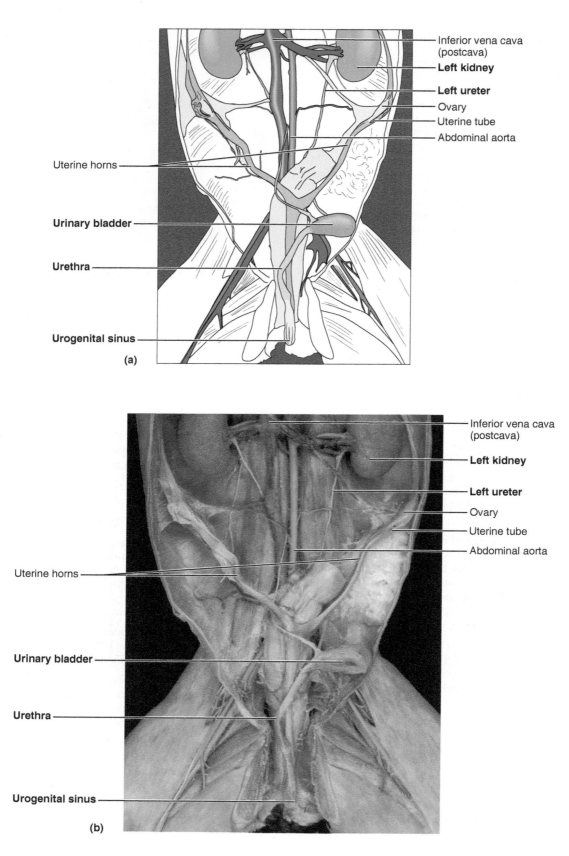

Figure 28.8 Urinary system of the female cat. (Reproductive structures are also indicated.) **(a)** Diagram. **(b)** Photograph of female urogenital system.

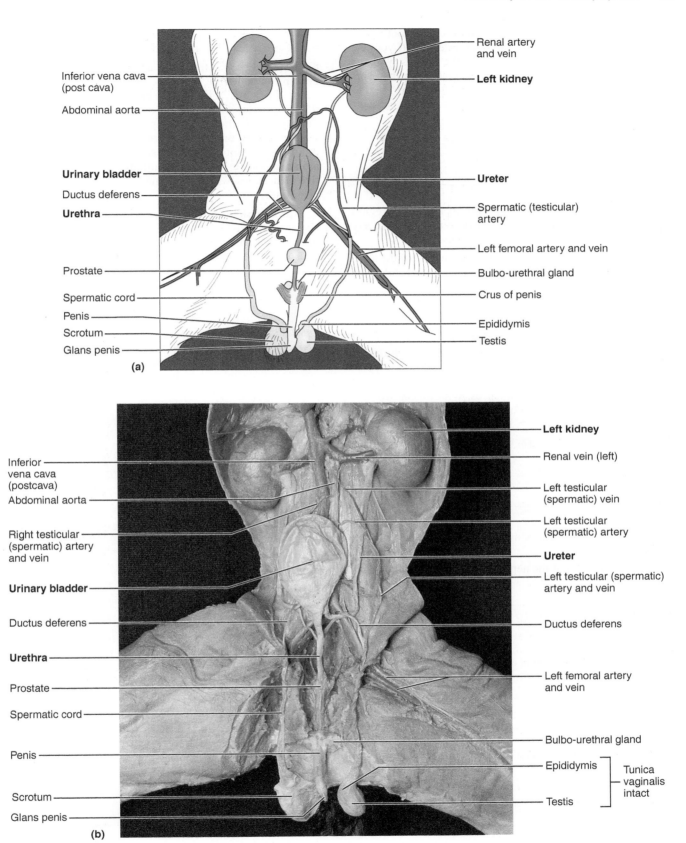

Figure 28.9 Urinary system of the male cat. (Reproductive structures are also indicated.) **(a)** Diagram. **(b)** Photograph of male urogenital system.

5. Trace the **ureters** to the **urinary bladder,** a smooth muscular sac located superiorly to the small intestine. If your cat is a female, be careful not to confuse the ureters with the uterine tubes, which lie superior to the bladder in the same general region (Figure 28.8). Observe the sites where the ureters enter the bladder. How would you describe the entrance point anatomically?

_____

_____

6. Cut through the bladder wall, and examine the region where the **urethra** exits to see whether you can discern any evidence of the *internal urethral sphincter.*

7. If your cat is a male, identify the prostate (part of the male reproductive system), which encircles the urethra distal to the neck of the bladder (Figure 28.9). Notice that the urinary bladder is somewhat fixed in position by ligaments.

8. Using a probe, trace the urethra as it exits the bladder. In the male, the urethra goes through the prostate and enters the penis. In the female cat, the urethra terminates in the **urogenital sinus,** a common chamber into which both the vagina and the urethra empty. In the human female, the vagina and the urethra have separate external openings. At this time, do not perform dissection to expose the urethra along its entire length, you might damage the reproductive structures, which you may study in a separate exercise (Exercise 29).

9. To complete this exercise, observe a cat of the opposite sex. Before you leave the laboratory, follow the boxed instructions to prepare your cat for storage and clean the area (page 219). ■

GROUP CHALLENGE

Urinary System Sequencing

Arrange the following sets of urinary structures in the correct order for the flow of urine, filtrate, or blood. Work in small groups, but refrain from using a figure or other reference to determine the sequences. Within your group, assign a facilitator and a recorder. The facilitator will list each term in a given question set on a separate sticky note. All members of the group will discuss the correct order for the structures and arrange them accordingly. After all of the sticky notes have been arranged, the recorder will write down the terms in the appropriate order.

1. renal pelvis, minor calyx, renal papilla, urinary bladder, ureter, major calyx, and urethra

_____

_____

2. distal convoluted tubule, ascending limb of the nephron loop, glomerulus, collecting duct,

descending limb of the nephron loop, proximal convoluted tubule, and glomerular capsule

_____

_____

3. segmental artery, afferent arteriole, cortical radiate artery, glomerulus, renal artery, interlobar

artery, and arcuate artery _____

_____

4. arcuate vein, inferior vena cava, peritubular capillaries, renal vein, interlobar vein, cortical

radiate vein, and efferent arteriole _____

_____

Anatomy of the Urinary System

Gross Anatomy of the Human Urinary System

1. Complete the following statements.

The kidney is referred to as an excretory organ because it excretes __1__ wastes. It is also a major homeostatic organ because it maintains the electrolyte, __2__, and __3__ balance of the blood.

Urine is continuously formed by the structural and functional units of the kidneys, the __4__, and is routed down the __5__ by the mechanism of __6__ to a storage organ called the __7__. Eventually, the urine is conducted to the body __8__ by the urethra. In the male, the urethra is __9__ centimeters long and transports both urine and __10__. The female urethra is __11__ centimeters long and transports only urine.

Voiding, or emptying the bladder, is called __12__. Voiding has both voluntary and involuntary components. The voluntary sphincter is the __13__ sphincter, composed of skeletal muscle. An inability to control this sphincter is referred to as __14__.

1. _____
2. _____
3. _____
4. _____
5. _____
6. _____
7. _____
8. _____
9. _____
10. _____
11. _____
12. _____
13. _____
14. _____

2. Which layer of support tissue holds the kidneys in the retroperitoneal position?

_____

3. Name the three structures that outline the triangular region of the bladder known as the trigone. _____

_____

4. Why is incontinence a normal phenomenon in the child under 1½ to 2 years old? _____

_____

In the adult what events may lead to incontinence? _____

_____

_____

5. Label the photograph of the kidney model by selecting the letter for the correct structure from the key below.

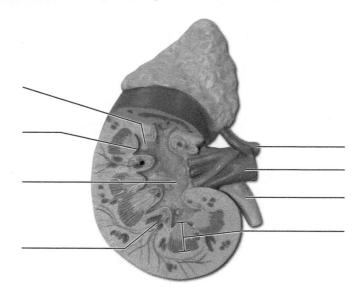

Key:

a.   minor calyx

b.   renal artery

c.   renal column

d.   renal papilla

e.   renal pelvis

f.   renal pyramids

g.   renal vein

h.   ureter

Gross Internal Anatomy of the Pig or Sheep Kidney

6. Match the appropriate structure in column B to its description in column A.

Column A

_____ 1.   smooth membrane that adheres tightly to the kidney surface

_____ 2.   portion of the kidney containing mostly collecting ducts

_____ 3.   superficial region of kidney tissue

_____ 4.   basinlike area of the kidney, continuous with the ureter

_____ 5.   a cup-shaped extension of the pelvis that encircles the apex of a pyramid

_____ 6.   area of cortical tissue running between the renal pyramids

Column B

a.   cortex

b.   fibrous capsule

c.   medulla

d.   minor calyx

e.   renal column

f.   renal pelvis

Functional Microscopic Anatomy of the Kidney and Bladder

7. Label the blood vessels and parts of the nephron by selecting the letter for the correct structure from the key below.

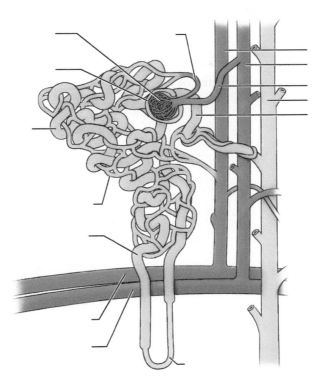

Key:

a. afferent arteriole

b. arcuate artery

c. arcuate vein

d. collecting duct

e. cortical radiate artery

f. cortical radiate vein

g. distal convoluted tubule

h. efferent arteriole

i. glomerular capsule

j. glomerulus

k. nephron loop—ascending limb

l. nephron loop—descending limb

m. peritubular capillary

n. proximal convoluted tubule

8. For each of the following descriptions of a structure, find the matching name in the question 7 key above.

_____ 1. capillary specialized for filtration

_____ 2. capillary specialized for resorption

_____ 3. cuplike part of the renal corpuscle

_____ 4. location of macula densa

_____ 5. primary site of tubular resorption

_____ 6. receives urine from many nephrons

9. Explain *why* the glomerulus is such a high-pressure capillary bed. _____

_____

10. What structural modification of certain tubule cells enhances their ability to resorb substances from the filtrate?

_____

11. Explain the mechanism of tubular secretion, and explain its importance in the urine formation process. _____

_____

_____

12. Compare and contrast the composition of blood plasma and glomerular filtrate. _____

_____

13. Describe the role of the juxtaglomerular complex. _____

_____

14. Label the drawing of the nephron using the key letters of the correct terms.

Key: a. granular cells

   b. cuboidal epithelium

   c. macula densa

   d. glomerular capsule (parietal layer)

   e. ascending limb of nephron loop

   f. glomerulus

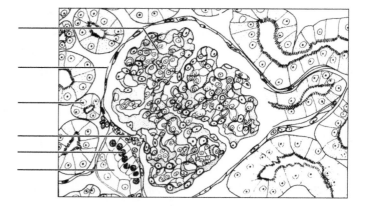

Dissection and Identification: The Urinary System of the Cat

15. How does the position of the kidneys in the cat differ from their position in humans? _____

_____

16. How does the site of urethral emptying in the female cat differ from its termination point in the human female?

_____

_____

17. What gland encircles the neck of the bladder in the male? _____

Is this part of the urinary system? _____ What is its function? _____

_____

18. ✚ A urinary tract infection (UTI) is an infection of any of the urinary tract structures: kidneys, ureter, bladder, or urethra. Considering the differences in the male and female anatomy and that the bacteria that cause UTIs are often found in the feces,

explain why females are more likely to contract a UTI. _____

_____

_____

19. ✚ Acute glomerulonephritis (GN) is sudden inflammation of the glomeruli, most commonly a result of a streptococcal infection that causes the body to attack its own tissue. The damage to the filtration membrane increases permeability of the membrane for

proteins and larger components. Describe the abnormal components of the urine for patients with GN. _____

_____

Anatomy of the Reproductive System

MATERIALS

- Three-dimensional models or large laboratory charts of the male and female reproductive tracts
- Prepared slides of sperm and of cross sections of the testis, penis, epididymis, seminal glands, uterine tube, uterus showing endometrium (secretory phase), and ovary
- Compound microscope
- Immersion oil
- Dissection animal, tray, and instruments
- Disposable gloves
- Safety glasses
- Embalming fluid
- Bone cutters
- Small metric rulers
- Organic debris container

LEARNING OUTCOMES

- ☐ Discuss the general function of the reproductive system.
- ☐ Identify the structures of the male and female reproductive systems on an appropriate model or image, and list the general function of each.
- ☐ Define *semen,* state its composition, and name the organs involved in its production.
- ☐ Trace the pathway followed by a sperm from its site of formation to the external environment.
- ☐ Define *erection* and *ejaculation*.
- ☐ Define *gonad.* Name the gametes and hormones of the testes and ovaries, indicating the cell types or structures responsible for their production.
- ☐ Describe the microscopic structure of the penis, seminal glands, epididymis, uterine wall, and uterine tube, and relate structure to function.
- ☐ Explain the role that the fimbriae and ciliated epithelium of the uterine tubes play in moving the egg from the ovary to the uterus.
- ☐ Identify the fundus, body, and cervical regions of the uterus.
- ☐ Define *endometrium, myometrium,* and *ovulation*.
- ☐ Describe the anatomy and discuss the reproduction-related function of female mammary glands.
- ☐ Identify the major reproductive structures of the male and female dissection animal.

PRE-LAB QUIZ

1. The essential organs of reproduction are the _____, which produce the sex cells.
   a. accessory male glands   c. seminal glands
   b. gonads   d. uterus
2. Circle the correct underlined term. The paired oval testes lie in the <u>scrotum</u> / <u>prostate</u> outside the abdominopelvic cavity, where they are kept slightly cooler than body temperature.
3. After sperm are produced, they enter the first part of the duct system, the:
   a. ductus deferens   c. epididymis
   b. ejaculatory duct   d. urethra
4. Circle the correct underlined term. The <u>acrosome</u> / <u>midpiece</u> of the sperm contains enzymes involved in the penetration of the egg.
5. The prostate, seminal glands, and bulbo-urethral glands produce

   _____.
   a. seminal fluid   c. urine
   b. testosterone   d. water
6. Circle the correct underlined term. The <u>interstitial endocrine cells</u> / <u>seminiferous tubules</u> produce testosterone, the hormonal product of the testis.

Text continues on next page.

7. The endocrine products of the ovaries are estrogen and:
   a. luteinizing hormone   c. prolactin
   b. progesterone   d. testosterone

8. The _____ is a pear-shaped organ that houses the embryo or fetus during its development.
   a. bladder   c. uterus
   b. cervix   d. vagina

9. Circle the correct underlined term. The endometrium / myometrium, or the thick mucosal lining of the uterus, has a superficial layer that sloughs off periodically.

10. Circle the correct underlined term. A developing egg is ejected from the ovary at the appropriate stage of maturity in an event known as menstruation / ovulation.

M ost organ systems of the body function from the time they are formed. However, the **reproductive system** begins its biological function, the production of off-spring, at puberty.

The essential organs of reproduction are the **gonads,** the testes and the ovaries, which produce the sex cells, or **gametes,** and the sex hormones. The reproductive role of the male is to manufacture sperm and to deliver them to the female reproductive tract. The female, in turn, produces eggs. The combination of sperm and egg produces a fertilized egg, which is the first cell of a new individual. Once fertilization has occurred, the female uterus provides a nurturing, protective environment in which the embryo, later called the *fetus,* develops until birth.

Gross Anatomy of the Human Male Reproductive System

The primary reproductive organs of the male are the **testes,** which produce sperm and the male sex hormones. All other reproductive structures are ducts or sources of secretions that help deliver the sperm safely to the body exterior or female reproductive tract.

ACTIVITY 1

Identifying Male Reproductive Organs

As the following organs and structures are described, locate them on **Figure 29.1** and then identify them on a three-dimensional model of the male reproductive system or on a large laboratory chart. ■

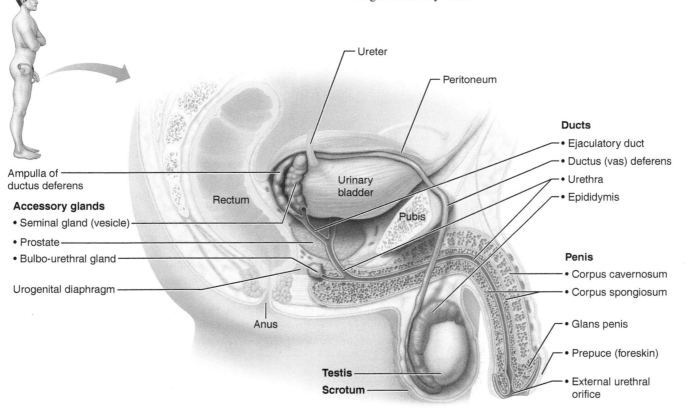

(a) Sagittal view

Figure 29.1 Reproductive organs of the human male. (a) Sagittal view.

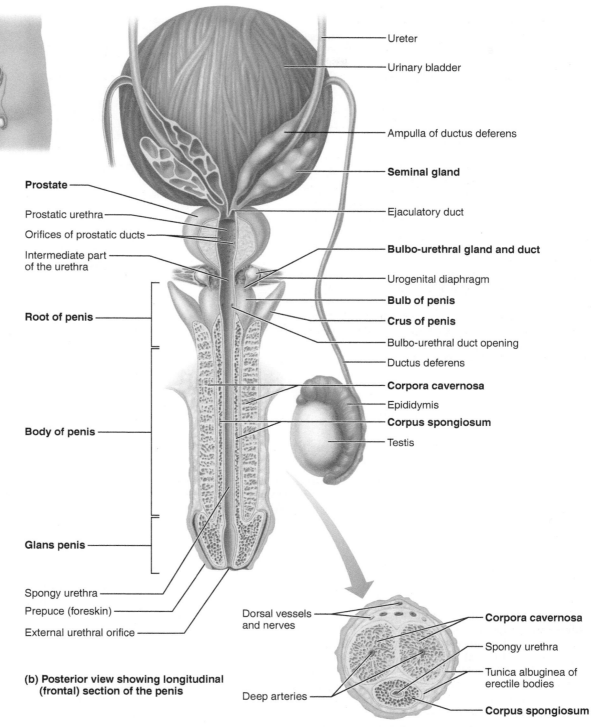

Ureter
Urinary bladder
Ampulla of ductus deferens
Seminal gland
Ejaculatory duct
Bulbo-urethral gland and duct
Urogenital diaphragm
Bulb of penis
Crus of penis
Bulbo-urethral duct opening
Ductus deferens
Corpora cavernosa
Epididymis
Corpus spongiosum
Testis

Prostate
Prostatic urethra
Orifices of prostatic ducts
Intermediate part of the urethra
Root of penis
Body of penis
Glans penis
Spongy urethra
Prepuce (foreskin)
External urethral orifice

(b) Posterior view showing longitudinal (frontal) section of the penis

Dorsal vessels and nerves
Corpora cavernosa
Spongy urethra
Tunica albuginea of erectile bodies
Corpus spongiosum
Deep arteries

(c) Cross section of the penis with dorsal surface at top

Figure 29.1 (*continued*) **(b)** Posterior view showing a longitudinal section of the penis. **(c)** Transverse section of the penis.

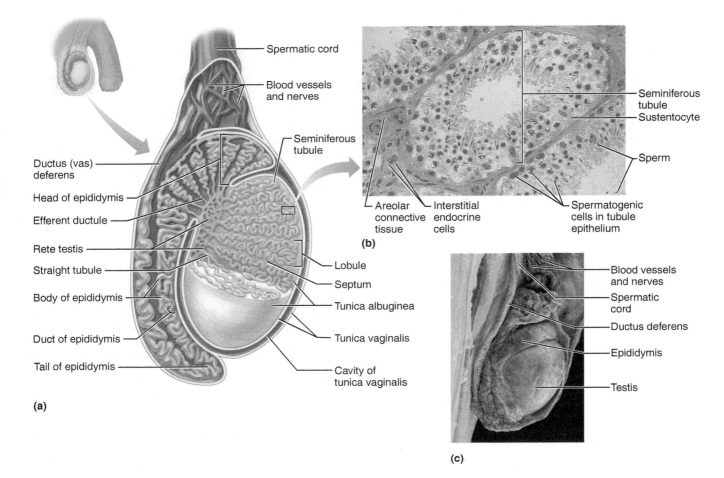

Figure 29.2 **Structure of the testis. (a)** Partial sagittal section of the testis and associated epididymis. **(b).** Photomicrograph of an active seminiferous tubule (215×). **(c)** External view of a testis from a cadaver; same orientation as in (a).

The paired oval testes lie in the **scrotum** outside the abdominopelvic cavity. The temperature there (approximately 94°F, or 34°C) is slightly lower than body temperature, a requirement for producing viable sperm.

The accessory structures forming the *duct system* are the epididymis, the ductus deferens, the ejaculatory duct, and the urethra. The **epididymis** is an elongated structure running up the posterior and lateral aspect of the testis and capping its superior aspect. The epididymis forms the first portion of the duct system and provides a site where immature sperm entering it from the testis complete their maturation process. The **ductus deferens,** or **vas deferens** (sperm duct), arches superiorly from the epididymis, passes through the inguinal canal into the pelvic cavity, and courses over the superior aspect of the urinary bladder. In life, the ductus deferens is enclosed along with blood vessels and nerves in a connective tissue sheath called the **spermatic cord (Figure 29.2)**. The terminus of the ductus deferens enlarges to form the region called the **ampulla,** which empties into the **ejaculatory duct.** During **ejaculation,** contraction of the ejaculatory duct propels the sperm through the prostate to the **prostatic urethra,** which in turn empties into the **intermediate part of the urethra** and then into the **spongy urethra,** which runs through the length of the penis to the body exterior.

The spermatic cord is easily palpated through the skin of the scrotum. When a *vasectomy* is performed, a small incision is made in each side of the scrotum, and each ductus deferens is cut through or cauterized. Although sperm are still produced, they can no longer reach the body exterior; thus a man is sterile after this procedure.

The *accessory glands* include the prostate, the paired seminal glands, and the bulbo-urethral glands. These glands produce **seminal fluid,** the liquid medium in which sperm leave the body. The location and secretion of the accessory glands are summarized in **Table 29.1**.

Semen consists of sperm and seminal fluid. Seminal fluid is overall alkaline, which buffers the sperm against the acidity of the female vagina.

The **penis,** part of the external genitalia of the male along with the scrotal sac, is the copulatory organ of the male. Designed to deliver sperm into the female reproductive tract, the penis consists of a **body,** or *shaft,* which terminates in an enlarged tip, the **glans penis** (see Figure 29.1a and b). The skin covering the penis is loosely applied, and it reflects

Table 29.1   Accessory Glands of the Male Reproductive System (Figure 29.1)

Accessory gland	Location	Secretion
Seminal glands	Paired glands located posterior to the urinary bladder. The duct of each gland merges with a ductus deferens to form the ejaculatory duct.	A thick, light yellow, alkaline secretion containing fructose and citric acid, which nourish the sperm, and prostaglandins for enhanced sperm motility. Its secretion has the largest contribution to the volume of semen.
Prostate	Single gland that encircles the prostatic urethra inferior to the bladder.	A milky, slightly acidic fluid that contains citric acid, several enzymes, and prostate-specific antigen (PSA). Its secretion plays a role in activating the sperm.
Bulbo-urethral glands	Paired tiny glands that drain into the intermediate part of the urethra.	A clear alkaline mucus that lubricates the tip of the penis for copulation and neutralizes traces of acidic urine in the urethra prior to ejaculation.

downward to form a circular fold of skin, the **prepuce,** or **foreskin,** around the proximal end of the glans. The foreskin may be removed in the surgical procedure called *circumcision*. Internally, the penis consists primarily of three elongated cylinders of erectile tissue, which engorge with blood during sexual excitement. This causes the penis to become rigid and enlarged so that it may more adequately serve as a penetrating device. This event is called **erection.** The paired dorsal cylinders are the **corpora cavernosa.** The single ventral **corpus spongiosum** surrounds the spongy urethra (Figure 29.1c).

WHY THIS **MATTERS** | Benign Prostatic Hyperplasia (BPH)

Benign has become synonymous with "noncancerous," and *hyperplasia* means an increase in the number of cells. In benign prostatic hyperplasia (BPH), an increased number of cells results in an enlarged prostate. As men age, normal hormonal changes can cause an increased proliferation of cells, which can enlarge the prostate and lead to urinary dysfunction. An enlarged prostate can also be a result of prostate cancer. Previously, an elevated blood level of prostate-specific antigen (PSA), a secretion of the prostate, was correlated to prostate cancer. However, research has shown that elevated PSA levels can have other causes, including BPH. A more accurate predictor of prostate cancer appears to be regular monitoring of the PSA level over time to see whether it continues to rise. ∎

ACTIVITY 2

Conducting a Microscopic Study of Selected Male Reproductive Organs

Testis

Each testis is covered by a dense connective tissue capsule called the **tunica albuginea** (literally, "white tunic"). Extensions of this sheath enter the testis, dividing it into a number of lobules, each of which houses one to four highly coiled

seminiferous tubules, the sperm-forming factories (Figure 29.2). The seminiferous tubules of each lobule converge to empty the sperm into a **straight tubule,** which in turn delivers its contents into another set of tubules, the **rete testis,** at the posterior side of the testis. Sperm traveling through the rete testis then enter the epididymis, located on the exterior aspect of the testis, as previously described. Lying between the seminiferous tubules and softly padded with connective tissue are the **interstitial endocrine cells,** which produce testosterone, the hormonal product of the testis.

1.   Obtain a slide of the testis and a microscope. Examine the slide under low power to identify the cross-sectional views of the cut seminiferous tubules. Then rotate the high-power lens into position, and observe the wall of one of the cut tubules. As you work, refer to Figure 29.2b to make the following identifications.

2.   Scrutinize the cells at the periphery of the tubule. The cells in this area are the **spermatogonia,** which undergo frequent mitoses to increase their number and maintain their population. About half of the spermatogonia's "offspring" become **primary spermatocytes,** spermatogenic cells that undergo meiosis, which leads to the formation of spermatids having half the usual genetic composition. The remaining daughter cells resulting from mitotic divisions of spermatogonia remain at the tubule periphery to maintain the germ cell line.

3.   Observe the cells in the middle of the tubule wall. There you should see a large number of spermatocytes that are obviously undergoing a nuclear division process. Look for coarse clumps of chromatin, or threadlike chromosomes, which have the appearance of coiled springs.

4.   Examine the cells at the tubule lumen. Identify the small round-nucleated spermatids, many of which may appear lopsided and look as though they are starting to lose their cytoplasm. See if you can find a spermatid embedded in an elongated cell type, a **sustentocyte,** or *Sertoli cell,* which extends inward from the periphery of the tubule. The sustentocytes nourish the spermatids as they begin their transformation into sperm. Also in the adluminal area, locate immature sperm, which can be identified by their tails. The sperm develop directly from the spermatids by the loss of extraneous cytoplasm and the development of a propulsive tail.

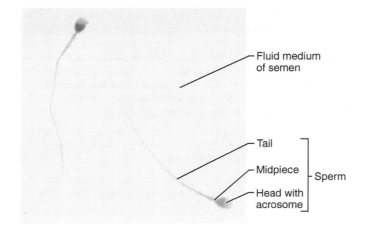

Figure 29.3 **Sperm in semen.** In addition to sperm, semen contains fluids secreted by the accessory glands (1160×)

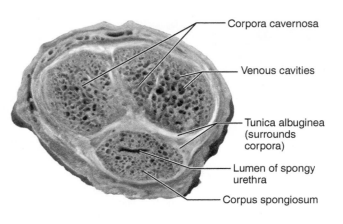

Figure 29.4 **Transverse section of the penis (2×)**

5.   Identify the testosterone-producing **interstitial endocrine cells** lying external to and between the seminiferous tubules. The interstitial endocrine cells produce testosterone, which stimulates sperm production.

6.   Obtain a prepared slide of human sperm, and view it with the oil immersion lens. Compare it to **Figure 29.3**. Identify the head (essentially the nucleus of the spermatid), acrosome, and tail regions. (The acrosome, which caps the nucleus anteriorly, contains enzymes involved in sperm penetration of the egg.)

Penis

Obtain a cross section of the penis. Scan the tissue under low power to identify the urethra and the cavernous bodies. Compare your observations to Figure 29.1c and **Figure 29.4**. Observe the lumen of the urethra carefully. What type of epithelium do you see?

_____

Explain the function of this type of epithelium.

_____

_____

_____

Seminal Gland

Obtain a slide showing a cross-sectional view of the seminal gland. Examine the slide at low magnification to get an overall view of the highly folded mucosa of this gland. Switch to higher magnification, and notice that the folds of the gland protrude into the lumen where they divide further, giving the lumen a honeycomb look **(Figure 29.5)**. Notice that the loose connective tissue lamina propria is underlain by smooth muscle fibers—first a circular layer, and then a longitudinal layer. Identify the glandular secretion in the lumen, a viscous substance that is rich in fructose and prostaglandins.

Epididymis

Obtain a cross section of the epididymis. Notice the abundant tubule cross sections resulting from the fact that the coiling epididymis tubule has been cut through many times in the specimen. Using Figure 29.2b and **Figure 29.6** as guides, look for sperm in the lumen of the tubule. Examine the composition of the tubule wall carefully. Identify the *stereocilia* of the pseudostratified columnar epithelial lining. These nonmotile microvilli absorb excess fluid and pass nutrients to the sperm in the lumen.

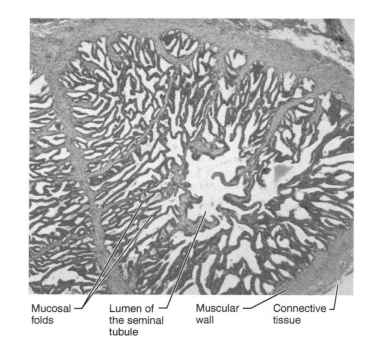

Mucosal folds | Lumen of the seminal tubule | Muscular wall | Connective tissue

Figure 29.5 **Histology of a seminal gland.** Cross-sectional view showing the elaborate network of mucosal folds (12×). The glandular secretion is seen in the lumen.

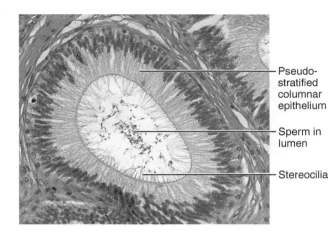

Pseudo-
stratified
columnar
epithelium

Sperm in
lumen

Stereocilia

Figure 29.6 Cross section of epididymis (185×)

Now identify the smooth muscle layer. What do you think the function of the smooth muscle is?

_____

_____

_____ ▬

Gross Anatomy of the Human Female Reproductive System

The **ovaries** are the primary reproductive organs of the female. Like the testes of the male, the ovaries produce gametes (eggs, or ova) and also sex hormones (estrogens and progesterone). The other accessory structures of the female reproductive system transport, house, nurture, or otherwise serve the needs of the reproductive cells and/or the developing fetus.

ACTIVITY 3

Identifying Female Reproductive Organs

As you read the descriptions of these structures, locate them on **Figure 29.7** and **Figure 29.8** and then on the female reproductive system model or large laboratory chart. ▬

External Genitalia

The **external genitalia (vulva)** consist of the mons pubis, the labia majora and minora, the clitoris, the external urethral and vaginal orifices, the hymen, and the greater vestibular glands. **Table 29.2** summarizes the structures of the female external genitalia (see also Figure 29.7). The diamond-shaped region between the anterior end of the labial folds, the ischial tuberosities laterally, and the anus posteriorly is called the **perineum.**

Internal Organs

The internal female organs include the vagina, uterus, uterine tubes, ovaries, and the ligaments and supporting structures

that suspend these organs in the pelvic cavity (Figure 29.8). The **vagina** extends for approximately 10 cm (4 inches) from the vestibule to the uterus superiorly. It serves as a copulatory organ because it receives the penis (and semen) during sexual intercourse. The vagina also provides a passageway for delivery of an infant and for menstrual flow. The pear-shaped **uterus,** situated between the bladder and the rectum, is a muscular organ with its narrow end, the **cervix,** directed inferiorly. The major portion of the uterus is referred to as the **body;** its superior rounded region above the entrance of the uterine tubes is called the **fundus.** A fertilized egg is implanted in the uterus, which houses the embryo or fetus during its development.

In some cases, the fertilized egg may implant in a uterine tube or even on the abdominal viscera, creating an **ectopic pregnancy.** Such implantations are usually unsuccessful and may even endanger the mother's life because the uterine tubes cannot accommodate the increasing size of the fetus. ✚

The **endometrium,** the thick mucosal lining of the uterus, has a superficial **functional layer,** or **stratum functionalis,** that sloughs off periodically (about every 28 days) in response to cyclic changes in the levels of ovarian hormones in the woman's blood. This sloughing-off process, which is accompanied by bleeding, is referred to as **menstruation,** or **menses.** The deeper **basal layer,** or **stratum basalis,** forms a new functional layer after menstruation ends.

The **uterine,** or **fallopian, tubes** are about 10 cm (4 inches) long and extend from the ovaries in the peritoneal cavity to the superolateral region of the uterus. The distal ends of the tubes are funnel-shaped and have fingerlike projections called **fimbriae.** Unlike the male duct system, there is no actual contact between the female gonad and the initial part of the female duct system—the uterine tube.

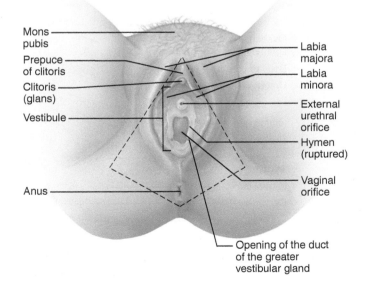

Mons pubis

Prepuce of clitoris

Clitoris (glans)

Vestibule

Anus

Labia majora

Labia minora

External urethral orifice

Hymen (ruptured)

Vaginal orifice

Opening of the duct of the greater vestibular gland

Figure 29.7 External genitalia (vulva) of the human female. (The region enclosed by dashed lines is the perineum.)

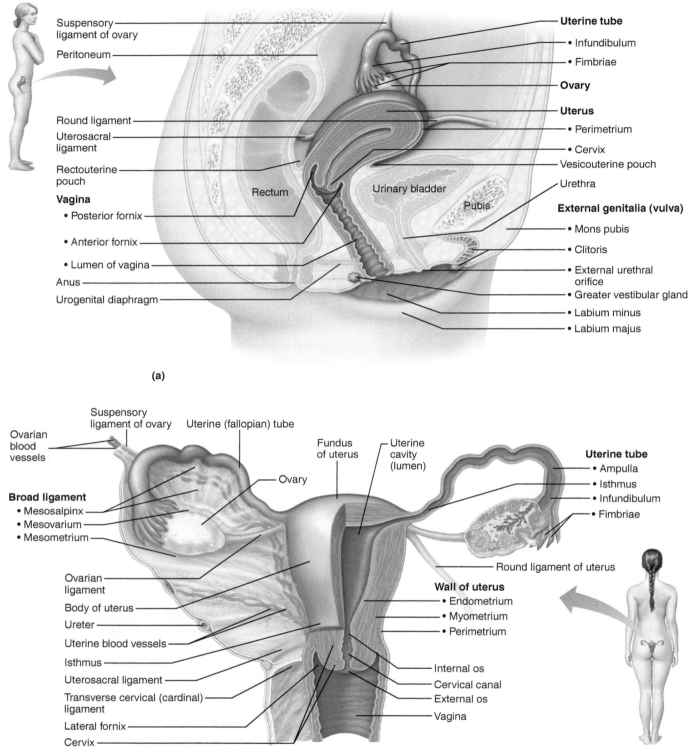

(a)

(b)

Figure 29.8 Internal reproductive organs of the human female. (a) Midsagittal section of the human female reproductive system. **(b)** Posterior view. The posterior walls of the vagina, uterus, uterine tubes as well as the broad ligament, have been removed on the right side to reveal the shape of the lumen of these organs.

Table 29.2   External Genitalia (Vulva) of the Human Female (Figure 29.7)

Structure	Description
Mons pubis	Rounded fatty eminence that cushions the pubic symphysis; covered with coarse pubic hair after puberty.
Labia majora (singular: *labium majus*)	Two elongated hair-covered skin folds that extend from the mons pubis. They contain sebaceous glands, apocrine glands, and adipose tissue. They are homologous to the scrotum.
Labia minora (singular: *labium minus*)	Two smaller folds located medial to the labia majora. They don't have hair or adipose tissue but they do have many sebaceous glands.
Vestibule	Region located between the two labia minora. From anterior to posterior, it contains the clitoris, the external urethral orifice, and the vaginal orifice.
Clitoris	Small mass of erectile tissue located where the labia minora meet anteriorly. It is homologous to the penis.
Prepuce of the clitoris	Skin folds formed by the union of the labia minora; they serve to hood the clitoris.
External urethral orifice	Serves as the outlet for the urinary system. It has no reproductive function in the female.
Hymen	A thin fold of vascular mucous membrane that may partially cover the vaginal opening.
Greater vestibular glands	Pea-sized mucus-secreting glands located on either side of the hymen. They lubricate the distal end of the vagina during coitus. They are homologous to the bulbo-urethral glands of males.

Because of this open passageway between the female reproductive organs and the peritoneal cavity, reproductive system infections, such as gonorrhea and other **sexually transmitted infections (STIs), also called sexually transmitted diseases (STDs),** can cause widespread inflammations of the pelvic viscera, a condition called **pelvic inflammatory disease (PID).** ✚

The internal female organs are all retroperitoneal, except the ovaries. They are supported and suspended somewhat freely by ligamentous folds of peritoneum. The supporting structures for the uterus, uterine tubes, and ovaries are summarized in **Table 29.3**.

Within the ovaries, the female gametes, or eggs, begin their development in saclike structures called *follicles*. The growing follicles also produce *estrogens*. When a developing egg has reached the appropriate stage of maturity, it is ejected from the ovary in an event called **ovulation.** The ruptured follicle is then converted to a second type of endocrine structure called a *corpus luteum,* which secretes progesterone and some estrogens.

The flattened almond-shaped ovaries lie adjacent to the uterine tubes but are not connected to them; consequently, an ovulated "egg," actually a secondary oocyte, enters the pelvic cavity. The waving fimbriae of the uterine tubes create fluid currents that, if successful, draw the egg into the lumen of the uterine tube. There the egg begins its passage to the uterus, propelled by the cilia of the tubal walls. The usual and most desirable site of fertilization is the uterine tube, because the journey to the uterus takes about 3 to 4 days and an egg is viable for up to 24 hours after it is expelled from the ovary. Thus, sperm must swim upward through the vagina and uterus and into the uterine tubes to reach the egg. This is an arduous journey, because they must swim against the downward current created by ciliary action—rather like swimming upstream!

Table 29.3   Supporting Structures for the Uterus, Uterine Tubes, and Ovaries (Figure 29.8)

Structure	Description
Broad ligament	Fold of the peritoneum that drapes over the superior uterus to enclose the uterus and uterine tubes and anchors them to the lateral body walls
Mesometrium	Portion of the broad ligament that supports the uterus laterally (mesentery of the uterus)
Round ligaments	Anchor the uterus to the anterior pelvic wall by descending through the mesometrium and the inguinal canal; attach to the skin of one of the labia majora
Uterosacral ligaments	Secure the inferior uterus to the sacrum posteriorly
Cardinal (transverse cervical) ligaments	Connect the cervix and vagina to the pelvic wall laterally
Mesosalpinx	Portion of the broad ligament that anchors the uterine tube (mesentery of the uterine tube)
Mesovarium	Posterior fold of the broad ligament that supports the ovaries (mesentery of the ovaries)
Suspensory ligaments	A lateral continuation of the broad ligament that attaches the ovaries to the lateral pelvic wall
Ovarian ligaments	Anchors the ovaries to the uterus medially and is enclosed within the broad ligament

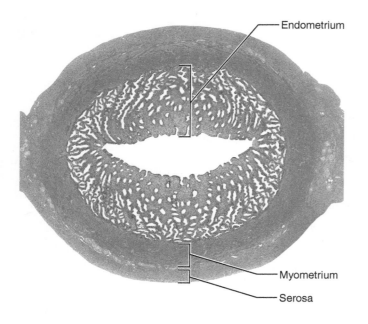

Figure 29.9 **Cross-sectional view of the uterine wall.**
The mucosa is in the secretory stage (2×).

ACTIVITY 4

Conducting a Microscopic Study of Selected Female Reproductive Organs

Wall of the Uterus

Obtain a cross-sectional view of the uterine wall. Identify the three layers of the uterine wall—the endometrium, myometrium, and serosa. Also identify the two strata of the **endometrium:** the functional layer and the basal layer. The latter forms a new functional layer each month. **Figure 29.9,** a photomicrograph that includes the secretory endometrium, will help in this study.

As you study the slide, notice that the bundles of smooth muscle are oriented in several different directions. What is the function of the **myometrium** (smooth muscle layer) during childbirth?

_____

_____

Uterine Tube

Obtain a prepared slide of a cross-sectional view of a uterine tube for examination. Notice that the mucosal folds nearly fill the tubule lumen **(Figure 29.10).** Then switch to high power to examine the ciliated secretory epithelium.

Ovary

Because many different stages of ovarian development exist within the ovary at any one time, a single microscopic preparation will contain follicles at many different stages of development **(Figure 29.11).** Obtain a cross section of ovary tissue, and identify the following structures.

Surface epithelium: Outermost layer of the ovary.

Primary follicle: One or a few layers of cuboidal cells surrounding the large central developing ovum, or **oocyte,** the immature egg.

Secondary follicles: Follicles consisting of several layers of granulosa cells surrounding the central developing ovum and beginning to show evidence of fluid accumulation and **antrum** (central cavity) formation.

Mature ovarian follicle: At this stage of development, the follicle has a large antrum containing fluid produced by the granulosa cells. The developing ovum is pushed to one side of the follicle and is surrounded by a capsule of several layers of granulosa cells called the **corona radiata** (radiating crown). When the immature ovum (secondary oocyte) is released, it enters the uterine tubes with its corona radiata intact. The connective tissue adjacent to the mature follicle forms a capsule that encloses the follicle and is called the **follicular theca.**

Corpus luteum: A solid glandular structure or a structure containing a scalloped lumen that develops from the ovulated follicle **(Figure 29.12).** ■

The Mammary Glands

The **mammary glands** exist within the breasts in both sexes, but they normally have a reproduction-related function only in females. Because the function of the mammary glands is to produce milk to nourish the newborn infant, their importance is more closely associated with events that occur when reproduction has already been accomplished. Periodic stimulation by the female sex hormones, especially estrogens, increases the size of the female mammary glands at puberty. During this period, the duct system becomes more elaborate, and fat is deposited—fat deposition is the more important contributor to increased breast size.

The rounded, skin-covered mammary glands lie anterior to the pectoral muscles of the thorax, attached to them by connective tissue. Slightly below the center of each breast is a pigmented area, the **areola,** which surrounds a centrally protruding nipple **(Figure 29.13).**

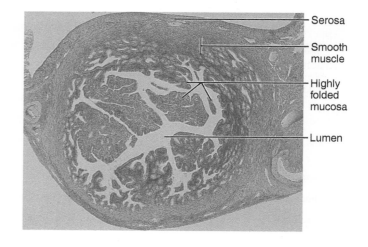

Figure 29.10 **Cross-sectional view of the uterine tube** **(10×).**

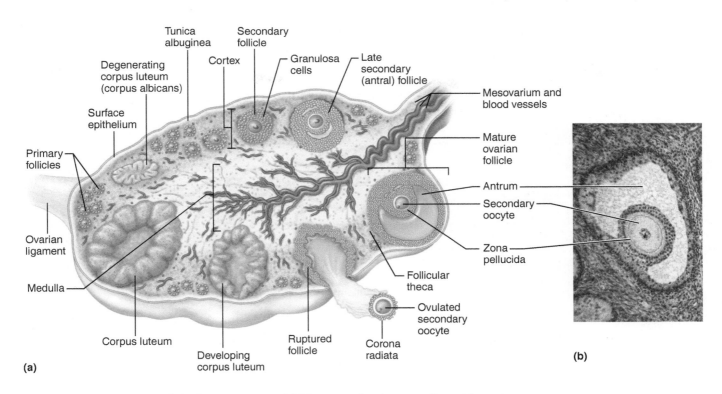

Figure 29.11 Anatomy of the human ovary. (a) The ovary has been sectioned to reveal the follicles in its interior. Note that not all the structures would appear in the ovary at the same time. **(b)** Photomicrograph of a mature ovarian follicle (115×).

Internally, each mammary gland consists of 15 to 25 **lobes** that radiate around the nipple and are separated by fibrous connective tissue and fat. Within each lobe are smaller chambers called **lobules,** containing the glandular **alveoli** that produce milk during lactation. The alveoli of each lobule pass the milk into a number of **lactiferous ducts,** which join to form an expanded storage chamber, the **lactiferous sinus,** as they approach the nipple. The sinuses open to the outside at the nipple. The description of mammary glands that we have just given applies only to nursing women or women in the last trimester of pregnancy.

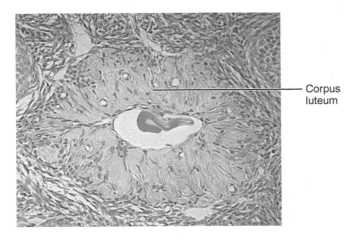

Figure 29.12 Glandular corpus luteum of an ovary. (170×).

DISSECTION AND IDENTIFICATION
The Reproductive System of the Cat

Don gloves and safety glasses, and obtain your cat, a dissection tray, and the necessary dissecting instruments. After you have completed the study of the reproductive structures of your specimen, observe a cat of the opposite sex. The following instructions assume that the abdominal cavity has been opened in previous dissection exercises. ■

ACTIVITY 5

Identifying Organs of the Male Reproductive System

Identify the male reproductive structures (refer to **Figure 29.14**).

1. Identify the **penis,** and notice the prepuce covering the glans penis. Carefully cut through the skin overlying the penis to expose the cavernous tissue beneath, then cross section the penis to see the relative positioning of the three cavernous bodies.

2. Identify the **scrotum,** and then carefully make a shallow incision through the scrotum to expose the **testes.** Notice that the scrotum is divided internally.

3. Lateral to the medial aspect of the scrotal sac, locate the **spermatic cord,** which contains the spermatic artery, vein, and nerve, as well as the ductus deferens, and follow it up through the inguinal canal into the abdominal cavity. It is not necessary to cut through the pubis; a slight tug on the spermatic cord in the scrotal sac region will reveal its position

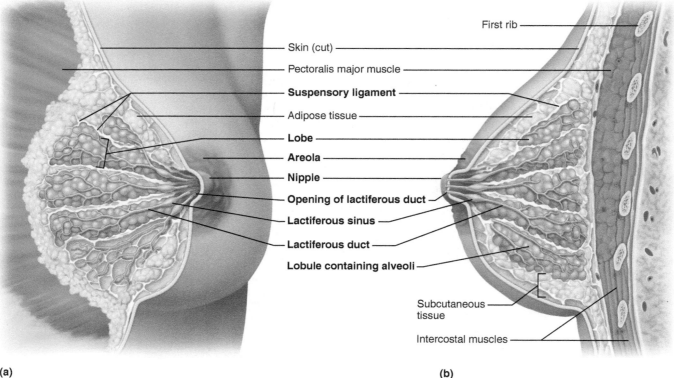

First rib

Skin (cut)

Pectoralis major muscle

Suspensory ligament

Adipose tissue

Lobe

Areola

Nipple

Opening of lactiferous duct

Lactiferous sinus

Lactiferous duct

Lobule containing alveoli

Subcutaneous tissue

Intercostal muscles

(a)   (b)

Figure 29.13 Anatomy of lactating mammary gland. (a) Anterior view of partially dissected breast. **(b)** Sagittal section of the breast.

in the abdominal cavity. Carefully loosen the spermatic cord from the connective tissue investing it, and follow its course as it travels superiorly in the pelvic cavity. Then, follow the **ductus deferens** as it loops over the ureter and then courses posterior to the bladder and enters the prostate. Using bone cutters, carefully cut through the pubic symphysis to follow the urethra.

4. Notice that the **prostate,** an enlarged whitish mass abutting the urethra, is comparatively smaller in the cat than in the human, and it is more distal to the bladder. In the human, the prostate is immediately adjacent to the base of the bladder. Carefully slit open the prostate to follow the **ductus deferens** to the urethra, which exits from the bladder midline. The male cat urethra, like that of the human, serves as both a urinary and a sperm duct. In the human, the ductus deferens is joined by the duct of the seminal gland to form the ejaculatory duct, which enters the prostate. Seminal glands are not present in the cat.

5. Trace the **urethra** to the proximal ends of the cavernous tissues of the penis, each of which is anchored to the ischium by a band of connective tissue called the **crus** of the penis. The crus is covered ventrally by the ischiocavernosus muscle, and the **bulbo-urethral gland** lies beneath it (Figure 29.14a).

6. Once again, turn your attention to a testis. Cut it from its attachment to the spermatic cord, and carefully slit open the **tunica vaginalis** capsule enclosing it. Identify the **epididymis**

running along one side of the testis. Make a longitudinal cut through the testis and epididymis. Can you see the tubular nature of the epididymis and the rete testis portion of the testis with the unaided eye? ■

ACTIVITY 6

Identifying Organs of the Female Reproductive System

Identify the female reproductive structures (refer to **Figure 29.15**).

1. Unlike the pear-shaped simplex, or one-part, uterus of the human, the uterus of the cat is Y-shaped (bipartite or bicornuate) and consists of a **uterine body** from which two **uterine horns** (cornua) diverge. Such an enlarged uterus enables the animal to produce litters. Examine the abdominal cavity, and identify the bladder and the body of the uterus lying just dorsal to it.

2. Follow one of the uterine horns as it travels superiorly in the body cavity. Identify the thin mesentery (the *broad ligament*), which helps anchor it and other reproductive structures to the body wall. Approximately halfway up the length of the uterine horn, it should be possible to identify the more important *round ligament,* a cord of connective tissue extending laterally and posteriorly from the uterine horn to the region of the body wall that would correspond to the inguinal region of the male.

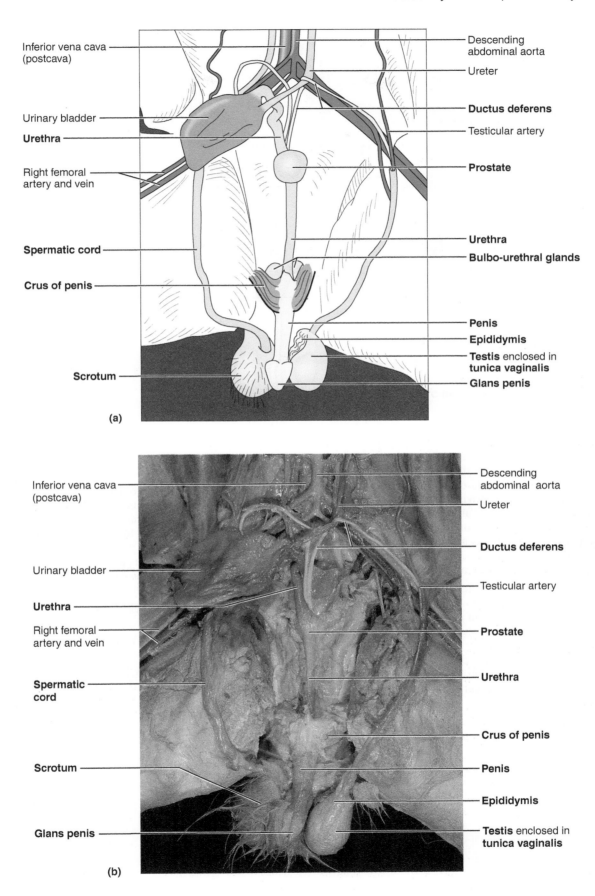

Inferior vena cava (postcava)

Urinary bladder

Urethra

Right femoral artery and vein

Spermatic cord

Crus of penis

Scrotum

(a)

Descending abdominal aorta

Ureter

Ductus deferens

Testicular artery

Prostate

Urethra

Bulbo-urethral glands

Penis

Epididymis

Testis enclosed in **tunica vaginalis**

Glans penis

Inferior vena cava (postcava)

Urinary bladder

Urethra

Right femoral artery and vein

Spermatic cord

Scrotum

Glans penis

(b)

Descending abdominal aorta

Ureter

Ductus deferens

Testicular artery

Prostate

Urethra

Crus of penis

Penis

Epididymis

Testis enclosed in **tunica vaginalis**

Figure 29.14 Reproductive system of the male cat. (a) Diagram. **(b)** Photograph.

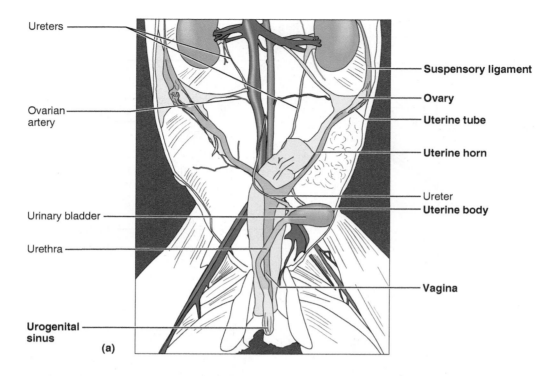

Ureters

Suspensory ligament

Ovary

Uterine tube

Ovarian
artery

Uterine horn

Ureter

Uterine body

Urinary bladder

Urethra

Vagina

Urogenital
sinus

(a)

Figure 29.15 Reproductive system of the female cat. (a) Diagram.

3. Examine the **uterine tube** and **ovary** at the distal end of the uterine horn just caudal to the kidney. Observe how the funnel-shaped end of the uterine tube curves around the ovary. As in the human, the distal end of the tube is fimbriated, or fringed, and the tube is lined with ciliated epithelium. The uterine tubes of the cat are tiny and much shorter than in the human. Identify the **ovarian ligament,** a short thick cord that extends from the uterus to the ovary and anchors the ovary to the body wall. Also observe the *ovarian artery* and *vein* passing through the mesentery to the ovary and uterine structures.

4. Return to the body of the uterus, and follow it caudally to the pelvis. Use bone cutters to cut through the pubic symphysis, cutting carefully so you do not damage the urethra deep to it. Expose the pelvic region by pressing the thighs dorsally. Follow the uterine body caudally to the vagina, and note the point where the urethra draining the bladder and the **vagina** enter a common chamber, the **urogenital sinus.** How does this anatomical arrangement compare to that seen in the human female?

_____

_____

5. On the cat's exterior, observe the **vulva,** which is similar to the human vulva. Identify the slim **labia majora** surrounding the urogenital opening.

6. To determine the length of the vagina, which is difficult to ascertain by external inspection, slit through the vaginal wall just superior to the urogenital sinus, and cut toward the body of the uterus with scissors. Reflect the cut edges, and identify the muscular cervix of the uterus. Measure the distance between the urogenital sinus and the cervix. Approximately how long is the vagina of the cat?

_____

7. To complete this exercise, observe a cat of the opposite sex.

8. Before you leave the laboratory, follow the boxed instructions to prepare your cat for storage and clean the area (page 219). ▪

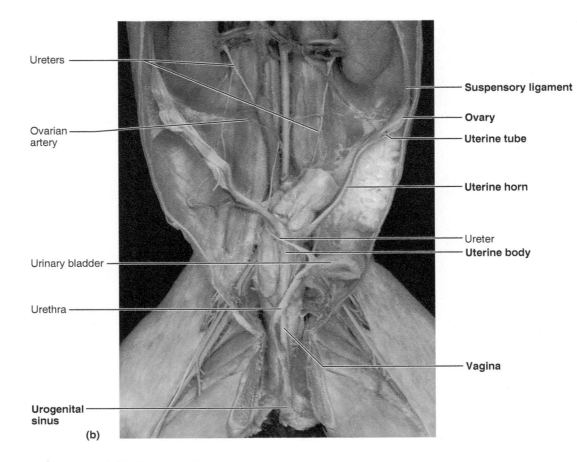

Ureters

Ovarian artery

Urinary bladder

Urethra

Urogenital sinus

(b)

Suspensory ligament

Ovary

Uterine tube

Uterine horn

Ureter

Uterine body

Vagina

Figure 29.15 *(continued)* **(b)** Photograph.

Anatomy of the Reproductive System

Gross Anatomy of the Human Male Reproductive System

1. List the two main functions of the testis: _____

and _____

2. Identify all indicated structures or portions of structures on the photo of the model of the male reproductive system below.

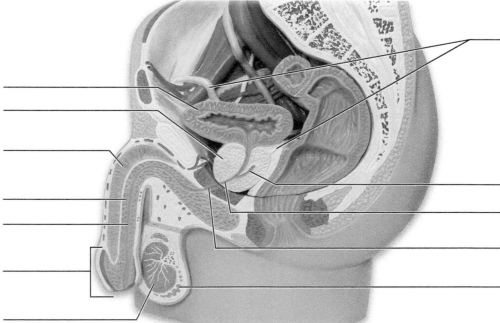

3. Why are the testes located in the scrotum rather than inside the ventral body cavity? _____

_____

WHY THIS MATTERS **4.** A screening tool for benign prostatic hyperplasia and prostate cancer includes palpation of the prostate. Explain

how this is accomplished. _____

_____

5. Would you expect a male with benign prostatic hyperplasia to have difficulty with ejaculation? Why or

why not? _____

_____

6. Match the terms in column B to the descriptive statements in column A.

Column A

_____ 1. copulatory organ/penetrating device

_____ 2. site of sperm/androgen production

_____ 3. muscular passageway conveying sperm to the
ejaculatory duct; in the spermatic cord

_____ 4. distal urethra that transports both sperm
and urine

_____ 5. sperm maturation site

_____ 6. location of the testis in adult males

_____ 7. loose fold of skin encircling the glans penis

_____ 8. portion of the urethra that is located in the
urogenital diaphragm

_____ 9. accessory gland that secretes fluid to cleanse
the urethra prior to ejaculation

_____ 10. accessory gland that secretes the largest
contribution to semen

Column B

a. bulbo-urethral glands

b. ductus (vas) deferens

c. epididymis

d. intermediate part of the urethra

e. penis

f. prepuce

g. scrotum

h. seminal gland

i. spongy urethra

j. testes

7. Describe the composition of semen, and name all structures contributing to its formation. _____

_____

_____

8. Of what importance is the fact that seminal fluid is alkaline? _____

_____

9. What structures compose the spermatic cord? _____

_____

Where is it located? _____

_____

10. Using the following terms, trace the pathway of sperm from the testes to the urethra: rete testis, epididymis, seminiferous
tubule, ductus deferens.

_____ → _____ → _____ → _____

Gross Anatomy of the Human Female Reproductive System

11. Name the structures composing the external genitalia, or vulva, of the female.

_____

12. On the photo of the model of the female reproductive system below, identify all indicated structures.

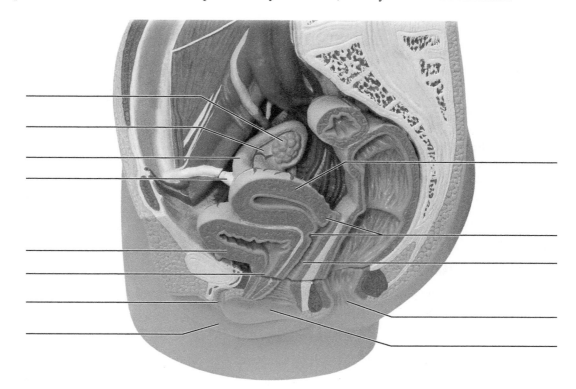

13. Identify the female reproductive system structures described below.

_____ 1. site of fetal development

_____ 2. copulatory canal

_____ 3. egg typically fertilized here

_____ 4. becomes erect during sexual excitement

_____ 5. glands homologous to the bulbo-urethral glands of the males

_____ 6. partially closes the vaginal opening; a membrane

_____ 7. produces oocytes, estrogens, and progesterone

_____ 8. fingerlike ends of the uterine tube

14. Do any sperm enter the pelvic cavity of the female? Why or why not? _____

_____

15. What is an ectopic pregnancy, and how can it happen? _____

_____

_____

_____

16. Put the following vestibular-perineal structures in their proper order from the anterior to the posterior aspect: vaginal orifice, anus, external urethral opening, and clitoris.

Anterior limit: _____ → _____ → _____ → _____

17. Name the male structure that is homologous to the female structures named below.

labia majora _____   clitoris _____

18. Assume that a couple has just consummated the sex act and that the male's sperm have been deposited in the woman's vagina. Trace the pathway of the sperm through the female reproductive tract.

_____

19. Define *ovulation*. _____

Microscopic Anatomy of Selected Male and Female Reproductive Organs

20. The testis is divided into a number of lobules by connective tissue. Each of these lobules contains one to four_____

_____, which converge on a _____ which, in turn, empties into a

tubular region at the testis called the _____.

21. What is the function of the cavernous bodies seen in the male penis? _____

_____

_____

22. Describe the arrangement of the layers of smooth muscle in the seminal gland. _____

_____

23. What is the function of the stereocilia exhibited by the epithelial cells of the mucosa of the epididymis? _____

_____

24. Name the three layers of the uterine wall from the inside out.

_____, _____, _____

Which of these is sloughed during menses? _____

Which contracts during childbirth? _____

25. Describe the epithelium found in the uterine tube. _____

_____

26. On the diagram showing the sagittal section of the human testis, correctly identify all structures provided with leader lines.

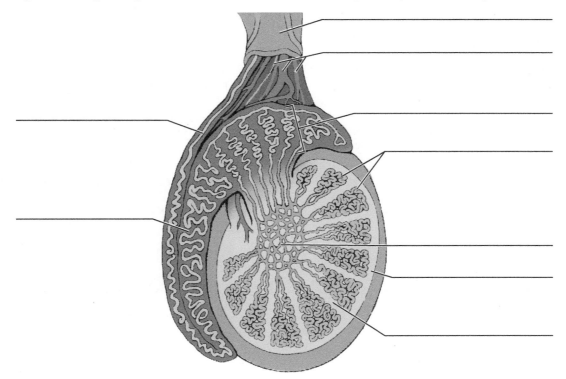

The Mammary Glands

27. Match the key term with the correct description.

_____ 1. gland that produces milk during lactation

_____ 2. subdivision of mammary lobe that contains alveoli

_____ 3. enlarged storage chamber for milk

_____ 4. duct connecting alveoli to the storage chambers

_____ 5. pigmented area surrounding the nipple

_____ 6. releases milk to the outside

Key:

a. alveolus

b. areola

c. lactiferous duct

d. lactiferous sinus

e. lobule

f. nipple

28. ✚ Cryptorchidism is failure of the testes to descend. Explain why this would cause sterility if not corrected. _____

_____

_____

29. ✚ Hysterectomy is a surgical removal of the uterus. It may or may not be accompanied by a salpingo-oophorectomy, removal of the uterine tubes and ovaries. Why would it be an advantage to leave the ovaries intact? _____

_____

_____

30. Using the key terms, correctly identify the breast structures on the following diagram.

> *Key:* a. adipose tissue
> b. areola
> c. lactiferous duct
> d. lactiferous sinus
> e. lobule containing alveoli
> f. nipple

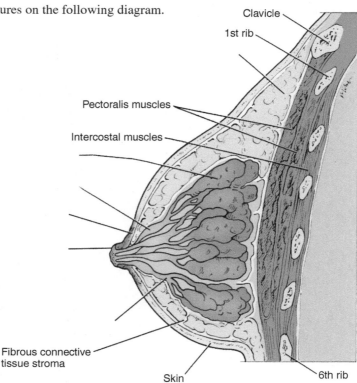

Clavicle

1st rib

Pectoralis muscles

Intercostal muscles

Fibrous connective
tissue stroma

Skin

6th rib

Dissection and Identification: The Reproductive System of the Cat

31. The female cat has a _____ uterus; that of the human female is _____.

Explain the difference in structure of these two uterine types. _____

_____

32. What reproductive advantage is conferred by the feline uterine type?

_____

33. Cite differences noted between the cat and the human relative to the following structures:

uterine tubes _____

site of entry of ductus deferens into the urethra _____

location of the prostate _____

seminal glands _____

urethral and vaginal openings in the female _____

_____

Surface Anatomy Roundup

MATERIALS

- Articulated skeletons
- Three-dimensional models or charts of the skeletal muscles of the body
- Hand mirror
- Stethoscope
- Alcohol swabs

LEARNING OUTCOMES

☐ Define *surface anatomy,* and explain why it is an important field of study; define *palpation.*

☐ Describe and palpate the major surface features of the cranium, face, and neck.

☐ Describe the easily palpated bony and muscular landmarks of the back, and locate the vertebral spines on the living body.

☐ List the bony surface landmarks of the thoracic cage, explain how they relate to the major soft organs of the thorax, and explain how to find the second to eleventh ribs.

☐ Name and palpate the important surface features on the anterior abdominal wall, and explain how to palpate a full bladder.

☐ Define and explain the following: *linea alba, umbilical hernia,* examination for an inguinal hernia, *linea semilunaris,* and *McBurney's point.*

☐ Locate and palpate the main surface features of the upper limb.

☐ Explain the significance of the cubital fossa, pulse points in the distal forearm, and the anatomical snuff box.

☐ Describe and palpate the surface landmarks of the lower limb.

☐ Explain exactly where to administer an injection in the gluteal region and in the other major sites of intramuscular injection.

PRE-LAB QUIZ

1. Why is it useful to study surface anatomy?
   a. You can easily locate deep muscle insertions.
   b. You can relate external surface landmarks to the location of internal organs.
   c. You can study cadavers more easily.
   d. You really can't learn that much by studying surface anatomy; it's a gimmick.
2. Circle the correct underlined term. <u>Palpation</u> / <u>Dissection</u> allows you to feel internal structures through the skin.
3. The epicranial aponeurosis binds to the subcutaneous tissue of the cranium to form the:
   a. mastoid process
   b. occipital protuberance
   c. true scalp
   d. xiphoid process
4. The _____ is the most prominent neck muscle and also the neck's most important landmark.
   a. buccinator
   b. epicranius
   c. masseter
   d. sternocleidomastoid
5. The three boundaries of the _____ are the trapezius medially, the latissimus dorsi inferiorly, and the scapula laterally.
   a. torso triangle
   b. triangle of ausculation
   c. triangle of back muscles
   d. McBurney's point
6. Circle True or False. The lungs do *not* fill the inferior region of the pleural cavity.

Text continues on next page.

7. Circle True or False. With the exception of a full bladder, most internal pelvic organs are not easily palpated through the skin of the body surface.

8. On the dorsal surface of your hand is a grouping of superficial veins known as the _____, which provides a site for drawing blood and inserting intravenous catheters.
   a. anatomical snuff box   c. radial and ulnar veins
   b. dorsal venous network   d. palmar arches

9. Circle True or False. To avoid harming major nerves and blood vessels, clinicians who administer intramuscular injections in the gluteal region of adults use the gluteus medius muscle.

10. The large femoral artery and vein descend vertically through the _____, formed by the border of the inguinal ligament, the medial border of the adductor longus muscle, and the medial border of the sartorius muscle.
    a. femoral triangle   c. medial condyle
    b. lateral condyle   d. quadriceps

urface anatomy is a valuable branch of anatomical and medical science. True to its name, **surface anatomy** does indeed study the *external surface* of the body, but more important, it also studies *internal* organs as they relate to external surface landmarks and as they are seen and felt through the skin. Feeling internal structures through the skin with the fingers is called **palpation** (literally, "touching").

Surface anatomy is living anatomy, better studied in live people than in cadavers. It can provide a great deal of information about the living skeleton (almost all bones can be palpated), and about the muscles and blood vessels that lie near the body surface. Furthermore, a skilled examiner can learn a good deal about the heart, lungs, and other deep organs by performing a surface assessment. Thus, surface anatomy serves as the basis of the standard physical examination. If you are planning a career in the health sciences or physical education, a study of surface anatomy will show you where to take pulses, where to insert tubes and needles, where to locate broken bones and inflamed muscles, and where to listen for the sounds of the lungs, heart, and intestines.

We will take a regional approach to surface anatomy, exploring the head first and proceeding to the trunk and the limbs. You will be observing and palpating your own body as you work through the exercise, because your body is the best learning tool of all. To aid your exploration of living anatomy, skeletons and muscle models or charts are provided around the lab so that you can review the bones and muscles you will encounter. Whenever possible, have a study partner assume the role of the subject for skin sites that you cannot reach on your own body.

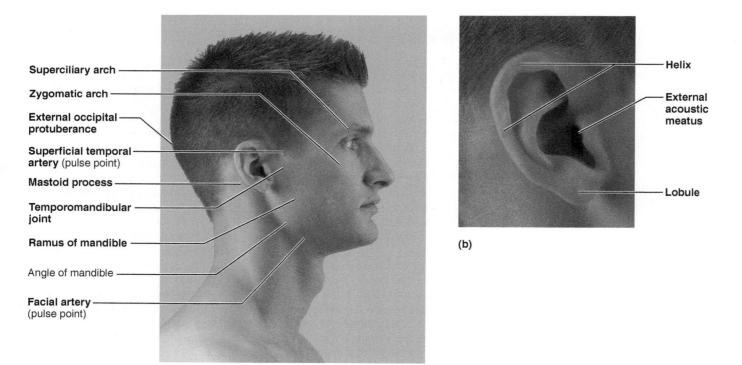

Figure 30.1 Surface anatomy of the head. (a) Lateral aspect. **(b)** Close-up of an auricle.

Palpating Landmarks of the Head

The head (**Figure 30.1** and **Figure 30.2**) is divided into the cranium and the face.

Cranium

1. Run your fingers over the superior surface of your head. Notice that the underlying cranial bones lie very near the surface. Proceed to your forehead and palpate the **superciliary arches** (brow ridges) directly superior to your orbits (Figure 30.1).

2. Move your hand to the posterior surface of your skull, where you can feel the knoblike **external occipital protuberance.** Run your finger directly laterally from this projection to feel the ridgelike *superior nuchal line* on the occipital bone. This line, which marks the superior extent of the muscles of the posterior neck, serves as the boundary between the head and the neck. Now feel the prominent **mastoid process** on each side of the cranium just posterior to your ear.

3. The **frontal belly** of the epicranius (see Figure 30.2) inserts superiorly onto the broad aponeurosis called the *epicranial aponeurosis* (Table 12.1, page 188) that covers the superior surface of the cranium. This aponeurosis binds tightly to the overlying subcutaneous tissue and skin to form the true **scalp.** Push on your scalp, and confirm that it slides freely over the underlying cranial bones. Because the scalp is only loosely bound to the skull, people can easily be "scalped" (in industrial accidents, for example). The scalp is richly vascularized by a large number of arteries running through its subcutaneous tissue. Most arteries of the body constrict and close after they are cut or torn, but those in the scalp are unable to do so because they are held open by the dense connective tissue surrounding them.

What do these facts suggest about the amount of bleeding that accompanies scalp wounds?

_____

_____

Face

The surface of the face is divided into many different regions, including the *orbital, nasal, oral* (mouth), and *auricular* (ear) areas.

1. Trace a finger around the entire margin of the bony orbit. The **lacrimal fossa,** which contains the tear-gathering lacrimal sac, may be felt on the medial side of the eye socket.

2. Touch the most superior part of your nose, its **root,** which lies between the eyebrows (Figure 30.2). Just inferior to this, between your eyes, is the **bridge** of the nose formed by the nasal bones. Continue your finger's progress inferiorly along the nose's anterior margin, the **dorsum nasi,** to the tip of the nose, the **apex.** Place one finger in a nostril and another finger on the flared winglike **ala** that defines the nostril's lateral border.

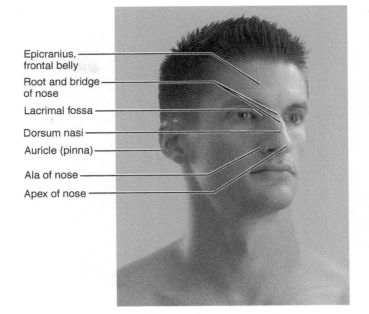

Epicranius, frontal belly
Root and bridge of nose
Lacrimal fossa
Dorsum nasi
Auricle (pinna)
Ala of nose
Apex of nose

Figure 30.2 Surface structures of the face.

3. Grasp your **auricle,** the shell-like part of the external ear that surrounds the opening of the **external acoustic meatus** (Figure 30.1). Now trace the ear's outer rim, or **helix,** to the **lobule** (earlobe) inferiorly. The lobule is easily pierced, and because it is not highly sensitive to pain, it provides a convenient place to obtain a drop of blood for clinical blood analysis. Next, place a finger on your temple just anterior to the auricle. There, you may be able to feel the pulsations of the **superficial temporal artery,** which ascends to supply the scalp (Figure 30.1).

4. Run your hand anteriorly from your ear toward the orbit, and feel the **zygomatic arch** just deep to the skin. This bony arch is easily broken by blows to the face. Next, place your fingers on the skin of your face, and feel it bunch and stretch as you contort your face into smiles, frowns, and grimaces. You are now monitoring the action of several of the subcutaneous **muscles of facial expression** (Table 12.1, page 188).

5. On your lower jaw, palpate the parts of the bony **mandible:** its anterior body and its posterior ascending **ramus.** Press on the skin over the mandibular ramus, and feel the **masseter muscle** bulge when you clench your teeth. Palpate the anterior border of the masseter, and trace it to the mandible's inferior margin. At this point, you will be able to detect the pulse of your **facial artery** (Figure 30.1). Finally, to feel the **temporomandibular joint,** place a finger directly anterior to the external acoustic meatus of your ear, and open and close your mouth several times. The bony structure you feel moving is the *condylar process of the mandible.* ■

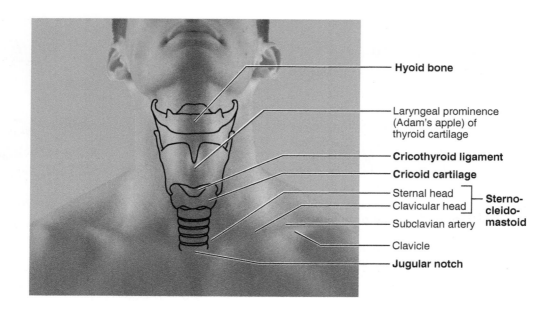

Figure 30.3 Anterior surface of the neck. A diagram of the underlying structures is superimposed on a photograph of the neck.

Palpating Landmarks of the Neck

Bony Landmarks

1.   Run your fingers inferiorly along the back of your neck, in the posterior midline, to feel the *spinous processes* of the cervical vertebrae. The spine of C_7 the *vertebra prominens,* is especially prominent.

2.   Now, beginning at your chin, run a finger inferiorly along the anterior midline of your neck **(Figure 30.3)**. The first hard structure you encounter will be the U-shaped **hyoid bone,** which lies in the angle between the floor of the mouth and the vertical part of the neck. Directly inferior to this, you will feel the **laryngeal prominence** (Adam's apple) of the thyroid cartilage. Just inferior to the laryngeal prominence, your finger will sink into a soft depression (formed by the **cricothyroid ligament**) before proceeding onto the rounded surface of the **cricoid cartilage.** Now swallow several times, and feel the whole larynx move up and down.

3.   Continue inferiorly to the trachea. Attempt to palpate the *isthmus of the thyroid gland,* which feels like a spongy cushion over the second to fourth tracheal rings (Figure 30.3). Then, try to palpate the two soft lateral *lobes* of your thyroid gland along the sides of the trachea.

4.   Move your finger all the way inferiorly to the root of the neck, and rest it in the **jugular notch,** the depression in the superior part of the manubrium between the two clavicles. By pushing deeply at this point, you can feel the cartilage rings of the trachea.

Muscles

The **sternocleidomastoid** is the most prominent muscle in the neck and the neck's most important surface landmark. You can best see and feel it when you turn your head to the side.

Obtain a hand mirror, hold it in front of your face, and turn your head sharply from right to left several times. You will be able to see both heads of this muscle, the **sternal head** medially and the **clavicular head** laterally (Figure 30.3). Several important structures lie beside or beneath the sternocleidomastoid:

- The *cervical lymph nodes* lie both superficial and deep to this muscle. Swollen cervical nodes provide evidence of infections or cancer of the head and neck.

- The *common carotid artery* and *internal jugular vein* lie just deep to the sternocleidomastoid, a relatively superficial location that exposes these vessels to danger in slashing wounds to the neck.

- Just lateral to the inferior part of the sternocleidomastoid is the large **subclavian artery** on its way to supply the upper limb. By pushing on the subclavian artery at this point, one can stop the bleeding from a wound anywhere in the associated limb.

- Just anterior to the sternocleidomastoid, superior to the level of your larynx, you can feel a carotid pulse—the pulsations of the **external carotid artery (Figure 30.4)**.

- The *external jugular vein* descends vertically, just superficial to the sternocleidomastoid and deep to the skin (Figure 30.4b). To make this vein "appear" on your neck, stand before the mirror, and gently compress the skin superior to your clavicle with your fingers.

Triangles of the Neck

The sternocleidomastoid muscles divide each side of the neck into the posterior and anterior triangles (Figure 30.4a).

1.   The **posterior triangle** is defined by the sternocleidomastoid anteriorly, the trapezius posteriorly, and the clavicle inferiorly. Palpate the borders of the posterior triangle.

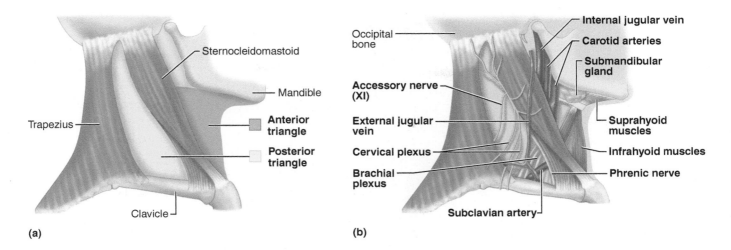

Figure 30.4 Anterior and posterior triangles of the neck. (a) Boundaries of the triangles. **(b)** Some contents of the triangles.

The **anterior triangle** is defined by the inferior margin of the mandible superiorly, the midline of the neck anteriorly, and the sternocleidomastoid posteriorly.

2. The contents of these two triangles include nerves, glands, blood vessels, and small muscles (Figure 30.4b). The posterior triangle contains the **accessory nerve** (cranial nerve XI), most of the **cervical plexus,** and the **phrenic nerve.** In the inferior part of the triangle are the **external jugular vein,** the trunks of the **brachial plexus,** and the **subclavian artery.** These structures are relatively superficial and are easily cut or injured by wounds to the neck.

3. In the neck's anterior triangle, important structures include the **submandibular gland,** the **suprahyoid** and **infrahyoid muscles,** and parts of the **carotid arteries** and **jugular veins** that lie superior to the sternocleidomastoid.

☐ Palpate your carotid pulse.

A wound to the posterior triangle of the neck can lead to long-term loss of sensation in the skin of the neck and shoulder, as well as partial paralysis of the sternocleidomastoid and trapezius muscles. Explain these effects.

_____

_____ +

ACTIVITY 3

Palpating Landmarks of the Trunk

The trunk of the body consists of the thorax, abdomen, pelvis, and perineum. The _back_ includes parts of all of these regions, but for convenience it is treated separately.

The Back

Bones

1. The vertical groove in the center of the back is called the **posterior median furrow (Figure 30.5)**. The _spinous_ _processes_ of the vertebrae are visible in the furrow when the spinal column is flexed.

☐ Palpate a few of these processes on your partner's back (C_7 and T_1 are the most prominent and the easiest to find).

☐ Also palpate the posterior parts of some ribs, as well as the prominent **spine of the scapula** and the scapula's long **medial border.**

    The scapula lies superficial to ribs 2 to 7; its **inferior angle** is at the level of the spinous process of vertebra T_7. The medial end of the scapular spine lies opposite the T_3 spinous process.

2. Now feel the **iliac crests** (superior margins of the iliac bones) in your own lower back. You can find these crests effortlessly by resting your hands on your hips. Locate the most superior point of each crest, a point that lies roughly halfway between the posterior median furrow and the lateral side of the body (Figure 30.5). A horizontal line through these two superior points, the **supracristal line,** intersects L_4 providing a simple way to locate that vertebra. The ability to locate L_4 is essential for performing a _lumbar puncture,_ a procedure in which the clinician inserts a needle into the vertebral canal of the spinal column directly superior or inferior to L_4 and withdraws cerebrospinal fluid.

3. The _sacrum_ is easy to palpate just superior to the cleft in the buttocks. You can feel the _coccyx_ in the extreme inferior part of that cleft, just posterior to the anus.

Muscles

The largest superficial muscles of the back are the **trapezius** superiorly and **latissimus dorsi** inferiorly (Figure 30.5). Furthermore, the deeper **erector spinae** muscles are very evident in the lower back, flanking the vertebral column like thick vertical cords.

1. Shrug your shoulders to feel the trapezius contracting just deep to the skin.

2. Feel your partner's erector spinae muscles contract and bulge as he straightens his spine from a slightly bent-over position.

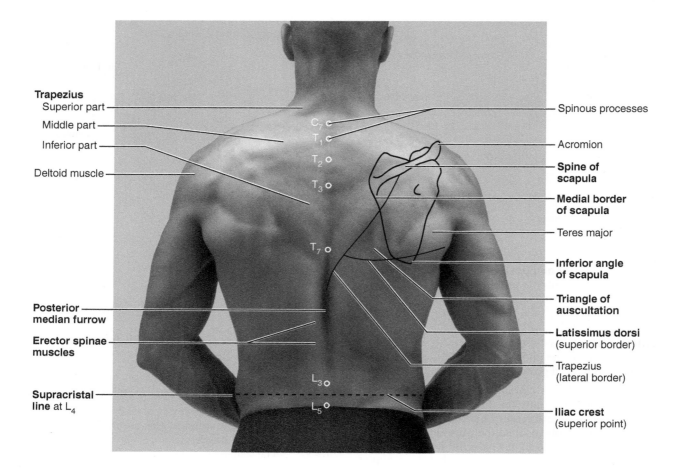

Trapezius
Superior part
Middle part
Inferior part

Deltoid muscle

Posterior
median furrow

Erector spinae
muscles

Supracristal
line at L₄

Spinous processes

Acromion

**Spine of
scapula**

**Medial border
of scapula**

Teres major

**Inferior angle
of scapula**

**Triangle of
auscultation**

Latissimus dorsi
(superior border)

Trapezius
(lateral border)

Iliac crest
(superior point)

C₇

T₁

T₂

T₃

T₇

L₃

L₅

Figure 30.5 Surface anatomy of the back.

The superficial muscles of the back fail to cover a small area of the rib cage called the **triangle of auscultation** (Figure 30.5). This triangle lies just medial to the inferior part of the scapula. Its three boundaries are formed by the trapezius medially, the latissimus dorsi inferiorly, and the scapula laterally. The physician places a stethoscope over the skin of this triangle to listen for lung sounds (*auscultation* = listening). To hear the lungs clearly, the doctor first asks the patient to fold the arms together in front of the chest and then flex the trunk.

What do you think is the precise reason for having the patient take this action?

_____

_____

3. Have your partner assume the position just described. After cleaning the earpieces with an alcohol swab, use the stethoscope to auscultate the lung sounds. Compare the clarity of the lung sounds heard over the triangle of auscultation to that over other areas of the back.

The Thorax
Bones

1. Start exploring the anterior surface of your partner's bony *thoracic cage* (**Figure 30.6** and **Figure 30.7**) by defining the extent of the *sternum.* Use a finger to trace the sternum's triangular *manubrium* inferior to the jugular notch, its flat *body,* and the tongue-shaped **xiphoid process.** Now palpate the ridgelike **sternal angle,** where the manubrium meets the body of the sternum. Locating the sternal angle is important because it directs you to the second ribs (which attach to it). Once you find the second rib, you can count down to identify every other rib in the thorax (except the first and sometimes the twelfth rib, which lie too deep to be palpated). The sternal angle is a highly reliable landmark—it is easy to locate, even in overweight people.

2. By locating the individual ribs, you can mentally "draw" a series of horizontal lines of "latitude" that you can use to map and locate the underlying visceral organs of the thoracic cavity. Such mapping also requires lines of "longitude," so let us construct some vertical lines on the wall of your partner's trunk. As he lifts an arm straight up in the air, extend a line

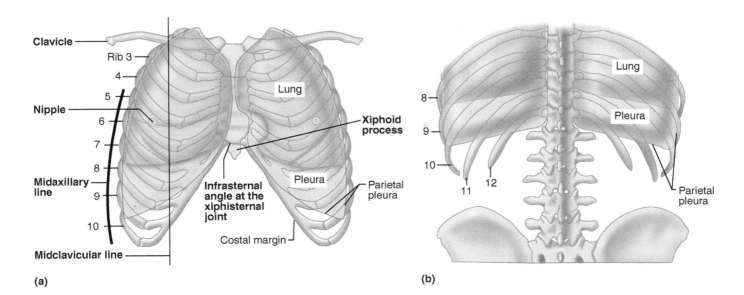

Clavicle

Rib 3

4

5

Nipple

6

7

8

Midaxillary line

9

10

Midclavicular line

Lung

Xiphoid process

Infrasternal angle at the xiphisternal joint

Pleura

Parietal pleura

Costal margin

(a)

Lung

8

Pleura

9

10

11   12

Parietal pleura

(b)

Figure 30.6 The bony rib cage as it relates to the underlying lungs and pleural cavities. Both the pleural cavities (blue) and the lungs (pink) are outlined. **(a)** Anterior view. **(b)** Posterior view.

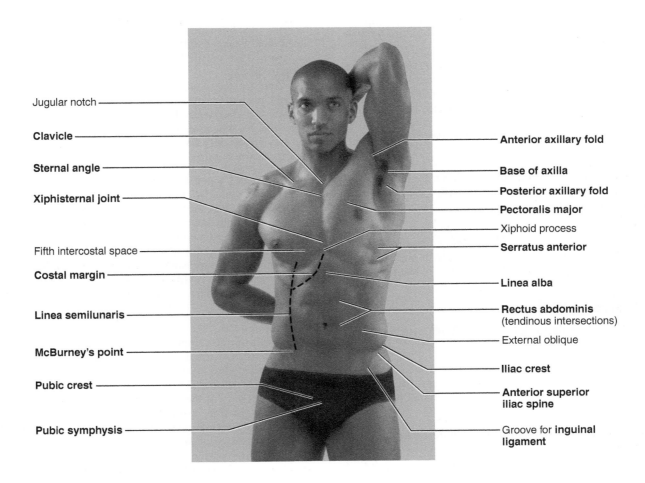

Jugular notch

Clavicle

Sternal angle

Xiphisternal joint

Fifth intercostal space

Costal margin

Linea semilunaris

McBurney's point

Pubic crest

Pubic symphysis

Anterior axillary fold

Base of axilla

Posterior axillary fold

Pectoralis major

Xiphoid process

Serratus anterior

Linea alba

Rectus abdominis (tendinous intersections)

External oblique

Iliac crest

Anterior superior iliac spine

Groove for **inguinal ligament**

Figure 30.7 The anterior thorax and abdomen.

inferiorly from the center of the axilla onto his lateral thoracic wall. This is the **midaxillary line** (Figure 30.6a). Now estimate the midpoint of his **clavicle**, and run a vertical line inferiorly from that point toward the groin. This is the **midclavicular line**, and it will pass about 1 cm medial to the nipple.

3.   Next, feel along the V-shaped inferior edge of the rib cage, the **costal margin**. At the **infrasternal angle**, the superior angle of the costal margin, lies the **xiphisternal joint**. The heart lies on the diaphragm deep to the xiphisternal joint.

4.   The thoracic cage provides many valuable landmarks for locating the vital organs of the thoracic and abdominal cavities. On the anterior thoracic wall, ribs 2–6 define the superior-to-inferior extent of the female breast, and the fourth intercostal space indicates the location of the **nipple** in men, children, and small-breasted women. The right costal margin runs across the anterior surface of the liver and gallbladder. Surgeons must be aware of the inferior margin of the *pleural cavities* because if they accidentally cut into one of these cavities, a lung collapses. The inferior pleural margin lies adjacent to vertebra T_{12} near the posterior midline (Figure 30.6b) and runs horizontally across the back to reach rib 10 at the midaxillary line. From there, the pleural margin ascends to rib 8 in the midclavicular line (Figure 30.6a) and to the level of the xiphisternal joint near the anterior midline. The *lungs* do not fill the inferior region of the pleural cavity. Instead, their inferior borders run at a level that is two ribs superior to the pleural margin, until they meet that margin near the xiphisternal joint.

5.   Let's review the relationship of the *heart* to the thoracic cage. The superior right corner of the heart lies at the junction of the third rib and the sternum; the superior left corner lies at the second rib, near the sternum; the inferior left corner lies in the fifth intercostal space in the midclavicular line; and the inferior right corner lies at the sternal border of the sixth rib.

Muscles

The main superficial muscles of the anterior thoracic wall are the **pectoralis major** and the anterior slips of the **serratus anterior** (Figure 30.7).

☐   Palpate these two muscles on your chest. They both contract during push-ups, and you can confirm this by pushing yourself up from your desk with one arm while palpating the muscles with your opposite hand. ▄

ACTIVITY 4

Palpating Landmarks of the Abdomen

Bony Landmarks

The anterior abdominal wall (Figure 30.7) extends inferiorly from the costal margin to an inferior boundary that is defined by several landmarks. Palpate these landmarks as they are described below.

1.   **Iliac crest:** Locate the iliac crests by resting your hands on your hips.

2.   **Anterior superior iliac spine:** Representing the most anterior point of the iliac crest, this spine is a prominent landmark. It can be palpated in everyone, even those who are overweight. Run your fingers anteriorly along the iliac crest to its end.

3.   **Inguinal ligament:** The inguinal ligament, indicated by a groove on the skin of the groin, runs medially from the anterior superior iliac spine to the pubic tubercle of the pubis.

4.   **Pubic crest:** You will have to press deeply to feel this crest on the pubis near the median **pubic symphysis.** The **pubic tubercle,** the most lateral point of the pubic crest, is easier to palpate, but you will still have to push deeply.

Inguinal hernias occur immediately superior to the inguinal ligament and may exit from a medial opening called the **superficial inguinal ring.** To locate this ring, one would palpate the pubic tubercle (**Figure 30.8**). An inguinal hernia in a male can be detected by pushing into the superior inguinal ring. ✚

Muscles and Other Surface Features

The central landmark of the anterior abdominal wall is the *umbilicus* (navel). Running superiorly and inferiorly from the umbilicus is the **linea alba** (white line), represented in the skin of lean people by a vertical groove (see Figure 30.7). The linea alba is a tendinous seam that extends from the xiphoid process to the pubic symphysis, just medial to the rectus abdominis muscles (Table 12.3, page 193). The linea alba is a favored site for surgical entry into the abdominal cavity because the surgeon can make a long cut through this line with no muscle damage and minimal bleeding.

Several kinds of hernias involve the umbilicus and the linea alba. In an **acquired umbilical hernia,** the linea alba weakens until intestinal coils push through it just superior to the navel. The herniated coils form a bulge just deep to the skin.

Another type of umbilical hernia is a **congenital umbilical hernia,** present in some infants. This type of umbilical hernia is seen as a cherry-sized bulge deep to the skin of the navel that enlarges whenever the baby cries. Congenital umbilical hernias are usually harmless, and most correct themselves automatically before the child's second birthday. ✚

1.   **McBurney's point** is the spot on the anterior abdominal skin that lies directly superficial to the base of the appendix (Figure 30.7). It is located one-third of the way along a line between the right anterior superior iliac spine and the umbilicus. Try to find it on your body.

McBurney's point is often the place where the pain of appendicitis is experienced most acutely. Pain at McBurney's point after pressure is removed (rebound tenderness) can indicate appendicitis. This is not a *precise* method of diagnosis, however.

2.   Flanking the linea alba are the vertical straplike **rectus abdominis** muscles (Figure 30.7). Feel these muscles contract just deep to your skin as you do a bent-knee sit-up (or as you bend forward after leaning back in your chair). In the skin of lean people, the lateral margin of each rectus muscle makes a groove known as the **linea semilunaris** (half-moon line). On your right side, estimate where your linea semilunaris crosses the costal margin of the rib cage. The *gallbladder* lies just deep to this spot, so this is the standard point of incision for gallbladder surgery. In muscular people, three horizontal grooves can be seen in the skin covering the rectus abdominis. These grooves represent the **tendinous intersections,** fibrous bands that subdivide the rectus muscle. Because of these

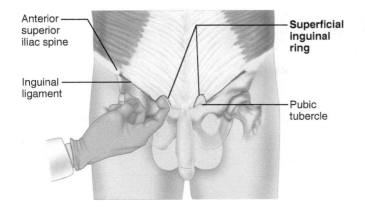

Figure 30.8 Clinical examination for an inguinal hernia in a male. The examiner palpates the patient's pubic tubercle, pushes superiorly to invaginate the scrotal skin into the superficial inguinal ring, and asks the patient to cough. If an inguinal hernia exists, it will push inferiorly and touch the examiner's fingertip.

Labels in figure: Anterior superior iliac spine; Inguinal ligament; **Superficial inguinal ring**; Pubic tubercle

subdivisions, each rectus abdominis muscle presents four distinct bulges. Try to identify these intersections on yourself or your partner.

3. The only other major muscles that can be seen or felt through the anterior abdominal wall are the lateral **external obliques.** Feel these muscles contract as you cough, strain, or raise your intra-abdominal pressure in some other way.

4. The anterior abdominal wall can be divided into four quadrants (Figure 1.8). A clinician listening to a patient's **bowel sounds** places the stethoscope over each of the four abdominal quadrants, one after another. Normal bowel sounds, which result as peristalsis moves air and fluid through the intestine, are high-pitched gurgles that occur every 5 to 15 seconds.

☐ Use the stethoscope to listen to your own or your partner's bowel sounds.

Abnormal bowel sounds can indicate intestinal disorders. Absence of bowel sounds indicates a halt in intestinal activity, which follows long-term obstruction of the intestine, surgical handling of the intestine, peritonitis, or other conditions. Loud tinkling or splashing sounds, by contrast, indicate an increase in intestinal activity. Such loud sounds may accompany gastroenteritis (inflammation of the GI tract) or a partly obstructed intestine. ✚

The Pelvis and Perineum

The bony surface features of the *pelvis* are considered with the bony landmarks of the abdomen (page 532) and the gluteal region (pages 536–537). Most *internal* pelvic organs are not palpable through the skin of the body surface. A full *bladder,* however, becomes firm and can be felt through the abdominal wall just superior to the pubic symphysis. A bladder that can be palpated more than a few centimeters above this symphysis is retaining urine and dangerously full, and it should be drained by catheterization. ▬

ACTIVITY 5

Palpating Landmarks of the Upper Limb

Axilla

The **base of the axilla** is the groove in which the underarm hair grows (Figure 30.7). Deep to this base lie the axillary *lymph nodes* (which swell and can be palpated in breast cancer), the large *axillary vessels* serving the upper limb, and much of the brachial plexus. The base of the axilla forms a "valley" between two thick, rounded ridges, the **axillary folds.** Just anterior to the base, clutch your **anterior axillary fold,** formed by the pectoralis major muscle. Then grasp your **posterior axillary fold.** This fold is formed by the latissimus dorsi and teres major muscles of the back as they course toward their insertions on the humerus.

Shoulder

1. Again locate the prominent spine of the scapula posteriorly (Figure 30.5). Follow the spine to its lateral end, the flattened **acromion** on the shoulder's summit. Then, palpate the **clavicle** anteriorly, tracing this bone from the sternum to the shoulder (Figure 30.5 and **Figure 30.9**). Notice the clavicle's curved shape.

2. Now locate the junction between the clavicle and the acromion on the superolateral surface of your shoulder, at the **acromioclavicular (AC) joint.** To find this joint, thrust your arm anteriorly repeatedly until you can palpate the precise point of pivoting action.

3. Next, place your fingers on the **greater tubercle** of the humerus. This is the most lateral bony landmark on the superior surface of the shoulder. It is covered by the thick **deltoid muscle,** which forms the rounded superior part of the shoulder. Intramuscular injections are often given into the deltoid, about 5 cm (2 inches) inferior to the greater tubercle (Figure 30.17a, page 538).

Arm

Remember, according to anatomists, the arm runs only from the shoulder to the elbow, and not beyond.

1. In the arm, palpate the humerus along its entire length, especially along its medial and lateral sides.

2. Feel the **biceps brachii** muscle contract on your anterior arm when you flex your forearm against resistance. The medial boundary of the biceps is represented by the **medial bicipital groove** (Figure 30.9). This groove contains the large *brachial artery,* and by pressing on it with your fingertips you can feel your *brachial pulse.* Recall that the brachial artery is the artery routinely used in measuring blood pressure with a sphygmomanometer.

3. All three heads of the **triceps brachii** muscle (lateral, long, and medial) are visible through the skin of a muscular person **(Figure 30.10)**.

Elbow Region

1. In the distal part of your arm, near the elbow, palpate the two projections of the humerus, the **lateral** and **medial epicondyles** (Figures 30.9 and 30.10). Midway between the

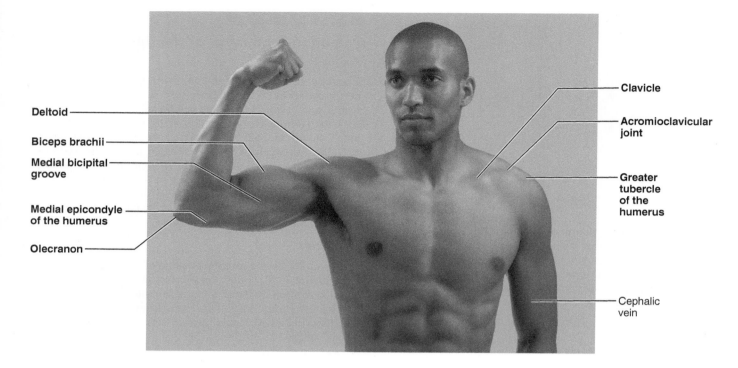

Deltoid

Biceps brachii

Medial bicipital groove

Medial epicondyle of the humerus

Olecranon

Clavicle

Acromioclavicular joint

Greater tubercle of the humerus

Cephalic vein

Figure 30.9 Shoulder and arm.

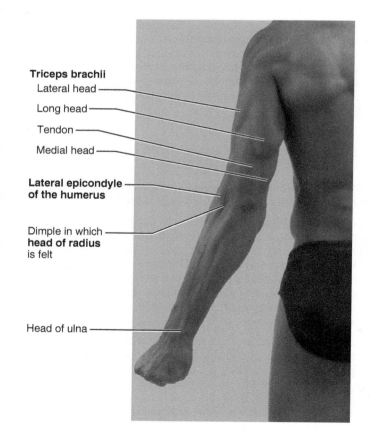

Triceps brachii
Lateral head

Long head

Tendon

Medial head

Lateral epicondyle of the humerus

Dimple in which **head of radius** is felt

Head of ulna

Figure 30.10 Surface anatomy of the upper limb, posterior view.

epicondyles, on the posterior side, feel the **olecranon,** which forms the point of the elbow.

2. Confirm that the two epicondyles and the olecranon all lie in the same horizontal line when the elbow is extended. If these three bony processes do not line up, the elbow is dislocated.

3. Now feel along the posterior surface of the medial epicondyle. You are palpating your ulnar nerve.

4. On the anterior surface of the elbow is a triangular depression called the **cubital fossa (Figure 30.11).** The triangle's superior *base* is formed by a horizontal line between the humeral epicondyles; its two inferior sides are defined by the **brachioradialis** and **pronator teres** muscles (Figure 30.11b). Try to define these boundaries on your own limb. To find the brachioradialis muscle, flex your forearm against resistance, and watch this muscle bulge through the skin of your lateral forearm. To feel your pronator teres contract, palpate the cubital fossa as you pronate your forearm against resistance. (Have your partner provide the resistance.)

Superficially, the cubital fossa contains the **median cubital vein** (Figure 30.11a). Clinicians often draw blood from this superficial vein and insert intravenous (IV) catheters into it to administer medications, transfused blood, and nutrient fluids. The large **brachial artery** lies just deep to the median cubital vein (Figure 30.11b), so a needle must be inserted into the vein from a shallow angle (almost parallel to the skin) to avoid puncturing the artery. Tendons and nerves are also found deep in the fossa (Figure 30.11b).

5. The median cubital vein interconnects the larger **cephalic** and **basilic veins** of the upper limb. These veins are visible through the skin of lean people (Figure 30.11a). Examine your arm to see whether your cephalic and basilic veins are visible.

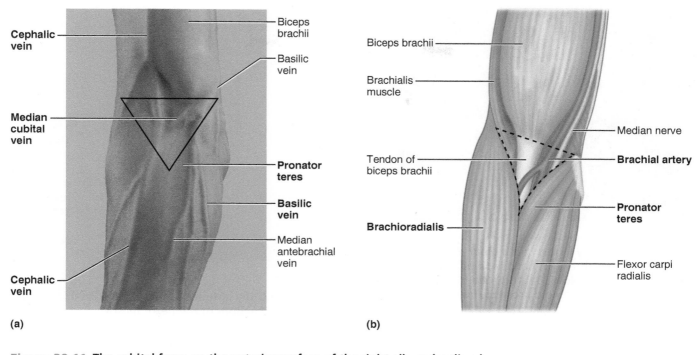

Figure 30.11 The cubital fossa on the anterior surface of the right elbow (outlined by the triangle). (a) Photograph. **(b)** Diagram of deeper structures in the fossa.

Forearm and Hand

The two parallel bones of the forearm are the medial *ulna* and the lateral *radius.*

1. Feel the ulna along its entire length as a sharp ridge on the posterior forearm (confirm that this ridge runs inferiorly from the olecranon). As for the radius, you can feel its distal half, but most of its proximal half is covered by muscle. You can, however, feel the rotating **head** of the radius. To do this, extend your forearm, and note that a dimple forms on the posterior lateral surface of the elbow region (Figure 30.10). Press three fingers into this dimple, and rotate your free hand as if you were turning a doorknob. You will feel the head of the radius rotate as you perform this action.

2. Both the radius and ulna have a knoblike **styloid process** at their distal ends. Palpate these processes at the wrist **(Figure 30.12)**. Do not confuse the ulnar styloid process with the conspicuous **head of the ulna,** from which the styloid process stems. Confirm that the radial styloid process lies about 1 cm (0.4 inch) distal to that of the ulna.

Colles' fracture of the wrist is an impacted fracture in which the distal end of the radius is pushed proximally into the shaft of the radius. This sometimes occurs when someone falls on outstretched hands, and it most often happens to elderly women with osteoporosis. Colles' fracture bends the wrist into curves that resemble those on a fork. ✚

Can you deduce how physicians use palpation to diagnose Colles' fracture?

_____

_____

3. Next, feel the major groups of muscles within your forearm. Flex your hand and fingers against resistance, and feel the anterior *flexor muscles* contract. Then extend your hand at the wrist, and feel the tightening of the posterior *extensor muscles.*

4. Near the wrist, the anterior surface of the forearm reveals many significant features **(Figure 30.13)**. Flex your fist against resistance; the tendons of the main wrist flexors will bulge the skin of the distal forearm. The tendons of the **flexor carpi radialis** and **palmaris longus** muscles are most obvious. The palmaris longus, however, is absent from at least one arm in 30% of all people, so your forearm may exhibit just one prominent tendon instead of two. The **radial artery** lies just lateral to (on the thumb side of) the flexor carpi radialis tendon, where the pulse is easily detected (Figure 30.13). Feel your radial pulse here. The *median nerve,* which innervates the thumb, lies deep to the palmaris longus tendon. Finally, the **ulnar artery** lies on the medial side of the forearm, just lateral to the tendon of the **flexor carpi ulnaris.** Locate and feel your ulnar arterial pulse (Figure 30.13).

5. Extend your thumb and point it posteriorly to form a triangular depression in the base of the thumb on the back of your hand. This is the **anatomical snuff box (Figure 30.14)**. Its two elevated borders are defined by the tendons of the thumb extensor muscles, **extensor pollicis brevis** and **extensor pollicis longus.** The radial artery runs within the snuff box, so this is another site for taking a radial pulse. The main bone on the floor of the snuff box is the scaphoid bone of the wrist, but the radial styloid process is also present here. (If displaced by a bone fracture, the radial styloid process will be felt outside the snuff box rather than within it.) The "snuff box" took its name from the practice of putting snuff (tobacco for sniffing) in this hollow before lifting it up to the nose.

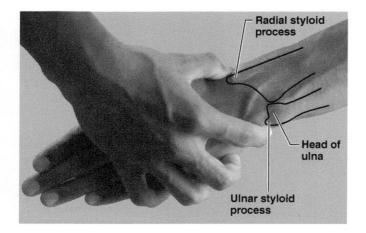

Figure 30.12 **A way to locate the ulnar and radial styloid processes.** The right hand is palpating the left hand in this picture. Note that the head of the ulna is not the same as the ulnar styloid process. The radial styloid process lies about 1 cm distal to the ulnar styloid process.

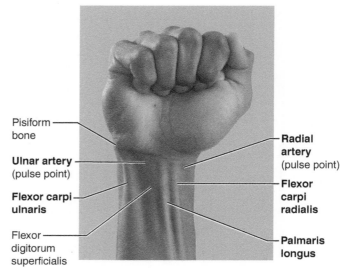

Figure 30.13 **The anterior surface of the distal forearm and fist.** The tendons of the flexor muscles guide the clinician to several sites for taking the pulse.

6. On the dorsal surface of your hand, observe the superficial veins just deep to the skin. This is the **dorsal venous network,** which drains superiorly into the cephalic vein. This venous network provides a site for drawing blood and inserting intravenous catheters and is preferred over the median cubital vein for these purposes. Next, extend your hand and fingers, and observe the tendons of the **extensor digitorum** muscle.

7. The anterior surface of the hand also contains some features of interest **(Figure 30.15)**. These features include the *epidermal ridges* (fingerprints) and many **flexion creases** in the skin. Grasp your **thenar eminence** (the bulge on the palm that contains the thumb muscles) and your **hypothenar eminence**

(the bulge on the medial palm that contains muscles that move the little finger). ▬

ACTIVITY 6

Palpating Landmarks of the Lower Limb
Gluteal Region

Dominating the gluteal region are the two *prominences* (cheeks) of the buttocks. These are formed by subcutaneous fat and by the thick **gluteus maximus** muscles **(Figure 30.16)**. The midline groove between the two prominences

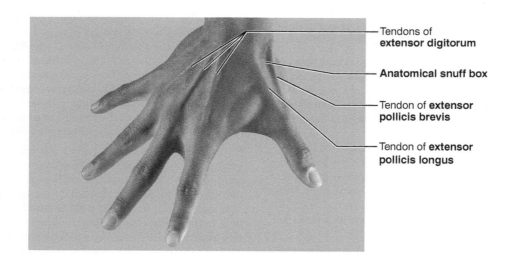

Figure 30.14 **The dorsal surface of the hand.** Note especially the anatomical snuff box and dorsal venous network.

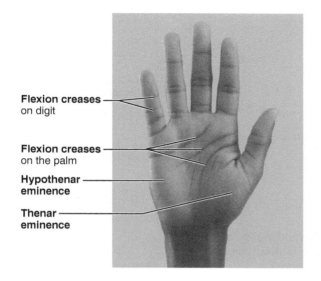

Flexion creases
on digit

Flexion creases
on the palm

Hypothenar
eminence

Thenar
eminence

Figure 30.15 The palmar surface of the hand.

is called the **natal cleft** (*natal* = rump), or **gluteal cleft.** The inferior margin of each prominence is the horizontal **gluteal fold,** which roughly corresponds to the inferior margin of the gluteus maximus.

1. Try to palpate your **ischial tuberosity** just above the medial side of each gluteal fold (it will be easier to feel if you sit down or flex your thigh first). The ischial tuberosities are the robust inferior parts of the ischial bones, and they support the body's weight during sitting.

2. Next, palpate the **greater trochanter** of the femur on the lateral side of your hip (Figure 30.16). This trochanter lies just anterior to a hollow and about 10 cm (one hand's breadth, or 4 inches) inferior to the iliac crest. To confirm that you have found the greater trochanter, alternately flex and extend your thigh. Because this trochanter is the most superior point on the lateral femur, it moves with the femur as you perform this movement.

3. To palpate the sharp **posterior superior iliac spine** (Figure 30.16), locate your iliac crests again, and trace each to its most posterior point. You may have difficulty feeling this spine, but it is indicated by a distinct dimple in the skin that is easy to find. This dimple lies two to three fingerbreadths lateral to the midline of the back. The dimple also indicates the position of the *sacroiliac joint,* where the hip bone attaches to the sacrum of the spinal column. You can check *your* "dimples" out in the privacy of your home.

The gluteal region is a major site for administering intramuscular injections. When such injections are given, the health care provider must take extreme care to avoid piercing the major nerve that lies just deep to the gluteus maximus muscle. Can you guess what nerve this is?

This nerve is the thick *sciatic nerve,* which innervates much of the lower limb. Furthermore, the needle must avoid the gluteal nerves and gluteal blood vessels, which also lie deep to the gluteus maximus.

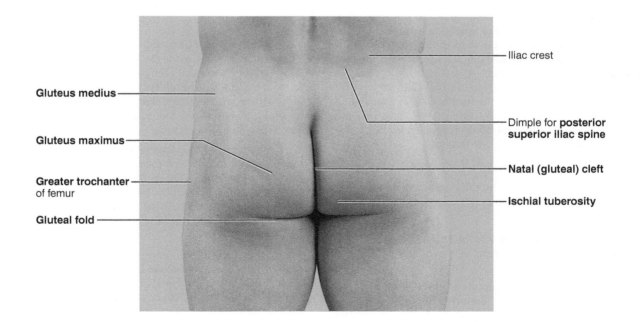

Gluteus medius

Gluteus maximus

Greater trochanter
of femur

Gluteal fold

Iliac crest

Dimple for **posterior**
superior iliac spine

Natal (gluteal) cleft

Ischial tuberosity

Figure 30.16 The gluteal region. The region extends from the iliac crests superiorly to the gluteal folds inferiorly. Therefore, it includes more than just the prominences of the buttock.

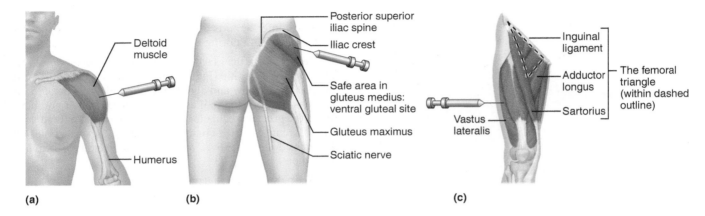

Figure 30.17 Three major sites of intramuscular injections. (a) Deltoid muscle of the arm (for injection volumes of less than 1 ml). **(b)** Ventral gluteal site (gluteus medius). **(c)** Vastus lateralis in the lateral thigh. The femoral triangle is also shown.

To avoid harming these structures, the injections are most often applied to the **gluteus** medius (not maximus) muscle superior to the cheeks of the buttocks, in a safe area called the **ventral gluteal site (Figure 30.17b).** To locate this site, mentally draw a line laterally from the posterior superior iliac spine (dimple) to the greater trochanter; the injection would be given 5 cm (2 inches) superior to the midpoint of that line. Another safe way to locate the ventral gluteal site is to approach the lateral side of the patient's left hip with your extended right hand (or the right hip with your left hand). Then, place your thumb on the anterior superior iliac spine and your index finger as far posteriorly on the iliac crest as it can reach. The heel of your hand comes to lie on the greater trochanter, and the needle is inserted in the angle of the V formed between your thumb and index finger about 4 cm (1.5 inches) inferior to the iliac crest.

Gluteal injections are not given to small children because their "safe area" is too small to locate with certainty and because the gluteal muscles are thin at this age. Instead, infants and toddlers receive intramuscular shots in the prominent **vastus lateralis** muscle of the thigh.

Thigh

Much of the femur is clothed by thick muscles, so the thigh has few palpable bony landmarks (**Figure 30.18** and **Figure 30.19**).

1. Distally, feel the **medial** and **lateral condyles of the femur** and the **patella** anterior to the condyles (Figure 30.18b).

2. Next, palpate your three groups of thigh muscles—the **quadriceps femoris muscles** anteriorly, the **adductor muscles** medially, and the **hamstrings** posteriorly (Figures 30.18a and 30.19). The **vastus lateralis,** the lateral muscle of the quadriceps group, is a site for intramuscular injections. Such injections are administered about halfway down the length of this muscle (Figure 30.17c).

3. The anterosuperior surface of the thigh exhibits a three-sided depression called the **femoral triangle** (Figure 30.18a). As shown in the figure (Figure 30.17c), the superior border of

this triangle is formed by the **inguinal ligament,** and its two inferior borders are defined by the **sartorius** and **adductor longus** muscles. The large *femoral artery* and *vein* descend vertically through the center of the femoral triangle. To feel the pulse of your femoral artery, press inward just inferior to your midinguinal point (halfway between the anterior superior iliac spine and the pubic tubercle). Be sure to push hard, because the artery lies somewhat deep. By pressing very hard on this point, one can stop the bleeding from a hemorrhage in the lower limb. The femoral triangle also contains most of the *inguinal lymph nodes,* which are easily palpated if swollen.

Leg and Foot

1. Locate your patella again, then follow the thick **patellar ligament** inferiorly from the patella to its insertion on the superior tibia (Figure 30.18b). Here you can feel a rough projection, the **tibial tuberosity.** Continue running your fingers inferiorly along the tibia's sharp **anterior border** and its flat **medial surface**—bony landmarks that lie very near the surface throughout their length.

2. Now, return to the superior part of your leg, and palpate the expanded **lateral** and **medial condyles of the tibia** just inferior to the knee. You can distinguish the tibial condyles from the femoral condyles because you can feel the tibial condyles move with the tibia during knee flexion. Feel the bulbous **head of the fibula** in the superolateral region of the leg (Figure 30.18c).

3. In the most distal part of the leg, feel the **lateral malleolus** of the fibula as the lateral prominence of the ankle (Figure 30.18c and d). Notice that this lies slightly inferior to the **medial malleolus** of the tibia, which forms the ankle's medial prominence. Place your finger just posterior to the medial malleolus to feel the pulse of your *posterior tibial artery.*

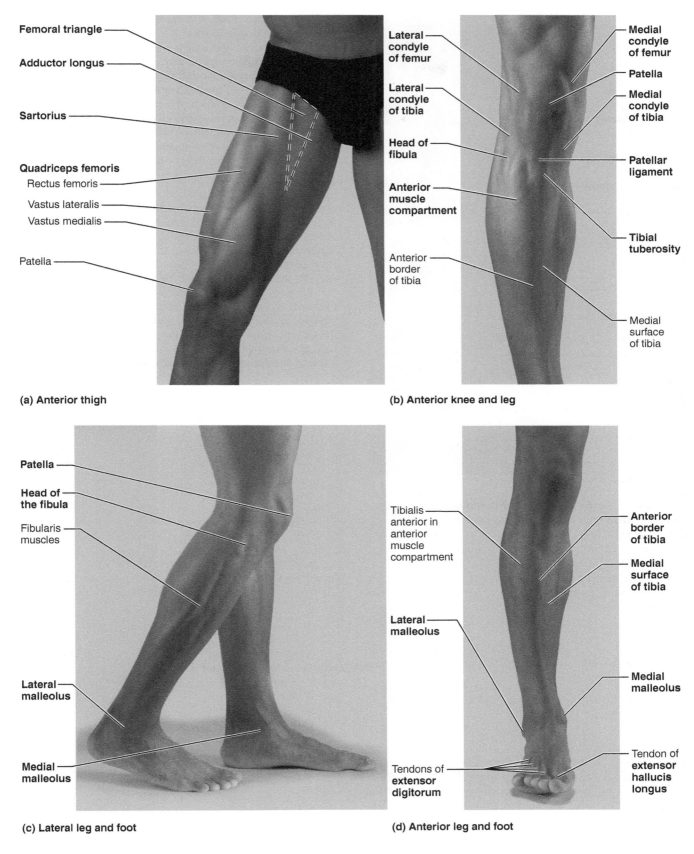

(a) Anterior thigh

Femoral triangle

Adductor longus

Sartorius

Quadriceps femoris
Rectus femoris
Vastus lateralis
Vastus medialis

Patella

(b) Anterior knee and leg

Lateral condyle of femur

Lateral condyle of tibia

Head of fibula

Anterior muscle compartment

Anterior border of tibia

Medial condyle of femur

Patella

Medial condyle of tibia

Patellar ligament

Tibial tuberosity

Medial surface of tibia

(c) Lateral leg and foot

Patella

Head of the fibula

Fibularis muscles

Lateral malleolus

Medial malleolus

(d) Anterior leg and foot

Tibialis anterior in anterior muscle compartment

Lateral malleolus

Tendons of extensor digitorum

Anterior border of tibia

Medial surface of tibia

Medial malleolus

Tendon of extensor hallucis longus

Figure 30.18 Anterior surface of the lower limb.

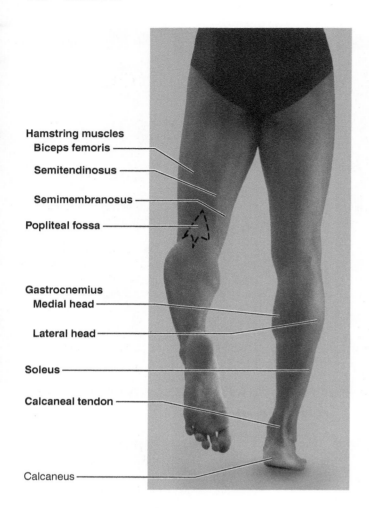

Hamstring muscles
Biceps femoris

Semitendinosus

Semimembranosus

Popliteal fossa

Gastrocnemius
Medial head

Lateral head

Soleus

Calcaneal tendon

Calcaneus

Figure 30.19 Posterior surface of the lower limb. Notice the diamond-shaped popliteal fossa posterior to the knee.

4. On the posterior aspect of the knee is a diamond-shaped hollow called the **popliteal fossa** (Figure 30.19). Palpate the large muscles that define the four borders of this fossa: the **biceps femoris** forming the superolateral border, the **semitendinosus** and **semimembranosus** defining the superomedial border, and the two heads of the **gastrocnemius** forming the inferior border. The main vessels to the leg, the *popliteal artery* and *vein,* lie deep within this fossa. To feel a popliteal pulse, flex your leg at the knee, and push your fingers firmly into the popliteal fossa. If a physician is unable to feel a patient's popliteal pulse, the femoral artery may be narrowed by atherosclerosis.

5. Observe the dorsum (superior surface) of your foot. You may see the superficial **dorsal venous arch** overlying the proximal part of the metatarsal bones (Figure 30.18d). This arch gives rise to both saphenous veins (the main superficial veins of the lower limb). Visible in lean people, the *great saphenous vein* ascends along the medial side of the entire limb (Figure 24.9, page 406). The *small saphenous vein* ascends through the center of the calf.

As you extend your toes, observe the tendons of the **extensor digitorum longus** and **extensor hallucis longus** muscles on the dorsum of the foot. Finally, place a finger on the extreme proximal part of the space between the first and second metatarsal bones. Here you should be able to feel the pulse of the **dorsalis pedis artery.** ■

Surface Anatomy Roundup

_____ 1. A blow to the cheek is most likely to break which superficial bone or bone part? (a) superciliary arches, (b) mastoid process, (c) zygomatic arch, (d) ramus of the mandible

_____ 2. Rebound tenderness (a) occurs in appendicitis, (b) is whiplash of the neck, (c) is a sore foot from playing basketball, (d) occurs when the larynx falls back into place after swallowing.

_____ 3. The anatomical snuff box (a) is in the nose, (b) contains the radial styloid process, (c) is defined by tendons of the flexor carpi radialis and palmaris longus, (d) cannot really hold snuff.

_____ 4. Some landmarks on the body surface can be seen or felt, but others are abstractions that you must construct by drawing imaginary lines. Which of the following pairs of structures is abstract and invisible? (a) umbilicus and costal margin, (b) anterior superior iliac spine and natal cleft, (c) linea alba and linea semilunaris, (d) McBurney's point and midaxillary line, (e) lacrimal fossa and sternocleidomastoid

_____ 5. Many pelvic organs can be palpated by placing a finger in the rectum or the vagina, but only one pelvic organ is readily palpated through the skin. This is the (a) nonpregnant uterus, (b) prostate, (c) full bladder, (d) ovaries, (e) rectum.

_____ 6. A muscle that contributes to the posterior axillary fold is the (a) pectoralis major, (b) latissimus dorsi, (c) trapezius, (d) infraspinatus, (e) pectoralis minor, (f) a and e.

_____ 7. Which of the following is _not_ a pulse point? (a) anatomical snuff box, (b) inferior margin of mandible anterior to masseter muscle, (c) center of distal forearm at palmaris longus tendon, (d) medial bicipital groove on arm, (e) dorsum of foot between the first two metatarsals

_____ 8. Which pair of ribs inserts on the sternum at the sternal angle? (a) first, (b) second, (c) third, (d) fourth, (e) fifth

_____ 9. The inferior angle of the scapula is at the same level as the spinous process of which vertebra? (a) C_5, (b) C_7, (c) T_3, (d) T_7, (e) L_4

_____ 10. An important bony landmark that can be recognized by a distinct dimple in the skin is the (a) posterior superior iliac spine, (b) ulnar styloid process, (c) shaft of the radius, (d) acromion.

_____ 11. A nurse missed a patient's median cubital vein while trying to withdraw blood and then inserted the needle far too deeply into the cubital fossa. This error could cause any of the following problems, _except_ this one: (a) paralysis of the ulnar nerve, (b) paralysis of the median nerve, (c) bruising the insertion tendon of the biceps brachii muscle, (d) spurting of blood from the brachial artery.

_____ 12. Which of these organs is almost impossible to study with surface anatomy techniques? (a) heart, (b) lungs, (c) brain, (d) nose

_____ 13. A preferred site for inserting an intravenous medication line into a blood vessel is the (a) medial bicipital groove on arm, (b) external carotid artery, (c) dorsal venous network of hand, (d) popliteal fossa.

_____ 14. One listens for bowel sounds with a stethoscope placed (a) on the four quadrants of the abdominal wall; (b) in the triangle of auscultation; (c) in the right and left midaxillary line, just superior to the iliac crests; (d) inside the patient's bowels (intestines), on the tip of an endoscope.

_____ 15. A stab wound in the posterior triangle of the neck could damage any of the following structures _except_ the (a) accessory nerve, (b) phrenic nerve, (c) external jugular vein, (d) external carotid artery.

16. ✚ What procedure requires locating the supracristal line? _____

_____

What disease is this procedure used to detect? _____

17. ✚ Describe the procedure used to detect a full urinary bladder. _____

_____

18. ✚ A patient is experiencing mastoiditis. Where would you expect the inflammation to be located? _____

_____

Credits

Index